D0880906

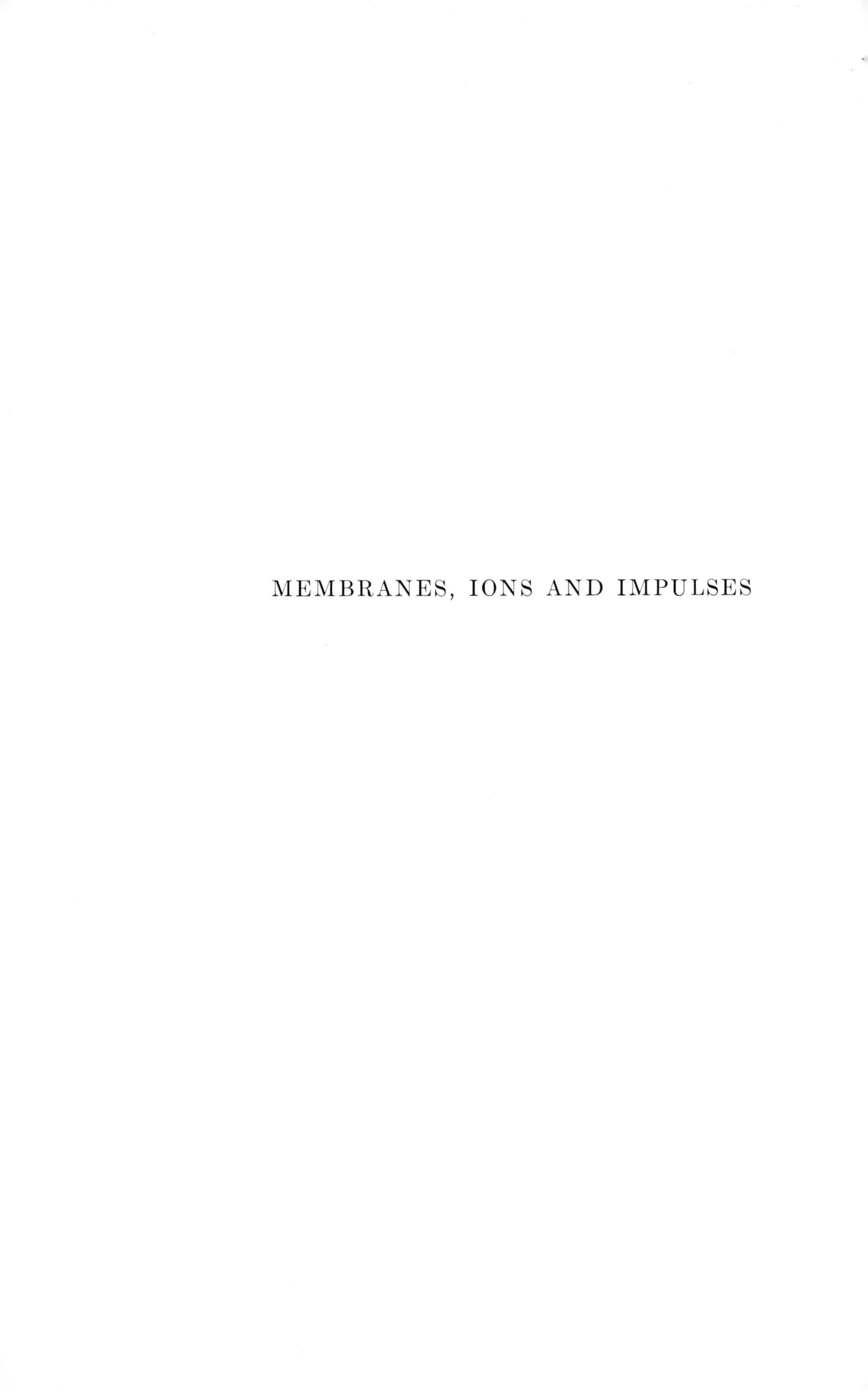

MEMBRANES, IONS AND IMPULSES

Membranes, Ions

KENNETH S. COLE

and Impulses

A Chapter of Classical Biophysics

1968

BERKELEY AND LOS ANGELES

UNIVERSITY OF CALIFORNIA PRESS

BIOPHYSICS SERIES
Cornelius A. Tobias, Editor

Volume 1. MEMBRANES, IONS, AND IMPULSES

Kenneth S. Cole
Laboratory of Biophysics, NINDB
National Institutes of Health, Bethesda
Division of Medical Physics
University of California, Berkeley

University of California Press
Berkeley and Los Angeles, California
Cambridge University Press
London, England
Library of Congress Catalog Card Number: 67-24121
Printed in the United States of America

E L I Z A B E T H

LIBBY · MOM · LIZ

MY WIFE

1902 · 1932 · 1966

Preface

This book is not the result of extensive, careful planning or leisurely writing. It is in fact the exception to a deliberate decision, made long ago, never to write any book. I had watched other authors with mixed admiration and horror, and I could not imagine myself matching either their effort or their achievement. Yet I have been completely surprised to find that I had been composing the story of this epoch of science for about thirty years.

I began the organization with a course at Columbia University, continuing and compacting it for the Priestley Lectures at Pennsylvania State University. To this background both progress and bewilderment were added for the course at the University of Chicago. Then after the pieces of the puzzle fell together—except for the still unknown core—I gave very condensed versions at the University of Pennsylvania and at Yale, with even smaller fragments elsewhere. But I can now see that much of the opinion and the comment, spoken and unspoken and in written and mental notes, would have remained in that form but for my invitation to give a seminar at the University of California in Berkeley, and for my appointment there as Regents' Professor.

I knew that I would be talking mostly of an era that was only a preliminary to the future. I, among others, had found it a notable era for many reasons, and it had a definite beginning and a definite end. I had known of the beginning from Höber and Fricke. The end was very neatly made official by the Nobel awards to Hodgkin, Huxley, and Eccles in 1963 as I was finishing Part III in this account. I had been closer to more of the work than anyone else who might write it down for those who cared to read about it, so I started to put my notes into the form of this book.

The book is little more and no less than the Berkeley seminar. It is in no way intended to be a definitive work—complete and impersonal, accurate, consistent, and authoritative. It is only an introduction to and a survey of what I have found the most interesting, challenging—and frustrating—steps in a single restricted field of science. The book is essentially pedagogical. I hope to teach some a little of the substance of physical sciences and their application to the problems of living cell membranes. Through the physical approach I hope to introduce others, in their own language, to the success and the difficulty of using elementary concepts and to the challenges of a small, fundamental part of biology and medicine. Perhaps of most importance is the hope that I may help some of our future scientists, teachers, and investigators, and even others, to learn of the art, the joys, and the despair of curiosity and exploration—to learn that efforts are often in the wrong direction and that without the many small steps there would not be much progress. As I have now completed my task, except for the thousand trivia demanded by convention, I find my fears are all realized but my hopes are only a little less.

The book is a product of the help and patience of many—family, teachers, friends, colleagues, collaborators, chiefs, and students. I wish it were a better tribute to them.

Contents

Introduction

Life, electricity, and the relations between them have long been subjects of intense curiosity and often of acrimonious speculation. After the Egyptian records of the electric catfish *Malapterurus* about 2600 B.C. (Fritsch, 1887) there was a long and diverse history of electrical phenomena in living things. This was brought to a focus in 1790 by Galvani and then, along with the surge of physical sciences, it spread out into two primary channels. The first, that of a description of living tissues—particularly human skin—as electrical conductors, followed upon the appearance of Ohm's law in 1827. But this was premature and soon came to a frustrated halt (Gildemeister, 1928b).

The channel that was to become electrophysiology advanced through electrical stimulation of tissues and into the production of electrical currents by injury and activity—sometimes leading and always promptly following new instruments (Brazier, 1959). Du Bois-Reymond (1848–1860) made major advances in the facts while Helmholtz (1853) clearly recognized the physical problems and principles but could do little with them at the time. In a contribution to potential theory, Helmholtz showed that a distribution of potential over a surface could not define a unique internal distribution of sources and sinks. His work, with an extension by Hermann (1877), is discussed in some detail by Lorente de Nó (1947). The controversy then was between alteration and preexistence—whether the sources of the injury and the action currents were created at the time they appeared or had been present all the time in the resting tissues. In the same period the core conductor model for a nerve fiber was started on a curious development (Taylor, 1963). It began with investigations of analogues and by 1905 it was represented by the same partial differential equation that Kelvin (1855) had used for the theory of the first Atlantic cable. Then came the idea of Hermann (1905) that in impulse propagation current flow from an active region stimulated the resting region ahead. In marked contrast to these simple physical ideas, the all-or-none concept had to be created and finally proved for muscle and nerve by Lucas (1909) and Adrian (1912). From here, classical electrophysiology grew and developed into an imposing structure.

In the onward course of physics and chemistry, Kohlrausch (1876) and Arrhenius (1887) saw the fundamentals of electrolytes and they produced an experimental method and the interpretation of electrolytic conduction—both of which soon came to be widely used in the investigation of biological systems. Nernst and Boltzmann formulated the energetics of electrolytes and Nernst almost continuously attempted to apply his ideas to tissues and cells. These ideas Julius Bernstein put together into his remarkably complete and durable membrane hypothesis of the role of potassium ions.

BERNSTEIN'S MEMBRANE HYPOTHESIS

Both the passive and the active electrical aspects of biology were considered by Bernstein in his membrane hypothesis. This consisted of three propositions: (1) Living cells are composed of an electrolytic interior surrounded by a thin membrane relatively impermeable to ions. (2) There is an electrical difference of potential across this membrane at rest. (3) During activity the ion permeability of the membrane increases in such a way as to reduce this potential to a comparatively low value.

The first of Bernstein's three hypotheses may have developed quite early (1868). He believed that the interior of a living cell was an electrolyte in which ions were free to move much as they did outside the cell. But at its surface the cell was to be separated from its environment by a membrane only slightly permeable to ions. By the time of his definitive statement, Bernstein (1902) took full advantage of Nernst's work (1888, 1889). He postulated that the membrane permeability was for the potassium ion and that a Nernst diffusion potential was to be found across the resting membrane. When the membrane became active this selective permeability for potassium was lost as the membrane became highly permeable to all ions. The diffusion potential all but disappeared as the "excess negative ions inside the membrane joined with excess positive outside." By 1912 the capillary electrometer and the string galvanometer had stimulated a large amount of electrobiology and in this Bernstein (1912) found more support for his postulates, and published his work in more detail.

At that time, over fifty years ago, the physical structure of a living cell was wholly unknown and, in particular, the existence of an enveloping membrane was only a hypothesis. Thus Bernstein's work must be looked upon as a classic. It was not only a starting point for what we now know of membranes, ions, and impulses but it has also been a clear guide along the path. The application of concepts of physics along with the developing power of physical techniques and the exploitation of uniquely favorable living cells was this time to meet with a continuing success.

This volume is a recounting and a commentary for this past half century. It tells of electrical descriptions of cell cytoplasm and of the exquisite thinness of cell membranes with unexplained hints as to their structure. The sheer size of the giant axons of the lowly North Atlantic squids, *Loligo pealii* and *forbesi*, made them vulnerable to new physical attacks. Measures were made of the ion permeabilities of resting and active nerve membranes. As the phenomena of excitation began to connect with the structure of neurophysiology, there came an observation which it was beyond Bernstein's hypothesis to explain. The resolution of this difficulty, anticipated by Overton (1902), then led to the description of the physical behavior of the ions as they crossed the squid membrane. This membrane had to yield the facts as to how many sodium and potassium ions it let through, and when. These facts have most of the complexity needed to correlate and contain the vast body of classical electrophysiology. They also have a complexity that seems to deny the possibility of any single elementary explanation. Yet in these facts and in the more recent experiments on other membranes a unity and a simplicity have appeared which suggest that an understanding of how ions get through membranes to make nerve impulses may not be far away.

Even though the Bernstein membrane hypothesis is now importantly inadequate, the new added factor is most gratifyingly of the same kind as that in the original daring application of physical science to biology. It seems most appropriate to relate the development of the half century shown in the frontispiece to this starting point (see next page).

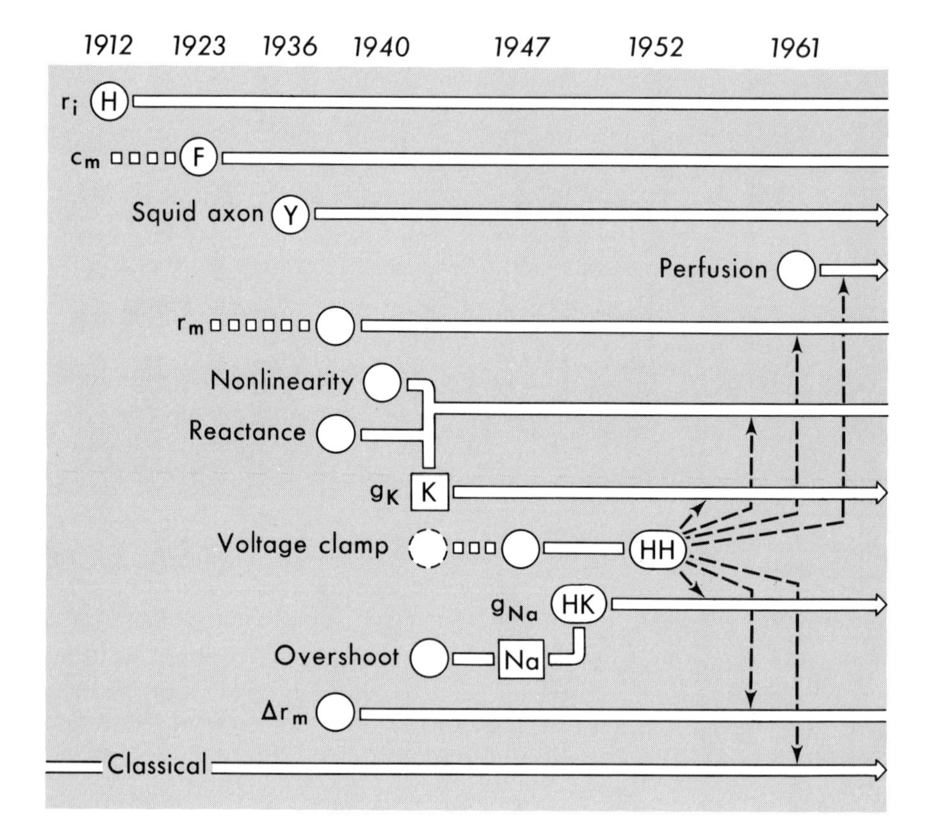

A chart of the development of membranes, ions, and impulses. Höber(H) found an electrolytic cytoplasm r_i. Fricke(F) demonstrated a thin membrane c_m. Young(Y) introduced the squid axon. The resistance of its membrane r_m allowed an ion permeability and the decrease during activity Δr_m indicated an ionic mechanism. The nonlinearity and anomalous reactance could be attributed to the potassium ion [K] and its conductance g_K. The overshoot of the action potential was attributed to the sodium ion [Na] and its conductance g_{Na} by Hodgkin and Katz (HK.) The voltage clamp led to a quantitative statement of the ionic properties of the squid axon membrane by Hodgkin and Huxley (HH.) These equations included much of the other available information for the squid and suggestions for much of classical electrophysiology, as indicated by the dashed lines. As the applications to the classical and to other membranes continued, the search for the fundamental mechanisms of the ionic conductances was intensified and broadened by internal perfusion of the squid axon.

PART I

Linear and Passive Cell Properties

Contents to Part I

Introduction

Bernstein's basic speculation was that living cells were contained by a membrane with a low permeability to ions. This required that the ions of an electric current pass mostly around the cells of tissues and in cell suspensions. One of the few items of general agreement among the measurements of electrical resistance, following Kohlrausch (1876), was that living tissues had considerably higher resistance than dead tissues, body fluids, or artificial media (McClendon, 1929; Cole, 1933). In further partial confirmation of the presence of a poorly conducting membrane, it was shown, by Stewart (1899) and others, that the conductance of blood decreased as the corpuscle concentration was increased. This was much as would be expected if the corpuscles were effectively nonconducting. Analytical expressions for the relation between conductance and concentration, such as that of Stewart (1899), were entirely without theoretical foundation. But the conclusion was supported by the analogue experiments of Oker-Blom (1900) on sand suspensions.

The next question was whether the entire corpuscle contributed to this high effective resistivity of the blood cells or if indeed the interior was a reasonably good electrolytic conductor and only the still hypothetical membrane had the necessary high resistance. Here again we are in debt to Nernst, for it was in his laboratory that the first answer was found, and—as Höber led me to suspect—it may be that Nernst provided both the question and the way to the answer.

With three remarkable experiments, Höber (1910, 1912, 1913) showed that the red cell membrane was a very poor conductor while the interior was in the range of biological electrolytes. For example, at a frequency (f) of 1 kilocycle/sec (kc) the resistivity of a mass of red cells centrifuged in sucrose was 1200 ohm (Ω) cm. At a frequency of 10 megacycles/sec (Mc) this mass was equivalent to a 0.4 percent NaCl solution with a resistivity of about 200 Ωcm. Furthermore, after hemolysis and membrane destruction by saponin, the resistivity not only remained unchanged at high frequency but now showed this same low value at the low frequency. It was then to be concluded that the presence or absence of the cell membranes made no difference at 10 Mc and this resistivity was a measurement of the cell interior. Höber found this internal resistivity to be in the range of that of 0.1 to 0.4 percent NaCl. It was further to be concluded that the unique high resistance of intact cells at 1 kc required the membranes to have relatively high resistance. This work, as I have been told by his son, was tremendously gratifying to Bernstein as the first direct confirmation of his membrane hypothesis.

Here, and probably for the first time, the concept of a membrane capacity and the variation of its reactance with frequency were introduced. The capacity C of a nonconductor between two electrodes is defined by the charges $\pm Q$ on the electrodes, and a potential difference V between them, as $C = Q/V$. As the potential changes, a charging current I flows, $I = dQ/dt = C\,dV/dt$, and for an alternating, sinusoidal potential, $V = \bar{V} \sin \omega t$ where $\omega = 2\pi f$,

$$I = \bar{I} \cos \omega t = C\omega \bar{V} \cos \omega t.$$

Corresponding to the definition of a resistance by Ohm's law, $R = V/I$, there is here the reactance $X = \bar{V}/\bar{I} = 1/C\omega$. Now, with the confidence and brilliance of hindsight, the membranes of the red cell mass might be expected to have a reactance considerably higher than 1200 Ωcm at 1 kc, and about as much below 200 Ωcm at 10 Mc. Then, taking a geometric mean of these frequencies, we might hope that the membranes would have a mean reactance of 1500 Ωcm at 100 kc or 10^{-9} farad (F) for a centimeter cube. Assuming the red cells to be conveniently deformed into cubes, 5.5 μ ($5.5 \cdot 10^{-4}$ cm) on a side, this capacity is that of 3600 membranes in series, each with an area of 1 cm^2. This conclusion that the living red cell membrane had a capacity of $3.6 \cdot 10^{-6}$ F/cm^2—of the order of 1 μF/cm^2—was obviously possible a decade before it was published, but I have no hint that Nernst, or even Fricke, made such an estimate.

Then came the momentous work of Osterhout (1922) on the conductance of the marine kelp, *Laminaria*, at 1 kc. After death, this plant showed about the same conductance as sea water. The relatively low conductance in life was attributed largely to the low ion permeability of the cell membranes, and the variations of tissue conductance were explained as variations of the permeability. I was very pleased to assist Blinks (1928) in support of this interpretation. We measured the limiting resistance of the kelp at high frequencies and found it to be constant at about that of sea water and independent of the changes at low frequencies. In this same period Crile and his associates (1922) had followed the lead of Osterhout and made extensive 1 kc measurements on mammalian organs, particularly brain and liver. These they correlated with histological and clinical observations.

After World War I, Philippson (1921), a Belgian banker and radio engineer, measured packed blood cells, guinea pig muscle and liver, and potato (resting and germinating) by a vacuum tube ammeter–voltmeter method, over the frequency range from 500 c to 3 Mc. The ratio of the peak amplitudes $\bar{I}$ and $\bar{V}$ now became a vector impedance. The amplitude, $|Z| = \bar{V}/\bar{I}$, varied with frequency as in figure 1:1a.

Philippson interpreted these data in terms of the circuit of figure 1:1b, in which R_i represented the resistance of the cell interior and R_m and C_m the resistance and capacity of the membrane, ignoring any current flow around and between the cells. The impedance data were extrapolated to zero frequency to give $R_i + R_m = R_0$ and to infinite frequency to give $R_i = R_\infty$. In particular, Philippson confirmed Höber's conclusions by finding an internal resistivity for the red cell of about 3.5 times that of plasma. Then C_m was computed to give a much more specific model than was possible from Höber's experiments.

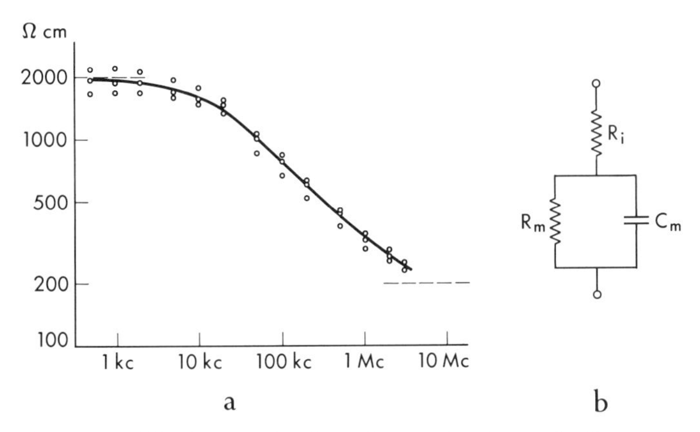

FIG. 1:1. (a) The specific impedance of guinea pig liver in Ω cm as a function of the alternating current measuring frequency. (b) The equivalent circuit proposed for several tissues. C_m and R_m are attributed to the cell membrane capacity and resistance and R_i to the cytoplasm resistance.

The concept of an impedance as a vector on the R,X plane and techniques for analysis of the alternating current behavior of circuits were well established and were used by Philippson. The complex impedance $Z = R + jX$ with the imaginary operator $j = \sqrt{-1}$ was soon found more convenient and came into common usage. This technique may first be illustrated with a resistance, a capacity and an inductance (L) in series (Page and Adams, 1931). By definition $V_R = RI$, $dV_C/dt = I/C$ and $V_L = L\,dI/dt$ and for the series circuit $V_R + V_C + V_L = V$. Differentiations and substitutions give

$$L\frac{d^2I}{dt^2} + R\frac{dI}{dt} + \frac{1}{C}I = \frac{dV}{dt}.$$

For applied potentials $\bar{V}\cos\omega t$ and $\bar{V}\sin\omega t$ the currents I' and I'' given by

$$L\frac{d^2I'}{dt^2} + R\frac{dI'}{dt} + \frac{1}{C}I' = \frac{d}{dt}[\bar{V}\cos\omega t]$$

and

$$L\frac{d^2I''}{dt^2} + R\frac{dI''}{dt} + \frac{1}{C}I'' = \frac{d}{dt}[\bar{V}\sin\omega t].$$

Adding these for the total current, $I = I' + jI''$, gives

$$L\frac{d^2I}{dt^2} + R\frac{dI}{dt} + \frac{1}{C}I = \frac{d}{dt}[\bar{V}\,e^{j\omega t}]$$

since

$$e^{j\Phi} = \cos \Phi + j \sin \Phi.$$

For $I = \bar{I}\, e^{j\omega t}$ we have

$$\left[-L\omega^2 + Rj\omega + \frac{1}{C} \right] \bar{I}\, e^{j\omega t} = j\omega \bar{V}\, e^{j\omega t}$$

and the impedance

$$Z = \bar{V}/\bar{I} = R + j\omega L + 1/j\omega C.$$

Here

$$X_L = \omega L \text{ and } X_C = -1/\omega C$$

are the inductive and capacitive series reactances and the circuit reactance $X = X_L + X_C$ to give in general $Z = R + jX$. Here the impedance vector

$$Z = |Z|\, e^{j\Phi}$$

has the amplitude $|Z| = \sqrt{R^2 + X^2}$ and the direction, or phase angle, Φ, given by $\tan \Phi = X/R$.

Returning to Philippson's circuit of figure 1:1b the impedance

$$Z = R_i + \cfrac{1}{\cfrac{1}{R_m} + j\omega C_m}$$

or

$$Z = \frac{R_0 + j\omega\tau R_\infty}{1 + j\omega\tau} \tag{1:1}$$

where $\tau = R_m C_m$. Since he measured only $|Z|$ and not Φ or R and X, we derive

$$|Z|^2 = \frac{R_0{}^2 + \omega^2\tau^2 R_\infty{}^2}{1 + \omega^2\tau^2}$$

then

$$C_m = \frac{1}{\omega R_m} \sqrt{\frac{R_0{}^2 - |Z|^2}{|Z|^2 - R_\infty{}^2}} \tag{1:2}$$

could be computed from the data. It had been assumed, as above, that C_m was an ideal capacity, independent of frequency, but this analysis of the data showed that it decreased with increasing frequency. Furthermore, this capacity was well approximated by the empirical expression

$$C_m = \bar{C}_m \omega^{-\alpha}$$

where $\bar{C}_m$ is the value at $\omega = 1$ and α is a constant, 0.12 for red cells, 0.55 for the guinea pig tissues, and 0.75 for potato. Philippson then noted that this behavior along with a resistive component, $R(\omega)$, such that $R(\omega)C(\omega)\omega = \tan \phi_m = m$, a constant, was a characteristic of metal electrodes in electrolytes. He further invoked this constant phase angle analogy as a basis for believing that this "polarization capacity" also arose from the movement of ions across the cell surface—a hope that was to persist for two decades.

Membrane Capacity

CELL SUSPENSIONS

Fricke's Discovery

The hypothesis that a living cell had a well-conducting interior surrounded by a relatively impermeable, poorly conducting region had by 1921 been well supported by measurements on red blood cells and several tissues. The interiors showed specific resistances of the order of a hundred ohm cm but there were no similar estimates for the properties of the bounding membrane region. All that seemed to be established (cf. McClendon, 1926; Gildemeister, 1928b; Cole, 1933) was that various living cells had capacitive properties correlated with the presence of a membrane, that the capacitive reactance became negligible above a few megacycles, and that at low frequencies, such as 1 kc, the high resistances to be attributed to high membrane resistances were interesting indices of some physiological changes.

At this time Hugo Fricke was brought to the Cleveland Clinic, primarily in the hope that he could help Crile understand his tissue conductance measurements (Crile et al., 1922). Although he was principally interested in radiological physics, Fricke had, just before the summer of 1923, which I spent in his laboratory, determined and published the capacity of the red blood cell membrane as 0.81 $\mu F/cm^2$. This figure of about 1 $\mu F/cm^2$ has been so confirmed and refined, extended and approximated for membranes of the red cells and almost all other living cells, as to become a biophysical constant. But Fricke also assumed that the membrane might be an oil with a dielectric constant of 3, and so estimated that the capacity might correspond to a membrane thickness of 33 Å. This was the first indication that the membrane was of molecular dimensions, and it is within a factor of only 2 or 3 of the best estimates available even in 1967.

Fricke's original work has seemed to me to be not only the beginning of an important era of biophysics but also an achievement of a magnitude that I have come to respect more and more over the past two score years. With a characteristic independence, he developed an original potential theory for the capacity and resistance of suspensions of spherical or spheroidal cells (Fricke, 1924a,b, 1925a–c). With Morse (1925a), he combined the purely physical theory and experiment to provide a major advance in biology.

Potential theory is a part of classical mathematics and mathematical physics which is treated in many texts (Page and Adams, 1931; Jeans, 1927). It covers a wide variety of problems which are stated in the form of general partial differential equations and are solved generally and particularly in many different ways. We will be concerned primarily with electrostatic problems of potential and charge distributions in nonconductors, with "volume conductor" problems of the distributions of potential and current density in conductors, and with kinetic problems, related to or including diffusion of ions. These problems are stated by the Laplace, Poisson or continuity equations, and we will need answers, usually relatively simple and approximate, in the appropriate coordinate systems, Cartesian, spherical, or cylindrical.

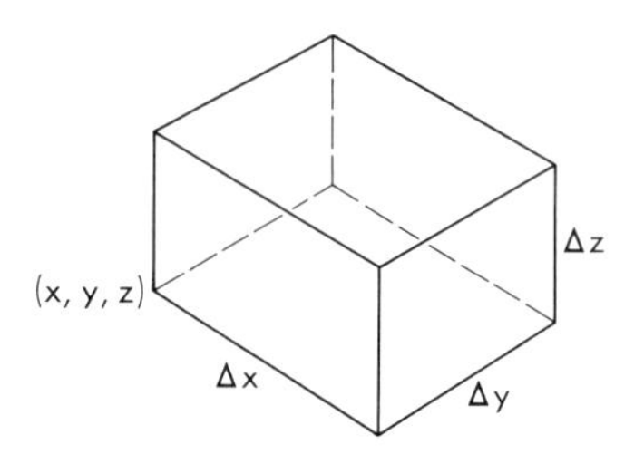

Fig. 1:2. The elementary volume for the derivation of the Laplace equation in rectangular coordinates.

For the "volume conductor" problem in Cartesian coordinates we consider the element of volume $\Delta x \Delta y \Delta z$ of figure 1:2 in a medium of uniform conductivity $\Lambda(\Omega^{-1}\text{cm}^{-1})$ at a point x,y,z. For a potential V the components of the field are

$$X = -\partial V/\partial x, \qquad Y = -\partial V/\partial y, \qquad Z = -\partial V/\partial z$$

and the components of the current density (A/cm^2) are

$$i_x = \Lambda X, \qquad i_y = \Lambda Y, \qquad i_z = \Lambda Z.$$

The current entering a face $\Delta y \Delta z$ at x is

$$i_x(x)\Delta y \Delta z = \Lambda X(x)\Delta y \Delta z$$

and that leaving at $x + \Delta x$ is

$$i_x(x + \Delta x)\Delta y \Delta z = \Lambda X(x + \Delta x)\Delta y \Delta z.$$

The net inflow

$$\Delta i_x = \Lambda[X(x) - X(x + \Delta x)]\Delta y \Delta z$$
$$= -\Lambda \frac{\partial X}{\partial x} \Delta x \Delta y \Delta z$$

and with similar expressions for Δi_y, Δi_z, the total current

$$I = \Lambda \left[\frac{\partial^2 V}{\partial x^2} + \frac{\partial^2 V}{\partial y^2} + \frac{\partial^2 V}{\partial z^2} \right] \Delta x \Delta y \Delta z.$$

For a charge density of q,

$$I = \frac{\partial q}{\partial t} \Delta x \Delta y \Delta z$$

and

$$\frac{\partial q}{\partial t} = \Lambda \left[\frac{\partial^2 V}{\partial x^2} + \frac{\partial^2 V}{\partial y^2} + \frac{\partial^2 V}{\partial z^2} \right] \qquad (1:3)$$

which is the equation of continuity. For the electrostatic problem

$$-\frac{4\pi q}{\epsilon} = \frac{\partial^2 V}{\partial x^2} + \frac{\partial^2 V}{\partial y^2} + \frac{\partial^2 V}{\partial z^2} \qquad (1:4)$$

where q is the free charge density, we have the Poisson equation. In the absence of current sinks and sources or of free charge these become the Laplace equation

$$\frac{\partial^2 V}{\partial x^2} + \frac{\partial^2 V}{\partial y^2} + \frac{\partial^2 V}{\partial z^2} = 0. \qquad (1:5)$$

A more general derivation gives this Laplacian operator in vector forms

$$\text{div grad } V = 0 \qquad \text{or} \qquad \nabla^2 V = 0 \qquad (1:6)$$

from which transformations can be made to any particular coordinate system.

In addition to the many and powerful analytical solutions there are certainly uncounted procedures and devices for obtaining numerical and graphical solutions of all manner of potential theory problems. Some of these have been found quite helpful in a number of membrane and cell problems.

Fricke (1925a) first placed a nonconducting sphere in a medium of conductivity Λ and an original uniform field X. The potential will be symmetrical about the axis of X, and the Laplace equation in the polar coordinates is then

$$\nabla^2 V = \frac{1}{r^2}\frac{\partial}{\partial r}\left(r^2 \frac{\partial V}{\partial r}\right) + \frac{1}{r^2 \sin\theta}\frac{\partial}{\partial\theta}\left(\sin\theta \frac{\partial V}{\partial\theta}\right) = 0.$$

The boundary conditions are

$$\frac{\partial V}{\partial x} \to -X \text{ as } x \to \pm\infty \text{ and } \frac{\partial V}{\partial r} = 0 \text{ for } r = a.$$

Of the many possible solutions only

$$V = A\, r \cos\theta + B \cos\theta / r^2$$

will satisfy them with

$$A = -X \text{ and } B = \frac{a^3}{2} A = -\frac{a^3 X}{2}.$$

At the surface,

$$V = -\frac{3Xa}{2} \cos\theta,$$

and for a simple capacity c_3 at the surface and the conducting interior at $V = 0$, the electrostatic energy of a sphere

$$u = 1/2 \int c_3 V^2\, dS = \frac{9 c_3 X^2 \pi a^4}{2} \int_0^{\pi/2} \cos^2\theta \sin\theta\, d\theta$$

$$= 3/2\ c_3 X^2 \pi a^4.$$

For n spheres in a unit cube the energy is assumed equal to that of the suspension,

$$nu = 3/2\ c_3 X^2 \pi a^4 n = U = 1/2\ C X^2$$

$$\text{or } c_3 = \frac{4}{9} \cdot \frac{1}{a\rho} \cdot C \qquad (1{:}7)$$

where C is the capacity of the suspension and ρ is the volume concentration of the spherical cells. This most ingenious and novel solution is certainly an excellent approximation. Since it assumes that the original field is also the effective field for each cell it does not take into account the interaction between cells, which was later found to be significant above volume concentrations of 20–30 percent. A measurement of dog corpuscles (Fricke, 1925a,b) at $\rho = 21$ percent gave $C = 129$ pF and, for an equivalent radius $a = 2.6\ \mu$, equation 1:7 gives the membrane capacity $c_3 = 1.0\ \mu\text{F/cm}^2$.

The potential V_1 outside a sphere with an appreciable conductance Λ_2 in a medium of conductance Λ_1 is satisfied by the same form of equation,

$$V_1 = A_1 r \cos\theta + B \cos\theta / r^2.$$

The potential inside is now of the form

$$V_2 = A_2 r \cos\theta,$$

and the boundary condition at $r = a$ is $\Lambda_1 \partial V_1/\partial r = \Lambda_2 \partial V_2/\partial r$. Again $A_1 = -X$, but

$$B = -\frac{1 - \Lambda_2/\Lambda_1}{2 + \Lambda_2/\Lambda_1} a^3 A_1$$

and

$$A_2 = \frac{3}{2 + \Lambda_2/\Lambda_1} A_1$$

to give

$$V_1 = -\left[1 + \frac{1 - \Lambda_2/\Lambda_1}{2 + \Lambda_2/\Lambda_1} \cdot \frac{a^3}{r^3}\right] Xr\cos\theta$$

$$V_2 = -\frac{3}{2 + \Lambda_2/\Lambda_1} Xr\cos\theta.$$

This result is both interesting and useful because it can be shown that the added external potential is that of a central dipole with a moment

$$M = X\frac{1 - \Lambda_2/\Lambda_1}{2 + \Lambda_2/\Lambda_1} a^3.$$

Several of the solutions are shown in figure 1:3 as the paths of current flow, everywhere perpendicular to the equipotentials. These were obtained by a procedure of L. J. Savage. The flow around a conducting sphere was shown to be the same as that around a smaller nonconducting sphere, and all the patterns were obtained by scaling the single solution for the flow around a nonconducting sphere.

By an again original but approximate calculation, Fricke (1924a) gave a conductance for a suspension Λ as

$$\frac{1 - \Lambda/\Lambda_1}{3} = \rho\frac{1 - \Lambda_2/\Lambda_1}{2 + \Lambda_2/\Lambda_1}$$

and for a nonconducting sphere, $\Lambda_2 = 0$,

$$\rho = 2(1 - \Lambda/\Lambda_1)/3.$$

By substituting this in equation 1:7 we have Fricke's favorite result

$$c_3 = \frac{2C}{3a(1 - \Lambda/\Lambda_1)}. \tag{1:8}$$

However, it can be shown that equation 1:8 is rather better because in it the two major approximations quite neatly cancel out!

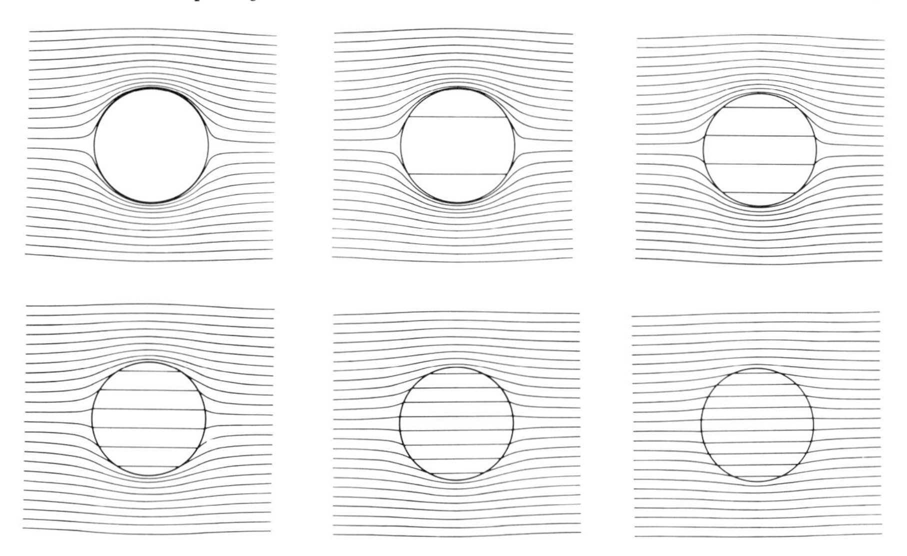

FIG. 1:3. Lines of current flow around spheres placed in a uniform field. The approximate ratios of the conductances of sphere to medium from left to right, top, 0, 0.02, 0.08 and bottom, 0.18, 0.35, 0.6.

Fricke, with Morse (1925a), also built and painstakingly calibrated and corrected all components of a Wheatstone bridge from 800 c to $4\frac{1}{2}$ Mc. Fricke (1925a) measured the parallel capacity at 87 kc for suspensions of dog blood at various concentrations. A modification of equation 1:8 derived for a suspension now gave

$$c_3 = 0.81\ \mu\text{F/cm}^2$$

and this is an achievement of truly historic importance. It was in the first place amazingly large by comparison with man-made condensers. But then, on the basis of Overton's lipid solubility theory of membrane permeability, Fricke assumed a dielectric constant $\epsilon = 3$ as characteristic of oils. The formula for a parallel plate condenser

$$c_3 = \epsilon/4\pi\delta$$

gave $\delta = 33$ Å—about the length of a single molecule of 20 to 30 carbon atoms. Although the value of ϵ has been and still is an unresolved problem (Danielli, 1935, and others), Fricke certainly gave the first, and probably still the best, indication of the molecular dimension of the membrane of an intact, functional, living cell.

Fricke and Morse (1925a) measured suspensions over the entire frequency range and presented the results in terms of the equivalent

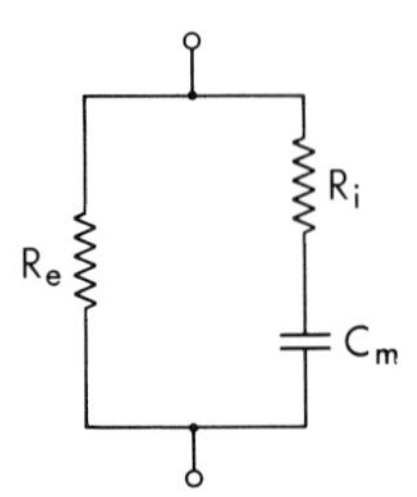

FIG. 1:4. An equivalent circuit for a red blood cell suspension. R_e represents the resistance of the lines of current that pass around the cells, R_i the lines that go to and through the cells, and C_m the capacity of the membranes.

circuit of figure 1:4. As an example, a suspension of calf cells, for which $\rho = 46$ percent, gave

$$R_e = 191\ \Omega\text{cm}; \qquad R_i = 370\ \Omega\text{cm}; \qquad C_m = 146\ \text{pF/cm}.$$

The components of such circuits were identified with the physical parameters of the suspensions (Cole, 1928a) and produced excellent agreements between the calculated and the observed values. This showed that the assumption c_3 independent of frequency was at least approximately satisfied and reasonably consistent with Philippson's result for packed red cells. The extrapolated high frequency resistance of the above example, 124 Ωcm, gave an internal resistance that was equivalent to 0.17 percent NaCl. This is close to Höber's mean value and fully confirms that his frequency was high enough to approximate the infinite frequency asymptote of resistance.

Subsequent work by Fricke and many others has both considerably improved, extended, and qualified the theory and produced so wide a variety of experiments on many different forms as to make a membrane capacity (within a factor of 2) of 1 μF/cm^2 a physical constant of most living cells. However, an impressive array of chemical, mechanical, optical, x-ray, and electron microscope experiments on cell components and models has made a membrane thickness of nearly 100 Å seem more probable than Fricke's estimate of 33 Å.

More on red cells

Along with his theory for a suspension of spherical cells, Fricke (1924b) developed the more complicated theory for spheroids as a better approximation to the shape of the biconcave mammalian red blood cells. Spheroids and ellipsoids are rather nice forms to deal with because, like spheres, they have a uniform internal field, and if a geometric axis is in the direction of the applied field, the inside field is in the same direction. The treatment is more elaborate, using confocal coordinates and expressing solutions in ellipsoidal harmonics to give the contribution for a single principal orientation as an elliptic integral. The applied field at each cell was then modified to include an average of the effects of the other cells in the suspension.

For ellipsoids with axes a, b, c, the integral

$$L_a = \int_0^\infty \frac{d\lambda}{(a^2 + \lambda)\sqrt{(a^2 + \lambda)(b^2 + \lambda)(c^2 + \lambda)}} \tag{1:9}$$

is evaluated for the a axis parallel to the field. The conductance of this oriented suspension is

$$\Lambda = \Lambda_1 + \frac{\rho}{1 - \rho} \cdot \frac{2(\Lambda_2 - \Lambda_1)}{2 + abc\, L_a[(\Lambda_2/\Lambda_1) - 1]}$$

with similar integrals and conductance for the other principal orientations. The effect of random orientation was then obtained by averaging the principal contributions for the three axes. The results were given for various ratios of internal and external resistivities, r_2 and r_1, and axial ratios of prolate and oblate spheroids as a form factor γ in the equation

$$\frac{1 - r_1/r}{\gamma + r_1/r} = \rho \frac{1 - r_1/r_2}{\gamma + r_1/r_2}.$$

It must have been a disappointment for Fricke to find that O. Wiener (1912), although confirming his work, had done essentially the same problem in his analysis of form birefringence. Fricke applied his results to the data of Oker-Blom on sand—using a high axial ratio for an oblate spheroid. Also, in 1925, Fricke developed the corresponding equation for the capacity of a suspension of spheroidal cells,

$$c_3 = \frac{C}{\alpha a(1 - r_1/r)}$$

where a is the major axis of the spheroid and α is a form factor. From this he obtained a reasonable agreement with experiments on dog cells with an axial ratio of 4.25.

For spheres the general resistance equation now reduced to

$$\frac{1 - r_1/r}{2 + r_1/r} = \rho \frac{1 - r_1/r_2}{2 + r_1/r_2} \tag{1:10}$$

This equation had long been well known in theoretical physics, appearing in Poisson's theory of induced magnetism, the Clausius-Mossotti theory for the dielectric constant, and the Lorenz-Lorentz index of refraction theory. Fricke and Morse (1925b) now showed that it applied to cream with $r_2 = \infty$ for butter fat concentrations up to 62 percent with an accuracy of 0.5 percent.

In this same period McClendon (1926) developed a Wheatstone bridge and made measurements on red cells and a few other tissues. He packed beef red cells to transparency in a homemade vacuum centrifuge and was able to balance them in the bridge with the circuit of figure 1:1 at frequencies from 1 kc to 1 Mc. The value of $R_i = 200\ \Omega\text{cm}$ was quite reasonable. He felt justified, as had Philippson, in assuming no leakage between cells. Then for an effective cell dimension of 3 μ he derived a value of $c_3 = 9\ \mu\text{F/cm}^2$ from $C_m = 1{,}500$ pF/cm, and similarly he found a membrane resistance of 0.11 Ωcm^2 from $R_m = 675\ \Omega\text{cm}$.

Fricke's analysis and measurement were so extensive and so careful as to make a major mistake seem highly improbable. By comparison, McClendon's conclusions based on an ingenious, but intuitive, analysis were much less impressive. I, for one, was unable to explain the order of magnitude difference in the membrane capacity until K. S. Johnson called my attention to the work of Zobel (1923) on equivalent circuits. In particular it is a simple matter to obtain the general relations between the elements of the circuits of figures 1:1 and 1:4 in order that the two circuits have the same complex impedance at all frequencies as given by equation 1:1. McClendon's data for figure 1:1 are then as well represented by figure 1:4 with $R_e = 875\ \Omega\text{cm}$, $R_i = 260\ \Omega\text{cm}$ and $C_m =$ 890 pF/cm. By McClendon's argument this produces 5.3 $\mu\text{F/cm}^2$, and a crude consideration of the high leakage represented by R_e reduces the membrane capacity by another factor of 2. Thus it was an important mistake of interpretation to assume a negligible current flow between the cells. I found this a far more satisfactory conclusion, and also a lesson in both the virtues of analysis and the hazards of guessing (Cole, 1928a).

Spheres and Marine Eggs

I realized soon after my first experience with the beautiful eggs of the sea urchin *Arbacia punctulata* at Woods Hole (Rogers and Cole, 1925) that it should be rather easy and simple, as well as interesting, to measure their membrane capacity. The experiment should be easier than for the red blood cells which were ten times smaller, and the interpretation should be simpler because the eggs were spheres instead of biconcave disks. After the joint efforts of F. K. Richtmyer and W. J. V. Osterhout produced support, I looked for some way to avoid the many and considerable difficulties of a Wheatstone bridge.

Calculations from the data of Fricke and Morse showed that ammeter–voltmeter impedance measurements could be interpreted by the method of Philippson, equivalent to equation 1:2, to give the circuit of figure 1:4, and then the results of Fricke and Morse. The equipment, consisting of only an oscillator and two vacuum thermocouples, was highly satisfactory and the experiments went well; troubles began in understanding the data they gave (Cole, 1928b).

The calculation of $1/C\omega$ similar to equation 1:2, showed that C changed as about $C(\omega) = \bar{C}\omega^{-1/2}$. As Philippson had noted, a variable resistance $R(\omega)$ had been found to accompany such a capacity in other phenomena, such as electrode polarization, but he had then failed to include $R(\omega)$ in his analysis. As he also noted, such systems often gave $R(\omega)C(\omega)\omega = m$, a constant. Not only was this an uncomfortable assumption but there was no basis to estimate the constant, m, between the expected limits of zero and minus infinity. An estimate at the characteristic frequency of about 350 kc gave a membrane capacity of the order of 1 μF/cm^2—the first indication that any other cell might have a membrane similar to the red cell.

After considerable effort, no way had been found to determine $R(\omega)$ and $C(\omega)$ from measurements of $|Z(\omega)|$ alone, and this approach was abandoned. It was only a slight consolation to find later that this effort had been contemporary—as so often happens—with the successful solution now known as one form of the Kramers-Kronig integrals (Kronig, 1926). Here I might have computed the phase angle,

$$\Phi(\omega) = \frac{2}{\pi}\int_0^\infty \frac{|Z(\nu)|}{\nu^2 - \omega^2}\, d\nu.$$

These powerful results have become widely known (Cole and Cole, 1941; Bode, 1945, among others) and special techniques have been developed in many forms for their application in communications and optical problems (Tuttle, 1958).

The Fricke theory for the resistance and capacity of a suspension of spheres was the basis on which I had expected to interpret the *Arbacia* egg data. In reviewing the many derivations of the resistance equation I was particularly pleased with that given by Maxwell (1873, Art. 314). He placed a sphere of radius b containing a uniform suspension of n spheres, each of radius a and resistance r_2, in a medium of resistance r_1, into an infinite medium with a resistance r_1 and an initial uniform field. The potential, V, at an angle θ and a large distance, R, will be

$$V = A\left[R + \frac{1 - r_1/r_2}{2 + r_1/r_2} \cdot \frac{na^3}{R^2}\right] \cos\theta$$

where A is proportional to the applied field. Maxwell then replaced the spherical suspension with a sphere of uniform resistance r, such that V was unchanged,

$$V = A\left[R + \frac{1 - r_1/r}{2 + r_1/r} \cdot \frac{b^3}{R^2}\right] \cos\theta$$

to obtain

$$\frac{1 - r_1/r}{2 + r_1/r} = \rho \frac{1 - r_1/r_2}{2 + r_1/r_2}. \qquad (1{:}10)$$

This we will now call the Maxwell equation.

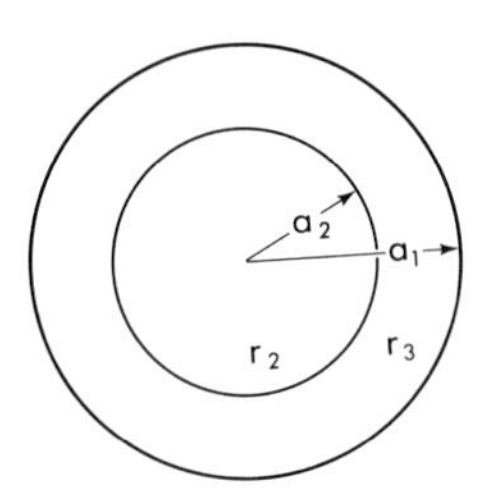

FIG. 1:5. Equivalent conductance of a stratified sphere, of radius a_1, in a medium of resistivity r_1 with a layer of resistivity, r_3 over a core of radius a_2 and resistivity r_2.

But I was surprised and even more pleased to find in his preceding Article 313 that Maxwell had obtained the equivalent resistance $\hat{r}_2$ of the two-phase sphere with a core of resistance r_2 and radius a_2 covered by a shell of resistance r_3 and outer radius a_1 shown in figure 1:5. There is now one more equation and one more boundary condition than in the problem of equation 1:10 and figure 1:3 and these determine the two additional constants for the potentials. The algebra then produced,

$$\hat{r}_2 = r_2 \frac{(2r_2 + r_3)a_1^3 + (r_2 - r_3)a_2^3}{(2r_2 + r_3)a_1^3 - 2(r_2 - r_3)a_2^3}. \qquad (1{:}11)$$

Danzer (1934), Fricke (1953b), and Pauly and Schwan (1959),

among probably many others, have carried through various calculations for suspensions with all of this complication of detail.

For $a_2 = a_1 - \delta = a - \delta$, and with the approximations $\delta \ll a$ and $r_3 \gg r_2$ I found (Cole, 1928a)

$$\hat{r}_2 = r_2 + r_3\delta/a$$

and replacing $r_3\delta$ by z_3, a surface impedance for a unit area in Ωcm^2,

$$\hat{r}_2 = r_2 + z_3/a. \tag{1:12}$$

This useful relation was almost obvious—after it was known. The internal field is uniform to give a uniform current density i_x and a potential at the inside of the bounding layer in figure 1:6,

$$V_2 = r_2 i_x x = r_2 i_x \, a \cos\theta.$$

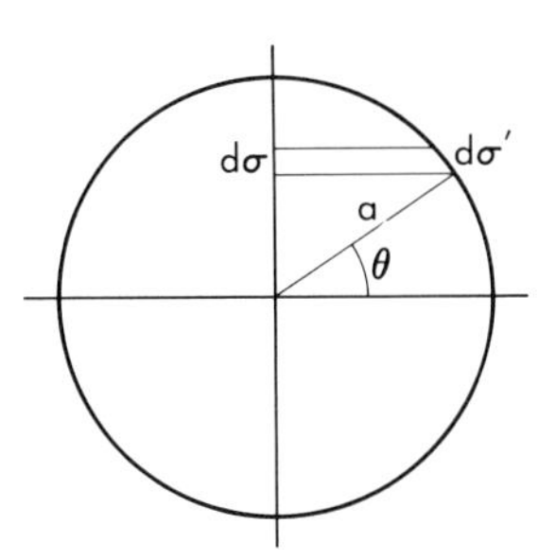

FIG. 1:6. Derivation of the equivalent resistivity of a membrane covered sphere or cylinder of radius a. $d\sigma$ is an element of area normal to the field $d\sigma'$ is its intersection with the surface.

The element of current in the tube of area $d\sigma$ crosses a surface area $d\sigma'$ to give a potential across it of

$$V_1 - V_2 = z_3 i_x \, d\sigma/d\sigma' = z_3 i_x \cos\theta$$

so that

$$V_1 = (r_2 + z_3/a) i_x \, a \cos\theta.$$

Since the potential at the surface of a uniform sphere with an equivalent resistance $\hat{r}_2$ is

$$V_1 = \hat{r}_2 i_x \, a \cos\theta$$

we again have

$$\hat{r}_2 = r_2 + z_3/a.$$

Returning to the impedance of a suspension as expressed by equation 1:10, and replacing r_2 by $\hat{r}_2$ as above and r by z, the impedance of the suspension is

$$z = r_1 \frac{(1 - \rho)r_1 + (2 + \rho)(r_2 + z_3/a)}{(1 + 2\rho)r_1 + 2(1 - \rho)(r_2 + z_3/a)}. \tag{1:13}$$

If now $z_3 \to \infty$ for $\omega \to 0$ and $z_3 \to 0$ for $\omega \to \infty$, the corresponding resistances are

$$r_0 = \frac{2 + \rho}{2(1 - \rho)} r_1; \; r_\infty = \frac{(1 - \rho)r_1 + (2 + \rho)r_2}{(1 + 2\rho)r_1 + 2(1 - \rho)r_2} r_1.$$

We identify r_0 with R_e of figure 1:4 as before from the current passing around the cells. Also

$$R_i = \frac{(1 - \rho)(2 + \rho)}{9\rho} r_1 + \frac{(2 + \rho)^2}{9\rho} r_2$$

so this circuit element is nicely separated into the outer and inner components of resistance to the additional current passing through the cells. The intuitive interpretation of the circuit is this time justified. However, a circuit of the type of figure 1:1b may also be an adequate equivalent of a suspension as pointed out by Fatt (1964). By comparison with equation 1:12, C_m is replaced by the equivalent impedance for the cells; R_m is for the paths around the cells, whereas R_i expresses the series resistance of the rest of the medium.

The interpretation is continued with the assumption that the membrane has a pure capacity $z_3 = 1/j\omega\, c_3$. More algebra gives

$$C = \frac{9\rho}{(2 + \rho)^2} a\, c_3$$

and in resistivity terms,

$$c_3 = \frac{2C}{(2 + r_1/r_0)(1 - r_1/r_0)a}.$$

These relations are a significant improvement upon the original equation 1:7 and lead to a red cell membrane capacity closer to 1 μF/cm^2.

At about this stage I became acquainted with the work of Carter (1925) and the representation of two-terminal impedances on the complex plane. The consideration of two frequency-dependent reactances in a resistance network was more general than was to be found useful for some time to come. But Carter introduced this problem in terms of the theory of a complex variable as applied to a resistance circuit with a single pure reactance. Any such circuit has an impedance of the form

$$Z = \frac{R_0 + j\omega\tau R_\infty}{1 + j\omega\tau} = R_\infty + \frac{R_0 - R_\infty}{1 + j\omega\tau}$$

and on the complex plane $Z = R + jX$, figure 1:7, these series components are

$$R = R_\infty + \frac{R_0 - R_\infty}{1 + \omega^2\tau^2}\,; \qquad X = -\frac{(R_0 - R_\infty)\omega\tau}{1 + \omega^2\tau^2}$$

shown as A and B in figure 1:7. Elimination of $\omega\tau$ between R and X gives the semicircle C of radius $(R_0 - R_\infty)/2$ with its center at $(R_0 + R_\infty)/2$, which is the locus of all values of R and X as ω goes from 0 to ∞. The maximum value $\bar{X} = -(R_0 - R_\infty)/2$, and the corresponding value $\bar{R} = (R_0 + R_\infty)/2$, appear at $\bar{\omega}\tau = 1$; this point defines the characteristic frequency $\bar{f} = 1/2\pi\tau$ of the circuit.

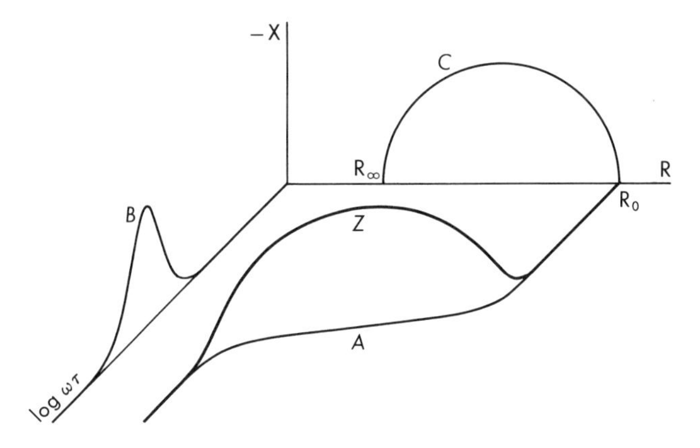

FIG. 1:7. Representation of the impedance Z of a system with a capacitive reactance and time constant τ. Curves A and B are the projections on the horizontal and a vertical plane to give the resistance and the reactance components R and $-X$ as functions of frequency $\omega/2\pi$. Curve C is the projection on the complex impedance plane, R vs. $-X$.

The data of Fricke and Morse were plotted as the impedance locus of figure 1:8 and, although they could be approximated by the semicircle required by theory, there was a systematic deviation; and a circular arc with a depressed center was a better representation. This reminded me not only of the small difference found for blood by Philippson and in a similar calculation of the data of Fricke and Morse (Cole, 1928b), but also of the work reviewed by Gildemeister (1928a) on human skin; he found only a straight line, representing a constant phase angle polarization element, on the impedance locus.

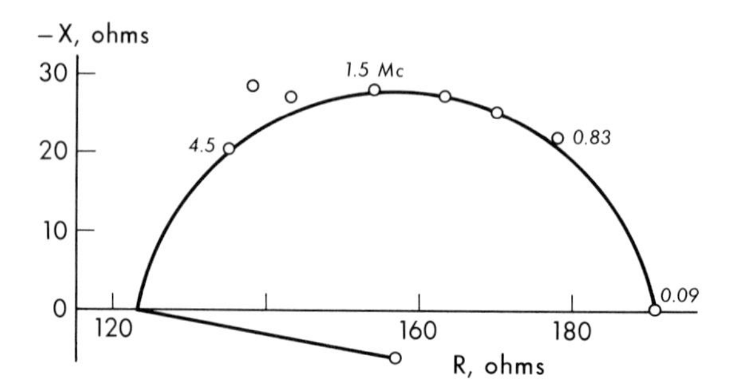

FIG. 1:8. Complex impedance plane representation, R vs. $-X$, of the original data for a calf red blood cell suspension.

A constant phase angle was assumed for $z_3 = r_3 + jx_3$ in equation 1:13 in the form $x_3 = mr_3$, so $\tan \phi = m$. Direct algebraic manipulations showed (Cole, 1928a) that as x_3 varied from 0 to ∞ the locus was again a circle—but with the depressed center as in figure 1:9. The half angle between the radii to R_0 and R_∞ was now ϕ as given above. It was also found that $|z_3|$ was proportional to u/v where u and v are the distances of the corresponding impedance Z from R_∞ and R_0.

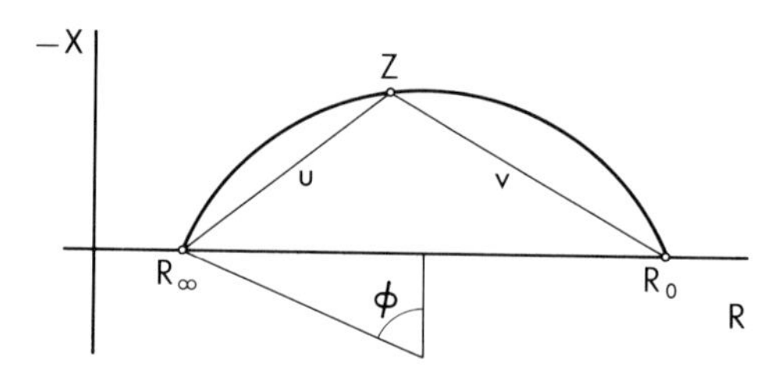

FIG. 1:9. Complex impedance plane representation of a system containing an element of constant phase angle ϕ. The absolute impedance of this element is computed at each frequency from the chords u and v.

I was somewhat chagrined, later, to find that Campbell (1911) had dealt quite generally with such problems. He expressed the impedance Z of all circuits with a single variable element x in the form

$$Z = \frac{a + bx}{c + dx}. \tag{1:14}$$

This is the bilinear transformation which is well known in complex variable theory to convert any circle in the x plane into a circle in the Z plane. So I should have known without labor that for a straight line element, x, whether $\phi = 90°$ or less, the Z locus was a circle. Furthermore, Campbell gave the coefficients of the transformation as transformer constants for the two circuits we have been considering—among others. Mechanical linkages have long been made to perform such transformations, the best known being that invented by C. N. Peaucellier in 1864. This and other linkages designed to produce a straight line from a circle, the inverse of our problem, are given by Kempe (1877). Since then it has become far easier to buy a pair of nonlinear potentiometers with outputs

$$R - R_{\infty} = (R_0 - R_{\infty}) \tanh \theta$$
$$X = (R_0 - R_{\infty}) \operatorname{sech} \theta$$

for shaft rotations, $\theta = \alpha \log \omega\tau$, given by a linear potentiometer. Through operational amplifiers these outputs may then be combined on an $X - Y$ plotter.

As these manipulations of constant phase angle membrane systems were made and published, the conventions and definitions were changed continually. This was so frustrating that it led Rothschild (1946) to assemble the whole as a self-consistent and orderly structure!

E. Newton Harvey had told me about the big white sea urchin, *Hipponoë* now *Tripneustes*, which he had found "full of eggs" in Bermuda during the Christmas holidays. After scouting all of the echinoderms there and finding that the breeding season of this urchin started about the end of September, I went back the next fall to take advantage of yet another form, and also because this particular one produced up to 100 cc of eggs from a single specimen. The job was not easy, even with the help of H. J. Curtis, because of emergencies which were beyond the facilities of the Station, but neither was it dull. The first satisfactory run gave a membrane capacity of nearly 1 $\mu F/cm^2$, but with the utterly unexpected phase angle $\phi = 90°$, as close as could be measured. This became very disturbing as the complete series of unfertilized egg suspensions gave an average of 0.87 $\mu F/cm^2$; and each one with the 90° phase angle as in figure 1:10. Why was *Hipponoë* so different from *Arbacia?* Was I only unlucky to have used an inadequate method on *Arbacia* and an unduly complicated one in Bermuda? But there were other questions as well as some progress (Cole, 1935).

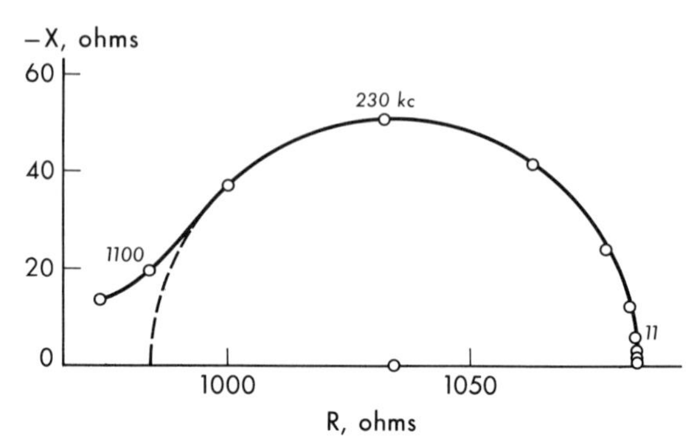

FIG. 1:10. Complex impedance locus, resistance vs. reactance, for a suspension of sea urchin (*Hipponoë*) eggs.

The high-frequency data gave indication of another dispersion region for unfertilized eggs and a clear departure above the characteristic frequency in the fertilized eggs. These results are not yet explained and they prevent a good estimate for the internal resistivity, which appears to be quite high—ten or more times that of sea water.

The resistance equations for nonconducting spheres had been accurately checked with cream (Fricke and Morse, 1925b) and, less completely, for nonconducting spheroids with red blood cells at low frequencies. But the volume concentrations of the egg suspensions had to be measured to assure the validity of the sphere equation for them. Of several methods, the only satisfactory one was a conductance titration with iso-osmotic sucrose. This gave concentrations for the *Hipponoë* egg suspensions that were all within ± 2 percent of that resulting from the assumption that both fertilized and unfertilized eggs were effectively nonconducting at low frequencies in equation 1:10. As a consequence it had to be assumed that the membranes were nonconducting within the accuracy of either the volume concentration or the resistance measurements. It was not until some time later that this result was used to set a lower limit on the membrane resistance, as shown in figure 1:28.

Perhaps for no good reason except that it could be done, I measured the membrane capacity of osmotically swollen unfertilized eggs. Probably I expected an increase of capacity as the membrane became thinner on stretching, so I was completely unprepared to find that the capacity decreased linearly with the increase of area, figure 1:11. Iida (1943b) also obtained this result for another urchin egg. At long last an explanation was given by Rothschild (1957) that the normal corrugations of the membrane were only flattened in swelling without stretching and thinning. This possibility has important but unexplored implications.

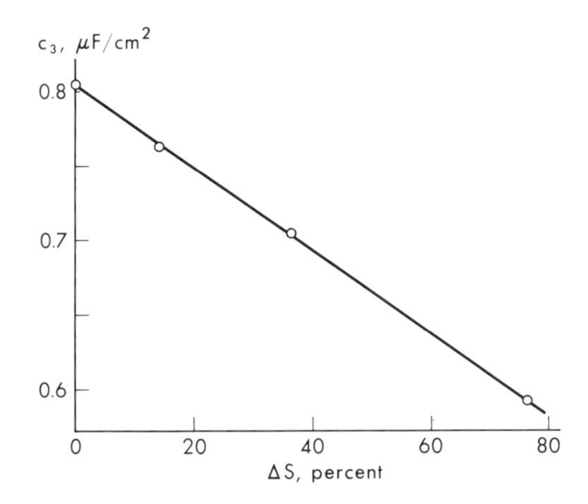

FIG. 1:11. Decrease of sea urchin (*Hipponoë*) egg membrane capacity, c_3 with increase of membrane area ΔS by osmotic swelling.

Although not *Arbacia*, an echinoderm egg had proven even more interesting and useful than I had hoped. The fact of a 90° phase angle membrane was new, apparently unique, and certainly provocative. But probably as important was the fact that experience and an abundance of eggs encouraged the replacement of the original bubbler measuring cell for the suspensions by one of a buret form (Cole, 1928b, 1935). The first of these cells required over 5 cc of suspension, but later this amount was reduced to 1.5 cc, 0.8 cc, 0.2 cc and finally to but a single egg.

It now became possible, and far more than a matter of mere curiosity and tidiness, to return to *Arbacia* for better information on the nature and extent of its differences from *Hipponoë*. My brother, Robert H. Cole, joined me again to become the guests of Fricke and Curtis. They gave us the nighttime use of their complete laboratory, including bridge and resonance equipment covering the range from 1 kc to 16 Mc. Although the oyster beds of Cold Spring Harbor had been dredged for starfish, we found a few females that had escaped. Then as the urchins came in season they were shipped from Woods Hole. The eggs from only a single female of either *Asterias* or *Arbacia* were required for the buret type conductivity cell with a volume of 1.5 cc.

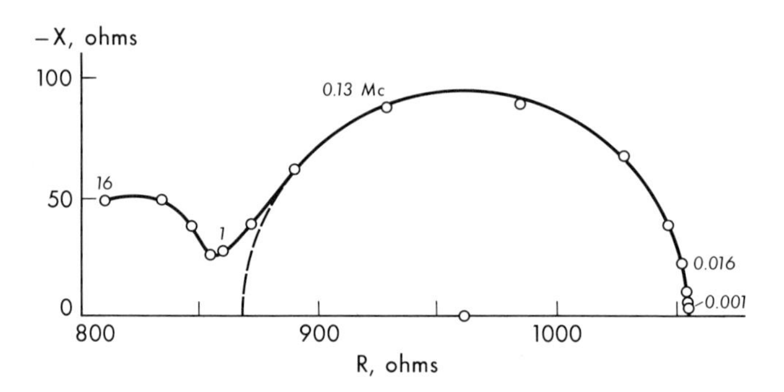

FIG. 1:12. Complex impedance locus for starfish (*Asterias*) egg suspension.

The data such as figure 1:12 on the unfertilized *Asterias* eggs (Cole and Cole, 1936a) gave an average membrane capacity of 1.1 $\mu F/cm^2$ with a phase angle of 90° between 8 kc and 256 kc. The extended high frequency range covered more than half of another dispersion region, with a characteristic frequency of 12 Mc. This could be extrapolated to give an internal resistivity of 4.5 times that of sea water. As the frequency decreased below 8 kc the capacity increased significantly above that to be expected for the simple model. However, for a purely reactive membrane it was useful to turn from the impedance of the suspension to its admittance, $Y = (1/R_P) + j\omega C_P$, where R_P and C_P were the parallel resistance and capacity. There is then a single linear relation between C_P and $1/R_P$ from $R_P = R_\infty$, $C_P = 0$, at high frequency and $R_P = R_0$, $C_P = C_0$ at low frequency as found for the considerable range shown in figure 1:13. But the clear break at C_0 indicates a different phenomenon. Following Fricke and Curtis (1936, 1937) in their interpretation of similar data on red cells and colloids, this was attributed to a tangential surface conductance; but an analysis of it was long in coming (Schwartz, 1962). The *Asterias* experiments raised new problems as they apparently put *Arbacia* in a minority position that seriously needed reexamination.

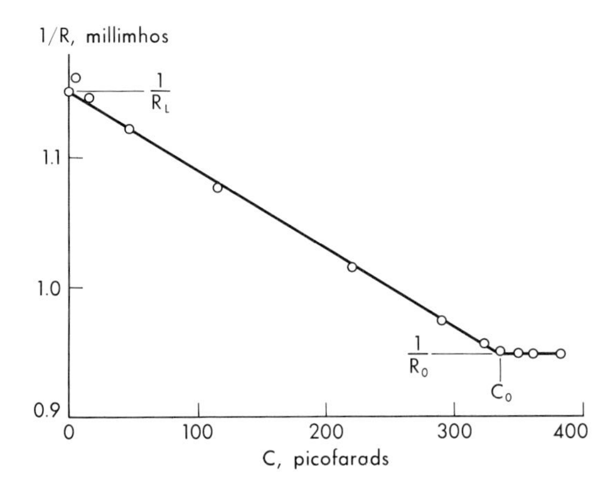

FIG. 1:13. An inverse representation of starfish (*Asterias*) egg suspension data, parallel conductance $1/R$, vs. parallel capacity C to show the added capacitive component and the interpolation for the membrane capacity C_0 at zero frequency.

When *Arbacia* eggs (Cole and Cole, 1936b) became available, their electrical behavior did not differ in any major way from those of the Bermuda urchin and the starfish eggs. The membrane capacity averaged 0.74 μF/cm^2 again with a 90° phase angle over the considerable frequency range from 4 kc to 250 kc. And, as for the other forms, there was evidence of another dispersion region at both extremes of frequency. A less satisfactory high frequency extrapolation than for *Asterias* gave an internal resistivity of about four times that of sea water. Such extrapolations could not be taken seriously without a defensible mechanism for the high-frequency dispersion. Since red cells without nuclei had not shown this characteristic, it was attractive to attribute it to the egg nuclei. This postulate produced nuclear membrane capacities of the order of 0.1 μF/cm^2 which seemed rather small—but why must a nuclear membrane be the same as a plasma membrane? On the other hand this frequency range was that of dielectric dispersion for some proteins (Oncley, 1943). But the red cells had high protein contents, too. And so the data were not only unexplained but, much more regrettable, a unique tool for the investigation of the interior of a living cell could not be used effectively.

The three echinoderm egg experiments were powerful and were highly satisfactory in many ways. They allowed the use of simple theory because they were such beautiful spheres—without the complex geometry of a red cell—and they cooperated by generally excellent support of the theory. They further cooperated by producing for the first time the simple, 90°, static membrane capacity that Fricke had postulated and interpreted as a dielectric capacity—without even a trace of polarization impedance.

The first *Arbacia* conclusions about the membrane impedance were now quite thoroughly repudiated and corrected. But I was thoroughly confused and probably not a little defensive. I did not really believe that the eggs in my first experiments were in as bad shape as I suggested later (Cole, 1938), and it was most unreasonable for those eggs to have become so similar to all of the tissues that were such a problem. I have no idea how an acceptable answer ultimately appeared, but I do remember opening urchin after urchin, not only to find females in those days before electric sexing, but also to find enough females with enough eggs to give usable concentrations in the more than 5 cc volume of the measuring cell. With experience, I had learned to admire the uniformity of the eggs from a single specimen, but I had also come to respect the individuality of the urchins as expressed in the several characteristics of their eggs. The key quantity in the theory and interpretation was not that just the membrane capacity c_3 be independent of frequency, but that the cell radius a must also be included to make the product c_3a a constant. The finding of a 90° phase angle for the eggs of single urchins was a tribute to the uniformity of this product, but, since the distribution of a is very narrow, the distribution of c_3 must be similarly narrow. It was known by then that the capacity distribution among urchins was rather broad, as are the egg radii, but there was no known interdependence. So the theory of impedance for distributions of c_3a needed to be considered. And this might account for the small but definite phase difference which Fricke (1933) reported for rabbit red cells which had been made spherical with lecithin, figure 1:14.

Fertilization

There had been clear indications of a 90° membrane capacity of 2–3 $\mu F/cm^2$ which appeared at low frequencies after fertilization of the *Hipponoë* and *Arbacia* eggs. This is one of the few major physio-

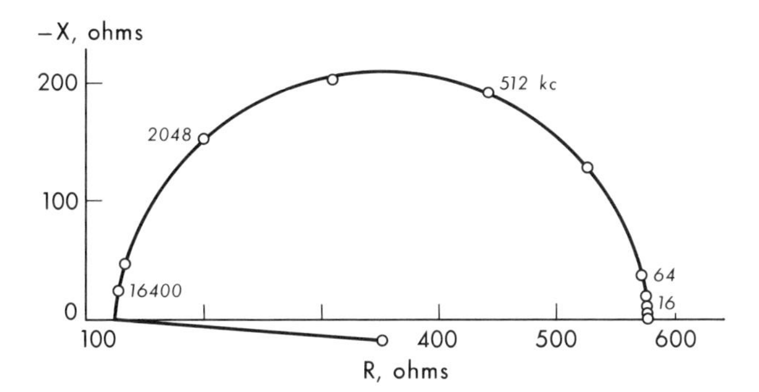

FIG. 1:14. Complex impedance locus, resistance vs. reactance, for a lecithinated rabbit red blood cell suspension.

logical changes of capacity and I hope that it may be a significant factor in understanding the dramatic phenomenon of fertilization.

The *Hipponoë* measurements made on but two suspensions of fertilized eggs were in good agreement at more than twice the average membrane capacity of the unfertilized eggs. As for *Hipponoë*, the *Arbacia* egg membrane capacity increased on fertilization to an average of 3.1 $\mu F/cm^2$, figure 1:15. These increases were confirmed on several other forms, first by Iida (1943a) and more recently by Hiramoto (1959).

For both *Hipponoë* and *Arbacia* in the 64 kc to 2 Mc region there was also an essentially identical departure from the low-frequency

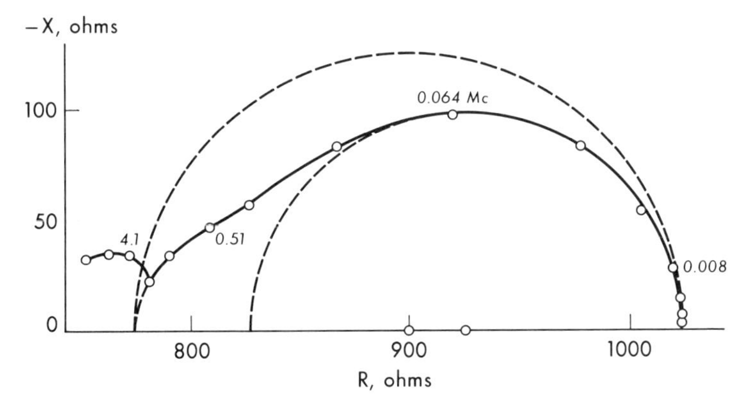

FIG. 1:15. A complex impedance locus for a suspension of fertilized *Arbacia* eggs showing evidence of two time constants.

behavior, as in figure 1:15. The ingenious explanation was proposed—and illustrated (Cole, 1937)—that the plasma membrane capacity remained unchanged as a second high-capacity barrier was produced at the outer boundary of the fertilization membrane. One of the serious difficulties with this idea was that the analysis was inadequate and the conclusions were found to be self-inconsistent!

Cole and Spencer (1938) continued this investigation with *Arbacia* eggs lightly centrifuged into a 0.8 cc measuring cell in order to reduce the handling. In addition to the routine impedance vs. frequency measurements, aliquots of the suspension were counted and the diameters of the plasma and fertilization membranes were measured for 50 eggs. The impedance loci were almost perfect semicircles for both unfertilized and fertilized eggs, to give average membrane capacities of 0.86 $\mu F/cm^2$ and 3.3 $\mu F/cm^2$ respectively. The nonconducting volume concentration ρ_0 was calculated from r_0 and r_1 on the assumption of a nonconducting membrane in equation 1:10. This agreed to within a few percent with the plasma membrane enclosed volume ρ_p, obtained by counts and diameters while the corresponding volume concentration enclosed by the fertilization membrane was 30 percent larger, figure 1:16.

This careful, tedious work made it quite evident that the plasma

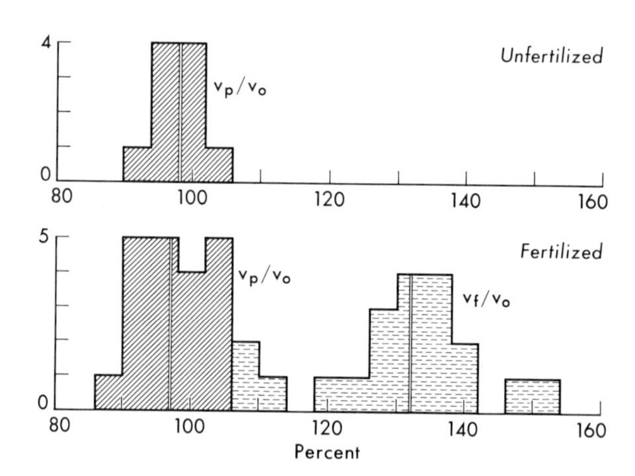

Fig. 1:16. Comparison of the non-conducting volume v_o from electrical measurements with the plasma volume v_p for unfertilized *Arbacia* eggs (upper) and with v_p and the fertilization membrane volume v_f (below). The ordinates are the number of determinations for ratios in 5 percent intervals.

membrane was effectively nonconducting in both unfertilized and fertilized eggs, that the conductivity of the fertilization membrane was not appreciably different from sea water, and that the capacity of the plasma membrane increased nearly fourfold on fertilization, figure 1:17. Another result was the strong suggestion that instead of both plasma and fertilization membranes being present on each and every egg, there had been a significant fraction of unfertilized eggs in the previous measurements on fertilized suspension. This was certainly no great credit and I did not relish the need to admit it in trying to set the record straight (Cole, 1938). Beyond the immediate additional warning of the hazards of inhomogeneity, the indication of a considerable change in the membrane structure on fertilization was an important challenge (Rothschild, 1956).

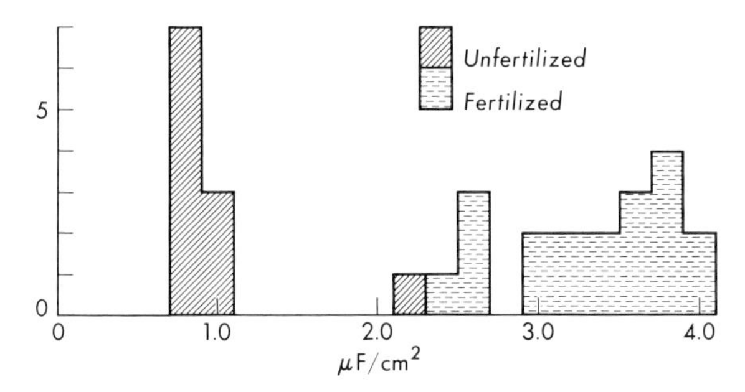

Fig. 1:17. Comparison of the membrane capacities of unfertilized and fertilized *Arbacia* eggs. The ordinates are the number of results within intervals of 0.1 $\mu F/\text{cm}^2$.

Alberto Monroy has however given me more details of the surface phenomena in the fertilization process of echinoderm eggs. The membranes of the cortical granules fuse with the plasma membrane lying immediately above as a pinocytosis process in reverse. After their interiors communicate with the external sea water, an average of 30,000 granules, 0.8 μ in diameter, then increase the effective surface area to 4.5 times that of the 74 μ unfertilized egg. This is not unreasonably different from the nominal capacity increase of 3×. Further, at the early stages of this exteriorization, a connecting channel of radius a can be expected to have a resistance of at least

$$R_g = r_1/2\pi a$$

which may complicate the measured impedance. As seen from

figure 1:18, there would be two dispersion regions. The capacity $C_u + C_g$ will be measured at sufficiently low frequencies but for $\omega \gg 1/R_gC_g$ only C_u will be important. This explanation for data such as figure 1:15 could only apply for a few minutes after fertilization. So it is improbable, although it would be much more palatable than the alternative that I had accepted.

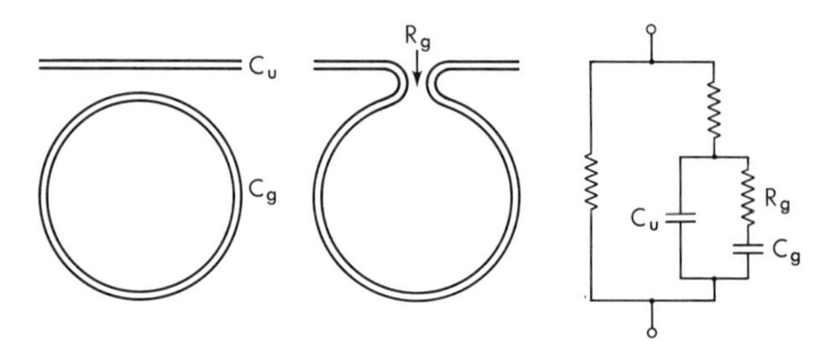

FIG. 1:18. A possible explanation for the increase of the plasma membrane capacity on fertilization. The plasma and cortical granule membranes are indicated at the left for the unfertilized egg and in the center shortly after fertilization. An equivalent circuit diagram on the right shows the plasma membrane capacity C_u and the granule capacity C_g in series with the resistance of the channel, R_g.

Further Work on Suspensions

Within a few years, Fricke (1931) was able to state that the red cells of man, calf, dog, rabbit, chicken, and turtle had this same membrane capacity of 0.8 $\mu F/cm^2$. He also stated that this capacity did not depend upon the composition of the suspending solution and that it was independent of the measuring frequency. On a similar basis he, with Curtis (1934a,b; 1935), came up with more and similar facts on these and other living cells.

Velick and Gorin (1940) extended the evaluation of the Fricke integrals, 1:9, to calculate the low-frequency resistance of ellipsoids. These they applied to most ingenious measurements on flow-oriented avian red blood cells. Expressing their results as

$$r = r_1 \frac{1 - \rho + \gamma\rho}{1 - \rho}$$

where γ is the form factor, they showed that by successive dilutions of a concentrated suspension ρ, v to a final ρ', v'

$$\frac{v}{v'} = \frac{\rho'\gamma}{(r/r_1) - 1} + \rho'.$$

Plots of v/v' vs. $1/[(r/r_1) - 1]$ then gave constant values of γ up to $\rho = 50$ percent. I have been told that this procedure discriminates against low values of ρ to approximate the data at the high ρ for which the theory is inadequate, but I have no proof of this.

It is now convenient—although probably not entirely correct as to reactance—to present the impedance of a dilute suspension in the general form

$$z = r_1 \frac{(1 - \rho)r_1 + (\gamma + \rho)(r_2 + z_3/a)}{(1 + \gamma\rho)r_1 + \gamma(1 - \rho)(r_2 + z_3/a)}.$$

The theory has also been carried through for both spheres (Schwan and Morowitz, 1962) and ellipsoids (Fricke, 1953b), with a finite membrane thickness and for several layered membranes (Fricke, 1955), but I do not know that these more general calculations have been found useful.

It is to be noted that the simple suspension theory becomes questionable for volume concentrations of about 50 percent and is almost certainly inadequate at 70 percent, as found for muscle. The theory must be replaced by a new approach for tissues and packed cells occupying nearly 100 percent of the available volume.

In further investigations on red cells, Schwan (1963) has found electrode effects at high volume concentrations. For osmotically lysed cell suspensions, Schwan and Carstensen (1957) also obtained a 90° phase angle. This was attributed to a more nearly spherical form than in normal cells, where the phase difference may be explained by the two relaxation times corresponding to the two principal axes (Fricke, 1953a).

Fricke continued work on suspensions, and with Curtis he found 0.6 $\mu F/cm^2$ for a yeast (Fricke and Curtis, 1934b) and 1.0 $\mu F/cm^2$ for leucocytes (Fricke and Curtis, 1935). He had obtained an uncertain result with a bacterium (1933) in which the cellulose cell wall may have been a complicating factor. As was found for *Nitella* (Curtis and Cole, 1937; Tasaki, unpublished) and *Chara* (Gaffey and Mullins, 1958), this also has the conductance characteristics of an ion exchanger in two other bacteria (Carstensen et al., 1965). Here the thin layer approximation of equation 1:12 is not valid and equation 1:11 is needed.

Since the Second World War, Schwan (1957) and the group with him have worked with small cells and organelles, to find them included in the earlier generalization with 1 $\mu F/cm^2$ membranes. First came the bacterium *E. coli*, with 0.7 $\mu F/cm^2$ (Fricke et al., 1956), then mitochondrium, 0.5–1.1 $\mu F/cm^2$ (Pauly et al., 1960); later the pleuro-pneumonia-like organism with a diameter of 0.25 μ gave 1.3 $\mu F/cm^2$ (Schwan and Morowitz, 1962).

TISSUE IMPEDANCES AND CELL PHASE ANGLES

Following the fiasco of the sea urchin egg experiments and the consequent development of the representation and interpretation of impedances on the complex plane, the published data on various tissues were reviewed and recalculated and bridge measurements were begun at frequencies below 90 kc. The approximations to circular arc loci led to the conclusion that constant phase angle elements could at least be used to represent the available data over considerable parts of the frequency ranges used, figure 1:19 (Cole,

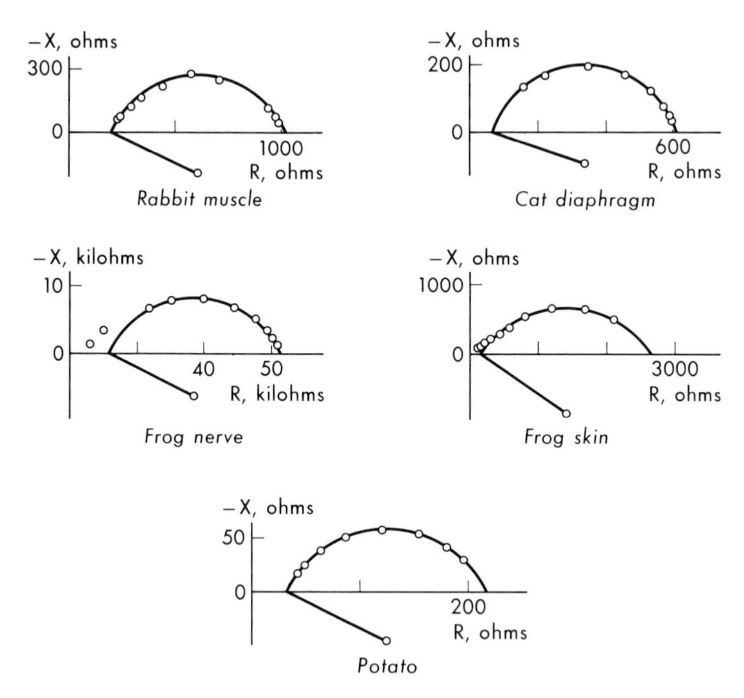

FIG. 1:19. Some early impedance measurements on tissues shown on the complex plane.

1932). In contrast to the only slight departure from 90° that appeared in the red cell experiments, these other data gave a range of considerably smaller phase angles ϕ as given in Table 1:1 (Cole, 1933).

TABLE 1:1

Tissue	Phase angle ϕ
Frog sciatic nerve (Lullies, 1930)	64°
Rabbit muscle (Fricke, 1931)	65°
Human skin (various)	~55°
Valonia (Blinks, 1926)	55°
Frog skin	55°
Cat diaphragm	71°
Potato	64°
Laminaria	78°

Aside from the fact that low-frequency data on human skin gave no indication of either a characteristic frequency or a zero frequency resistance, the most marked departure from a circular arc appeared in the high-frequency results on nerve. I find it interesting to note that this worrisome anomaly was later found to be a necessary consequence of the longitudinal electrode arrangement.

In the meantime we continued with other tissues. *Laminaria* seemed particularly important because of the need to understand further Osterhout's (1922) very extensive work at 1 kc. Although this kelp was best collected by diving at chilly dawn in Long Island Sound, Bruce M. Hogg and I (unpublished) repeated some of the more dramatic experiments and fully confirmed Osterhout's results. Furthermore, we found a rather constant phase angle of about 78°, figure 1:20, and that the impedance at 1 kc was near the zero frequency extrapolation R_0, while the infinite frequency extrapolation R_∞ remained nearly fixed at the sea water and dead tissue value. Aside from this, all that could be seen was that the impedance arcs enlarged and shrank with the 1 kc resistances, which was later shown to be explained by the change of membrane ion permeability that Osterhout had proposed.

Polarization

Electrode polarization had been interesting to Fricke (1924a), but apparently it was not until his only published experiment on a tissue, an unspecified rabbit muscle, that he became seriously con-

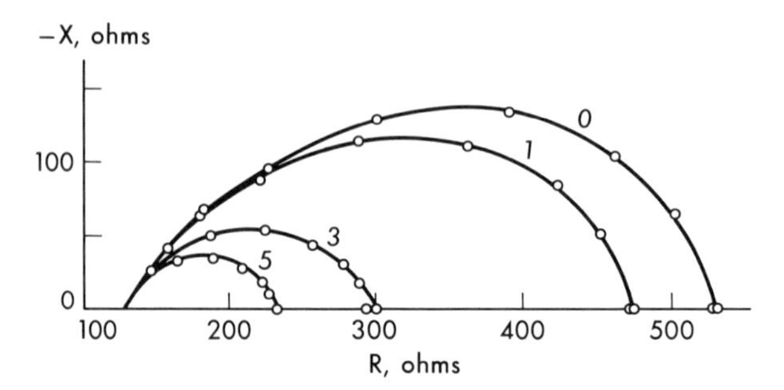

FIG. 1:20. Impedance measurements on the kelp, *Laminaria*, shown on the complex plane, before 0, and at hours indicated after replacing the external sea water with iso-osmotic NaCl.

cerned about it as a membrane property. Here he found (Fricke, 1931) that the muscle data could be represented by a membrane impedance of a resistance R and a capacity C in series. These were both frequency dependent to give a constant phase angle

$$C\,\omega\,R = m = \tan\phi$$

where $m = 0.46$, and with the capacity variation

$$C \sim \omega^{-x}; \qquad x = 0.28.$$

He did not estimate this capacity for a unit area of muscle membrane but considered this behavior to be due to polarization at the cell surface and to be more usual than the static capacity of the red cell membranes.

Fricke (1932) then published a theory of electrolytic polarization based on Fourier integrals, giving the frequency exponent for a constant phase angle element as expressed by x and m above.

The Fourier integrals, or Fourier transforms, are an important classical part of mathematics, physics, and engineering. Although largely supplanted by other more powerful and general methods, the Laplace transform in particular, the Fourier transform is still widely applicable. It also has an attractive conceptual simplicity in presenting the relations between the steady state frequency phenomena that we have been considering and the transient behaviors as a function of time, such as the potential V after a sudden change of current I. This real time analysis seems less suspect to many than a complex variable which in turn may be preferable to a contour integral. The literature on Fourier methods for both analytical and numerical functions is tremendous, but at least an introduction is provided by Weber (1954).

For many of our simple problems, the procedure and the collection of integrals by Campbell and Foster (1931; hereafter cited as CF) is entirely adequate. Following them, the cause $I(t)$ may be expressed as a continuous spectrum of frequencies $\bar{I}(\omega)$ with $0 \leq \omega \leq \infty$ given by

$$\bar{I}(p) = \int_0^\infty I(t)e^{pt}\,dt$$

where $p = j\omega$. For this example, $I(t) = 0$, $t < 0$, and $I(t) = I_0$, $t > 0$, and

$$\bar{I}(p) = I_0/p \qquad \text{(CF No. 415).}$$

Thus a current step is equivalent to the continuous application of all frequencies, with amplitudes inversely proportional to the frequency. Similarly, for a very short pulse of current,

$$Q_0 = \int_0^\infty I(t)\,dt$$

and all frequencies are of equal amplitude (CF No. 403.1). It is then no surprise to find that for $I(t) = I_0 \exp(-\lambda t)$ the frequency spectrum

$$\bar{I}(p) = I_0/(p + \lambda) \qquad \text{(CF No. 438)}$$

includes both the step and the pulse as limits for λ small and large. If now the impedance $Z(p)$ of the circuit is known,

$$\bar{V}(p) = Z(p)/\bar{I}(p)$$

and the synthesis of these amplitudes for all frequencies is given by

$$V(t) = \int_0^\infty \bar{V}(p)e^{pt}\,dp.$$

As they should, $\bar{V}(p)$ given by Q_0, V_0/p or $V_0/(\lambda + p)$ produce the pulse, step, or exponential transient for $V(t)$. If now we apply a step of current to a resistance R and capacity C in parallel,

$$\bar{I}(p) = I_0/p;\ Z(p) = \frac{1}{C} \cdot \frac{1}{p + 1/RC}$$

and

$$V(t) = RI_0[1 - \exp(-t/RC)]. \qquad \text{(CF No. 448)}$$

Conversely a time response may be analyzed by many procedures to produce the equivalent frequency characteristics. One numerical method has been discussed and applied in biological situations by Teorell (1946). This very old, poor man's approach is both nice and effective. It had been partially mechanized some time before, but this is indeed ancient history. The Fast Fourier Transform, outlined by Cooley and Tukey (1965), revolutionized these procedures and was rapidly translated into computer languages.

We now apply a step current to an electrolytic polarization impedance $Z(p)$, and accept Fricke's ad hoc assumption that $V(t) = V_0 t^{-\alpha}$ to find that

$$Z(p) = \frac{\bar{V}(p)}{\bar{I}(p)} = \frac{V_0}{I_0}\Gamma(\alpha - 1)/p^{(\alpha-1)} \qquad \text{(CF No. 516)}$$

where Γ is the gamma function. Then for

$$Z = Z_0(j\omega)^{(1-\alpha)} \tag{1:15}$$

we have the constant phase angle tan $X/R = j^{(1-\alpha)}$ and the absolute impedance

$$|Z| = \sqrt{R^2 + X^2} \sim \omega^{(1-\alpha)} \tag{1:16}$$

to give

$$\phi = \alpha\pi/2 \tag{1:17}$$

and

$$X = -\frac{1}{C\omega} = R \tan \phi = Z_0\omega^{(-1+\alpha)}.$$

This answer was immediately available as the inverse of a Heaviside operational calculus expansion (Bush, 1929). A more powerful argument for the same result is found from the Kramers-Kronig integral, while a simpler but more subtle argument was given by Cole and Cole (1941). Since $j\omega$ appears because of the application of a cause of the form exp $j\omega t$ and only linear operations are used, any result must be expressed as a function of $(j\omega)$. Consequently, if the phase angle given by $j^{-\alpha}$ is constant, the impedance variation must be given by $\omega^{-\alpha}$ since it is required that $Z = Z_0(j\omega)^{-\alpha}$, and Fricke's relation follows.

Fricke showed that the available experimental data on electrode polarization as well as his results for a rabbit muscle followed this theoretical relation as well as could be expected. It has since received many and various experimental confirmations although, as everyone freely admits, no fundamental explanation for either the constant phase angle or the frequency dependence of the resistance or capacitive components for such an impedance has yet been found.

There was the surviving hope that a polarization impedance might be some expression of behavior of ions in or near cell membranes. For $\alpha = 1$ and $\phi = 90°$ the membrane would have the properties of a perfect static capacity in which ion movement and leakage current had no part. So this might be an expression of a completely ion-impermeable membrane. At the other extreme, with $\alpha = 0$ and $\phi = 0$, the polarization impedance came to be a pure resistance which might correspond to a relatively complete membrane permeability for all ions. In between these two extremes lay the case of a selective permeability. A membrane permeable to a single ion would then be a reversible electrode for that ion and, according to the theory of Warburg (1899), the diffusion polarization of such an electrode gave $\alpha = 0.5$ and $\phi = 45°$.

Fricke (1931) ascribed the constant $\phi < \pi/2$ for a rabbit muscle to an electrolytic polarization resulting from a current flow across the fiber membranes. By 1933 he suggested that the slight departure from $\phi = 90°$ for red cell membranes might be a polarization effect while that for tissues could be a result of heterogeneity. But by 1934 the enormous increases of capacities observed with Curtis for yeast below 16 kc, and later the smaller capacity increases of white cells (1935), were ascribed to electrolytic polarization membrane current flow phenomena along with similar effects from saponin treated red cells. Then similar observations on suspensions of nonconducting particles placed the primary emphasis on the tangential phenomena of electrophoresis and surface conductance (Fricke and Curtis, 1937).

Muscle

With no basis for any real optimism, Bozler and I undertook to see what we could find out about the frog sartorius muscle and rigor (Bozler and Cole, 1935). In an extending retrospect this effort continues to have been a fortunate one. The work went well, both in experiment and in analysis, but—more importantly—it has been so often and so much questioned and extended that it is now largely supplanted by the results of powerful and sophisticated attacks (Fatt, 1964). It is not at all unrealistic to suspect that all this progress may soon lead to a concept of electromechanical coupling that will outdate everything that has gone before!

It was a simple matter to calculate the impedance of a suspension of parallel circular cylinders measured with current flow perpendicular to the cylinder axes. The problem is the solution of the Laplace equation 1:6 in a plane,

$$\nabla^2 V = \frac{\partial^2 V}{\partial \rho^2} + \frac{1}{\rho} \cdot \frac{\partial V}{\partial \rho} + \frac{1}{\rho^2} \cdot \frac{\partial^2 V}{\partial \theta^2} = 0$$

and is analogous to that for spheres. Similarly following the procedure used by Maxwell for spheres and described by Cole and Curtis (1936), the resistivity r of the suspension of uniform cylinders, resistivity r_2 and volume concentration ρ, in a medium of resistivity r_1, is now

$$\frac{1 - r_1/r}{1 + r_1/r} = \rho \frac{1 - r_1/r_2}{1 + r_1/r_2}. \qquad (1{:}18)$$

Again the equivalent resistivity $\hat{r}_2$ of the cylinder of radius a with a surface membrane impedance z_3 is

$$\hat{r}_2 = r_2 + z_3/a$$

and for a pure capacity $z_3 = 1/j\omega c_3$ the capacity of the suspension C gives

$$c_3 = \frac{C}{(1 + r_1/r_0)(1 - r_1/r_0)a}. \qquad (1{:}19)$$

As had happened before, I found later that Rayleigh (1892) had given equation 1:18—but as only a first approximation for a rectangular array of cylinders that he treated in detail!

With a developmental model of the Wheatstone bridge to come, measurements were made from 1 kc to 1 Mc. The impedance loci for the muscles were well represented by circular arcs giving a phase angle of 70°, both relaxed and contracted as shown in figure 1:21.

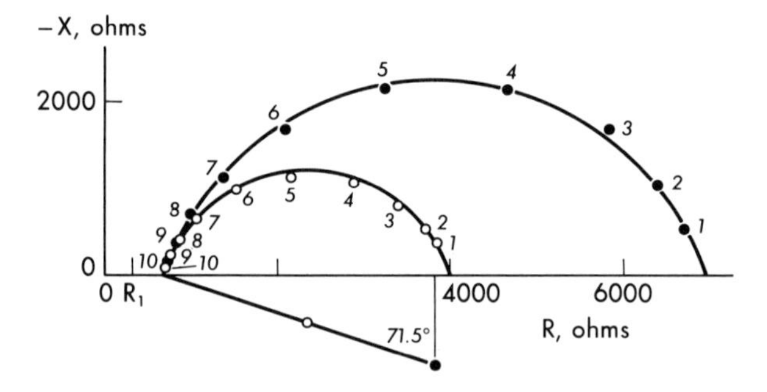

FIG. 1:21. Transverse impedance of frog sartorius muscle on the complex plane. Solid circles are before and open circles after the muscle was in rigor. The ratio between successive frequencies is $10^{1/3}$ over the range from 1.08 kc to 1.08 Mc. R_1 is the resistance for Ringer's solution.

The infinite frequency extrapolation r_∞ remained unchanged, so the internal resistivity r_2 of the fibers was thus presumably constant. Without a simpler way to determine the volume concentration ρ, the r_∞ was also measured with two dilutions of the outside Ringer's solution by iso-osmotic sugar solution. These were self-consistent, figure 1:22, for $\rho = 77$ percent and $r_2 = 260$ Ωcm or about three times the resistance of Ringer's solution.

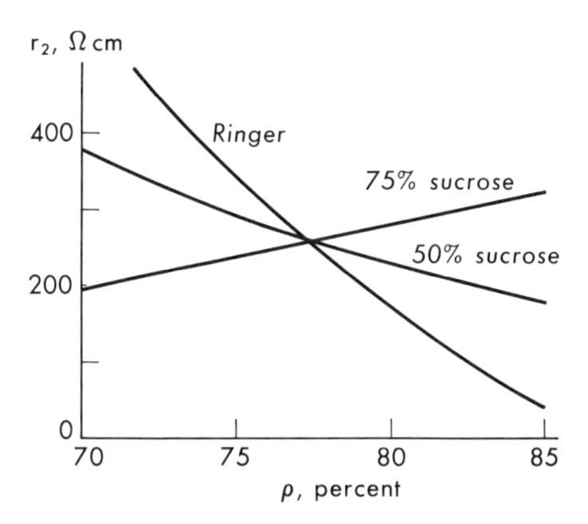

FIG. 1:22. The relations between the myoplasm resistivity and the fiber volume concentrations as calculated from the high frequency impedance data on a muscle in Ringer's solution and two dilutions with iso-osmotic sucrose.

When the experiments were well along, it belatedly occurred to me that the complex plane circle was not at all an adequate demonstration that the membrane impedance was of the form $\bar{z}(j\omega)^{-\alpha}$. Although it did show that $j^{-\alpha}$ was indeed constant, there was no experimental reason to believe that the frequency parameter appeared as $\omega^{-\alpha}$. The direct solutions

$$|z_3| = ar_0 \frac{1 - (r_1/r_0)^2}{1 - r_\infty/r_0} \cdot \frac{u}{v} \qquad (1:20)$$

as found earlier (Cole, 1928a) and illustrated in figure 1:9, were run through rapidly. To my considerable relief they produced the typical behavior of figure 1:23 with values of α in the neighborhood of 0.79. There was some question as to how well the 1 mm diameter electrodes first used conformed to the assumptions of the theory, but later measurements with 15 mm long electrodes did not show a significant difference (Cole and Curtis, 1936).

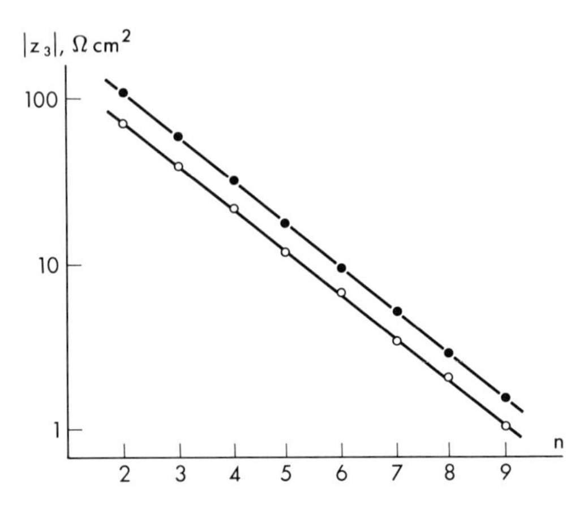

FIG. 1:23. The absolute values of the muscle fiber membrane impedance $|z_3|$ as calculated from the data of Fig. 1:21.

Among other things, these sartorius measurements considerably strengthened the conclusion that such a tissue behaved as if it had uniform fibers with a polarization impedance at the membrane. Furthermore, the capacity equivalent of this impedance was a few $\mu F/cm^2$. The interpretation of the low-frequency data was fuzzy, to say the least, but there was the clear implication in both these data and the membrane impedance behavior at higher frequencies that a membrane capacity might be increasing in rigor.

The characteristics of normal muscle were confirmed (Guttman, 1939) and, with the techniques of analysis that had by then become available, the effect of narcotics was found to be an alteration of the membrane resistance.

Further measurements and more care led to further complications, some of which are yet to be resolved. I suggested that Schwan extend the transverse sartorius measurements to lower frequencies to see if there was anything as startling as the squid axon had presented. He essentially confirmed the earlier results and did find an additional capacitive dispersion region at about 100 c (Schwan, 1954).

On the other hand, between 1947 and 1958 longitudinal and transient measurements led to membrane capacities of 5 $\mu F/cm^2$ and more for sartorius, and up to 40 $\mu F/cm^2$ in some crab muscle fibers. Most recently Fatt (1964) resurrected the transverse technique to confirm the high frequency capacity of 2.3 μF/cm^2 with a constant phase angle of about 72°. He also made a detailed analysis of the low-frequency dispersion and ascribed it to the sarcotubular system.

Other Tissue Impedances

There have been many measurements of many tissues under many conditions. These go back—probably almost continuously—to about the announcement of Ohm's law in 1827. They include plant and animal tissues; *in situ* and excised; living, surviving, injured, and dead, under all manner of physiological, pharmacological, and pathological conditions. The techniques are somewhat less numerous—using direct current and alternating currents, time transients, and frequency variations, each almost as soon as the new approach made its appearance in the physical sciences. In a large majority of these measurements the geometry of the current flow and/or the electrode characteristics have been serious problems, as have been the interpretations of the results.

I had been interested in Sapegno's (1930) early work on the longitudinal resistance of muscle as a function of electrode separation, and very puzzled by the finite resistance extrapolated at zero length. The explanation was that he had used several pronged fork electrodes which were pushed through the muscle, and the high current density in the tissue at and near the small electrode surfaces had contributed a resistance that was independent of separation of the forks. This idea of a resistance associated with but one electrode has since been applied quite usefully many times and in many different ways.

The impedance of the lungs has an important bearing on the distribution of cardiac potentials which concerned F. N. Wilson. The measurements had to be made *in situ*, and the concept of a single electrode was developed from potential theory (Kaufman and Johnston 1943). The resistance R between two electrodes in a large conductor of specific resistance r is approximated by $R = r[\gamma_1 + (1/x) + \gamma_2]$ (Mason and Weaver, 1929), where x is the electrode separation and γ_1, γ_2 are the form factors for the electrodes. For a sphere of radius a,

$$dR = \frac{r}{4\pi x^2}\,dx; \qquad R = \frac{r}{4\pi}\int_a^\infty \frac{dx}{x^2} = \frac{r}{4\pi a}$$

so $\gamma = 1/4\pi a$ while other geometries require more analytical power.

The lung impedance at electrocardiac frequencies approximated that of other adjacent organs in which the membrane-surrounded volume was rather close to the air and membrane-surrounded volume in the lung. This technique and result were criticized, modified, and further applied by Schwan and Kay (1956, 1957).

Electrode polarization has been a major hazard in measurements of impedance (Schwan, 1963) and the corresponding transients. It is conspicuous at low frequencies and long times, particularly so for small imbedded single and pair electrode arrangements. Many of these difficulties are avoided by four-electrode techniques with one pair for a constant current supply and the second pair for potential measurement. These have been variously applied—such as for measurements on skin (Barnett 1938), for pulse recording and analysis (Nyboer, 1950, 1960), and for brain and torso impedances (Schmitt, unpublished). Present-day instrumentation has so simplified this procedure as to encourage its wider use.

A systematic investigation of mammalian tissue impedances was undertaken by Rajewsky and his colleagues (Rajewsky, 1938). Since that time more tissues have been added (Spector, 1956) and the frequency ranges have been extended until Schwan (1963) listed results on nine tissues between 100 c to 10^{10} c. In general, detailed interpretations, such as given by Cole (1940, 1941b) and Cole and Curtis (1950), were rather uncertain for some of these tissues so the data were presented as specific resistances and dielectric constants. However, Schwan (1957; Schwan and Cole, 1960) did recognize three dispersion regions and their origins, figure 1:24:

A, about 10^3 c, surface conductance and/or nonlinearity,
B, about 10^6 c, membrane capacities, and
C, about 10^{10} c, electrolyte and protein.

A particularly obvious theoretical—as well as practical—difficulty is the interpretation of most tissue measurements because of the usually high volume concentration of membrane covered cells. An extreme case is that of centrifuged red blood cells. The assumption of very thin intercellular layers and appropriate measurements may lead to an interesting and useful indication of a membrane conductance.

The almost universal finding of rather constant phase angles ϕ that were less than 90° for everything having a considerable spread of cell sizes contrasted sharply with the 90° phase angle given by marine egg suspensions. Since the marine eggs most closely con-

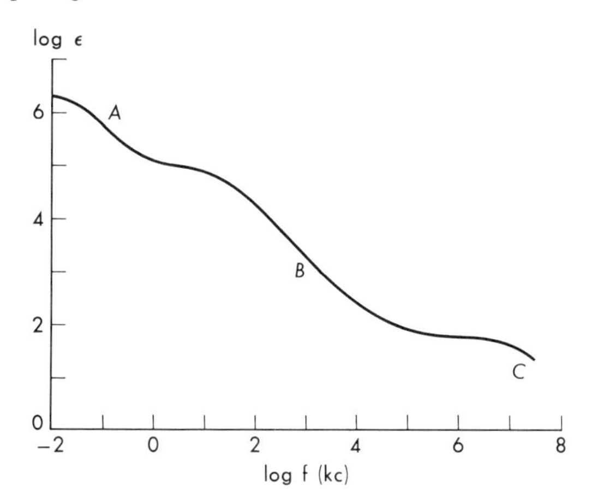

FIG. 1:24. A representation of the principal dispersions of tissues in terms of their effective dielectric constants ϵ as a function of frequency. Based on red blood cell suspensions, the *B* region is explained by the cell membrane capacity. The *A* dispersion probably represents the relaxation of a surface conductance on these cells or of ion conductance in an excitable membrane. The *C* dispersion is mostly the dielectric relaxation of intracellular proteins.

formed to the theoretical assumption of uniformity, it became obvious that the theory would have to be modified if it were to conform to the realities of experiments on other cells and tissues.

Statistical Distribution

The problem of the impedance of a suspension of cells in which the membrane capacities c_3 and/or the cell radii a were not uniform was first considered for the muscle model (Cole and Curtis, 1936). The Maxwell method shows that

$$\frac{z - r_1}{z + r_1} = \sum \rho_i \frac{r_2 - r_1 + jx_i}{r_2 + r_1 + jx_i}$$

where $x_i = -1/(c_3a)_i\omega$ and ρ_i is the volume concentration of the cells having the product $(c_3a)_i$. By equation 1:18 we know that the impedance of a suspension of uniform cylindrical cells with a membrane impedance z_3 is

$$\frac{z - r_1}{z + r_1} = \rho \frac{r_2 - r_1 + z_3}{r_2 + r_1 + z_3}.$$

After considerable manipulation and replacement of the sum by the limiting integral we have

$$z_3 = (r_1 + r_2)(1 - g)/g$$

where

$$g = \int_0^\infty \frac{G(s)}{1 + js}\, ds \tag{1:21}$$

and $s = x/(r_1 + r_2)$. $G(s)\, ds$ is the distribution function which is the fractional volume concentration of those cells in the range between s and $s + ds$.

At that time it was not yet known how to find $G(s)$ corresponding to either an analytical or a numerical function z_3. It was, however, much too simple to set up arbitrary block distributions of c_3a which were logarithmically symmetrical about a central value $\overline{c_3a}$ and work out the consequences on an adding machine. As was to be anticipated, distributions could be found to approximate almost any value of α (equation 1:17) in the frequency range $\bar{f}/10$ to $10\bar{f}$. But, as could also have been predicted, any such finite distribution failed, more or less noticeably, to approximate a constant phase angle for z_3 at frequencies above and below those given by the least and the largest values of c_3a assumed in the distribution. This same problem had been stated long before by Wagner (1913), who gave a few calculations for some Gaussian distributions. These calculations were essentially completed by Yager (1936) and showed that even this distribution was attenuated so rapidly at both extremes as to also fail to approximate the most general empirical finding—that of a constant phase angle—at the extremes of the frequency range. However, it seemed not at all unlikely that an observed z_3 could be reasonably represented by some distribution of c_3a. This problem was considered more generally and more powerfully by Weinberg and Householder (1941) who showed that the solution was that of the Stieltjes integral which appears in equation 1:21.

If indeed these biological cell suspensions and tissues were only aggregations of cells with static, 90° phase angle membrane capacities, it was probably important—and certainly interesting—to find the principal or central value of c_3. This most probable component $\bar{c}_3$ was also the most important at the characteristic frequency $\bar{f} = \varpi/2\pi$, and was given by

$$\bar{c}_3 = \frac{1 - r_\infty/r_0}{a r_0 \varpi (1 - r_1/r_0)(1 + r_1/r_0)}$$

rather than by equation 1:19. When we apply this analysis to the early muscle data (Bozler and Cole, 1935) and assume a frog sartorius muscle fiber diameter $2a = 100\ \mu$, we find an average value for $\bar{c}_3$ of 1.0 μF/cm^2 with no systematic differences between normal muscle, muscle in rigor, and muscles soaked in iso-osmotic sugar solutions. Later and less complete measurements gave a value for $\bar{c}_3$ of 1.9 μF/cm^2 (Cole and Curtis, 1936).

Another interesting consequence of this approach was that the phase angle of red blood cell suspensions was closer to 90° than could be accounted for by the normal broad distribution of cell radii. Thus it was necessary to invoke an inverse relation between membrane capacity and cell size!

A distribution was thus a simple, possible explanation for the polarization impedance that had become rather usual. It was also obviously a factor to consider, and one of at least some importance, after it was finally recognized by such circuitous fumbling. The distributions that would fit polarization or other observed impedances were not known nor was there then anything but a cut and try process for finding them.

However, for muscle, Fatt (1964) found measurements showing that the maximum ratio of fiber diameters was only 3 to 1 and he showed that this was far too small to account for the observed phase angle. Thus, in this case at least, it appears again that a nice and an obvious explanation was too fragile to survive except on the basis of an untested assumption. So we still need something like a constant phase angle for a single muscle fiber membrane.

SINGLE CELLS

Although the explanation of a distribution of capacities, or of dimensions, or particularly of their product, might be a sufficient explanation of a constant phase angle in suspensions, it became obvious that the only proof of its necessity would have to come from single cell measurements. Blinks (1936) had started with work on *Valonia* with both external and internal electrodes and on *Nitella*. He showed that these cells both had capacities of about 1 μF/cm^2 that decreased for increasing frequencies. Our next preparations were *Nitella* and the squid giant axon that was destined to play so important a role beginning in 1936. The first transverse measurements, interpreted by the Rayleigh (1892) theory for a rectangular array of cylinders, gave phase angles of certainly less than 90° along with negligible membrane conductances and rather erratic internal conductances.

During the winter *Nitella* was used both for its own sake and because of its similarity in size to the squid axon. It did give (Curtis and Cole, 1937) the capacity of 0.94 $\mu F/cm^2$ with a phase angle of 80°, for one of the first single cell measurements, but it could not give an indication of membrane conductance. Rather, the work emphasized what has since been recognized as an ion exchanger characteristic of the cellulose sheath. Thus with the sheath on a glass rod the transverse impedance and the two phase cylinder theory gave a sheath resistivity relatively independent of the concentration of the surrounding medium. This was later confirmed by Gaffey and Mullins (1958), and Tasaki (unpublished).

The summer of 1937 brought the transverse measurements of the squid axon with $c_3 = 1.3\ \mu F/cm^2$ and $\phi = 80°$ for the first single animal cell phase angle less than 90°. (Curtis and Cole, 1938). Although it was still possible to have a distribution of capacities over the 1 cm length of axon and 1.5 mm of circumference, it was at least not now necessary to assume a distribution among a population of cells.

Once again there was no indication of a membrane conductance. With this came the most disconcerting observation. The axons deteriorated, over several hours, beyond their ability to conduct an impulse. But in this period there was no significant change of either the membrane capacity or the phase angle. There had remained the haunting hope that these might be the characteristics of a polarization caused by an ion current flow across the membrane and so were in some way related to the activity and life of a cell. Quite apart from my own vague doubts, this considerably encouraged some of my less intimate colleagues in their opinion that I was not very bright to waste so much time and effort on a purely physical property of a membrane that was so obviously not related to the physiological, and important, function of axons.

Then, too, single marine eggs from several species when squeezed into a glass capillary, had close to a 90° phase angle (Cole and Curtis, 1938a). Here it had been hoped that the current flow between the egg membrane and the glass wall might be negligible, and so allow a measure of the membrane conductance at low frequency. Unfortunately the membrane capacity derived on this basis was much larger than that found from suspensions. The assumption of a negligible membrane conductance was far more satisfactory because it again produced capacities close to 1 $\mu F/cm^2$ from data such as in figure 1:25.

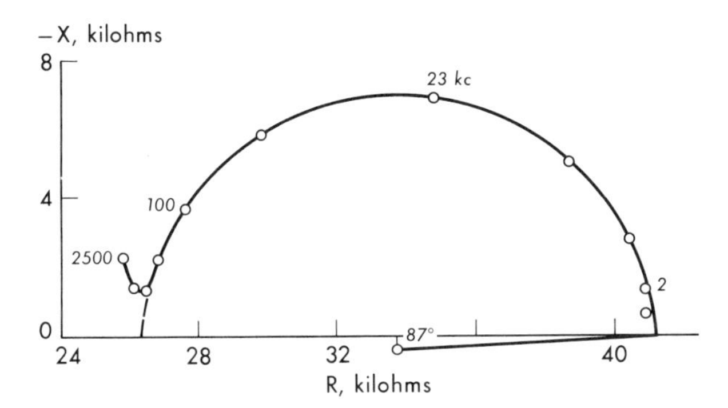

FIG. 1:25. Complex impedance behavior of a single *Arbacia* egg in a narrow glass capillary.

Cole and Guttman (1942) used single frog eggs in a short cylindrical channel between two large electrode chambers. This was not a simple geometry, and we had to be quite careful since we needed absolute as well as relative values of the egg parameters. The convergence resistance near the end of the hole had long been an important practical problem in mercury standards of resistance, and narrow limits had been set by potential theory (Jeans, 1927), but it was encouraging to confirm these conclusions experimentally. I could neither find nor develop a useful theory for the resistance of a spherical cell in a cylindrical hole; finally it was necessary to rely entirely on an analogue analysis which was probably at least as accurate and convenient as a theory. A series of small glass spheres gave an effective volume for the hole and showed that the usual simple Maxwell formula was adequate up to an equivalent volume concentration of 50 percent. One result from these experiments was a capacity of 2.0 $\mu F/cm^2$ with $\phi = 86°$ and internal resistivities of about 570 Ωcm. The same problem came up for the Coulter counter for suspensions of small cells. Gregg and Steidley (1965) also resorted to an analogue in the absence of anything more than a very approximate analysis.

Rothschild (1946) avoided all of this trouble by measuring salmon and trout eggs in a cubical cell, thus conforming to the rectangular array of spheres which Rayleigh (1892) had calculated. From this he found for unfertilized and fertilized trout eggs $c_3 = 0.58\ \mu F/cm^2$, $\phi = 83°$, $r_2 = 180\ \Omega cm$ approximately.

Quite recently direct measurements of the squid axon membrane between internal and external electrodes have given capacities of about $0.7\ \mu F/cm^2$, with constant phase angles in the region of 75° to 80° over the range from 10 to 70 kc (Taylor and Chandler, 1962).

NATURE OF THE MEMBRANE CAPACITY

Polarization Impedance

Both Hermann (1905) and Cremer (1900) were aware of the electrotonic spread in nerves, and sought to explain its spatial and temporal characteristics by a cable with high surface resistance and capacity (Taylor, 1963). This they found in the polarization of a metal-electrolyte interface that eventually evolved into the passivated iron analogue of Ostwald (1900) and Lillie (1936). Such electrode polarizations have been widely and intensively investigated and are still without a satisfactory explanation of their impedance characteristics, particularly the common finding of a constant phase angle. The appearance of the frequency dependent capacity and the consequent constant phase angle in living cell membranes without evidence of another conduction path long encouraged the assumption of an electrode process. The membrane capacity of a microfarad/cm^2, although large enough to satisfy the requirements of Hermann and Cremer, was almost never found in polarization systems. These latter systems usually gave tens of microfarad/cm^2, sometimes one or two orders of magnitude more. Furthermore they were highly sensitive to composition and state of the surface and to the composition and concentration of the electrolyte. Also, this parameter always had a negative temperature coefficient; while it is negligible for the membrane of the red cell (Schwan, 1948) and muscle fiber (Fatt, 1964), and for at least one membrane, the squid axon, the capacity increases with temperature (Taylor and Chandler, 1962).

Although polarization remains as at least a possible membrane mechanism, it was improbable and has since become unnecessary as a characteristic of ion transport.

Membranes as Dielectrics

The conclusions from tissues and single cells and the forthcoming evidence that this capacity and phase angle of membranes might be relatively independent of their conductance led me to consider living membranes primarily as passive nonconducting dielectrics with constant phase angles from 90° down to perhaps as low as 60°. This was in contrast to the ion conduction phenomenon of electrolytic polarization as originally proposed by Philippson, used by Gildemeister to explain the behavior of human skin, and used by Fricke who demonstrated it for the red cell and a muscle.

The transient currents in many dielectrics (Cole and Cole, 1942) had long been known to follow a power law, $i(t) = \bar{i}(t/\tau)^{-\alpha}$, after a potential was applied. But there were two difficulties. For the initial charge to remain finite at short times it is necessary for $\alpha < 1$, while the total charge can remain finite as $t \to \infty$ only for $\alpha > 1$. Thus the corresponding impedance element of equations 1:15, 1:16, 1:17, $Z = \bar{Z}(j\omega\tau)^{\alpha-1}$, could not be used without some restrictions, and it was necessary to turn to Debye's theory (1929) for the dielectric behavior of a dilute suspension of free dipoles. Here the complex dielectric constant is

$$\epsilon^* = \epsilon_\infty + (\epsilon_0 - \epsilon_\infty)/(1 + j\omega\tau) = \epsilon' - j\epsilon''$$

where ϵ_∞ represents the deformation polarization at infinite frequency, the dipoles rotate with a time constant τ, and contribute the low-frequency polarization $\epsilon_0 - \epsilon_\infty$. This is again a semicircle A on the complex plane ϵ' vs. ϵ'' figure 1:26. However, enough examples have been found to justify the use of the empirical expression of the form $1/[1 + (j\omega\tau)^\alpha]$. This also gives a circular arc but with a depressed center B, which is well correlated with liquids and solids in which considerable molecular interaction is to be expected (Cole and Cole, 1941). Much work over many years on usual solid insulators gave $\phi < 90°$ at high frequency and showed no inclination to bend over—much less extrapolate to a finite ϵ_0—as in C. Then for the older dielectrics with considerable loss, τ and ϵ_0 may be thought to become quite large even though ϕ may be degrees or tens of degrees. This process was illustrated for β-lactoglobulin by Shaw et al. (1944). At low concentrations in glycine-water mixtures ϵ^* approximated a Debye dispersion but α, τ and ϵ_0 increased with higher concentrations until ϵ^* approached the straight line C of figure 1:26 for the liquid crystal. Somewhat similarly, as halowax was cooled

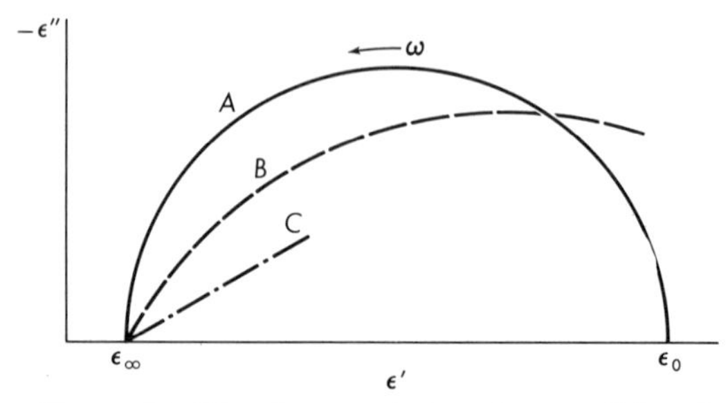

FIG. 1:26. The characteristics of some dielectrics on the complex dielectric plane; ϵ', the conservative storage factor, vs. ϵ'', the dissipation component. The solid semicircle A shows the effect of rotation of dipole molecules in low concentration. The circular arc B shows the behavior of some dipolar substances in high concentration or as a liquid. The linear representation, C on this plane, has been found for some solids.

α rose to above 0.8 at -100°C (Field, 1945). Furthermore, some of the newer synthetic high polymers, such as polystyrene, may have so small a value of ϵ'' and so little a variation of ϵ' over an enormous frequency range as to give little basis to extrapolate ϵ_∞. They thus approach the perfect classical dielectric with ϵ independent of frequency.

Cell membrane impedance measurements are restricted to a relatively small frequency range because of the difficulties as well as the complications and ambiguities at both high and low extremes. But there are no certain indications that they depart significantly from $(j\omega\tau)^{-\alpha}$ as determined by equations such as 1:15 for figure 1:9. Thus C_0 and τ are too large to estimate and $C_\infty \approx 0$, figure 1:27 (Cole, 1947). If we allow a 10 percent error in the extrapolation as $\omega \to \infty$ and assume $\epsilon_\infty = 1$, we would conclude that the mean value of 1 μF/cm^2 corresponds to $\epsilon = 10$. Then we arrive at a membrane with a thickness of 100 Å and a highly condensed structure of at least a liquid far below some critical point, and perhaps a solid.

This approach and the conclusions as presented are not only obviously highly speculative but may well come to be naïve or absurd as more is known of the structure of cell membranes and the dielectric behavior of known molecules and their interactions.

Probably more than forty years ago, Irving Wolff told me, in connection with his electrode polarization work (Wolff, 1926), that a constant phase required the energy dissipated per cycle to be

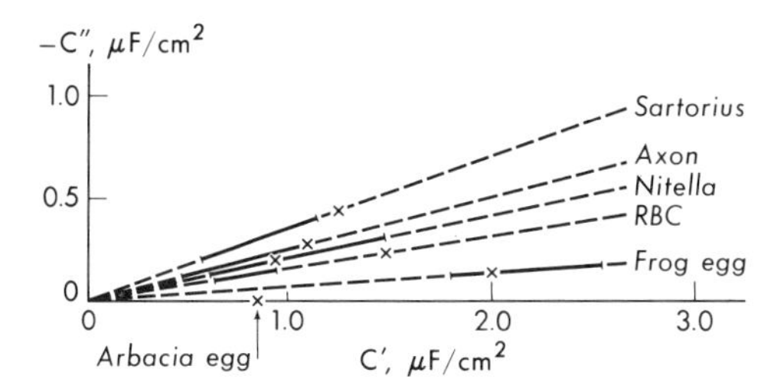

FIG. 1:27. The characteristics of some cell membranes shown on the complex capacity plane. The solid lines represent the range of measurements, the dashed lines are the constant phase angle extrapolations and the crosses are the values at 1 kc.

proportional to average energy stored independent of the frequency. This is obvious from Bode's (1935) expression for the impedance in energy terms, equation 1:43, but a molecular basis has not been found. The empirical impedance function has been shown to be equivalent to a continuous distribution of relaxation times both sharper and broader than a Gaussian expression (Cole and Cole, 1941). But this is only an alternative expression of our ignorance of the mechanism. The most direct and encouraging theoretical approach has been that of Fuoss and Kirkwood (1941), who considered the interactions of polar groups in polyvinyl chloride to obtain at least an approximation to a constant phase angle element. The recent developments and present status of dielectric dispersions, particularly as cooperative phenomena, were reviewed by R. H. Cole (1965). But then Havriliak and Negami (1966) produced an even more involved and general empirical relation to express the behavior of sixteen plastics.

SURFACE IMPEDANCE

Long ago, Fricke and Curtis (1935, 1936) found truly enormous capacities for red cell suspensions as they ventured below 1 kc, with the necessary resistance variation practically too small to measure. This was completely foreign to the transverse dielectric behavior and was attributed at first to a conduction polarization. However, a tangential or surface phenomena explanation became necessary to explain the similar phenomena found for suspensions of Pyrex spheres (Fricke and Curtis, 1937), and was strongly suggested as the basis for the even more dramatic impedance characteristics of cotton fibers (Murphy, 1929) and gelatin (Fricke and Parker, 1940). The interpretation of the *Asterias* egg measurements (Cole and Cole, 1936a) was highly satisfactory except that the capacity became too large and continued to increase as the frequency was decreased. This anomaly appeared most strikingly when the data were plotted on the complex admittance plane, figure 1:13, and the unknown surface mechanism was invoked as a possible explanation. The data of Goldman (1943) on artificial membranes showed a similar behavior between different electrolytes. But there had been the completely devastating inductive reactance for the squid axon which ion alterations could make larger or reduce and change to a capacitive reactance. The reasons, to be given later, for ascribing this to an ion flow across the membrane seem now to have been adequately substantiated. The capacitive anomaly found for muscle by Schwan (1954) in the 100 c range might have had either a transverse or a tangential origin. In spite of our high hopes that it was similar to the squid, Moore and Shirer (unpublished) found that this effect was only somewhat modified by external potassium, not at all by calcium, and never did the reactance become inductive! And now Fatt (1964) ascribes it to the sarcotubular structure.

Additional data on suspensions of glass and polystyrene spheres were not to be interpreted on anything but a tangential basis. Then came the empirical generalization that the time constants of these phenomena were proportional to the square of the principal dimension of the particle involved (Schwan et al., 1962). This alone should have been an adequate hint to almost anyone that the phenomena were diffusion defined. It should have been obvious—as had been emphasized by Onsager (1945)—that for such behavior with a diffusion coefficient D,

$$D\tau/L^2 = \text{constant} \qquad (1{:}22)$$

where τ is the time constant of the process and L is the limiting dimension. But it remained for Schwartz (1962) to appreciate this possibility and to put it into quantitative form. Taking an approximation to the complete electrodiffusion equations (cf. Planck, Debye), he calculated a motion of the adsorbed charge density σ around a spherical particle of radius a. This had a characteristic frequency corresponding to a time constant

$$\tau = a^2/2ukT$$

and an increase of the dielectric constant of the sphere of

$$\Delta\epsilon = ae\sigma/ekT.$$

The time constant as calculated by equation 1:22 was in excellent agreement with experiment, whereas the calculated dispersion was qualitatively satisfactory but significantly different from the results of the difficult experiments.

SUMMARY

Fricke's original theory and experiment on red blood cells have been a spectacular confirmation of Bernstein's hypotheses that the cell interior is an electrolyte with a poorly permeable surface membrane. They further, and for the first time, characterized this membrane by a capacity of 1 $\mu F/cm^2$ and as being of molecular dimensions. Subsequent developments of theory and further experiments on other cells have confirmed and extended the original conclusions. All of the work suggested that the ionic permeability—if any—was small while the membrane might have the dielectric characteristics that could be ascribed to a liquid or solid layer with strong intermolecular forces. Apparently quite apart from such phenomena involving the thickness of the membrane, a time dependent conductance along the surface of the membrane was measured and calculated reasonably well.

Membrane Conductance

In almost all of the work discussed so far it has been sufficient to consider the cell membranes as nonconductors. The measurements of Fricke and Morse (1925b) on cream gave ample confirmation of the Maxwell theory considerably beyond the analogous equations for other phenomena. No serious discrepancy had appeared in the various and numerous measurements of blood and other small cell suspensions. Increasingly careful marine egg experiments were portrayed as in figure 1:28 showing measured volumes in suspensions to be practically identical with the nonconducting volume produced by impedance measurements. Similar experiments but with single spherical and cylindrical cells did not require or even suggest an appreciable membrane conductance.

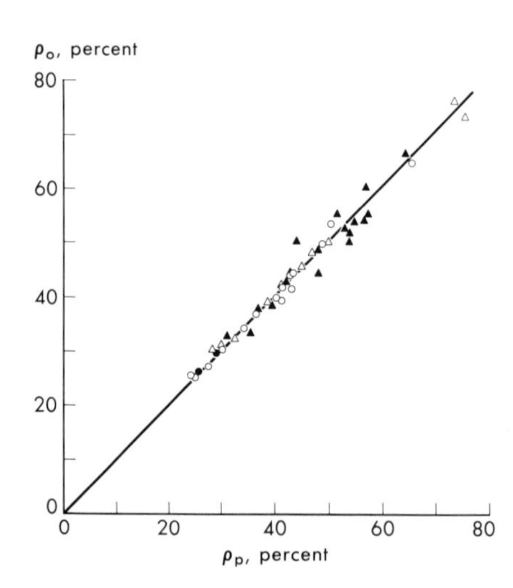

Fig. 1:28. Relation between the volume concentration of sea urchin egg suspensions ρ_o as calculated from electrical measurements on the assumption of a non-conducting membrane, and ρ_p as measured by sucrose dilution or by count and diameter. Circles for *Hipponoë* and triangles for *Arbacia*, open for unfertilized and filled for fertilized. The line is for the equality of the concentration estimates.

This absence of conductance had appeared rather reasonable for the protection of the intricacies of the living cell against the ions of its inanimate environment. But it did seem to require that no large amount of metabolites enter or leave the cells in ionic form, and some attempts were made to construct molecular transport models. Still at least some slight ion permeability was necessary unless Bernstein's second proposition was wrong. As he pictured it, potassium ions could and did diffuse out of the cell and the excess of negative ions left behind made the potential of the interior increasingly negative. This process would continue until an equilibrium was reached in which the inward force of the electric field X on the ions e balanced the outward diffusion of the ions down the concentration gradient dn/dx. Then at each point

$$0 = D\, dn/dx + uneX$$

where D is the diffusion coefficient and u the mechanical mobility. Integrating from outside, $n(0)$, to inside, $n(\delta)$, concentrations and using the Einstein (1905) relation $D = ukT$, the potential

$$E = -\int_0^\delta X\, dx = \frac{kT}{e} \ln \frac{n(0)}{n(\delta)}. \qquad (1{:}23)$$

This is the equilibrium potential as given by Nernst and in the form usually used by chemists. Physicists prefer the equivalent expression,

$$n(\delta) = n(0) \exp(-Ee/kT) \qquad (1{:}24)$$

given by Boltzmann.

Of electric potentials there had long been the tremendous amount of evidence—mostly as bioelectric injury and action currents—that had led Bernstein to his hypotheses. It had been well demonstrated that whatever the source of energy, it could produce injury currents for hours and days. But if Bernstein was right, potassium ions must be able to leave the cells to maintain these currents. Then away from equilibrium

$$I_K = uekT \frac{dn}{dx} - une^2X \qquad (1{:}25)$$

and the potential difference across the membrane,

$$V = -\int X\, dx = E_K + I_K \int dx/ue^2n. \qquad (1{:}26)$$

And further

$$I_K = G_K(V - E_K) \qquad (1{:}27)$$

since the specific conductance within the membrane is ue^2n and the conductance G_K of the whole membrane is the reciprocal of the integral. If, then, there is to be a current, not only must $V \neq E_K$

but also u and n must both be >0 everywhere. The membrane conductance is then the measure of the ionic permeability in that it is the flow of charge I_K per unit of the driving force $V - E_K$. Not only was such a finite permeability necessary to establish an equilibrium potential and to maintain injury currents, but it should be directly measurable as a current for an imposed change of potential—a conductance.

EARLY ESTIMATES

Apparently the first reliable indications of membrane conductances were obtained by Blinks on *Valonia* (1930a) and *Nitella* (1930b). The *Valonia* measurements were obtained between a chronic capillary impaling the cell, and the exterior sea water; they ranged from 3000 Ωcm^2 for a freshly prepared cell to the order of 10^5 Ωcm^2 after apparent recovery, with a nominal value of 10^4 Ωcm^2. The *Nitella* work, although less direct, was an interesting example of experimental and analytical ingenuity. The resistances were measured first between two external electrodes on a normal cell, figure 1:29,

$$\frac{R_e(R_i + 2R_m)}{R_e + R_i + 2R_m}$$

second after the cell was killed,

$$\frac{R_e R_i}{R_e + R_i}$$

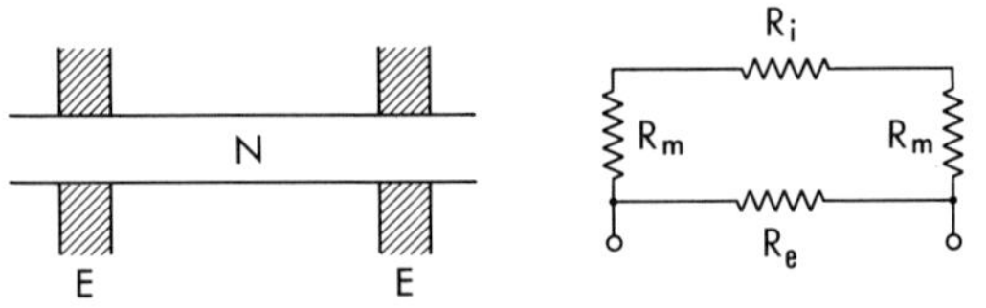

FIG. 1:29. Schematic arrangement, left, of *Nitella* cell N between two electrodes E. The equivalent circuit, right, shows the external, internal, and membrane resistances represented by R_e, R_i, and R_m.

and finally after the cellulose envelope was inflated with air, R_e. These then gave r_3 from $1 - 7 \cdot 10^5$ Ωcm^2 and an average of $2.5 \cdot 10^5$ Ωcm^2.

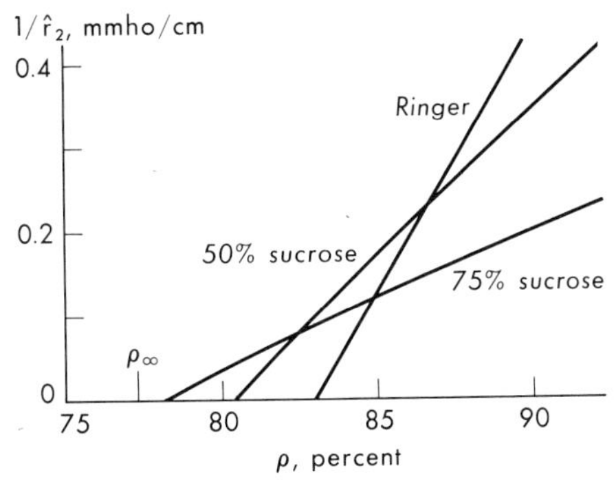

FIG. 1:30. An estimate of frog sartorius muscle fiber membrane resistance. The lines each show possible combinations of equivalent, myoplasm plus membrane, resistivities $\hat{r}_2$ and volume concentrations ρ for the indicated external solutions.

A rather different attempt was made in connection with transverse impedance measurements on sartorius muscle (Bozler and Cole, 1935). The external Ringer's was successively diluted by 50 percent and 75 percent iso-osmotic sugar solution, primarily to estimate the fiber volume concentration. However, the most self-consistent values were obtained, figure 1:30, for a membrane resistance of 36 Ωcm^2. A still later use of the single frog egg (Cole and Guttman, 1942) gave an average membrane resistance of 170 Ωcm^2. Such a very low resistance—at least as compared to large plant cells—led to serious consideration of the theoretical possibilities for measuring such membrane resistances in cell suspensions (Cole, 1937). We have from equation 1:13,

$$r_0 = r_1 \frac{(1-\rho)r_1 + (2+\rho)(r_2 + r_3/a)}{(1+\rho)r_1 + 2(1-\rho)(r_2 + r_3/a)}$$

but if r_3 were infinite,

$$r_0 = r_1 \frac{2 + \rho_0}{2(1 - \rho_0)}$$

where ρ_0 is the equivalent nonconducting volume concentration. For $r_3 \gg r_1 a$ and $r_3 \gg r_2 a$ we have

$$(\rho - \rho_0)/\rho = 3r_1 a/r_3. \qquad (1{:}28)$$

Thus ρ would have to be measured to one percent for sea urchin eggs in sea water to show a difference between $r_3 = 11$ Ωcm^2 and $r_3 = \infty$. This showed that extremely accurate values of ρ were necessary to detect a finite value of r_3 unless r_1 or a were quite large. Furthermore, the similar derivation for a transverse measurement of cylindrical cells, such as *Nitella* and a squid axon, led to the conclusion (Curtis and Cole, 1937, 1938) that the accuracy of the squid axon measurements would not permit the calculation of a finite membrane conductance if the resistance were more than 3 Ωcm^2. Thus it was possible that during deterioration the membrane resistance decreased from a high, but then unknown, value for reasonably normal activity, down to a point still somewhat higher than 3 Ωcm^2 at which the axon became inexcitable. Then, continuing its downward course, the resistance carried an increasing current that finally became apparent above the fluctuations encountered in experiments lasting more than a few hours. This was the first, and still somewhat vague, realization of the role of the membrane conductance.

These calculations and observations had shown that the larger the cell the higher the membrane resistance that could be measured or, qualitatively, that the current path over the cell surface had to be long if any useful fraction of the current were to cross the membrane. Thus, in an entirely coincidental preparation for the advent of the squid giant axon, theoretical attention was turned to longitudinal impedance for use with muscle and nerve measurements (Cole and Curtis, 1936).

CABLE THEORY AND EXPERIMENTS

The treatment of a nerve fiber as a cable was an old and well worked field. In the face of complexities, a number of early estimates, particularly on whole nerve, for a membrane conductance as a measure of ion permeability were more expressions of hope than of fact (Rosenberg and Schnauder, 1923; Lullies, 1930; Rushton, 1927, 1934; Hill, 1932; Danielli, 1939).

As a simple start, I first went through an analysis for a single fiber with the assumption of a negligible membrane conductance, equation 1:33. When this was not only complicated but useless, the more general problem was solved for a finite membrane conductance with arbitrary electrode widths and separation. The steady state resistance came from equation 1:34. For very narrow electrodes, a distance x apart, this was

$$R = \frac{r_e r_i}{r_e + r_i} x + \frac{r_e^2}{r_e + r_i} \lambda[1 - \exp(-x/\lambda)] \qquad (1{:}29)$$

where λ is the characteristic length as given by Rushton (1934). For small electrode separations, $x \ll \lambda$, this resistance extrapolated to $R = r_e x$, which is the external resistance. This shows that in this range the membrane resistance prevents any significant current flow into and along the internal path r_i. At the other extreme, for $x \gg \lambda$

$$R = \frac{r_e r_i}{r_e + r_i} x + \frac{r_e^2 \lambda}{r_e + r_i}.$$

Here the two electrode regions do not interfere with each other, the impedance is that of the external and internal paths in parallel plus the additional terms representing the extra resistance encountered by the internal current.

The wide electrode prediction seemed too complicated for any practical use. Attempts to understand the effect of electrode width on muscle measurements were also quite futile. Some data were taken on muscle with narrow electrodes; these indicated that there was no reactive element in the direction of the fiber axis either in the myoplasm or as a tangential surface reactance (Cole and Curtis, 1936). They further gave a basis for a characteristic length, and it should then have been possible to make an estimate of the membrane resistance. But we concluded that the comparison of single fiber calculations with whole muscle or nerve measurements was much too uncertain to encourage this effort.

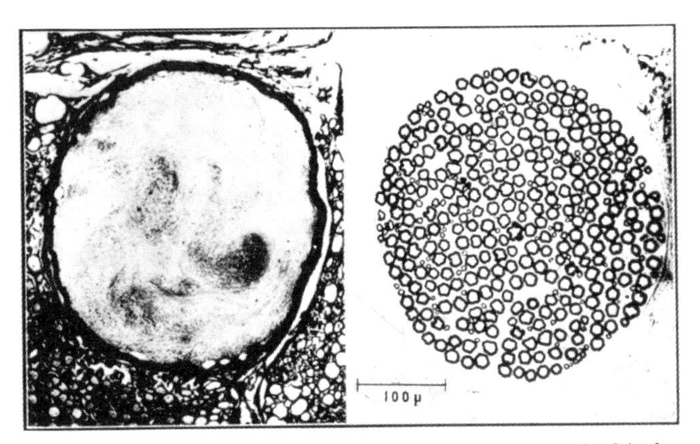

FIG. 1:32. A comparison between a giant axon in the hindmost stellar nerve of *Loligo*, left, and the many, more usual, small axons in a rabbit leg nerve, right. The figure is by J. Z. Young (1951).

At this point J. Z. Young (1936, 1951) and his enthusiastic promotion of the squid and its giant axons, figures 1:31 and 1:32 were encountered at Cold Spring Harbor. Our first attempts in 1936 to use the axon were worthless, but we were convinced that we could not ignore this opportunity. Then as we were measuring the membrane capacity, I made an elaborate but mostly futile effort to get the membrane resistance by direct current measurements between narrow electrodes.

With the conductance increase during a propagating impulse in *Nitella* during 1937–1938 and the similar result for the squid axon in the summer of 1938, the need for a resting conductance became much more immediately important. This was particularly true for the squid axon, and attention was turned again to cable theory.

Infinite Cable Theory

The cable theory that has now become the backbone for the analysis of cylindrical cells, both near their resting conditions and in activity, has had a long and erratic development (Taylor, 1963). Its early form was as a model for a polarizable wire in an electrolyte which, in turn, had been evolved in connection with muscle problems and was known as a "Kernleiter." This was investigated in elaborate detail by Weber (1873). He started with the Laplace equation, equation 1:6, expressed in the cylindrical coordinates

$$\frac{\partial^2 V}{\partial \rho^2} + \frac{1}{\rho} \cdot \frac{\partial V}{\partial \rho} + \frac{1}{\rho^2} \cdot \frac{\partial^2 V}{\partial \theta^2} + \frac{\partial^2 V}{\partial x^2} = 0.$$

To express the potential distribution given by a small source of current at the surface of the core conductor he developed some methods of solution that have become parts of standard Bessel function theory. But the theory has been evolved and used most extensively for the simpler problems of axial symmetry with V independent of θ,

$$\frac{\partial^2 V}{\partial \rho^2} + \frac{1}{\rho} \cdot \frac{\partial V}{\partial \rho} + \frac{\partial^2 V}{\partial x^2} = 0. \qquad (1:30)$$

As one example, Weinberg (1941), who noted the impact of Weber's biological incentive on mathematics, considered the effect of the thickness of the acid layer around an iron wire on the speed of the wave of activity in this nerve analogue. Also Lorente de Nó (1947) has published extensive calculations both for cylindrical and for plane geometries. In the course of his work, Weinberg calculated a Green's function for a narrow ring Δx at the surface of the wire, and in extrapolating to an infinite medium he came to the conclusion that the distribution was not exponential in x. This should perhaps be investigated further to estimate the effect of a large external conducting medium, as Clark and Plonsey (1966) did for a propagating impulse.

In order to deal with the attenuation and distortion in the first Atlantic cable, Lord Kelvin (1855) simplified the statement of the problem further to make transient and sinusoidal solutions possible. He assumed that the inside and outside potentials were independent of the radial distance ρ and were functions of x alone, and he represented the insulation by a capacity and leakage conductance. Almost everyone presents this in his own way—so here is mine:

V_i and V_e, inside and outside membrane potentials,
i_i and i_e, inside and outside longitudinal currents,
r_i and r_e, inside and outside longitudinal resistances, Ω/cm.
$V = V_i - V_e$, membrane potential difference,
i_m, radial membrane current, A/cm,
g_m and z_m, membrane admittance, mho/cm, and membrane impedance, Ωcm.
Then

$$dV_e/dx = -r_e i_e, \quad dV_i/dx = -r_i i_i$$

and in the absence of external and internal sources or sinks

$$di_e/dx = -di_i/dx = i_m.$$

Differentiation and substitution give

$$\frac{d^2V}{dx^2} = (r_e + r_i)i_m$$

for steady state. When V and i_m change with time this becomes

$$\frac{\partial^2 V}{\partial x^2} = (r_e + r_i)i_m \qquad (1:31)$$

which is the general cable equation. It is valid for any membrane behavior as indicated in figure 1:33 so long as r_e and r_i are truly linear.

When further the membrane is linear

$$\frac{\partial^2 V}{\partial x^2} = (r_e + r_i)g_m V$$

and for the conventional cable, with $g_m = pc_m + 1/r_m$ and p representing $\partial/\partial t$,

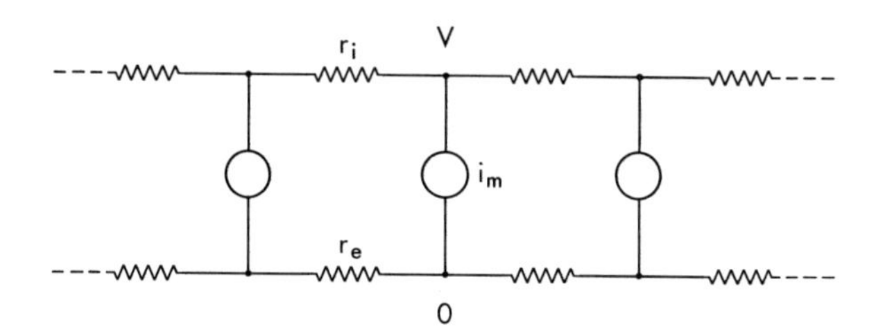

FIG. 1:33. Schematic representation of a general cable in which the nature of the membrane conductance current i_m is not specified.

$$\frac{\partial^2 V}{\partial x^2} = \frac{r_e + r_i}{r_m} V + (r_e + r_i)c_m \frac{\partial V}{\partial t}. \tag{1:32}$$

which is shown in figure 1:34.

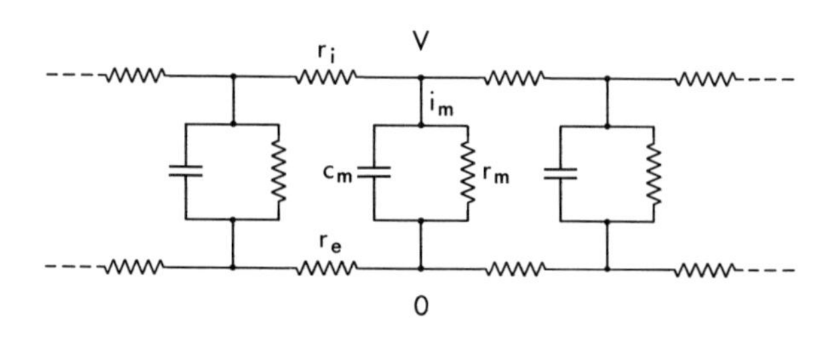

$$i_m = c_m \frac{\partial V}{\partial t} + \frac{V}{r_m} = \frac{1}{r_i + r_e} \cdot \frac{\partial^2 V}{\partial x^2}$$

FIG. 1:34. Schematic representation of a cable with linear membrane capacity and resistance.

For a pure capacity without leakage

$$\frac{\partial^2 V}{\partial x^2} = (r_e + r_i)c_m \frac{\partial V}{\partial t} \tag{1:33}$$

which is the conventional cable equation. For a leaky cable and steady state direct current,

$$\frac{d^2 V}{dx^2} = \frac{r_e + r_i}{r_m} \cdot V. \tag{1:34}$$

In this latter case a solution is

$$V = A \exp(-x/\lambda) + B \exp(x/\lambda) \tag{1:35}$$

where $\lambda = \sqrt{r_m/(r_e + r_i)}$ for the characteristic length. For $V = V_0$ at $x = 0$, $A = V_0$. Also $i_e = -i_i = i$ with

$$\frac{dV}{dx} = -(r_e + r_i)i.$$

At $x = 0$, $i = I_0 = V_0/\lambda(r_e + r_i)$

and $$Z_0 = V_0/I_0 = \sqrt{(r_e + r_i)z_m}. \qquad (1:36)$$

Circuit analysis procedures give an interesting alternative derivation (Zobel, 1923). Let the impedance of the infinite cable be Z_0, figure 1:35a. Now add a short equivalent section of length $2\,\Delta x$, figure 1:35b. Since the cable was already infinite, this addition will leave the input impedance unchanged for Δx sufficiently small. This leads to $Z_0^2 = (r_e + r_i)z_m + (r_e + r_i)^2\Delta x^2$ and the same result as before, equation 1:36, for $\Delta x \to 0$.

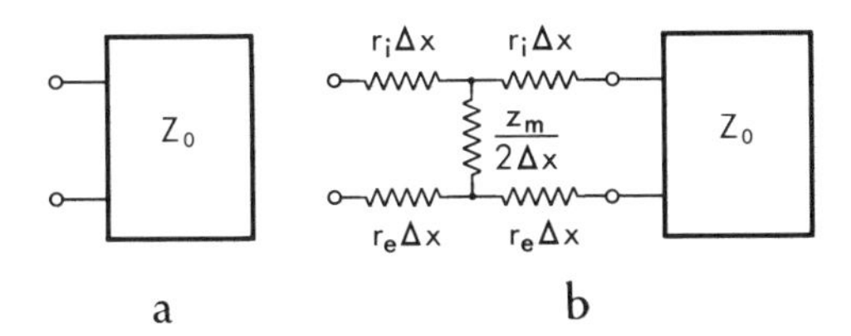

FIG. 1:35. Scheme for derivation of the impedance of a semi-infinite cable Z_0 using the fact that the input impedance with the added elements at the right is also Z_0 for Δx small.

For the submarine cable, λ is many miles, but for some nerves under some conditions it may be of the order of a millimeter. The giant axons of the *Loligo* squid vary from an average diameter of 500 μ for *pealeii* at Woods Hole to perhaps 700 μ for *forbesi* at Plymouth, and the Humboldt current *Dosidicus gigas* may be 2 mm in diameter, so the Kelvin assumption is not necessarily a good approximation. Returning to equation 1:29 with $V_e = 0$ everywhere and $V = V_i = V_0$ at $x = 0$ and in steady state at $\rho = a$,

$$\frac{1}{r_2} \cdot \frac{\partial V}{\partial \rho} = \frac{1}{r_m} V$$

we find $V = \Sigma\, A_n J_0(\rho/\lambda_n)$ where J_0 is a Bessel function.

In our case this may be simplified (Carslaw and Jaeger, 1947, Section 78) to give

$$V(a, x) = V_0 \sum_{n=1}^{\infty} A_n \exp(-x/\lambda_n)$$

with

$$A_n = 2h/a(h^2 + 1/\lambda_n^2), \; h = r_2/r_m.$$

For $\lambda = a$, $A_1 = 0.88$ and $\lambda_1 = 0.94\ \lambda$ to give errors of 12 percent and 6 percent respectively. These decrease rapidly for $\lambda > a$ and may usually be ignored (Taylor 1963).

Falk and Fatt (1964) went on to consider and calculate more complicated problems. For example, an angular dependence of potential and current around the circumference of the membrane was included as a closer approximation to their experimental situation.

Finite Electrodes and Separations

Simple and direct as these analyses and experiments may now seem and trivial as they may already appear to some, they were quite undreamed of in 1936 or even in 1938 when the membrane resistance of the squid axon became obviously important—not to say imperative. The result of a rather general development for outside electrodes was presented in 1936 (Cole and Curtis) as shown in figure 1:36 in which three regions are considered. By symmetry,

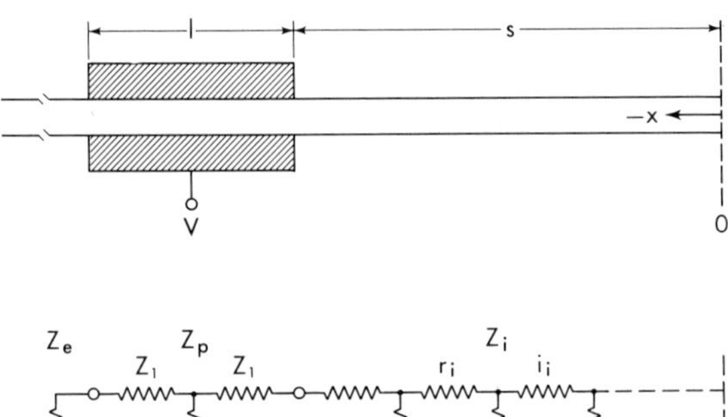

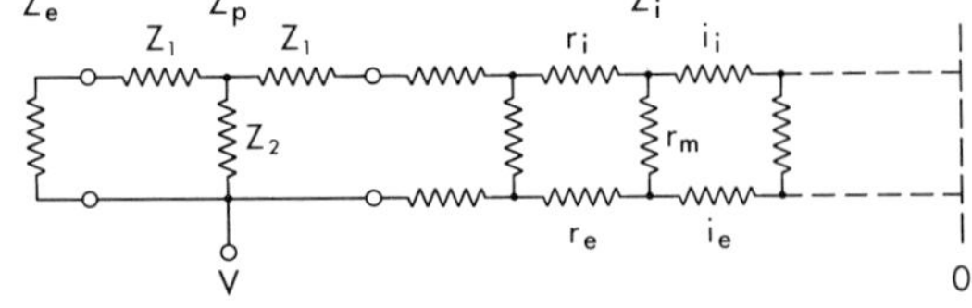

Fig. 1:36. Impedance of an axon as measured longitudinally. Schematic arrangement of one half, above, and equivalent circuit, below.

only one side of the center of the interpolar stretch need be considered. Here $V_e = V_i = 0$ at $x = 0$. On each side at $x = s$ the electrode region extends from $s \leq x \leq s + l$, and the extrapolar region is then beyond, $x \geq s + l$. For the interpolar region to be considered, first an equivalent solution of equation 1:34 is used:

$$V = A \sinh x/\lambda + B \cosh x/\lambda. \qquad (1:37)$$

And since $V = 0$ for $x = 0$, $B = 0$. Here also $i_e + i_i = i_0$ applied. At each point

$$\frac{dV_e}{dx} = -r_e i_e, \quad \frac{dV_i}{dx} = -r_i i_i$$

and with $V = V_i - V_e$,

$$\frac{dV}{dx} = r_e i_e - r_i i_i = \frac{A}{\lambda} \cosh x/\lambda.$$

So

$$A = \lambda \, (r_e i_e - r_i i_i)/\cosh x/\lambda$$

and

$$V = (r_e i_e - r_i i_i)\lambda \tanh x/\lambda. \qquad (1:38)$$

Also

$$V_e = -r_e \int_0^x i_e \, dx, \quad V_i = -r_i \int_0^x i_i \, dx$$

and

$$V = \int_0^x (-r_i i_i + r_e i_e) \, dx$$

$$= -r_i i_0 \, x + (r_e + r_i) \int_0^x i_e \, dx$$

$$= -r_i i_0 \, x - \frac{r_e + r_i}{r_e} V_e$$

or

$$V_e = -\frac{r_e r_i}{r_e + r_i} i_0 \, x - \frac{r_e}{r_e + r_i} V. \qquad (1:39)$$

Conventional circuit analysis (Guillemin, 1935) then produces the T equivalent for the polar region as shown, where

$$Z_1 = \sqrt{r_i/g_m} \tanh (l/2)\sqrt{r_i/g_m} \text{ and}$$
$$Z_2 = \sqrt{r_i/g_m}/\sinh l\sqrt{r_i/g_m}.$$

As shown before, the extrapolar region has the impedance

$$Z_e = \sqrt{(r_e + r_i)/g_m}.$$

At the junction of the interpolar and polar regions

$$i_i = V/Z_{pe}$$

where for the combined impedance of the polar and extrapolar regions

$$Z_{pe} = Z_1 + \frac{(Z_1 + Z_e)Z_2}{Z_1 + Z_2 + Z_e}.$$

Then

$$i_i Z_{pe} = (r_e i_e - r_i i_i)\lambda \tanh x/\lambda.$$

From this

$$i_i = \frac{r_e\lambda \tanh x/\lambda}{Z_{pe} + (r_e + r_i)\lambda \tanh x/\lambda} i_0$$

and

$$i_e = \frac{Z_{pe} + r_i\lambda \tanh x/\lambda}{Z_{pe} + (r_e + r_i)\lambda \tanh x/\lambda} i_0.$$

By equation 1:38,

$$V = \frac{i_0 r_e Z_{pe} \lambda \tanh x/\lambda}{Z_{pe} + (r_e + r_i)\lambda \tanh x/\lambda}$$

and by equation 1:39

$$V_e = \frac{r_e r_i}{r_e + r_i} i_0 x + \frac{r_e^2}{r_e + r_i} i_0 \frac{Z_{pe} \lambda \tanh x/\lambda}{Z_{pe} + (r_e + r_i)\lambda \tanh x/\lambda}.$$

For the impedance from $x = -s/2$ to $x = s/2$

$$Z = 2V_0/i_0$$

$$= \frac{r_e r_i}{r_e + r_i} s + \frac{2r_e^2}{r_e + r_i} \cdot \frac{\lambda}{(r_e + r_i)\lambda/Z_{pe} + \coth s/2\lambda}. \qquad (1:40)$$

Because of its generality this expression has an obvious complexity that could promise only an unpleasant and fruitless analysis—both in the abstract and in the analysis of any data that might be forthcoming. However, as the electrode length $l \to 0$, figure 1:37a,

$$Z_{pe} \to Z_e = \sqrt{(r_e + r_i)/g_m}$$

and

$$Z = \frac{r_e r_i}{r_e + r_i} s + \frac{r_e^2}{r_e + r_i} \cdot \lambda[1 - \exp(-s/\lambda)]. \qquad (1:29)$$

But the first and poorly designed experiments on the squid axon could not exploit the obvious simplicity of this infinitesimal electrode theory. On the other hand, for infinite electrode regions, figure 1:37b

$$Z_{pe} \to Z_p = \sqrt{r_2/g_m}$$

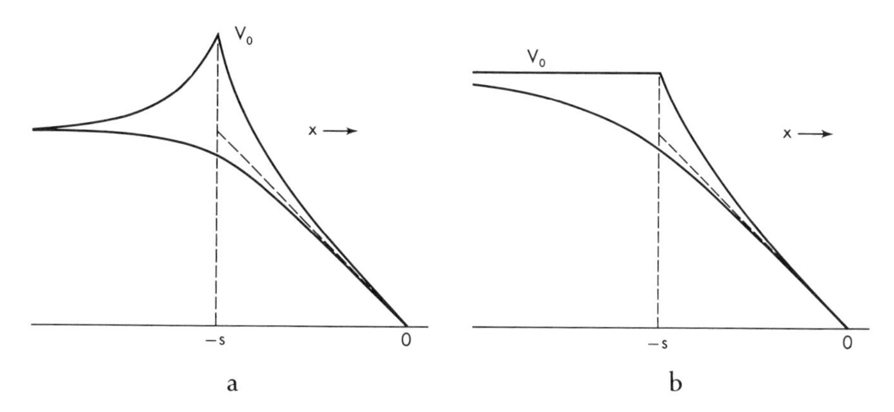

FIG. 1:37. Potential distributions, along half of an axon for an infinitesimal length of electrode, *a*, and an infinite electrode, *b*. For each the external potential is above and the internal, below.

to give

$$Z = \frac{r_e r_i}{r_e + r_i} s + \frac{2r_e^2}{r_e + r_i} \cdot \frac{\lambda}{\sqrt{(r_e + r_i)/r_i} + \coth s/2\lambda}. \quad (1:41)$$

The complexity of this relation between the impedance and the electrode separation was also too discouraging to me to even allow consideration of the use of infinite electrodes.

Squid Axon

A. L. Hodgkin insisted that the direct current resistance vs. axon length data between long electrodes should be interpretable in spite of the considerable complication of the theory, so we went ahead with the experiment as shown schematically in figure 1:38a to obtain the typical data as published (Cole and Hodgkin, 1939), and given in figure 1:38b. Hodgkin proved his contention by fitting the data as shown by the theory curve. Equation 1:41 is written in the form

$$R = ms + y(s)$$

where $m = r_e r_i/(r_e + r_i)$ and, for s large, $y \to y_\infty$, are obtained directly from the graph of experimental data. Defining $(h + 1)^2 = (r_e + r_i)/r_i$ gives

$$\frac{y}{y_\infty} = \frac{2 + h}{1 + h + \coth h(ms/y_\infty)}.$$

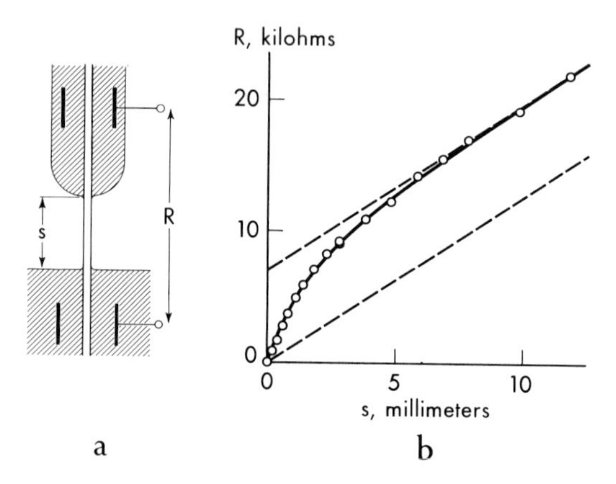

Fig. 1:38. (a) Schematic arrangement for measuring a squid axon resistance R with a length s between upper and lower sea water electrodes. (b) Data from one axon with upper dashed line extrapolating the final slope back to the vertical axis. The intercept is the contribution of the membrane resistance. The lower dashed line is resistance to be found for a negligible membrane resistance at 100 kc.

The plot of y/y_∞ vs. ms/y_∞ was compared with the family of curves

$$\frac{2 + h}{1 + h + \coth hx}$$

for various values of h. From the best values of h, m, and y_∞, all of the axon parameters were determined. These gave a nominal value of 1000 Ωcm^2 for the membrane resistance of a superior axon and less for those in poorer condition. This figure has most surprisingly—and for no yet apparent reason—come to be quite representative of the many other membrane resistances since measured. The axoplasm resistance averaged 1.4 times that of sea water—considerably lower than the four times average that Curtis and I had produced. An interesting and discouraging observation was that the resistance R remained essentially unchanged up to as high currents as could be conveniently measured—thus providing no indication of non-linearity.

The choice of an effective zero of length for the electrode separation and for the extraneous electrode resistance was a constant and unresolved problem. The best answer seemed to be to measure the impedance of the axon and models at a frequency sufficiently high to make the membrane impedance negligible. With the Wheatstone bridge available this was no problem, and it remained only to make measurements in the upper frequency range showing that 100 kc gave an essentially infinite frequency extrapolation, as in figure 1:38b.

This led later to a convenient and powerful method for determining the axon characteristics from the measurements of the zero and infinite frequency resistances R_0 and R_∞ at a series of electrode separations which were essentially measured indirectly by R_∞ rather than directly by a length.

For long electrodes we have again

$$R_0 = \frac{r_e r_i}{r_e + r_i} s + \frac{2r_e^2}{r_e + r_i} \cdot \frac{\lambda}{\sqrt{(r_e + r_i)/r_i} + \coth s/2\lambda}. \quad (1:41)$$

Now let $X = R_\infty = \dfrac{r_e r_i}{r_e + r_i} s$ and $Y = R_0 - R_\infty$.

As $X \to \infty$,

$$Y \to \overline{Y} = \frac{2r_e^2\lambda}{(r_e + r_i)[\sqrt{(r_e + r_i)/r_i} + 1]}$$

and as $X \to 0$,

$$\frac{Y}{\overline{Y}} \approx \frac{\sqrt{(r_e + r_i)/r_i} + 1}{2\lambda} \cdot \frac{r_e + r_i}{r_e r_i} \cdot X.$$

As this initial slope is extrapolated to $\overline{Y}$ it intersects at

$$\overline{X} = \frac{2r_e r_i \lambda}{(r_e + r_i)[\sqrt{(r_e + r_i)/r_i} + 1]}$$

and then the entire curve is given by

$$\frac{Y}{\overline{Y}} = \frac{B + 1}{B + \coth [(X/\overline{X})/(B + 1)]} \quad (1:42)$$

for $B = \sqrt{(r_e + r_i)/r_i}$.

The data are most conveniently plotted as log R_∞ vs. log $(R_0 - R_\infty)$ as shown in figure 1:39. From this $\overline{Y}$ and X are easily determined to give $r_e/r_i = \overline{Y}/\overline{X}$ and so to determine B. To my complete surprise, and for reasons still unknown, the maximum variation between the curves of equation 1:42 is at most a few percent over the range of r_1/r_2 from 0.5 to 8.0 and of $X/\overline{X}$ from 0.1 to 5.0. Thus the best fit of the experimental points to a single master curve is all that is needed to determine $\overline{X}$ and $\overline{Y}$. These with $r_e r_i/(r_e + r_i)$ from R_∞ vs. s completely determine r_e, r_i, and r_m by means of coefficients that are functions of r_e/r_i only. The plotted points of figure 1:39 are all of the data taken for other purposes in nine experiments including two on *Ommestrephes*.

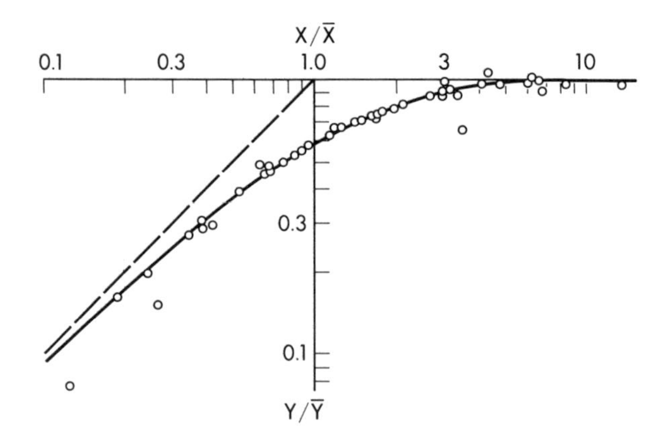

FIG. 1:39. An analysis of axon parameters from extrapolated resistances R_0 and R_∞ at zero and infinite frequencies. $X = R_\infty$, $Y = R_0 - R_\infty$ are plotted and fitted to the master curve as shown. The points are for nine experiments on axons from two species.

The axon parameters, sometimes determined from as few as three points, are given in Table 1:2.

TABLE 1:2

Experiment	Axon	Medium	r_2 — Ωcm	r_4 — Ωcm²
41-6-1	L	SW	33.4	2,700
-5		6xK	32.6	490
-11-1	L	SW	31.6	1,570
-2		Ca-free	31.6	270
-3		SW	31.6	810
-15-1	O*	SW	48.	7,000
-16-1	O*	SW	62.	10,000
-17-1	L	SW	37.7	5,400
-3		5xCa	28.9	27,100

L = *Loligo*, O = *Ommestrephes;* SW = sea water, 6xK = six times [K] of SW, 5x Ca = five times [Ca] of SW; Ca free = SW made up without Ca.

* The O axon diameters are lost and the values given have not been checked.

Conclusions

The squid axon experiments had been very rewarding as the first reliable measurements of a membrane resistance for an animal cell. They had shown that careful—although relatively simple—physical theory, experiment, and analysis could produce a physical membrane parameter, and we already had reason to expect that it was of considerable physiological significance. In the first place it gave a means and a measure of ion movement across the membrane. The value of this resistance was so high that the sensitivity of the suspension measurement—which had produced the membrane capacities—was quite inadequate for this purpose. The decrease of resistance with deterioration, perhaps to the point of conduction failure, without appreciable change of membrane capacity was a most attractive reason to believe that this was a dielectric capacity and not a polarization capacity arising from ion movement across the membrane. The effects on the membrane resistance found later for external ion changes might be expected to relate to physiological changes. Since these, too, were without any considerable effect on the capacity, it seemed probable that the capacity represented an entirely passive aspect of the membrane, with permeability and function expressed in the resistance (Cole, 1940).

INDUCTIVE REACTANCE

In connection with the original determination of squid axon membrane resistance it was reassuring, as well as helpful, to make high-frequency bridge measurements of the longitudinal impedance of an axon. We had found that the reactance became negligible above 100 kc, so we used this frequency as a reasonable approximation to infinite frequency. However, since frequency had long been our most powerful parameter and the axon and the bridge were connected, it was inevitable that we go to lower frequencies and make a complete series of measurements over the available range. Below 200 c the bridge could not be balanced, and only after reviewing the data did I think to add capacity to the unknown arm and obtain a balance. This showed that the capacity of the preparation was negative, and negative it continued to be down to the lower limit of 30 c. Another axon gave less blatant but, nonetheless, overtly similar characteristics requiring an inductive reactance.

Generalized Coordinates

The suggestion of an inductive reactance anywhere in the system was shocking to the point of being unbelievable. It seemed to be certainly a property of the axon, figure 1:40 (Cole and Baker, 1941).

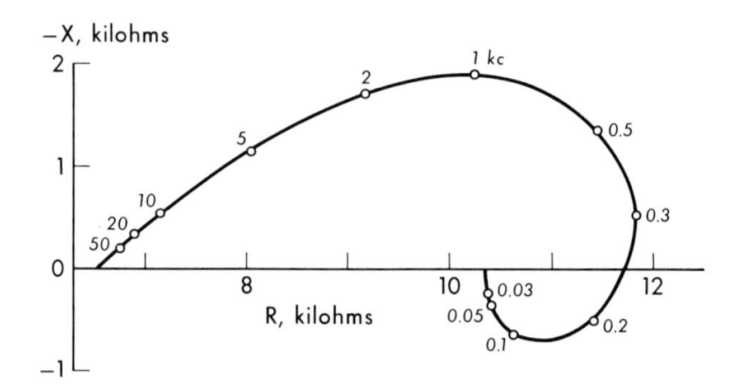

Fig. 1:40. Longitudinal impedance of a squid axon between distant electrodes shown on the complex plane.

The use of a Wien bridge for an oscillator had emphasized that the transfer characteristics of a capacity–resistance four-terminal network were similar to those of a two-terminal resonant circuit. Such a possibility was ruled out by the generalized coordinate treatment of any two-terminal resistance–inductance–capacity circuit by Bode (1935). In this he used the matrix formulation to determine the potential difference v and current i in each element of an arbitrary network. On the definitions of r, c, and l given by $v = ri$, $i = c\, dv/dt$, and $v = l\, di/dt$, the average rate of energy dissipation evaluated in the entire network was $F = \Sigma\, ri^2$. Similarly, the average potential energy $U = \frac{1}{2}\, \Sigma\, cv^2$ and the average kinetic energy $T = \frac{1}{2}\, \Sigma\, li^2$. In these calculations it was not possible to rely on the conventional complex operator exp $j\omega t$ since the operations are not linear, and the real operator exp $(j\omega t)$ + exp $(-j\omega t)$ was used instead. Then Bode showed that for an average current flow I into the network, the terminal impedance was given by

$$Z = R + jX = 2[F + j\omega(T - U)]/I^2 \qquad (1:43)$$

corresponding to the Lagrangian result in classical mechanics.

This required that, for the circuit to have an inductive reactance, there must not only be inductances in the circuit, but also their average energy must exceed that of the capacities since F, U, and T

were not negative. It is interesting to note that this relation also shows that for a constant phase angle the dissipation per cycle must be proportional to the difference of the kinetic and potential average energies independent of the frequency.

The growth of the inductive reactance for the squid axon with increasing electrode separation to reach an asymptotic value for larger separations indicated that the reactance was not associated with current flow parallel to the axon axis. It was then most simply and certainly to be located as a transverse characteristic of the membrane.

The very idea of an inductive reactance in a membrane was not easy to accept. Hodgkin and Rushton (1946) thought it useful, whether true or false, and suggested that it might be a precursor of excitation. Lorente de Nó (1947) was more critical, rejecting it as a linear concept that had no place in such a nonlinear system. Hodgkin (1951) accepted the suggestions (Cole, 1941a) that an inductive reactance and the numerous oscillatory phenomena might have a common origin and that the reactance might result from changes of potassium permeability (Cole, 1947, 1949). I do not know that he has expressed himself in print, but Otto Schmitt has understandably objected to use of the term "inductance" and seemed to prefer that there was also not an inductive reactance in a membrane. In a review, Mauro (1961) accepted the term "anomalous reactance" as implying reservations on the underlying processes which produced them.

Membrane Impedance

Quite irrespective of early—and of course, subsequent—opinions, it seemed to be a fact that a squid giant axon could show an inductive reactance, and perhaps less certainly, but still quite probably, that this reactance was produced by the axon membrane. It then became necessary to interpret the experimental facts of the axial impedance measurement as membrane characteristics.

By retaining z_m in the original equation 1:41 we had (Cole and Baker, 1941) for the longitudinal impedance

$$Z = \frac{r_e r_i}{r_e + r_i} s + \frac{2r_e^2\lambda}{(r_e + r_i)[\sqrt{(r_e + r_i)/r_i} + \coth s/2\lambda]}$$

where $\lambda = \sqrt{z_m/(r_e + r_i)}$. Accepting the assumption that $z_m \to 0$ as $\omega \to \infty$ and $z_m \to r_m$ as $\omega \to 0$, we had

$$R_\infty = \frac{r_e r_i}{r_e + r_i} s:$$

$$R_0 = R_\infty + \frac{2r_e^{\,2}\lambda_0}{(r_e + r_i)[\sqrt{(r_e + r_i)/r_i} + \coth s/2\lambda_0]}$$

where $\lambda_0 = \sqrt{r_m/(r_e + r_i)}$.

Now a relative impedance was defined,

$$\overline{Z} = \frac{Z - R_\infty}{R_0 - R_\infty} = \frac{\sqrt{(r_e + r_i)/r_i} + \coth s/2\lambda}{\sqrt{(r_e + r_i)/r_i} + \coth s/2\lambda_0} \cdot \sqrt{\frac{z_m}{r_m}}.$$

Then for $s \gg \lambda_0$,

$$\overline{Z} = \sqrt{z_m/r_m} \text{ or } z_m = r_m\overline{Z}^2,$$

and since $\overline{Z}$ and z_m may both be expressed as vectors,

$$\overline{Z} = |\overline{Z}|e^{j\Phi}, \quad z_m = |z_m|e^{j\phi}$$

we have

$$|z_m| = r_m|\overline{Z}|^2, \quad \phi = 2\Phi. \tag{1:44}$$

Thus it is a simple matter to predict the cable impedance locus as half the phase angle and the square root of the membrane impedance vector. An ideal capacity–resistance membrane with the locus of figure 1:41 would then give the cable locus of figure 1:42 for comparison with that of the real axon. Conversely the axon membrane locus of figure 1:43 is attained by doubling the phase angle and squaring the measured impedance vector of figure 1:40.

With the membrane characteristics then available, and thought

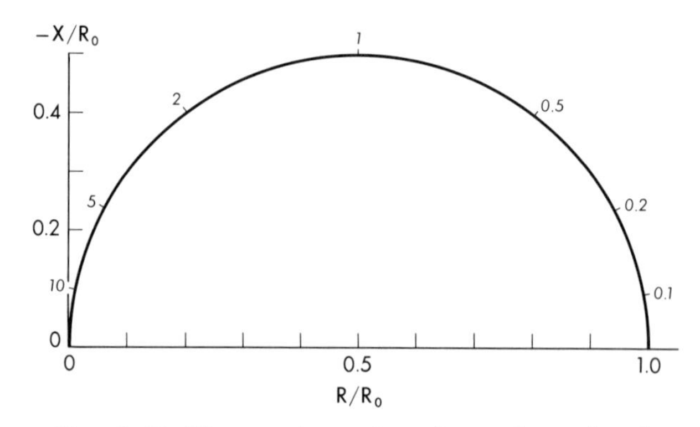

FIG. 1:41. The membrane impedance shown is calculated by squaring the impedance vector at each frequency for the cable. The frequencies are given in multiples of the characteristic frequency.

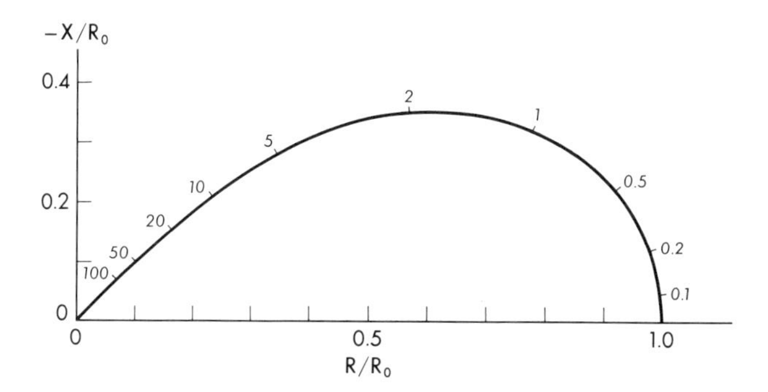

FIG. 1:42. Computed longitudinal impedance of an axon with a resistance-capacity membrane as it appears on the complex plane.

to be reasonably normal, there were certainly two frequency-dependent elements in the membrane—one capacitive and the other inductive. It had been shown by a topological analysis that there were eleven nonidentical circuits which could give any single impedance vs. frequency characteristic (Foster, 1932). Thus, with no structural guide, it seemed wise to make as general and noncommittal an analysis as possible.

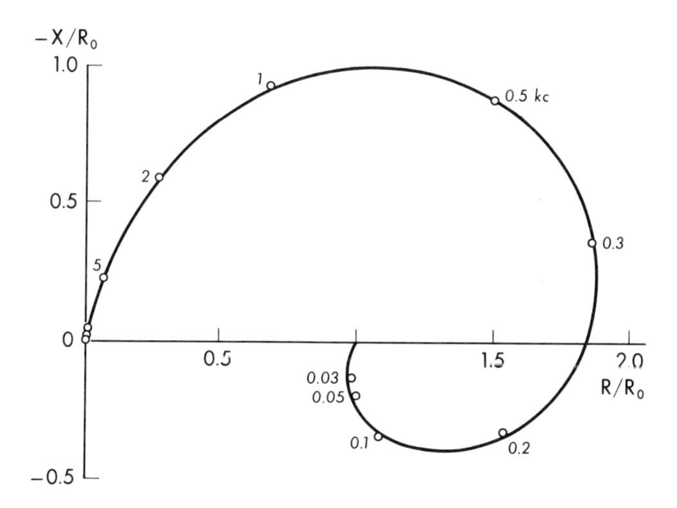

FIG. 1:43. Squid axon membrane impedance on the complex plane as calculated from longitudinal measurements by squaring. The curve is a pedal curve of an ellipse.

Pedal Curves

This approach had long been available in the work of Carter (1925), which gave a two-sheeted doughnut complex plane domain containing all the impedance loci, but it did not contribute materially toward the representation of the data. At this point, however, R. M. Foster gave me the apparently unpublished information that each of these loci was in fact a pedal curve of an ellipse with the parameters as shown in figure 1:44.

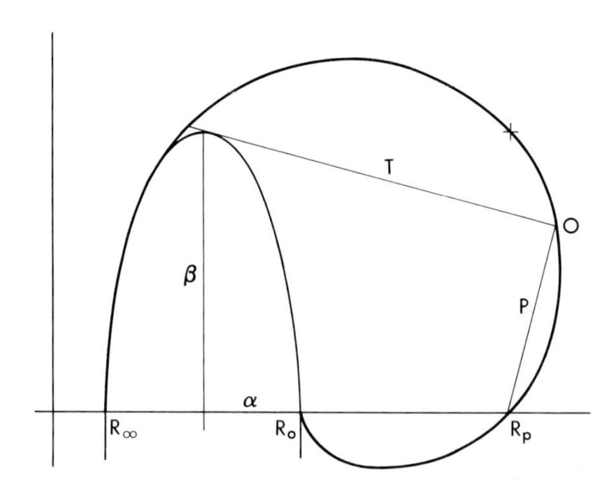

Fig. 1:44. A point O on the pedal curve is found by drawing the tangent T to the ellipse with half axes α, β and dropping the perpendicular P to the pedal point R_p.

In this figure the intersection of a tangent T to the ellipse, with semi axes α, β, and a perpendicular P to it from the fixed pedal point R_P trace the pedal curve O. For an impedance in biquadratic form,

$$Z = \frac{Ap^2 + Bp + C}{Dp^2 + Ep + F} \tag{1:45}$$

the geometric parameters are given by

$$\begin{aligned} R_\infty &= A/D \text{ at } \omega = \infty \\ R_0 &= C/F \text{ at } \omega = 0 \\ R_P &= B/E \text{ at } \omega^2 = (BF - CE)/(BD - AE) \end{aligned}$$

and

$$Z = R_P + j\beta \text{ at } \omega^2 = F/D \tag{1:46}$$

is among other helpful relations that determine the other frequency locations along the pedal curve locus. Although the curves are easily drawn from point by point constructions, an alternate construction as an envelope gives a smooth curve more easily (Salmon, 1879). These pedal curves have been found to be highly informative representations of two reactance circuits, and a number were published without disclosing their construction. One particularly interesting and helpful family of these loci, figure 1:45, was given for the conventional circuit of figure 1:46. The resistance R was main-

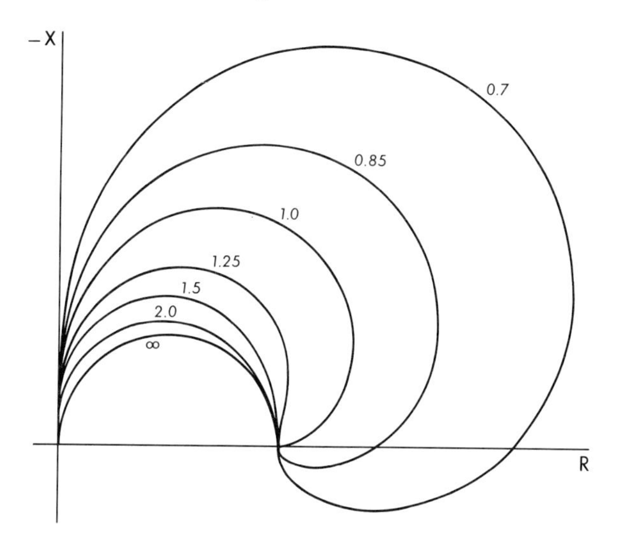

FIG. 1:45. Complex impedance loci for the capacity-inductance membrane circuit with a constant resistance and the indicated damping coefficients.

tained constant as the damping coefficient, defined as $\eta = R\sqrt{C/L}$, was varied from 0.7 to ∞. The case of $\eta = 1.0$ is unique in that the locus is tangent to the resistance axis and, as shown by van der Pol (1937) and Bode (1938), the energies stored in the inductance and the capacity are equal for direct current flow. For $\eta = \sqrt{2}$ this circuit is underdamped but is so fast as to be considered optimum for many purposes. As we go on, $\eta = 2.0$ represents critical damping, and for $\eta = \infty$ the inductance is negligible and the locus reduces to the resistance–capacity semicircle.

This line of work has never been brought to the desired goal of matching the experimental data because of the loss angle attributed

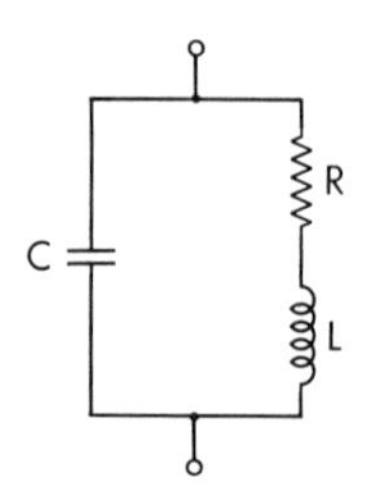

FIG. 1:46. A first approximation to a squid axon membrane equivalent circuit.

to the membrane capacity. It has not been found possible to derive the relations of equations 1:46 from equation 1:45; consequently the possibility of a similar construction for fractional powers of p has neither been derived nor arrived at empirically.

But ignoring the loss angle, it is almost obvious by inspection that the circuit of figure 1:46 can be a reasonable approximation to the membrane characteristic of figure 1:43. The equivalent inductance of 0.1 henry cm^2 is so extremely large as to warn that it is not of electromagnetic origin, and other possible explanations will be considered later.

EXTERNAL IONS

In the next step the effects of variations of the external potassium and calcium ions were undertaken (Cole and Marmont, 1942). It was found possible to maintain reasonably constant impedances and excitation thresholds for several hours with the interpolar region in water-saturated air rather than the oxygenated petroleum oil as used before. This also contributed considerably to the speed and convenience of the experiments. Although the low-frequency characteristics in sea water were somewhat variable, they usually

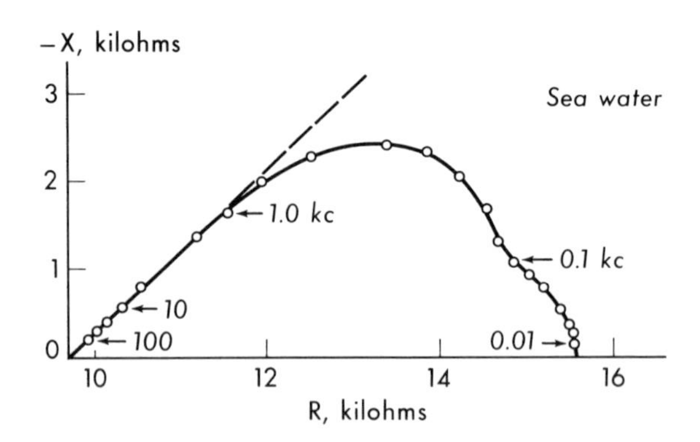

FIG. 1:47. Longitudinal complex plane impedance for a typical squid axon in sea water.

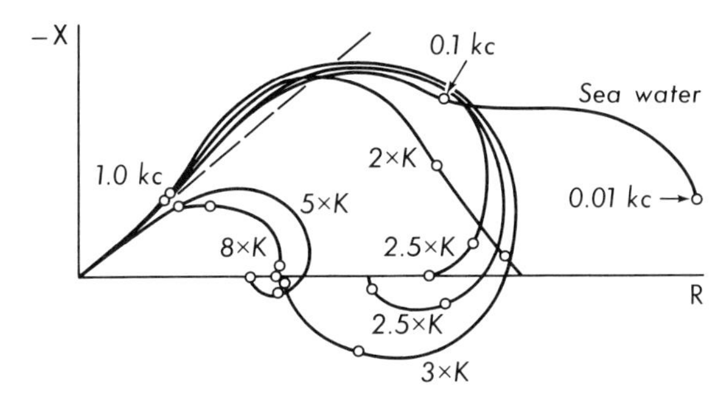

FIG. 1:48. Longitudinal complex plane impedances for a squid axon in sea water and for increased external potassium concentrations.

gave an extra, small, low-frequency capacitive reactance of figure 1:47 rather than the inductive component that had been thought of before as the normal behavior in sea water.

The axial impedances for increasing external potassium ion concentrations shown in figure 1:48 suggest that there may have been an increased potassium ion concentration in the oil preparation. Increased inductive and capacitive reactances and decreased zero-frequency resistance in the absence of external calcium are clearly properties of a rather good tuned circuit, figures 1:49, 1:50. But the large low-frequency capacitive component produced by high (10x) external calcium seemed to add a considerable complication to the problem of interpretation. It was then thought that both the inductive and the capacitive components might be present in the membrane at all times in addition to that of the usual capacity of 1 μF/cm^2. It was supposed (Cole and Marmont, 1942) that alterations of the resistive components might allow one or the other of

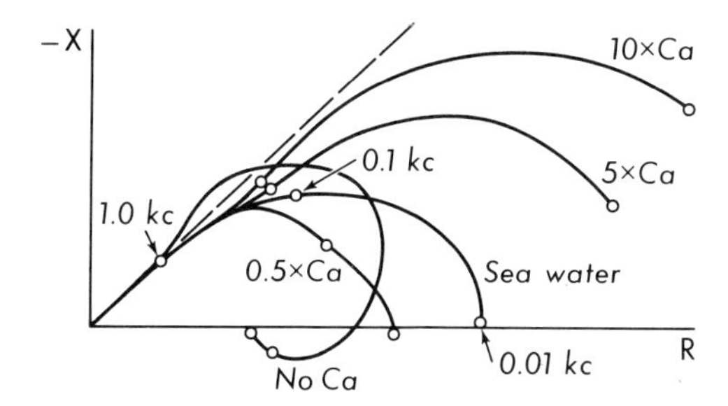

FIG. 1:49. Longitudinal complex plane impedances for a squid axon in sea water, and for increased and decreased external calcium concentrations.

these two low-frequency reactances to appear in measurements made at the boundaries. However, in an investigation of the topology of equivalent two-terminal networks Foster (1932) had found there were 113 to 122 equivalent three-reactance networks—depending on the exact definition of equivalence! Once again, without some guide as to structure, the problem of analysis had become even more hopelessly complex.

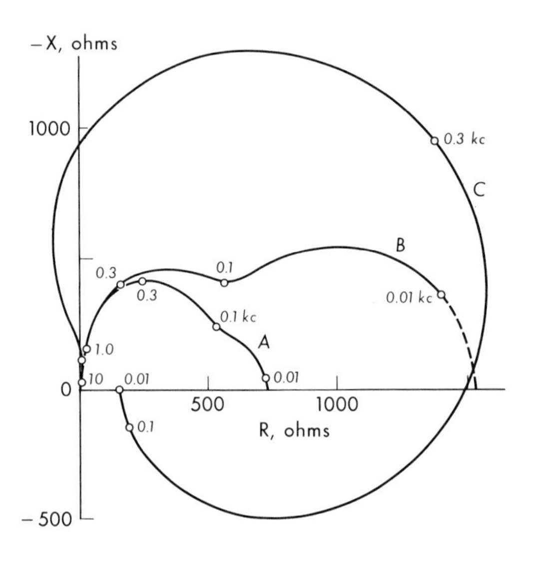

Fig. 1:50. Complex squid axon membrane impedance in sea water *A*, high calcium *B*, and low calcium *C*.

One simplification did, however, seem reasonable. It was possible to compute the membrane admittance as the reciprocal of the impedance. Then it was assumed as always before that the membrane capacity was directly between the internal and external electrolytes of the cell and was in parallel with the ion conductance machinery of the membrane. On this basis the constant phase angle admittance of the membrane capacity, as determined at high frequency and extrapolated to lower frequencies, could be subtracted to leave the conductive admittance. This rather tedious and somewhat unreliable procedure was carried out over some years to produce the better results shown in figure 1:51. Although it had become apparent that there might be three reactances in the membrane, their significance was not known and indeed is not yet fully understood.

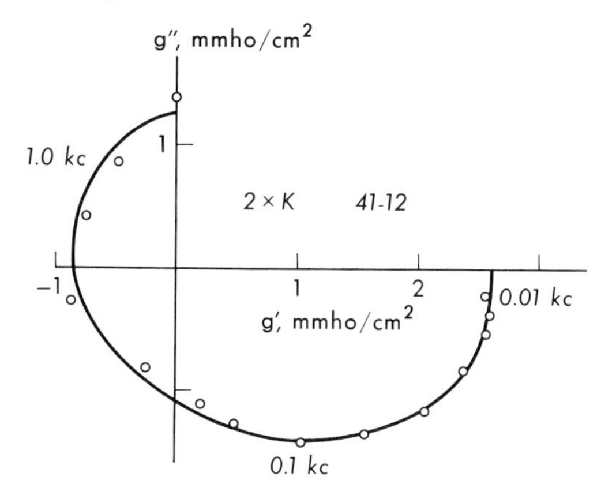

FIG. 1:51. Squid axon membrane conductance (admittance minus passive capacitive component) on complex plane in twice normal potassium.

SUMMARY

The measurements of cell membrane capacities had given no indication of an electrical leakage or an ion current. Several lines of evidence made it eminently desirable that there be such a thing but gave no suggestion of its magnitude. Following the lead of the large plant cells, the most timely advent of the squid giant axon allowed a sound measurement of a thousand ohms for a square centimeter of membrane. But in the process an impossible new fact emerged, an inductive reactance in the membrane. Both the resistance and this reactance were most encouragingly responsive to a few physiological changes.

Linear Parameters of Some Other Cells and Membranes

Membrane capacities have been relatively easier to determine and less interesting than membrane resistances. After measurements on the squid axon membrane had established a resistance of 1000 Ωcm^2, this parameter became increasingly accessible, particularly after World War II, by a variety of techniques, on many other membranes. The measurement of membrane resistance is now so commonplace and is being so widely interpreted that I can give only a cursory introduction to some of these developments and point to a partial summary of them (Cole, 1959).

For information on another single cell, Rita Guttman and I had turned to the frog egg (Cole and Guttman, 1942). This gave additional support for the constant phase angle membrane capacity, but it was also advantageous for a membrane resistance measurement. The accuracy of the effective volume concentration estimate, the relatively large diameter of the egg, and the high resistance of the medium all contributed to give the average of 170 Ωcm^2 for the surface membrane.

It had long been very frustrating that neither a membrane potential nor a membrane resistance was available for red blood cells, which have been so familiar and widely used and have such a relatively simple surface structure. So the achievement of a membrane resistance by Johnson and Woodbury (1964) is interesting. It is also gratifying because they avoided, with so small a cell, the difficulties that had prevented Curtis and me (1938a) from making the same measurement on the much more favorable marine eggs. After a cell was forced into a glass capillary, much more current flowed in the sea water between the membrane and the glass than flowed across the two membranes and the cytoplasm, and so prevented a membrane measurement. The red cell, however, was placed in a sucrose solution to block the shunt path after the sucrose at each end was replaced with electrolyte. The shunt path was of the order of 10 μ long and the diffusion coefficient about $10^{-5}cm^2/sec$, so the diffusion time $\tau \approx a^2/D \approx 25$ msec. This required prompt measurement after an abrupt washing away of the sucrose on both sides to give values of 10–15 Ωcm^2. This rather low resistance has been correlated with chloride permeability; it is a warning that the 1000 Ωcm^2 that had been found for many other resting membranes may not be an entirely universal constant.

The longitudinal resistance of isolated frog muscle fibers in air was also used by Tamasige (1950) to give an average membrane resistance of 10^4 Ωcm^2 at 18°C with a temperature coefficient of 6 percent per degree Celsius. Then the effects of changes of external ions were investigated (Tamasige, 1951), giving increased resistances for increases of calcium and magnesium and of potassium except for drastic reductions at above eight times normal.

Quite different, pioneering, and important investigations of the passive characteristics of the node of Ranvier of the frog myelinated axon have been led by Tasaki over many years and recently redetermined by him (Tasaki, 1955). In this delicate procedure two adjacent myelin internodes are placed in air gaps, and a three-electrode system is used to measure the central node. Current is passed from the node on one side through the central node to an external electrode. The potential is measured between the adjacent node on the other side and the central electrode. From this node a capacity of 3–7 $\mu F/cm^2$ and a resistance of 10–15 Ωcm^2 were found, both of which are somewhat extreme values. Additional data, assembled by Hodgkin (1951) and Stämpfli (1952), gave extremely interesting values for the internode. Here the myelin was found to have 5–10 as the dielectric constant and nearly 10^9 Ωcm for the resistivity, while the axoplasm was about 100 Ωcm. As Hodgkin (1951) pointed out, the lamellar structure of myelin from x-ray (Schmitt et al., 1941) and electron microscope (Fernandez-Moran, 1950) measurements gave about 250 layers each 80 Å thick. Then each of these further corresponded to a typical cell membrane with a capacity of about 0.6 $\mu F/cm^2$ and a resistance of 600 Ωcm^2.

Through much of the work on linear parameters, principal reliance has been on bridge measurements, particularly with the classical Wheatstone bridge. The requirement of a low power input and the need for high precision drove us (Cole and Curtis, 1937) to the noise level at the output and later to various filter arrangements to reduce this interference, especially at audio frequencies. The repeated observation that cell suspensions and tissues were noisier than their equivalent networks was very frustrating and aroused some curiosity. But nothing was done about it until Nightingale (1959) measured this excess noise between electrodes on humans. The artificial circuits gave a Nyquist-Johnson resistor noise (Nyquist, 1928), $\overline{V^2} = 4\ kTR$. This was independent of frequency and, by a Fourier synthesis (CF No. 403.1), corresponded to random pulses of potential. The excess noise, however, varied as $1/f$ which is the Fourier equivalent (CF No. 415) of random steps of potential. This same behavior is found in many physical, current-carrying, systems but in no case is it adequately explained. So, once again, the problems of physics may not be easier than those of biology—nor even much different!

Internal or transmembrane electrodes have added so much to the power of electrical investigations as to be revolutionary. Blinks (1930c) had allowed plant cells to heal around rather large capillaries and then Curtis and I (1940) and Hodgkin and Huxley (1939) had succeeded with the squid axon by measuring beyond the region affected by the insertion of the electrode.

A large proportion of the more recent work has taken advantage of the capillary intracellular microelectrode after it had been successfully used on frog muscle fibers by Ling and Gerard (1949). This I had attempted, following Robert Chambers, in 1924 (Cole, 1957) and some progress was made by Hogg, et al. (1934). However, with a tip diameter of a micron or less, most membranes form a satisfactory seal and even recover after the electrode is withdrawn. The almost unlimited possibilities for the measurement of potentials and for the introduction of current and foreign materials has aroused almost unlimited use of such microcapillaries.

After our many and futile efforts to find a marine egg membrane conductance I was especially pleased by the work of Tyler, et al. (1956) on the *Asterias* egg. The membrane was not easy to penetrate, and two microelectrodes had to be inserted before it could be proven that either one was inside. But with two internal electrodes, a membrane capacity of 0.5 $\mu F/cm^2$ was found to confirm the less direct measurements, and a resistance of 3000 Ωcm^2 was found for the first time in such an egg.

Among many other cytoplasmic membranes this two-needle procedure was applied to two gland and six egg cell nuclei (Loewenstein, 1964). The former membranes gave 15 mV potentials, $\sim 2\ \Omega cm^2$ resistances and about 500 $\mu F/cm^2$ capacities! On the other hand they found neither potential (<0.3 mV), nor resistance ($< .001\ \Omega cm^2$) nor capacity for the egg nuclear membranes. Both of these behaviors are so extreme as to deserve special attention.

"Electrical," as contrasted with "chemical," synapses were found by Furshpan and Potter (1959) to be low resistance couplings between the cells as shown in figure 1:52. Such "tight" or "electrotonic" junctions with low or negligible resistance between adjacent cells were soon found in other tissues. An analysis and summary were given by Bennett (1966). These electrical results together with electron microscopic evidence promise to be of great importance in an understanding of membrane ion permeabilities.

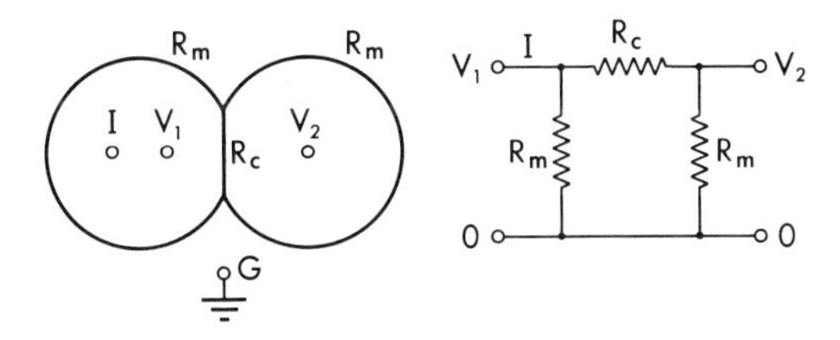

FIG. 1:52. Electronic coupling between cells. A current I into one cell, left, or an equivalent circuit, right, will produce a potential V_1 in it and V_2 in an adjacent cell. The ratio V_2/V_1, varies between zero and unity as the coupling resistance R_c ranges from much larger to much smaller than the external membrane resistances R_m.

CABLE IMPEDANCE

The cable theory and experiment have been developed and applied in several different ways to find the linear parameters of several nerve and muscle fibers. Rewriting the cable equation as

$$d^2V/dx^2 - R(Cp + G)V = 0 \qquad (1{:}32)$$

where $R = r_e + r_i$ and p replaces $\partial/\partial t$, the solution for $x \geq 0$ and an applied current I_0 at $x = 0$ gives for the transfer impedance

$$Z = V(x)/I_0 = \sqrt{R/(Cp + G)} \exp\left(-\sqrt{R(Cp + G)}x\right) \qquad (1{:}47)$$

with the characteristic impedance Z_0 at $x = 0$ as before, equation 1:36. Routine manipulations of the complex quantities yield the various algebraic forms appropriate for comparison with experiment, for example those by Schmitt (1955), Tasaki and Hagiwara (1957), Falk and Fatt (1964).

One approach has been used with power and elegance by Schmitt (1955) for both the squid axon and the frog sciatic nerve—first with "student servo" recording. By recording log $\bar{V}$ and Φ along the length of these preparations in the range of frequencies 40 c to 50 kc, Schmitt has shown the validity of the equations. Tasaki and Hagiwara (1957) have recorded $\bar{V}$ over the frequency range from 30 c to 2 kc for a single fiber in a toad sartorius muscle to show that it varies as $1/\sqrt{\omega}$ and to obtain the value 5–9 μF/cm^2. Falk and Fatt (1964) extended the range of these measurements to include 1 c to 10 kc. They chose to present their data on the cable complex plane of figure 1:42 rather than by equation 1:44 on the membrane plane of figure 1:41, which compresses the upper frequency part of the locus. With particular attention to stray capacities, they found the high-frequency capacity of 2.6 μF/cm^2 for frog sartorius, to confirm the earlier values, with a conductance corresponding to 3100 Ωcm^2. The values of 3.9 μF/cm^2 and 680 Ωcm^2 for crayfish were correlated with the extensive folding of these muscle fiber membranes. Below 500 c the data departed systematically from the simple representation of figure 1:39 as indicated in the locus of figure 1:53. The additional capacities of 4.1 μF/cm^2 for frog and 17 μF/cm^2 for crayfish with the respective resistances of 330 Ωcm^2 and 35 Ωcm^2 became evident below 500 c and were suggested as characteristics of the sarcoplasmic reticulum.

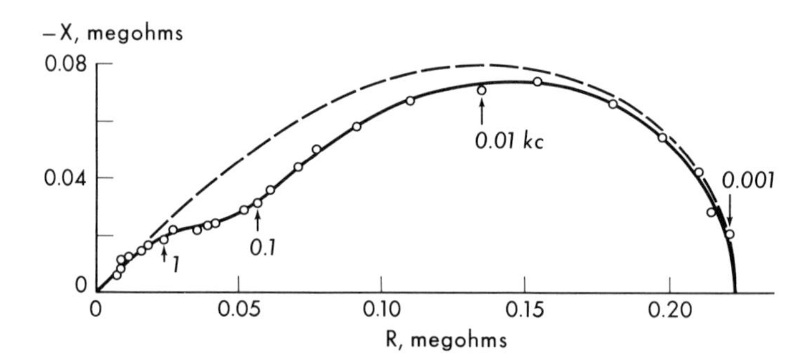

Fig. 1:53. Longitudinal complex impedance of a frog sartorius muscle fiber, circles, with the loci, dashed and solid, calculated for one and two time constants.

In sharp contrast to the conclusion of Fatt (1964) that distribution of fiber diameters could not account for the constant phase angle measurements of 70°, Falk and Fatt (1964) found no indication of significant difference in single fibers from the ideal of 90°. A reconciliation has not been arranged!

TISSUES

We were now in a position to try again to do something about packed cells and tissues. Since the days when we knew nothing about a membrane conductance, I have hoped for an impedance analysis that would start with completely packed cells and then give the effect of thin layers of intercellular electrolyte between membranes of adjacent cells.

The obvious model to use was the tetra kai decahedron of Kelvin (1904; D'Arcy Thompson, 1917). This beautiful form with all sides of the same length is made by slicing off the corners of a cube to give eight hexagonal faces and leave six square faces. It has a low surface area and it close packs in a cubic array. But after several attempts with cardboard and wire models, I did not find a simple way to compute the current flow through the assembly. I turned to the hexagonal packing of rhombic dodecahedra with only a slight promise of some success, although hexagonal packing is a possibility for cylindrical cells.

There were no useful experimental data. McClendon had low-frequency conductance measurements of centrifuged red cells but gave no details of his volume concentration data. Parpart and I packed red cells to far beyond Koeppe's criterion of transparency but we had no measure of volume concentration and were unable to reach a constant conductance. The only impression from these data was that the conductance was rather high, perhaps because the intercellular space was larger or the conductance in it was higher than expected. There are many other, usually inadequate, data on numerous tissues and a few on membranes one or two cells in thickness such as *Ulva* and guinea pig amnion (Silver et al., 1965).

But it was interesting to see what could be calculated for a cubic array of cells which are cubes. Let these cells be a cm on a side with an internal resistivity r_2 Ωcm, a membrane impedance z_3 Ωcm^2, and separated by the medium of thickness δ and resistivity r_1 Ωcm. Then using the planes of symmetry, which are of constant potential or not crossed by current, the problem was simplified to that of a single corner of a single cell as shown in figure 1:54a.

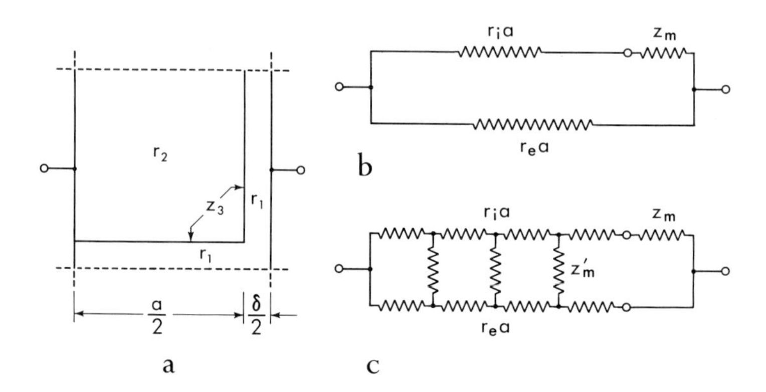

FIG. 1:54. (a) An analysis for the impedance of cubical cells, of edge a, separated by a medium of thickness δ. The circuit (b) approximates the behavior for the membrane impedance either very large or very small. In-between situations require the cable analysis indicated in (c)

Usually we approximate with $\delta \ll a$ and $r_2 \approx r_1$ but in careful detail even this problem is extremely complicated, so we have no idea of the effect of further and perhaps more drastic simplifications. Nonetheless, for a cm cube and for z_3 very large

$$Z \approx \frac{ar_1}{2\delta}\left(1 - \frac{a^2 r_1}{4\delta z_3}\right)$$

and for $z_3 = 1/j\omega c$ as $\omega \to 0$

$$R_0 \approx ar_1/2\delta.$$

On the other hand, for z_3 small

$$Z \approx r_2\left[1 - \frac{2\delta r_2}{ar_1} - \frac{2z_3}{ar_2}\right]$$

and if $z_3 = 1/j\omega c$ again and $\omega \to \infty$

$$R_\infty \approx r_2.$$

Both of these limiting impedances suggest the three element network of figure 1:54b, and the extrapolated resistances are obvious.

Somewhat in a spirit of adventure, we may go far beyond any probable range of validity for these approximations to see what happens for intermediate values of z_3. If the contributions of z_3 to each impedance are made equal then

$$z_3 = a^2 r_1/4\delta$$

and for $a = 40\ \mu$, $\delta = 1\ \mu$, $r_1 = 100\ \Omega$cm, $z_3 = 4\ \Omega\text{cm}^2$. Thus for such a tissue the approximations can only be trusted for a membrane impedance either much larger or much smaller than $4\ \Omega\text{cm}^2$. A membrane capacity of $1\ \mu\text{F/cm}^2$ for z_3 would have this reactance at about 40 kc to give a point of reference between R_0 and R_∞.

It is rather obvious, however, from figure 1:54a that the behavior of this cubical array should be much the same as for the Kelvin cable of figure 1:54c. Previously we used either infinite or infinitesimal electrode lengths but here the problem is worse with the electrode lengths about equal to their separation. Equation 1:40 now becomes

$$Z = \frac{r_e r_i}{r_e + r_i} a + \frac{2r_e^2}{r_e + r_i} \frac{\lambda z_m'}{(r_e + r_i)\lambda + z_m' \coth a/2\lambda}$$

for a single whole cell with

$$\begin{aligned} r_e &= r_1/2a\delta\ \Omega/\text{cm},\ r_i = r_2/a^2\ \Omega/\text{cm} \\ z_m &= z_3/a\ \Omega\text{cm},\ z_m' = (z_3 + \delta r_1)/a^2\ \Omega \\ \lambda &= \sqrt{z_m/(r_e + r_i)},\ Z_{ep} = Z_1 = z_m'. \end{aligned}$$

This complicated expression extrapolates to the previous limits as z_3 approaches 0 and ∞. The membrane part of it which varies between 0 and 1 may be rewritten

$$\frac{1}{2\theta^2 + \theta/\tanh\theta}$$

where $\theta = a/2\lambda$. Then for $z_3 = 1/j\omega c$ and the real part of the denominator near 2, we find from Kennelly (1921) that

$$z_3 = a^2 r_1/12\delta.$$

This also reasonably confirms the more primitive conclusion.

An analysis of this type may be applied to the double layer structure of an amnion. Here we find $\delta/a \approx 0.4$ with $c = 10\ \mu\text{F/cm}^2$. It will be interesting to find how well these values correspond to the histology of this membrane, both as to the volume concentration of the cells and the crinkles and wrinkles of the plasma membrane which are suggested and which we have come to expect. However, the measured value of R_∞ seems unreasonably large. Perhaps there is a series resistance associated with the membrane capacity—as has been suspected for other cells and demonstrated for the squid axon.

It is also possible to predict some properties of packed red blood cells. Robertson (1960) gives $\delta \approx 100$ Å, Johnson and Woodbury (1964) give $r_3 = 15\ \Omega\text{cm}^2$. Then with $r_1 = r_2 = 100\ \Omega\text{cm}$, and $2a = 4.5\ \mu$, $\lambda = 5.5\ \mu$, and $\theta = 0.2$. As a consequence the effect of the membrane conductance is less than 10 percent and the equivalent circuit, figure 1:54b is a good approximation. With $z_3 = 1/jc_3\omega$, the center of the dispersion region should then be about 300 kc.

TRANSIENT RESPONSE

The behavior of the cable for steps or pulses of current or potential requires a return to the circuit of figure 1:34 and the partial differential equation

$$c_m \frac{\partial V}{\partial t} + \frac{1}{r_m} V = i_m = \frac{1}{r_e + r_i} \frac{\partial^2 V}{\partial x^2}. \qquad (1:32)$$

The equation may be integrated directly by the powerful and elegant classical methods of partial differential equations. The transient solution after an applied step of current at $x = 0$ was given, graphed, and tabulated by Hodgkin and Rushton (1946)—after a six-year diversion—and with several graphs by Lorente de Nó (1947). They used an operational method and noted that the solution could be obtained by Fourier transforms. The answer to the immediate problem as stated in equation 1:48 is given by Campbell and Foster (1931) Pairs I, No. 818.1, 812.5 and II, No. 5, and when interpreted means

$$V_m = \frac{r_e \lambda I_0}{4} \{e^X[1 - \text{erf}(X/2\sqrt{T} + \sqrt{T})] - e^{-X} 1[+ \text{erf}(X/2\sqrt{T} - \sqrt{T})]\} \qquad (1:48)$$

where $X = x/\lambda$, $T = t/\tau_m$, $\lambda = \sqrt{r_m/(r_e + r_i)}$, $\tau_m = r_m c_m$

and

$$\text{erf}\ z = \frac{2}{\pi} \int_0^z e^{-\omega^2}\, d\omega.$$

The similar function for the break of current has been tabulated and graphs presented for V_m as a function of X, T after current make and break.

The constant λ was determined by the exponential spatial distribution of the steady state potential. A constant, $y = r_e^2\lambda/2(r_e + r_i)$, was the ratio of the steady potential at the electrode to the current. Another constant, $m = r_e r_i/(r_e + r_i)$, was obtained from the ratio of the potential gradient at the center of the interpolar region to the current. The value of τ_m was more troublesome; seven methods were listed and tried at least once. The most popular method was given without theoretical justification and is still an interesting problem. The time to half maximum was found to be very nearly linear in x, and the factor $2\lambda/\tau_m$ was obtained from the slope of the plot x vs. $t_{1/2}$. The axon constants are then

$$r_2 = \pi a^2 m(1 + m\lambda/2y)$$
$$r_3 = 2\pi a\lambda m(2 + m\lambda/2y + 2y/m\lambda) \text{ and}$$
$$c_3 = \tau_m/r_3.$$

This transient approach to the parameters of a linear cable has been spectacularly popular. It relies on well-known, widely used, and fast experimental techniques, the analysis has been made simple and straightforward, and, to quote myself (1959), ". . . real time seems to inspire an instinctive faith that a complex number cannot." So only a few of the many results can be mentioned.

TABLE 1:3

TRANSIENT CABLE ANALYSIS OF MEMBRANE PARAMETERS

Fiber	Reference	c_3 μF/cm²	r_3 Ωcm²	r_2 Ωcm
Lobster axon	(1)	1.33	2300	60
Carcinus axon	(2)	1.1	7600	90
	(3)	1.2	8000	55
Frog muscle	(4, 5)	5–8	3000–4000	230
Crab muscle	(6)	40	100	70
Crayfish muscle	(7)	20	1000	125
Calf, sheep, and goat Purkinje fiber	(8, 9)	11–12	1200–2000	75–105

References: (1) Hodgkin and Rushton (1946), (2) Hodgkin (1947), (3) Katz (1948), (4) Katz (1948), (5) Fatt and Katz (1951), (6) Fatt and Katz (1953), (7) Fatt and Ginsborg (1958), (8) Weidmann (1952), (9) Coraboeuf and Weidmann (1954).

Nerve membrane figures had come to be conventional, but those for muscles challenged this complacency. Although forewarned by our results and Schwan's at low frequency, I did not find a basis for such an explanation of these transient measurements. However.

Falk and Fatt (1964) demonstrated most dramatically that an interpretation cannot be much better than the experimental facts. They showed two relaxation times with the power of impedance measurements (figure 1:53), and identified the dispersion at about 8 c as predominant in the usual transient procedure. Also, to be even more convincing, they took transient records on time bases of 5 and 25 msec to show the high-frequency dispersion which only slightly perturbed the 100 msec record. As in other cases, for example Cole and Curtis (1939) and Cole (1965), a single record on a log time scale would be adequate and more revealing. This should be expected from sinusoidal work where log frequency has long been in common use. Such a radical innovation would increase both the speed and power of the transient method, but it seems difficult to abandon the familiar linear time scale.

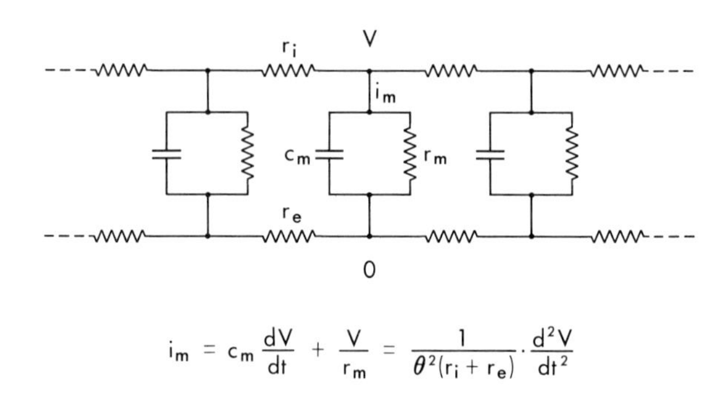

$$i_m = c_m \frac{dV}{dt} + \frac{V}{r_m} = \frac{1}{\theta^2(r_i + r_e)} \cdot \frac{d^2V}{dt^2}$$

FIG. 1:55. A linear cable and the equation for a constant signal propagating at a constant speed θ.

One departure from the linear time scale analysis was used effectively by Adrian and Peachey (1965). The cable impedance is

$$Z_0 \sqrt{(r_e + r_i)r_m/(1 + pc_mr_m)}$$

as shown in figure 1:55. At high frequencies $p \gg 1/c_mr_m$ and for $r_e \ll r_i$,

$$Z_0(p) \rightarrow \sqrt{r_i/pc_m}$$

to give the 45° approach to the resistance axis. The transform is then (CF 522)

$$Z_0(t) = \sqrt{r_i\, t/\pi c_m}.$$

Adrian and Peachey obtained this approximation from the Hodgkin and Rushton complete solution, equation 1:48, and then confirmed it experimentally by plotting V_0 vs. $\sqrt{t}$ up to 20 msec for frog slow muscle fibers. This gave a mean membrane capacity of 2.5 μF/cm^2.

The potential or current response of a complex conducting system is not necessarily restricted to the application of a current or a potential of sinusoidal, step, or pulse form. Although these forms may be particularly easy to generate or apply, or the responses especially accurate or simple to observe or analyze, some other applied waveform may be far more appropriate. Exponentials in general are interesting and useful but Tasaki and Hagiwara (1957) made a novel application for the analysis of cable parameters from a propagating action potential.

For an impulse speed θ, $x = \theta t$ and equation 1:32 becomes

$$\frac{1}{\theta^2(r_e + r_i)} \frac{d^2V}{dt^2} - c_m \frac{dV}{dt} - \frac{V}{r_m} = 0. \qquad (1:49)$$

Everyone has long known that the beginning, the foot, of an action potential is usually an exponential

$$V = \bar{V} \exp(t/\tau)$$

where $\bar{V}$ may be about a half of the spike and $t \leq 0$. If also r_m is very high in this region, from Cole and Curtis (1938b)

$$1/\tau = \theta^2(r_e + r_i)c_m$$

and c_m can be found. In this way Tasaki and Hagiwara hesitatingly arrived at an average of 7 μF/cm^2—because $r_e + r_i$ was uncertain—while Falk and Fatt (1964) suggest that the appropriate values of 200 Ωcm and 100 μ lead to 2 μF/cm^2, as is to be expected at the corresponding frequency of 800 c.

The problems were much more difficult in the darkness and complexity of the spinal cord. But Eccles (Brock et al., 1952) succeeded first with a single microelectrode in a motor horn cell and then with the double barreled electrode. This latter at least assured that both electrodes were in the same cell at the same time. The two circuits were coupled at the tips by current flow and by capacity in the shanks but they again gave about 1 μF/cm^2 and 1 kΩcm^2 at rest, on the assumption of reasonable values for the effective membrane area. Most interesting analyses of the complicated soma-dendritic system were made by Rall (1959, 1960). He was able to reduce a branching dendritic tree to an equivalent cable which was fed from an equivalent spherical soma. In spite of the obvious complexity he was able to give an analysis procedure for the passive characteristics of both components. The further experiments and results have been far reaching into an understanding of excitation and inhibition—as presented by Eccles (1964), Frank and Fuortes (1961), and Curtis (1963).

INTRACELLULAR PARAMETERS

It was of first importance for Bernstein's hypothesis to establish that all cell interiors were reasonably like ordinary biological electrolytes. Beginning half a century ago, conductances have been measured with frequencies high enough to penetrate the interiors of many living cells and, as has been given in various summaries (Cole, 1941b; Cole and Curtis, 1950; Cole, 1959), they have certainly supported the hypothesis. Except for cells in fresh water, the specific resistances have averaged a few times those of the cell environments but these ratios have seemed to be somewhat erratic rather than showing any regularity or obvious pattern. On the other hand, these resistivities are almost all within a factor of five of 100 Ωcm—which leads me to believe that intracellular structures and processes survive and operate most successfully in media of corresponding concentrations.

The problems of such measurements and their interpretations are far from simple for the inside of a cell. But the solutions of these problems have more than a negligible importance because they are among the most informative and definitive that can be made on cells that are alive and functional.

It is far from certain that there is any such thing as a standard value for the resistivity of the axoplasm of the squid giant axon. But with the vast amount and high significance of the other data on this axon, this resistivity has a considerable importance. The earlier estimates that were discussed in quite another connection (Cole and Moore, 1960) suggested a compromise of about 30 Ωcm or about 50 percent more than sea water at laboratory temperatures. But then two experiments with R. E. Taylor (unpublished) on extruded axoplasm agreed on a resistivity equal to that of sea water. So far my efforts to encourage less casual measurements have been endorsed but not implemented.

A delightful and much more sophisticated approach to intracellular conductance has been used for a *Sepia* axon by Hodgkin and Keynes (1953). They introduced a spot of radioactive potassium and followed it as it spread by diffusion and traveled in an electric field as shown in figure 1:56. Analysis showed that the potassium mobility was essentially the same by both diffusion and electric transport, and close to the value in water. This made it quite improbable that the tagged potassium combined to any considerable extent with units of higher molecular weight—such as proteins—in the axoplasm. But, so far as I know, the possibility of such a very slow combination has not been excluded. At the other extreme, it has been shown (Hodgkin and Keynes, 1957) that what little calcium ion may enter an axon has so low a mobility in the axoplasm as to virtually require that it be present only in a combined form.

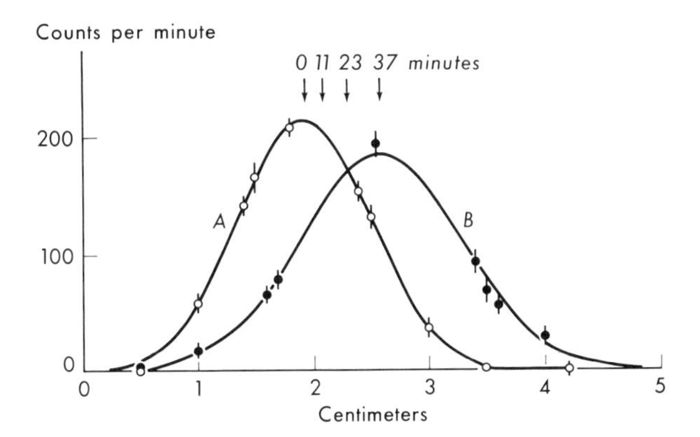

FIG. 1:56. Diffusion and electrical transport of radioactive potassium in a *Sepia* axon. The distribution of the tracer at the start is given by the open circles and curve *A*. After the electric field was applied, the maximum moved as indicated by the arrows. This transport gave a diffusion coefficient of $1.22 \cdot 10^{-5}$ cm²/sec. At the end the distribution, given by the solid circles and curve *B*, was lower and broader because of diffusion. These two measures gave coefficients of $1.3 \cdot 10^{-5}$ and $1.7 \cdot 10^{-5}$ cm^2/sec.

With the steadily increasing power of measurement and analysis, the dispersion region above that dominated by the plasma membranes has become increasingly useful.

I have thought that the nucleus modified the plasma membrane dispersions in marine eggs (Cole and Cole, 1936a,b). The analysis was uncertain and probably wrong (Cole, 1938). Measurements of a low resistance, 15 Ωcm² or much less, for some nuclear membranes (Loewenstein and Kanno, 1964) now suggest that this may be a general characteristic.

The theory of suspensions has been extended, in all of its complicated detail, by Fricke (1953a) and Pauly and Schwan (1959) to include the dielectric displacement of both the intercellular electrolyte and the protoplasm. These high-frequency dispersions have been measured to above 10^{10} c mostly by Schwan and colleagues. As had long been expected, the relaxation times of intracellular proteins (Oncley, 1942) participate, and it has been suggested that the fundamental properties of water are also involved (Schwan, 1957). The intracellular dielectric factor, at about 50, is considerably and very significantly below that of the aqueous medium for several red cells (Pauly, 1959) and tissues (Schwan, 1959). These values are reasonably explained mostly by protein content. The internal conductance was more of a problem in several tissues, and Pauly and Schwan (1966) considered it in detail for packed red cells. Seven species gave r_2 near 200 Ωcm while the inorganic ions predicted about 70 Ωcm for man. Several postulates did not explain the discrepancy.

The demands of biochemistry led to the isolation of large quantities of mitochondria. Suspensions of them have given an impedance dispersion region around 5 Mc, which leads to a membrane capacity of 1 $\mu F/cm^2$, as shown by Pauly et al. (1960). The higher frequency behavior is consistent with a similar capacity for the unit membranes of the lamellar internal structure.

SUMMARY

Bernstein's hypothesis of an electrolytic protoplasm has been found a good approximation, with specific resistances of the order of 100 Ωcm and some ion mobilities comparable to aqueous values. The contributions of intracellular proteins and mitochondria seem reasonable, while nuclei showed an unexpected high ion permeability.

General Summary

Following Fricke's discovery of a red cell membrane capacity of 1 $\mu F/cm^2$, a wide variety of measurements and appropriate analyses on many different living cells have established this value as a biological constant, with extreme values of about 0.5 and 2.0 $\mu F/cm^2$. This capacity seemed remarkably stable in spite of drastic physiological changes—except for an increase of two or three times for some marine eggs on fertilization. Demonstration of a membrane conductance representing an ion permeability came in due course with a rather surprising nominal value of 1000 Ωcm^2 for many cells at rest. Yet the range of variation was from 10 to 10^5 Ωcm^2 and this resistance showed a gratifying correlation with physiology that the capacity had not. It then became highly probable that the observed deviations from a perfect membrane capacity were not ion-conduction phenomena, but came in part from inadequate analysis of data for inhomogeneous cell populations, and in part from a dielectric loss in membrane structure.

The unexpected and unexplained complication of an inductive reactance made its appearance at low frequencies in the squid axon, as did another capacity in this and other cell membranes. The former seemed likely to be associated with ion transport through the membrane, and the latter from this or a conduction along the membrane surface, or the tubule system in muscle.

The cell interiors were found to be clearly electrolytic, with significant contributions by proteins.

References

Adrian, E. D., 1914. The all-or-none principle in nerve. J. Physiol., *47:* 460–474.

Adrian, R. H., and Peachey, L. D., 1965. The membrane capacity of frog twitch and slow muscle fibres. J. Physiol., *181:* 324–336.

Arrhenius, Sv., 1887. Über die Dissociation in wasser gelösten Stoffe. Zeit. physik. Chem., *1:* 631–648.

Barnett, A., 1938. The phase angle of normal human skin. J. Physiol., *93:* 349–366.

Bennett, M. V. L., 1966. Physiology of electrotonic junctions. Ann. N. Y. Acad. Sci., *137:* 509–539.

Bernstein, J., 1868. Über den zeitlichen Verlauf der negativen Schwankung des Nervenstroms. Arch. ges. Physiol., *1:* 173–207.
Bernstein, J., 1902. Untersuchungen zur Thermodynamik der bioelektrischen Ströme. I (Erster Theil). Arch. ges. Physiol., *92:* 521–562.
Bernstein, J., 1912. Elektrobiologie. Braunschweig, Wieweg und Sohn.
Blinks, L. R., 1928. High and low frequency measurements with *Laminaria.* Science, *68:* 235.
Blinks, L. R., 1930a. The direct current resistance of *Valonia.* J. Gen. Physiol., *13:* 361–378.
Blinks, L. R., 1930b. The direct current resistance of *Nitella.* J. Gen. Physiol., *13:* 495–508.
Blinks, L. R., 1930c. The variation of electrical resistance with applied potential. III. Impaled *Valonia ventricosa.* J. Gen. Physiol., *14:* 139–162.
Blinks, L. R., 1936. The effects of current flow in large plant cells. Cold Spring Harbor Symp. Quant. Biol., *4:* 34–42.
Bode, H. W., 1935. The driving point impedance function and its properties in terms of the energy functions. In Guillemin, E. A.: Communication networks. New York, Wiley, II, p. 226–229.
Bode, H. W., 1938. Impedance and energy relations in electrical networks. Physica, *5:* 143–144.
Bode, H. W., 1945. Network analysis and feedback amplifier design. N. Y., Van Nostrand, p. 151–157.
Bozler, E., and Cole, K. S., 1935. Electric impedance and phase angle of muscle in rigor. J. Cell. Comp. Physiol., *6:* 229–241.
Brazier, M. A. B., 1959. The historical development of neurophysiology. In Field, J. et al. (eds.): Handbook of physiology. Section 1: Neurophysiology. Washington, Am. Physiol. Soc., Vol. 1, p. 1–59.
Brock, L. G., Coombs, J. S., and Eccles, J. C., 1952. The recording of potentials from motoneurones with an intracellular electrode. J. Physiol., *117:* 431–460.
Bush, V., 1929. Operational circuit analysis. New York, Wiley.
Campbell, G. A., 1911. Cisoidal oscillations. Trans. Am. Inst. Elect. Eng., *30:* 873–909.
Campbell, G. A., and Foster, R. M., 1931. Fourier integrals for practical applications. Bell Tel. Sys. Monograph B-584. (Reprinted New York, Van Nostrand, 1948).
Carslaw, H. S. and Jaeger, J. C., 1947. Conduction of heat in solids. Oxford, Clarendon Press.
Carstensen, E. L., Cox, H. A., Jr., Mercer, W. B., and Natale, L. A., 1965. Passive electrical properties of microorganisms. I. Conductivity of *Escherichia coli* and *Micrococcus lysodeikticus.* Biophys. J., *5:* 289–300.
Carter, C. W., Jr., 1925. Impedance of networks containing resistances and two reactances. Bell Sys. Tech. J., *4:* 387–401.
Clark, J., and Plonsey, R., 1966. A mathematical evaluation of the core conductor model. Biophys. J., *6:* 95–112.
Cole, K. S., 1928a. Electric impedance of suspensions of spheres. J. Gen. Physiol., *12:* 29–36.
Cole, K. S., 1928b. Electric impedance of suspensions of *Arbacia* eggs. J. Gen. Physiol., *12:* 37–54.
Cole, K. S., 1932. Electric phase angle of cell membranes. J. Gen. Physiol., *15:* 641–649.
Cole, K. S., 1933. Electric conductance of biological systems. Cold Spring Harbor Symp. Quant. Biol., *1:* 107–116.

Cole, K. S., 1935. Electric impedance of *Hipponoë* eggs. J. Gen. Physiol., *18:* 877–887.
Cole, K. S., 1937. Electric impedance of marine egg membranes. Trans. Faraday Soc., *33:* 966–972.
Cole, K. S., 1938. Electric impedance of marine egg membranes. Nature, *141:* 79.
Cole, K. S., 1940. Permeability and impermeability of cell membranes for ions. Cold Spring Harbor Symp. Quant. Biol., *8:* 110–122.
Cole, K. S., 1941a. Rectification and inductance in the squid giant axon. J. Gen. Physiol., *25:* 29–51.
Cole, K. S., 1941b. Impedance of single cells. Tabulae Biologicae, *19* (Cellula, Pt. 2): 24–27.
Cole, K. S., 1947. Four lectures on biophysics. Rio de Janeiro, Institute of Biophysics, Univ. of Brazil.
Cole, K. S., 1949. Some physical aspects of bioelectric phenomena. Proc. Nat. Acad. Sci., U.S.A., *35:* 558–566.
Cole, K. S., 1957. Beyond membrane potentials. Ann. N. Y. Acad. Sci., *65:* 658–662.
Cole, K. S., 1959. The electric structure and function of cells. Proc. 1st Natl. Biophys. Conf., New Haven, Yale Univ. Press, p. 332–347.
Cole, K. S., 1965. Electrodiffusion models for the membrane of squid giant axon. Physiol. Rev., *45:* 340–379.
Cole, K. S., and Baker, R. F., 1941. Longitudinal impedance of the squid giant axon. J. Gen. Physiol., *24:* 771–788.
Cole, K. S., and Cole, R. H., 1936a. Electric impedance of *Asterias* eggs. J. Gen. Physiol., *19:* 609–623.
Cole, K. S., and Cole, R. H., 1936b. Electric impedance of *Arbacia* eggs. J. Gen. Physiol., *19:* 624–632.
Cole, K. S., and Cole, R. H., 1941. Dispersion and absorption in dielectrics. I. Alternating current characteristics. J. Chem. Physics, *9:* 341–351.
Cole, K. S., and Cole, R. H., 1942. Dispersion and absorption in dielectrics. II. Direct current characteristics. J. Chem. Physics, *10:* 98–105.
Cole, K. S., and Curtis, H. J., 1936. Electric impedance of nerve and muscle. Cold Spring Harbor Symp. Quant. Biol., *4:* 73–89.
Cole, K. S., and Curtis, H. J., 1937. Wheatstone bridge and electrolytic resistor for impedance measurements over a wide frequency range. Rev. Sci. Inst., *8:* 333–339.
Cole, K. S., and Curtis, H. J., 1938a. Electric impedance of single marine eggs. J. Gen. Physiol., *21:* 591–599.
Cole, K. S., and Curtis, H. J., 1938b. Electric impedance of *Nitella* during activity. J. Gen. Physiol., *22:* 37–64.
Cole, K. S., and Curtis, H. J., 1939. Electric impedance of the squid giant axon during activity. J. Gen. Physiol., *22:* 649–670.
Cole, K. S., and Curtis, H. J., 1950. Bioelectricity: Electric physiology. In Glasser, O. (ed.): Medical physics. Chicago, The Year Book Publishers, Inc., Vol. II, p. 82–90.
Cole, K. S., and Guttman, R. M., 1942. Electric impedance of the frog egg. J. Gen. Physiol., *25:* 765–775.
Cole, K. S., and Hodgkin, A. L., 1939. Membrane and protoplasm resistance in the squid giant axon. J. Gen. Physiol., *22:* 671–687.
Cole, K. S., and Marmont, G., 1942. The effect of ionic environment upon the longitudinal impedance of the squid giant axon. Fed. Proc., *1:* 15–16.
Cole, K. S., and Moore, J. W., 1960. Liquid junction and membrane potentials of the squid giant axon. J. Gen. Physiol., *43:* 971–980.

Cole, K. S., and Spencer, J. M., 1938. Electric impedance of fertilized *Arbacia* egg suspensions. J. Gen. Physiol., *21:* 583–590.

Cole, R. H., 1965. Relaxation processes in dielectrics. J. Cell. Comp. Physiol., *66:* (Suppl. 2), 13–20.

Cooley, J. W., and Tukey, J. W., 1965. An algorithm for the machine calculation of complex Fourier series. Math. Comp., *19:* 297–301.

Coraboeuf, E., and Weidmann, S., 1954. Temperature effects on the electrical activity of Purkinje fibres. Helv. Physiol. Acta, *12:* 32–41.

Crile, G. W., Hosmer, H. R., and Rowland, A. F., 1922. The electrical conductivity of animal tissues under normal and pathological conditions. Am. J. Physiol., *60:* 59–106.

Cremer, M., 1900. Ueber Wellen und Psuedowellen. Zeit. für Biol., *40:* 393–418.

Curtis, D. R., 1963. The pharmacology of central and peripheral inhibition. Pharmacol. Rev., *15:* 333–364.

Curtis, H. J., and Cole, K. S., 1937. Transverse electric impedance of *Nitella.* J. Gen. Physiol., *21:* 189–201.

Curtis, H. J., and Cole, K. S., 1938. Transverse electric impedance of the squid giant axon. J. Gen. Physiol., *21:* 757–765.

Curtis, H. J., and Cole, K. S., 1940. Membrane action potentials from the squid giant axon. J. Cell. Comp. Physiol., *15:* 147–157.

Danielli, J. F., 1935. The thickness of the wall of the red blood corpuscle. J. Gen. Physiol., *19:* 19–22.

Danielli, J. F., 1939. The resistance of nerve in relation to interpolar length. J. Physiol., *96:* 65–73.

Dänzer, H., 1934/35. Über das verhalten biologischer Korper bei Hochfrequenz. Ann. d. Physik, *21:* 783–790.

Debye, P., 1929. Polar molecules. New York, Chemical Catalog.

Du Bois-Reymond, Emil, 1848–1860. Untersuchungen über tierische Elekricität. Berlin, G., Reimer.

Eccles, J. C., 1964. The physiology of synapses. Berlin, Springer.

Einstein, A., 1905. Über die von der molekularkinetischen Theorie der Wärme geforderte Bewegung von in ruhenden Flüssigkeiten suspendierten Teilchen. Ann. d. Physik, *17:* 549–560.

Falk, G., and Fatt, P., 1964. Linear electrical properties of striatial muscle fibres observed with intracellular electrodes. Proc. Roy. Soc. (London) B, *160:* 69–123.

Fatt, P., 1964. An analysis of the transverse electrical impedance of striated muscle. Proc. Roy. Soc. (London) B, *159:* 606–651.

Fatt, P., and Ginsborg, B. L., 1958. The ionic requirements for the production of action potentials in crustacean muscle fibres. J. Physiol., *142:* 516–543.

Fatt, P., and Katz, B., 1951. An analysis of the end-plate potential recorded with an intra-cellular electrode. J. Physiol., *115:* 320–370.

Fatt, P., and Katz, B., 1953. The electrical properties of crustacean muscle fibres. J. Physiol., *120:* 171–204.

Fernández-Morán, H., 1950. Sheath and axon structures in the internode portion of vertebrate myelinated nerve fibers. An electron microscope study of rat and frog sciatic nerves. Exp. Cell Res., *1:* 309–337.

Field, R. F., 1945. The behavior of dielectrics over wide ranges of frequency, temperature and humidity. Unpublished manuscript.

Foster, R. M., 1932. Geometrical circuits of electrical networks. Trans. Am. Inst. Elec. Eng., *51:* 309–317.

Frank, K., and Fuortes, M. G. F., 1961. Excitation and conduction. Ann. Rev. Physiol., *23:* 357–386.
Fricke, H., 1923. The electric capacity of cell suspensions. Physic. Rev., *21:* 708–709.
Fricke, H., 1924a. A mathematical treatment of the electrical conductivity of colloids and cell suspensions. J. Gen. Physiol., *6:* 375–384.
Fricke, H., 1924b. A mathematical treatment of the electric conductivity and capacity of disperse systems. I. The electric conductivity of a suspension of homogeneous spheroids. Physic. Rev., *24:* 575–587.
Fricke, H., 1925a. The electric capacity of suspensions with special reference to blood. J. Gen. Physiol., *9:* 137–152.
Fricke, H., 1925b. The electric capacity of suspensions of red corpuscles of a dog. Physic. Rev., *26:* 137–152.
Fricke, H., 1925c. A mathematical treatment of the electric conductivity and capacity of disperse systems. II. The capacity of a suspension of conducting spheroids surrounded by a non-conducting membrane for a current of low frequency. Physic. Rev., *26:* 678–681.
Fricke, H., 1931. The electric conductivity and capacity of disperse systems. Physics, *1:* 106–115.
Fricke, H., 1932. The theory of electrolytic polarization. Phil. Mag. (7), *14:* 310–318.
Fricke, H., 1933. The electric impedance of suspensions of biological cells. Cold Spring Harbor Symp. Quant. Biol., *1:* 117–124.
Fricke, H., 1953a. The Maxwell-Wagner dispersion in a suspension of ellipsoids. J. Phys. Chem., *57:* 934–937.
Fricke, H., 1953b. The electric permittivity of a dilute suspension of membrane-covered ellipsoids. J. Appl. Physics, *24:* 644–646.
Fricke, H., 1955. The complex conductivity of a suspension of stratified particles of spherical or cylindrical form. J. Physic. Chem., *59:* 168–170.
Fricke, H., and Curtis, H. J., 1934a. Specific resistance of the interior of the red blood corpuscle. Nature, *133:* 651.
Fricke, H., and Curtis, H. J., 1934b. Electric impedance of suspensions of yeast cells. Nature, *134:* 102–103.
Fricke, H., and Curtis, H. J., 1935. Electric impedance of suspensions of leucocytes. Nature, *135:* 436.
Fricke, H., and Curtis, H. J., 1936. The determination of surface conductance from measurements on suspensions of spherical particles. J. Physic. Chem., *40:* 715–722.
Fricke, H., and Curtis, H. J., 1937. The dielectric properties of water dielectric interphases. J. Physic. Chem., *41:* 729–745.
Fricke, H., and Morse, S., 1925a. The electric resistance and capacity of blood for frequencies between 800 and 4½ million cycles. J. Gen. Physiol., *9:* 153–167.
Fricke, H., and Morse, S., 1925b. An experimental study of the electrical conductivity of disperse systems. I. Cream. Physic. Rev., *25:* 361–367.
Fricke, H., and Parker, E., 1940. The dielectric properties of the system gelatin-water. II. J. Physic. Chem., *44:* 716–726.
Fricke, H., Schwan, H. P., Li, K., and Bryson, V., 1956. A dielectric study of the low-conductance surface membrane in *E. coli.* Nature, *177:* 134–135.
Fritsch, G. T., 1887. Die elektrische Fische. Leipzig, Von Veit, Vol. 1, p. 4.
Fuoss, R. M., and Kirkwood, J. G., 1941. Electrical properties of solids. VII. Dipole moments in polyvinyl chloride—diphenyl systems. J. Am. Chem. Soc., *63:* 385–394.

Furshpan, E. J., and Potter, D. D., 1959. Transmission at the giant motor synapses of the crayfish. J. Physiol., *145:* 289–325.

Gaffey, C. T., and Mullins, L. J., 1958. Ion fluxes during the action potential in *Chara.* J. Physiol., *144:* 505–524.

Gildemeister, M., 1928a. Über elektrischen Widerstand Kapazität und Polarisation der Haut. Arch. ges. Physiol., *219:* 89–110.

Gildemeister, M., 1928b. Die passive-elektrischen Erscheinungen im Tier—und Pflanzenreich. In Bethe, A. (ed.) Handbuch d. normalen und pathologischen Physiologie Band VIII, Heft #2, p. 657–702.

Goldman, D. E., 1943. Potential, impedance, and rectification in membranes. J. Gen. Physiol., *27:* 37–60.

Gregg, E. C., and Steidley, K. D., 1965. Electrical counting and sizing of mammalian cells in suspension. Biophys. J., *5:* 393–405.

Guillemin, E. A., 1935. Communication networks. New York, Wiley, Vol. II, p. 178.

Guttman, R., 1939. The electrical impedance of muscle during the action of narcotics and other agents. J. Gen. Physiol., *22:* 567–591.

Havriliak, S., and Negami, S., 1966. A complex plane analysis of α-dispersions in some polymer systems. J. Polymer Sci. C, No. 14, p. 19.

Helmholtz, H., 1853. Ueber einige Gesetze der Vertheilung elektrischer Ströme in körperliche Leitern, mit Anwendung auf die thierischen elektrischen Versuche. Ann. Physik u. Chem., Ser. 2, *89:* 211–233, 353–377.

Hermann, L., 1877. Untersuchungen über die Entwicklung des Muskelstroms. Arch. ges. Physiol., *15:* 191–242.

Hermann, L., 1905. Beiträge zur Physiologie und Physik des Nerven. Arch. ges. Physiol., *109:* 95–144.

Hill, A. V., 1932. Chemical wave transmission in nerve. Cambridge, England, Univ. Press.

Hiramoto, Y., 1959. Electric properties of *Echinoderm* eggs. Embryologia, *4:* 219–235.

Höber, R., 1910. Eine Methode, die elektrische Leitfähigkeit im Innern von Zellen zu messen. Arch. ges. Physiol., *133:* 237–259.

Höber, R., 1912. Ein zweites Verfähren die Leitfähigkeit im Innern von Zellen zu messen. Arch. ges. Physiol., *148:* 189–221.

Höber, R., 1913. Messungen der innern Leitfähigkeit von Zellen. III. Arch. ges. Physiol., *150:* 15–45.

Hodgkin, A. L., 1947. The membrane resistance of a non-medullated nerve fibre. J. Physiol. *106:* 305–318.

Hodgkin, A. L., 1951. The ionic basis of electrical activity in nerve and muscle. Biol. Rev., *26:* 339–409.

Hodgkin, A. L., and Huxley, A. F., 1939. Action potentials recorded from inside a nerve fibre. Nature, *144:* 710–711.

Hodgkin, A. L., and Keynes, R. D., 1953. The mobility and diffusion coefficient of potassium in giant axons from *Sepia.* J. Physiol., *119:* 513–528.

Hodgkin, A. L., and Keynes, R. D., 1957. Movements of labelled calcium in squid giant axons. J. Physiol., *138:* 253–281.

Hodgkin, A. L., and Rushton, W. A. H., 1946. The electrical constants of a crustacean nerve fibre. Proc. Roy. Soc. (London) B, *133:* 444–479.

Hogg, B. M., Goss, C. M., and Cole, K. S., 1934. Potentials in embryo rat heart muscle cultures. Proc. Soc. Exp. Biol. and Med., *32:* 304–307.

Iida, T. T., 1943a. Changes of electric capacitance following fertilization in sea-urchin eggs. J. Fac. Sc. Imp. Univ. Tokyo (Sect. IV), *6:* 141–151.

Iida, T. T., 1943b. Effects of diluted sea water on membrane capacitance of sea-urchin eggs. J. Fac. Sc. Imp. Univ. Tokyo (Sect. IV), *6:* 165–173.
Jeans, J. H., 1927. The mathematical theory of electricity and magnetism. 5th edition, Cambridge, England, Univ. Press.
Johnson, S. L., and Woodbury, J. W., 1964. Membrane resistance of human red cells. J. Gen. Physiol., *47:* 827–837.
Katz, B., 1948. The electrical properties of the muscle fibre membrane. Proc. Roy. Soc. (London) B, *135:* 506–534.
Kaufman, W., and Johnston, F. D., 1943. The electrical conductivity of the tissues near the heart and its bearing on the distribution of the cardiac action currents. Am. Heart J., *26:* 42–54.
Kelvin, Lord (William Thompson), 1855. On the theory of the electric telegraph. Proc. Roy. Soc. London, *7:* 382–399.
Kelvin, Lord (William Thompson), 1904. Baltimore Lectures on: Molecular dynamics and the wave theory of light. London, C. J. Clay and Sons, Baltimore, Johns Hopkins Univ.
Kempe, A. B., 1877. How to draw a straight line; a lecture on linkages. London, Macmillan.
Kennelly, A. E., 1921. Tables of complex hyperbolic and circular functions. 2nd edition. Cambridge, Harvard Univ. Press.
Kohlrausch, F., 1876. Ueber das Leitungsvermögen der in Wasser gelösten Elektrolyte im Zusammenhang mit der Wanderung ihrer Bestandtheile. Nach. K. Ges. der Wiss. (Göttingen) *1876:* 213–244.
Kronig, R. de L., 1926. On the theory of dispersion of x-rays. J. Opt. Soc. Am., *12*, 547–557.
Lillie, R. S., 1936. The passive iron wire model of protoplasmic and nervous transmission and its physiological analogies. Biol. Rev., *11:* 181–209.
Ling, G., and Gerard, R. W., 1949. The normal membrane potential of frog sartorius fibers. J. Cell. Comp. Physiol., *34:* 383–396.
Loewenstein, W. R., 1964. Permeability of the nuclear membrane as determined with electrical methods. Protoplasmatologia, Handbuch der Protoplasmaforschung. Vienna, Springer, Bd. 5, T 2, p. 26–34.
Loewenstein, W. R., and Kanno, Y., 1964. Studies on an epithelial (gland) cell junction. I. Modifications of surface membrane permeability. J. Cell Biol., *22:* 565–586.
Lorente de Nó, R., 1947. A study of nerve physiology. Studies from the Rockefeller Institute. Vols. 131, 132. New York, Rockefeller Institute.
Lucas, Keith, 1908–1909. The "all or none" contraction of the amphibian skeletal muscle fibre. J. Physiol., *38:* 113–133.
Lullies, H., 1930. Über die Polarisation in Geweben. II. Mitteilung. Die Polarisation im Nerven. I. Arch. ges. Physiol., *225:* 69–86.
McClendon, J. F., 1926. Colloidal properties of the surface of the living cell. II. Electric conductivity and capacity of blood to alternating currents of long duration and varying in frequency from 266 to 2,000,000 cycles per second. J. Biol. Chem., *69:* 733–754.
McClendon, J. F., 1929. Polarization capacity and resistance of salt solutions, agar, erythrocytes, resting and stimulated muscle, and liver measured with a new Wheatstone bridge designed for electric currents of high and low frequency. Protopl., *7:* 561–582.
Mason, M., and Weaver, W., 1929. Electromagnetic field. Chicago, Univ. of Chicago Press, p. 241.

Mauro, A., 1961. Anomalous impedance, a phenomenological property of time varient resistance. An analytic review. Biophys. J., *1:* 353–372.
Maxwell, J. C., 1873. Treatise on electricity and magnetism. Oxford, Clarendon Press.
Murphy, E. J., 1929. Electrical conduction in textiles. II. Alternating current conduction in cotton and silk. J. Physic. Chem., *33:* 200–215.
Nernst, W., 1888. Zur Kinetik der in Lösung befindlichen Körper: Theorie der Diffusion. Zeit. physik. Chem., *2:* 613–637.
Nernst, W., 1889. Die elektromotorische Wirksamkeit der Ionen. Zeit. physik. Chem., *4:* 129–181.
Nightingale, A., 1959. "Background noise" in electromyography. Phys. in Med. and Biol. (London) *3:* 325–338.
Nyboer, J., 1950. Plethysmograph: impedance. In Glasser, O. (ed.) Medical physics. Chicago, Yearbook Publishers, Vol. 2, p. 736–743.
Nyboer, J., 1960. Plethysmograph: impedance. In Glasser, O. (ed.) Medical physics. Chicago, Yearbook Publishers, Vol. 3, p. 459–471.
Nyquist, H., 1928. Thermal agitation of electric charge in conductors. Physic. Rev., *32:* 110–113.
Oker-Blom, M., 1900. Thierische Säfte und Gewebe in physikalisch-chemischer Beziehung. Arch. ges. Physiol., *79:* 510–533.
Oncley, J. L., 1943. In Cohn, E. J. and Edsall, J. T. (eds.). Proteins, amino acids and peptides as ions and dipolar ions. New York, Reinhold, Chap. 22, p. 543–568.
Onsager, L., 1945. Theories and problems of liquid diffusion. Ann. N. Y. Acad. Sci., *46:* 241–265.
Osterhout, W. J. V., 1922. Injury, recovery and death in relation to conductivity and permeability. Philadelphia, Lippincott.
Ostwald, W., 1900. Periodische Erscheinungen bei der Auflösung des Chroms in Säuren, Parts I and II. Zeit. physik. Chem., *35:* 33–76, 204–256.
Overton, E., 1902. Beiträge zur allgermeinen Muskel-und Nervenphysiologie. Arch. ges. Physiol., *92:* 346–386.
Page, L., and Adams, N. I., 1931. Principles of electricity and magnetism. New York, Van Nostrand.
Pauly, H., 1959. Electrical conductance and dielectric constant of the interior of erythrocytes. Nature, *183:* 333–334.
Pauly, H., Packer, L., and Schwan, H. P., 1960. Electrical properties of mitochondrial membranes. J. Biophys. and Biochem. Cytol., *7:* 589–601.
Pauly, H. and Schwan, H. P., 1959. Über die Impedanz einer Suspension von kugelförmigen Teilchen mit einer Schale. Zeits. f. Naturfors., *14B:* 125–131.
Pauly, H., and Schwan, N. P., 1966. Dielectric properties and ion mobility in erythrocytes. Biophys. J., *6:* 621–639.
Philippson, M., 1921. Les lois de la resistance electrique des tissus vivants. Bull. Acad. roy. Belgique. Cl. Sci., *7:* 387–403.
van der Pol, B., 1937. New theorem on electrical networks. Physica, *4:* 585–589.
Rajewsky, B., 1938. Ultrakurzwellen, Ergebnisse der biophysikalischen Forschung, Bd. 1. Leipzig, Georg Thieme.
Rall, W., 1959. Branching dendritic trees and motoneuron membrane resistivity. Exp. Neurol., *1:* 491–527.
Rall, W., 1960. Membrane potential transients and membrane time constant of motoneurons. Exp. Neurol., *2:* 503–532.
Rayleigh, Lord (J. W. Strutt), 1892. On the influence of obstacles arranged in rectangular order upon the properties of a medium. Phil. Mag., *34:* 481–502.

Robertson, J. D., 1960. The molecular structure and contact relationships of cell membranes. Prog. Biophys., *10:* 343–418.
Rogers, C. G., and Cole, K. S., 1925. Heat production by the eggs of *Arbacia punctulata* during fertilization and early cleavage. Biol. Bull., *49:* 338–353.
Rosenberg, H., and Schnauder, F., 1923. Der scheinbare Widerstand verschieden langer Strecken und das Kernhüllenverhältnis des Froschnerven. Z. Biol., *78:* 175–194.
Rothschild, Lord, 1946. The theory of alternating current measurements in biology and its application to the investigation of the biophysical properties of the trout egg. J. Expt. Biol., *23:* 77–99.
Rothschild, Lord, 1956. Fertilization. New York, Wiley.
Rothschild, Lord, 1957. The membrane capacitance of the sea urchin egg. J. Biophys. Biochem. Cytol., *3:* 103–110.
Rushton, W. A. H., 1927. The effect upon the threshold for nervous excitation of the length of nerve exposed, and the angle between current and nerve. J. Physiol., *63:* 357–377.
Rushton, W. A. H., 1934. A physical analysis of the relation between threshold and interpolar length in the electric excitation of medullated nerve. J. Physiol., *82:* 332–352.
Salmon, G., 1876. A treatise on higher plane curves. (1934 photographic reprint of third edition New York, Stechert.)
Sapegno, E., 1930. Über die Impedanz und Kapazität des quergestreiften Muskels in Längs-und Querrichtung. Arch. ges. Physiol., *224:* 187–211.
Schmitt, F. O., Bear, R. S., and Palmer, K. J., 1941. X-ray diffraction studies on the structure of the nerve myelin sheath. J. Cell. Comp. Physiol., *18:* 31–42.
Schmitt, O. H., 1955. Dynamic negative admittance components in statically stable membranes. In Shedlovsky, T. (ed.). Electrochemistry in biology and medicine, New York, Wiley, p. 91–120.
Schwan, H., 1948. Die Temperaturabhängigkeit der Dielektrizitätskonstante von Blut bei Neiderfrequenz. Zeits. f. Naturfors., *3B:* 361–367.
Schwan, H., 1954. Die elekrischen Eigenschaften von Muskelgewebe bei Niederfrequenz. Zeits. f. Naturfors., *9B:* 245–251.
Schwan, H. P., 1957. Electrical properties of tissue and cell suspensions. In Lawrence, J. H. and Tobias, C. A. (eds.). Advances in biological and medical physics. New York, Academic Press, Vol. 5, p. 147–209.
Schwan, H. P., 1959. Alternating current spectroscopy of biological substances. Proc. Inst. Radio Eng., *47:* 1841–1855.
Schwan, H. P., 1963. Determination of biological impedances. In Nastuk, W. L., (ed.), Physical techniques in biological research, New York, Academic Press, Vol. 6, p. 323–407.
Schwan, H. P., and Carstensen, E. L., 1957. Dielectric properties of the membrane of lysed enthrocytes. Science, *125:* 985–986.
Schwan, H. P., and Cole, K. S., 1960. Bioelectricity: Alternating current admittance of cells and tissues. In Glasser, O. (ed.), Medical physics, Chicago, Yearbook Publishers, Vol. 3, p. 52–56.
Schwan, H. P. and Kay, C. F., 1956. Specific resistance of body tissues. Circulation Res., *4:* 664–670.
Schwan, H. P., and Kay, C. F., 1957. Capacitive properties of body tissues. Circulation Res., *5:* 439–443.
Schwan, H. P., and Morowitz, H. J., 1962. Electrical properties of the membranes of pleuropneumonia-like organism A 5969. Biophys. J., *2:* 395–407.
Schwan, H. P., Schwarz, G., Maczuk, J., and Pauly, H., 1962. On the low-frequency

dielectric dispersion of colloidal particles in electrolyte solution. J. Physic. Chem., *66:* 2626–2635.
Schwarz, G., 1962. A theory of the low-frequency dielectric dispersion of colloidal particles in electrolyte solution. J. Physic. Chem., *66:* 2636–2642.
Shaw, T. M., Jansen, E. F., and Lineweaver, H., 1944. The dielectric properties of β-lactoglobulin in aqueous glycine solutions and in the liquid crystalline state. J. Chem. Physics, *12:* 439–448.
Silver, G. A., Strauss, J., and Misrahy, G. A., 1965. Electrical impedance of isolated amnion. Biophys. J., *5:* 855–865.
Spector, W. S. (ed.), 1956. Handbook of Biological Data. Philadelphia, Saunders.
Stämpfli, R., 1952. Bau und Funktion isolierter markhaltiger Nervefasern. Ergebn. Physiol., *47:* 70–165.
Stewart, G. N., 1899. The relative volume or weight of corpuscles and plasma in blood. J. Physiol., *24:* 356–373.
Tamasige, M., 1950. Membrane and sarcoplasm resistance in an isolated frog muscle fibre. Annot. Zool. Japon., *23:* 125–134.
Tamasige, M., 1951. Effect of potassium ions upon the electrical resistance of an isolated frog muscle fibre. Annot. Zool. Japon., *24:* 141–149.
Tasaki, I., 1955. New measurements of the capacity and the resistance of the myelin sheath and the nodal membrane of the isolated frog nerve fiber. Am. J. Physiol., *181:* 639–650.
Tasaki, I., and Hagiwara, S., 1957. Capacity of muscle fiber membrane. Am. J. Physiol., *188:* 423–429.
Taylor, R. E., 1963. Cable theory. In Nastuk, W. L. (ed.), Physical techniques in biological research, New York, Academic Press, Vol. 6, Chap. 4, p. 219–262.
Taylor, R. E., and Chandler, W. K., 1962. Effect of temperature on squid axon membrane capacity. Biophys. Soc. Abst., TD 1.
Teorell, T., 1946. Application of "square wave analysis" to bioelectric studies. Acta Physiol. Scand., *12:* 235–254.
Thompson, D., 1917. On growth and form. Cambridge, University Press.
Tuttle, D. F., Jr., 1958. Network synthesis. New York, John Wiley, Vol. I, p. 804–819.
Tyler, A., Monroy, A., Kao, C. Y., and Grundfest, H., 1956. Membrane potential and resistance of the starfish egg before and after fertilization. Biol. Bull., *11:* 153–177.
Velick, S., and Gorin, M., 1940. The electrical conductance of suspensions of ellipsoids and its relation to the study of avian erythrocytes. J. Gen. Physiol., *23:* 753–771.
Wagner, K. W., 1913. Zur Theorie der unvollkommenen Dielektrika. Ann. d. Physik, *40:* 817–855.
Warburg, E., 1899. Ueber das Verhalten sogenannter unpolarisirbarer Elektroden gegen Wechselstrom. Ann. d. physik. Chem., *67:* 493–499.
Weber, E., 1954. Linear transient analysis, Vol. I. Lumped parameter two-terminal networks. New York, Wiley.
Weber, H., 1873. Ueber die stationären Strömungen der Electricität in Cylindern. J. Reine angew. Math., *76:* 1–20.
Weidmann, S., 1952. The electrical constants of Purkinje fibres. J. Physiol., *118:* 348–360.
Weinberg, A. M., 1941. Weber's theory of the kernleiter. Bull. Math. Biophys., *3:* 39–55.
Weinberg, A. M., and Householder, A. S., 1941. Statistical distribution of impedance elements in biological systems. Bull. Math. Biophys., *3:* 129–135.

Wiener, O., 1912. Die Theorie des Mischkörpers für das Feld der stationären Strömung. Abh. K. Sächs. Ges. Wiss., Math. Phys. Kl., *32:* 509–604.
Wolff, I., 1926. A study of polarization capacity over a wide frequency range. Physic. Rev., *27:* 755–763.
Yager, W. A., 1936. The distribution of relaxation times in typical dielectrics. Physics, *7:* 434–450.
Young, J. Z., 1936. Structure of nerve fibres and synapses in some invertebrates. Cold Spring Harbor Symp. Quant. Biol., *4:* 1–6.
Young, J. Z., 1951. Doubt and certainty in science. Oxford, Clarendon Press.
Zobel, O. J., 1923. Theory and design of uniform and composite electric wavefilters. Bell Sys. Tech. J., *2:* 1–46.

PART II

Nonlinear and Active Membrane Behaviors

Contents to Part II

Introduction

A friendly biologist told me long ago that a physicist, who should know something about cables, ought to be interested in nerve if only because the two were built in somewhat the same way and both carried electrical messages. Then some years later it became possible, as my friend and so many others had hoped and expected, to describe the squid axon at rest as a passive cable. Perhaps other axons and muscle cells might be somewhat similar.

Impulse Excitation and Propagation

CABLE VS. AXON

The squid axon as shown in figure 1:33 had a central conductor with a specific resistance of about 30 Ωcm or $r_i = 15$ kΩ/cm for the nominal 500 μ diameter. The insulation was not perfect—perhaps because of a dielectric loss as well as a leakage—but the nominal values of 1 μF/cm² and 1 kΩcm² then gave $c_m = 0.16$ μF/cm, and $r_m = 6.4$ kΩcm. The performance of this cable could now be worked out in detail, as had been done by Kelvin for the first Atlantic cable, as solutions of his basic equation 1:32,

$$c_m \frac{\partial V}{\partial t} = \frac{1}{r_e + r_i} \cdot \frac{\partial^2 V}{\partial x^2} - \frac{1}{r_m}. \tag{1:32}$$

As an illuminating example of cable behavior we may look at the spread and decay of potential after an initial pulse of current rather than the initial step already considered. This equation may be normalized to read

$$\frac{\partial V}{\partial T} = \frac{\partial^2 V}{\partial X^2} - V$$

where $T = t/\tau$ for $\tau = c_m r_m$ and $X = x/\lambda$ for $\lambda = \sqrt{r_m/(r_e + r_i)}$. This is a classical problem in the partial differential equations of mathematical physics. It is, however, better known as a statement of heat flow along a radiating wire than of a very inefficient cable. Carslaw and Jaeger (1957, Sect. 4.2) reduce the problem by substituting $V = U \exp(-T)$ to give

$$\frac{\partial U}{\partial T} = \frac{\partial^2 U}{\partial X^2}.$$

For a short pulse $U \cdot \Delta t = 1$ with Δt small, at $x = 0$, a solution is

$$U = \frac{1}{2\sqrt{\pi T}} \exp(-X^2/4T)$$

and so
$$V = \frac{1}{2\sqrt{\pi T}} \exp\,[(-X^2/4T) - T] \qquad (2:1)$$

as is also available by the Fourier and Laplace transform (Campbell and Foster, 1931, Pair 823).

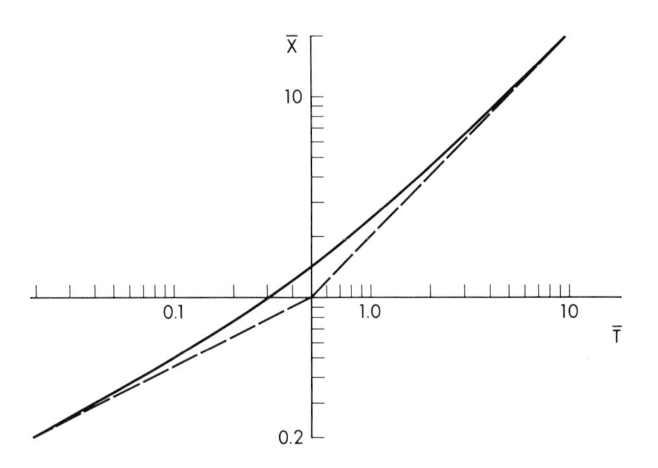

FIG. 2:1. Propagation in a passive axon after a short pulse of current. The maximum of potential arrives at a distance $\bar{X}$ given in units of λ at the time $\bar{T}$ in units of τ.

Taking $\bar{X}$ as the distance at which $\partial V/\partial T = 0$, the corresponding time, $\bar{T}$ is given by $\bar{X}^2 = 2\bar{T} + 4\bar{T}^2$. For $\bar{T} \ll 1$, $\bar{X}^2 \rightarrow 2\bar{T}$ and for $\bar{T} \gg 1$, $\bar{X} \rightarrow 2\bar{T}$, as shown in figure 2:1. Thus this solution starts out with the time to maximum proportional to the square of the distance, as for diffusion, heat conduction without radiation, or a cable without leakage, but for long times it closely approximates a constant speed $2\lambda/\tau$. For our standard squid axon in a large volume of sea water, $r_e \sim 0$, we have $\tau = 10^{-3}$ sec, $\lambda = 0.65$ cm, and the speed approaches the quite respectable value of 13 m/sec, comparable to the nerve impulse. However, the attenuation of the maximum membrane potential is given by exp $(-\bar{T}/\bar{X})$ and this is very severe —tenfold for less than a 3 cm length of axon. So it would require more than 100 volts to produce a millivolt at 6 inches away. Both the propagation and the attenuation are according to equation 2:1 illustrated in figure 2:2a in which the calculated responses for such a passive cable are at distances of one, two and three times the passive characteristic length λ from the initial short pulse. I have often remarked that the squid—and other—axons do behave in this thoroughly useless way so long as the initial potential change is only a few millivolts. But, in what Bullock (1959) called a revolution, it was found that such a primitive form of communication has been widely used in the nervous system. At short distances this "electrotonic spread" is probably the cheapest and most effective means available.

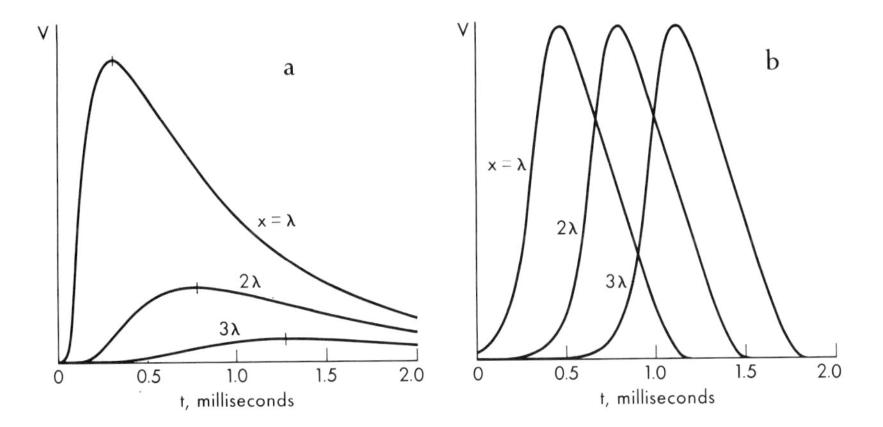

FIG. 2:2. (a.) Propagation in a passive axon after a short pulse of current. The potential is shown as a function of time at three distances. (b.) Propagation in an active axon after an adequate stimulus.

FIG. 2:3. Potential of a squid axon near a stimulating electrode and after a short shock. Below threshold a passive spread of potential is found, left. Above threshold this shock artifact is a little larger and is followed by the impulse, center. The passive response increases for a larger shock but the impulse remains essentially unchanged, right. Time in msec.

The needs for long distance communication are filled by the much more sophisticated and more intriguing performance of excitable axons. As shown in figure 2:3 by Brink (1951), when the initial V

exceeds 15–20 mV another potential appears as seen in figure 2:2b which is then independent in shape and velocity of the applied potential. It has a constant speed in a uniform axon, and for our nominal squid axon this is about 20 m/sec at 20°C. This is shown in figure 2:2b for comparison with the passive behavior of figure 2:2a.

Yet another striking and significant difference between cable and axon is found in the collision of two pulses coming from opposite directions (Tasaki, 1949, 1959). The cable is linear, superposition is the rule, the pulses add where they overlap, and eventually each continues as if there were no other. In the axon, figure 2:4, midway between the two oncoming impulses, they merge, on collision, into a single impulse. But this has almost the same shape and amplitude as each of the two components and sinks in place without a trace of either one continuing in either direction. So only one-way traffic is normally admitted to a single axon.

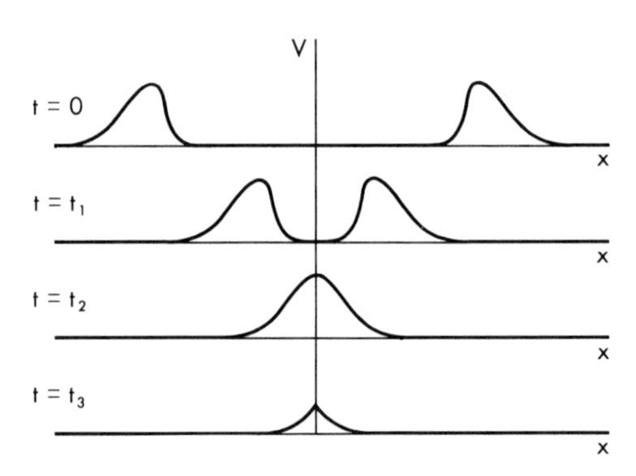

FIG. 2:4. Schematic description of impulse collision. Two impulses are shown headed at each other at the top. At the later time t_1 they are closer. At t_2 the rising phases of activity have merged and disappeared leaving only the two recovery processes. By the time t_3 the annihilation is almost complete.

The fundamental nature of the difference between the cable and the axon is both simple and obvious. The energy supplied to the passive cable is dissipated as the pulse is attenuated in both amplitude and speed. And since at least an internal resistance is unavoidable in a normal axon, the energy lost in it must be replaced by local sources along the axon if the amplitude and speed of the pulse are to remain constant. The problem is then the source of this local energy and the mechanism of its local release.

Decades ago I was brash enough to say at a Bell Telephone Laboratories' seminar that a nerve was merely a cable with a repeater at every molecule. Although I suspect that this is not literally true, at least submarine cables with repeaters every few miles are a good approximation and have now been in service for some years.

The performance of an axon in initiating and propagating an impulse was well summarized by the all-or-none law. By it the axon responds at each place and at every time to the maximum of its capacity at that place and time. The problem of this spectacular physiological behavior of an axon was thus somewhat better defined to become a more immediate goal.

SOME THEORIES OF EXCITATION AND PROPAGATION

Excitation and propagation had both been extensively and actively studied almost continuously since Galvani. The all-or-none nature of excitation, and propagation as a progressing excitation, had been early recognized for muscle by Lucas (1909), for nerve by Adrian (1914), and demonstrated directly by Adrian and Bronk (1928). Many models for electric excitation had been proposed. I reviewed some of these models as a part of the promotion of the first Cold Spring Harbor Symposium (Cole, 1933). As far as I have found, the first and, for two score years, the only attempt at a physical formulation of a mechanism of nerve excitation was begun by Nernst (1899, 1908). He showed that ion concentrations close to a selectively ion-permeable membrane would change according to the square root of time t after the application of constant current I. Then a critical change of an ion concentration to give excitation thus occurred by diffusion for $I\sqrt{t}$ = constant, and indeed there were more data to support this thesis than to oppose it. But it became quite certain that below some value of I excitation did not occur no matter how long t might be. Because of this, Nernst weakened his position by the ad hoc assumption of an "accommodation" such as by a slow increase of the threshold. This attempt to give a physical basis for excitation was continued by Hill (1910), who calculated ion concentration changes at boundaries a short distance apart. This modified the Nernst formula to give somewhat better agreement with the more and better experimental data for the longer stimuli that had become available.

The simplest and most generally useful model of excitation was in the form of first order kinetics as represented by a parallel resistance–capacity, RC network. This was extensively investigated by Lapicque (1907). In it the criterion of excitation was a threshold potential V_T. For a current I of very short duration t, $It = CV_T = Q_0$ and this was known as the constant quantity relation. At very long times, $I_0 = V_T/R$ was called the rheobase. The current duration of a threshold twice rheobase was defined as the chronaxie. I proposed a more general model (Cole, 1934) based on impedance measurements which were interpreted by a membrane characteristic of the form $z(\omega) = z_1(j\omega)z^{-\alpha}$. The Fourier transform required that $It^\alpha = \Gamma(1 + \alpha)V_T/z_1$ = constant at short times for a threshold change of potential V_T and a parallel membrane resistance R which then produced a rheobase $I_0 = V_T/R$. The chronaxie for such a model was more difficult to determine; the inversion and expansion of the Volterra integral equation was carried out by Davis (1936). Some impedance and excitation data agreed, at least to the extent that they gave values of α lying between 0.5, corresponding to the Nernst model, and 1.0 for the simple capacity, resistance model.

But many and more complicated experiments showed that no one of these simple models was adequate. In particular it was shown that a current increasing linearly with time at less than a "liminal gradient" would not stimulate. So something more had to be done. I had worked out a superposition correlation (Cole, 1933) of many of the excitation phenomena from the data of Bishop (1928) barely before Rashevsky (1933a) produced his two-factor formulation. At nearly the same time Monnier (1934) gave the same formulation in terms of an ingenious circuit, and soon Hill (1936) published his classic, detailed work on the same basis.

Two Factor Excitation

Hill's two variables were named the "local potential" V and the "threshold" U, and their behavior is given by

$$\begin{aligned} dV/dt &= bI - (V - V_0)/k \\ dU/dt &= (V - V_0)/\beta - (U - U_0)/\lambda. \end{aligned} \tag{2:2}$$

The resting values are V_0, U_0, the stimulus current is I; k and λ are time constants for excitation and accommodation, β/λ is the degree of accommodation, and $V = U$ is the condition for an impulse. In the usual case, λ may be several times longer than k. After a constant stimulus, V may increase rapidly to overtake U before it has risen appreciably and thus produce excitation. But for a slowly increasing stimulus, V may never catch U as they both increase indefinitely and so fail to excite. The ideal of a potential pursuing a fleeing threshold was indeed prophetic of the "phantom saddle point" to come (FitzHugh, 1955). Among the numerous other excitation phenomena, Hill calculated the effect of the accommodation in the time-honored strength-duration relation. Later, Le Fevre (1950) extended this to the limit as the accommodation time approached that for excitation. Even this interesting manipulation did not produce the sharp break between the constant quantity and the rheobase that was found experimentally (Cole, 1955). The use and success of this two factor approach did much to consolidate almost countless experiments into comparatively few and usually interesting numerical values (Katz, 1939).

A few years later in the confusing array of new experimental results, L. A. MacColl called my attention to Von Kármán's (1940) delightful Gibbs Lecture, "The Engineer Grapples with Nonlinear Problems," as a starting point for dealing with the obvious problems of membrane nonlinearity and instability. The field about which we had the most information and the best organization of information was that of subthreshold phenomena, and I first tried the new and intriguing topological and graphical methods on it.

Hill had given many analytical solutions to two-factor excitation problems, and Rushton (1937a) had produced a convenient and rapid graphical integration to give the time course of the approach to threshold. The utility of the two factor theory and its formulation as two first-order equations made it too attractive to avoid.

The variables were replaced by changes of excitation K and refractoriness or accommodation L; the threshold condition became $K - L \geq T > 0$ and, with $\beta = \lambda$, equations 2:2 are

$$\frac{dK}{dt} = -\frac{K}{k} + bI, \qquad \frac{dL}{dt} = \frac{K - L}{\lambda}$$

and the slope of the path on the K, L plane, figure 2:5,

$$\frac{dL}{dK} = \frac{K - L}{-(\lambda/k)K + b\lambda I}.$$

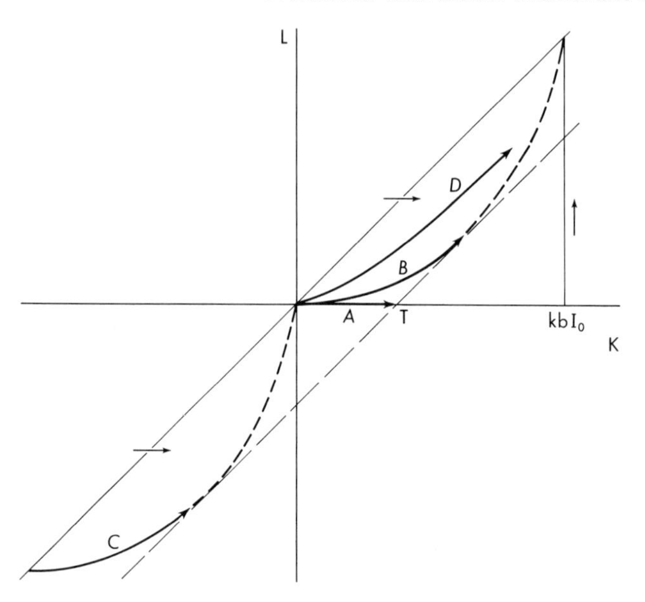

FIG. 2:5. Phase plane representation of two factor theory of excitation. K and L are the excitability and accommodation variables. The dashed line $K - L = T$ is the threshold. A short shock reaches threshold along path A. A constant current kbI_0 is just rheobase as path B touches the threshold line. After an anode break, threshold is reached along path C. A liminal gradient of current gives path D which becomes parallel to the threshold line.

This plane may now be covered with path directions at every point that may be connected up to give a complete path, or trajectory, from any initial condition. This was done by van der Pol (1926) by plotting the loci of a given slope which are here straight lines because the problem is trivial. Of these we are most interested in two, which I call orthoclines, for which

$$dK/dt = 0 \text{ and } dL/dt = 0.$$

These are given by $K = kbI$ and $K = L$, respectively and these with the threshold, $T = K - L$ are shown in figure 2:5. The trajectories A and B are for threshold constant stimuli of a very short duration and rheobase, C is for an anode break excitation, and D is for a linearly increasing stimulus in which K and L increase indefinitely without achieving threshold. However, this formulation was limited to the special, and perhaps extreme, case (Parrack, 1940) of 100 percent accommodation, and it did not produce the oscillatory phenomena for which there had been increasing evidence.

The Rashevsky, Monnier, and Hill theories, and a somewhat different approach by Blair (1932), were shown by Young (1937) to be included in the generalization

$$dK/dt = k_{11}K + k_{12}L + k_1 I$$
$$dL/dt = k_{21}K + k_{22}L + k_2 I \qquad (2{:}3)$$

where K, L, and I were again excitability, accommodation or inhibition, and stimulus, respectively, and were abstractions not defined physically. This formulation of excitation was discussed by Parrack (1940), and in more detail by Cole (1941). Hill had assumed the threshold to be given by a critical difference between K and L, and the substitution $T = K - L$ leads to

$$\frac{d^2T}{dt^2} - (k_{11} + k_{22})\frac{dT}{dt} + (k_{11}k_{22} - k_{21}k_{12})T$$
$$= (k_1 - k_2)\frac{dI}{dt} - [(k_{21} + k_{22})k_1 - (k_{12} + k_{11})k_2]I$$

which is elementary and well known. In the solution

$$\theta = A \exp(\lambda_1 t) + B \exp(\lambda_2 t)$$

the roots λ_1 and λ_2 might be real or complex conjugates depending on the coefficients. The latter possibility seemed improbable and had been arbitrarily excluded from consideration—regretfully by Monnier and Coppée (1939) and emphatically by Katz (1939) who later rescinded most graciously at Paris in 1949. But the experimental facts of excitation, the finding of an inductive reactance to allow an analogous oscillatory equivalent membrane circuit, and the appearance of oscillatory potentials (Arvanitaki, 1939; Cole and Curtis, 1941; Brink et al., 1946) all came to argue against the supposedly common sense restriction. And indeed the equivalent circuit (figure 1:45), with but the single ad hoc assumption of a threshold membrane potential, seems to have been the first model of excitation to be based upon measured, physical, membrane properties.

Other accommodations required $k_{21} \neq k_{22}$ and oscillations required a back coupling $k_{12} < 0$, so we return to the original general equations 2:3. On rewriting

$$K' = \sqrt{k_{21}}\, K,\ L' = \sqrt{k_{12}}\, L,\ t' = \sqrt{k_{12}k_{21}}\, t$$

$$\frac{dK'}{dt} = (k_{11}/\sqrt{k_{12}k_{21}})K' - L' + k_1'I$$

$$\frac{dL'}{dt} = K' - (k_{22}/\sqrt{k_{12}k_{21}})L' + k_2'I$$

and for threshold,

$$\sqrt{k_{12}}\,K' - \sqrt{k_{21}}\,L' \geq \sqrt{k_{12}k_{21}}\,T = T'$$

where k_{12} and k_{21} are now absolute values to make the radicals real. Then, dropping primes and dividing to obtain the slope of the path on the K, L plane,

$$\frac{dL}{dK} = \frac{K - (k_{22}/\sqrt{k_{12}k_{21}})L + k_2\,I}{-(k_{11}/\sqrt{k_{12}k_{21}})K - L + k_1\,I}.$$

The orthoclines, $dK/dt = 0$ and $dL/dt = 0$, are given by

$$K - (k_{22}/\sqrt{k_{12}k_{21}})L + k_2\,I = 0$$

and

$$-(k_{11}/\sqrt{k_{12}k_{21}})K - L + k_1\,I = 0.$$

These conditions are shown in figure 2:6 for one set of parameters:

$$k_{11} = 1\text{ msec}^{-1},\ k_{22} = k_{21} = 0.1\text{ msec}^{-1}$$

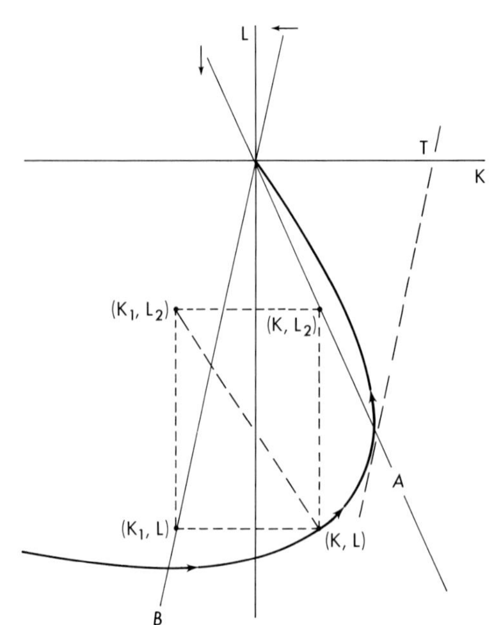

FIG. 2:6. Construction for tracing paths of excitation and accommodation according to the general formulation of the two factor theory. The perpendicular slopes are indicated on the orthoclines A and B and the slope at any other point, K, L, is found by the dashed line rectangle and diagonal.

with $k_1 = k_2 = 0$. In order to have an oscillatory solution it was necessary (Cole, 1941) for

$$(k_{11} - k_{22})^2 < 4k_{12}k_{21}$$

and the lowest value of k_{12} was chosen,

$$k_{12} = (k_{11} - k_{22})^2/4k_{21} = 2.0 \text{ msec}^{-1}.$$

Then, as shown

$$\frac{dL}{dK} = \frac{K - 0.22L}{-2.2K - L} \text{ and } L = 4.5K - T.$$

The coordinate transformations were so chosen that graphical solutions were readily run forward or backward from arbitrary initial or final conditions. For $\dot{L} = 0$ let $K_1 - 0.22L_1 = 0$. Then at a point K, L in figure 2:6,

$$\Delta L = [K - 0.22L - (K_1 - 0.22L_1)]\ \Delta t$$

and for $L_1 = L$,

$$\Delta L = (K - K_1)\ \Delta t.$$

Similarly, for $\dot{K} = 0$ let $2.2K_2 + L_2 = 0$ and

$$\Delta K = [-2.2K - L + (2.2K_2 + L_2)]\ \Delta t$$

and for $K_2 = K$,

$$\Delta K = (L_2 - L)\ \Delta t.$$

Then by similar triangles $\Delta L/\Delta K$ is perpendicular to the diagonal of the rectangle with sides $K - K_1$ and $L_2 - L$ in figure 2:6. Any trajectory is then followed piece by piece with compass or plastic ruler having a line perpendicular to an end. Many and varied exercises were worked out with this system. The representation was very satisfying and it was particularly important to show the conditions for oscillatory and non-oscillatory behavior.

There may be no such thing as a final word in any aspect of science. With the tremendous experience of the many intervening years, FitzHugh (1966) finally found it necessary to complete the Young formal theory, equations 2:3. He took it over the early barriers and into the regions of complex time constants and incomplete accommodation to get a better understanding of some recent investigations.

However, the two-factor variables were concepts without definition, and the parameters were defined only in terms of excitation itself. The complete absence of any physical significance for the variables or the parameters, and the inability to predict performance in any other terms than threshold were highly unsatisfactory.

Propagation

The abstract consolidation of the facts of the conduction of a nerve impulse had also not been impressive. Hermann had suggested (1899) that in the process of propagation part of the current into an excited region came through a passive region ahead and became an adequate stimulus for it. In this he, and also Cremer (1900), had independently developed the cable theory of Kelvin quite without any definitive evidence that the nerve had any characteristics in common with a submarine cable. Nonetheless, this local circuit theory was at least qualitatively satisfactory, and when backed by the physical concepts and theory of Kelvin, came to be generally accepted as a fundamental concept in the propagation of a nerve impulse. Since the physics of this process was relatively simple and straightforward, the only problems were: First, why should current flow along an axon between an excited and resting region? And, second, why should this current stimulate the resting region?

There were many attempts to provide mechanisms which would produce propagating impulses. These were, for me at least, culminated by Rashevsky's demonstration (1933b) that the entirely formal theory of excitation proposed by Blair (1932) could give reasonable propagation in a Kelvin cable-like nerve fiber. Later Rashevsky (1935) extended this approach to use his own and other more powerful expressions of the facts of excitation. Thus, impulse conduction was entirely understood, as a physical process, except for the cause and effect of excitation which were complete physical mysteries!

It had long been quite probable that bioelectric potentials had their origin in membranes, and it was becoming more likely that all membranes had a considerable electrical capacity. So these were two membrane properties that might change in excitation to produce propagation.

On the basis of the cable equation, Rushton (1937b) evolved a powerful theory for the initiation and propagation of an impulse. This assumed a sudden change of the membrane emf at a critical membrane potential as illustrated in figure 2:7a. Such an assumption was partly consistent with Bernstein's hypothesis that an increase of permeability essentially abolished the membrane potential. One of the most interesting conclusions was that a minimum length of axon had to be so excited before a sustained propagation could appear. Also the speed of uniform propagation was given entirely in terms of physical parameters which were all defined but not then all known. Another physical model was given by O. H. Schmitt (1937), who proposed that a membrane capacity suddenly increased at a threshold potential—a parametric capacity as it would now be called. This characteristic is shown in figure 2:7b. Schmitt demonstrated a model to illustrate the conclusions of his theoretical work on propagation.

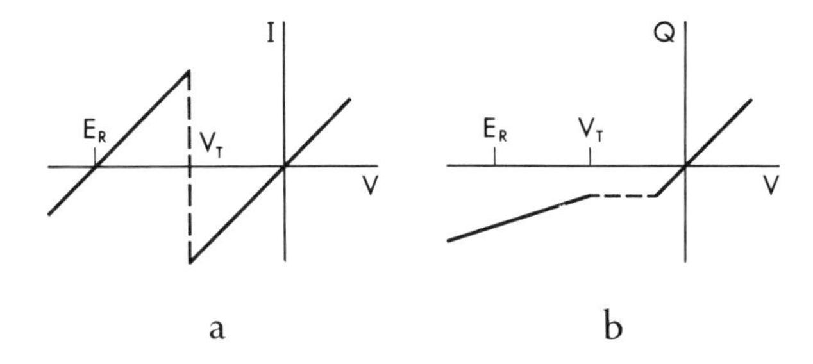

FIG. 2:7. Two theories of excitation. The resting membrane is represented by a negative emf E_R and a series resistance. In *a* the emf is reduced to zero at a threshold potential V_T. For *b* the membrane capacity is increased at threshold.

An ion permeability had not then been directly measured except for plant cells. But in retrospect it seems rather strange as well as unfortunate that a change of membrane conductance such as shown in figure 2:47 had not been made the basis of similar calculations. This would not only have explored another possibility but also would have conformed more closely to the Bernstein hypothesis. Yet this seems not to have been considered theoretically as a process of propagation, until by Offner et al. (1940) after it had been demonstrated as an experimental fact.

There was, however, another, and very interesting, general approach which dealt with relations between the known, physical parameters and function—quite independent of the unknown complications, such as the membrane conductance. Rosenblueth et al. (1948) and Hodgkin (1954) presented such an example that is worth reviewing in somewhat more adequate form. There was, and still is, good reason to trust the cable equation 1:31 so, in an extended medium,

$$\frac{\partial^2 V}{\partial x^2} = r_i i_m.$$

It is an experimental fact that there is a form for the membrane current, $i_m(V, t)$, by which a real axon propagates an impulse, $V(x, t)$, with a constant speed θ. We must then write $V = F(x - \theta t)$, and have equation 1:49,

$$r_i i_m = \frac{1}{\theta^2} \cdot \frac{\partial^2 F}{\partial t^2}.$$

The coefficients and variables are here for a unit length of axon and, replacing them by i_3 and r_2 for unit dimensions,

$$\frac{\theta^2}{a} r_2 = \frac{\partial^2 F}{\partial t^2} / i_3. \qquad (2:4)$$

If now F is the same function of i_3 for every axon and independent of its radius a we must have

$$\theta^2 r_2 / a = \text{constant}.$$

Under these assumptions the speed of the propagating impulse in an axon must vary directly as the square root of the diameter and inversely as the square root of the specific resistance of the media inside the membrane. Both of these relations are at least approximately true. However, without any indication that F and i were the same, it was thought by some that experiments (Pumphrey and Young, 1938) in which it was found that $\theta \sim a^{0.6}$ demonstrated the superiority of the axons over the simple-minded theorists.

Experimental Evidence

There had been some hints of changes in excitable tissues other than the production of a current, and later a potential, as a result of injury or activity, such as had led to Bernstein's hypothesis of an increase in ion permeability. There was the Ostwald-Lillie iron wire model in which the oxide layer membrane seemed to disappear entirely during the impulse, and it was too good an analogy to ignore. Then Blinks (1936a) had found a decrease of "polarizability" in activity. We had applied considerable 50 c potential to the bridge during transverse nerve and muscle impedance measurements (Cole and Curtis, 1936). In part, this was done to be sure that our routine measurements were independent of applied potential and, in part, just to see what would happen. In both preparations the parallel capacity and resistance both decreased near and somewhat above the threshold, as shown in figure 2:8 for a frog sciatic. A cat sciatic gave a low-frequency resistance which was reduced to half after removal of the sheath, but Lorente de Nó (1947) dismissed this with the comment that no proof of function was presented for the desheathed nerve.

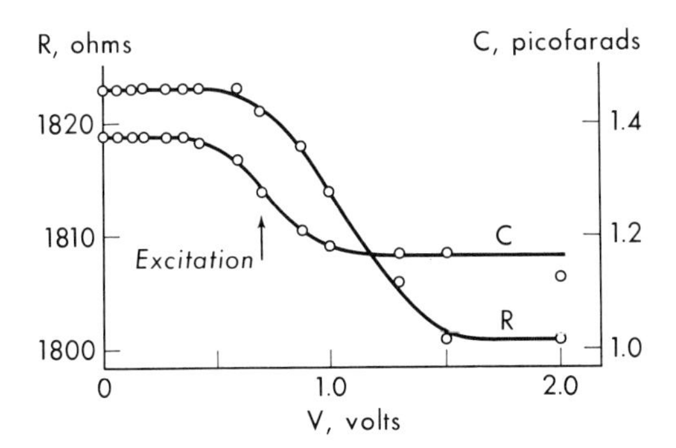

Fig. 2:8. Early measurements of the longitudinal impedance of a frog sciatic nerve at 50 c. The parallel resistance, R, and capacity, C, were found to decrease somewhat before and after the attached muscle first contracted.

MEMBRANE CONDUCTANCE IN EXCITATION

NITELLA

After the passive characteristics of *Nitella* and then the squid axon had been measured, we returned to measure the impedance of *Nitella* after it was stimulated and as activity was propagated between the narrow electrodes on each side of it, such as in figure 2:11 (Cole and Curtis, 1938b). At first only the maximum changes of impedance at one frequency were obtained during each impulse. Although the results were variable, these impedances clearly lay between two arcs coinciding with the rest locus at R_∞ and with the same phase angle. The decrease and return of the impedance were so slow that a movie of the bridge unbalance Lissajous figure along with the action potential gave a complete record at one frequency for each impulse.

The interpretation of the loci was not immediately obvious. Since a change of membrane capacity would only move the impedance along the rest locus, the effect of an increase of membrane conductance was tried. The impedance was given by equation 1:18, the cylinder equivalent of equation 1:13,

$$Z = r_1 \frac{(1 - \rho)r_1 + (1 + \rho)(r_2 + z_m/a)}{(1 + \rho)r_2 + (1 - \rho)(r_2 + z_m/a)} \qquad (2:5)$$

and for $1/z_m = 1/z_3 + 1/r_4$, where $z_3 = \bar{z}(j\omega)^{-\alpha}$, this reduced to

$$Z = \frac{a + br_4}{c + dr_4}$$

where a, b, c, and d are constants at one frequency. This is again a circular arc as for equation 1:14. By elimination of r_4 between the real and imaginary parts of Z, this arc was found to be tangent to the resistance axis at R_∞ for $r_4 = 0$ and ending at the rest frequency locus if $r_4 = \infty$ at rest. Furthermore, if the resistance decrease were the same at each frequency, the points of maximum change should form a smaller arc which has the same phase angle as the resting impedance and also passes through R_∞.

To a first approximation, the membrane resistance decreased from an effectively infinite value at rest to an average of 500 Ωcm^2; this change was independent of frequency between 100 c and 5 kc, figure 2:9a. However, the arcs during an impulse consistently fell on the low-frequency side of the predicted arc, to be interpreted as

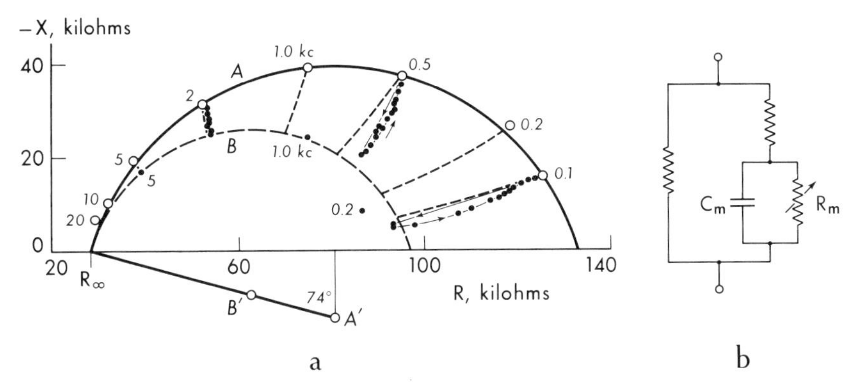

FIG. 2:9. (a.) Complex plane loci for transverse impedance of a *Nitella* cell. The open circles and the solid arc, *A*, were for the resting cell. The paths followed during the passage of an impulse are shown at three frequencies and maximum excursions for three more are presented as solid circles. A decrease of the membrane resistance alone would give the dashed circular arc *B* for the maxima and the connecting arcs to *A* at each frequency. (b.) Equivalent circuit for transverse impedance of *Nitella* in which the membrane resistance R_m decreases during an impulse while the membrane capacity C_m and the other resistances remain unchanged.

an average 15 percent maximum decrease of the membrane capacity below its rest value. These interpretations of the data were not accepted without some qualms. There were, after all, five known parameters in equation 2:5 r_1, r_2, the phase and amplitude of z_3, and r_4. All of these were obviously approximations to the complicated structure of the cell. The frequency-dependent impedance was presumably located at one or both of the protoplasm boundaries and there was no assurance that the leakage was in any way related to either of them. If but one boundary were primarily responsible, then either the internal or the external resistivities included the protoplasm; in either case the external resistivity was already known to include the complication of the cellulose sheath. But if both protoplasmic boundaries were making significant contributions, the equivalent circuit needed more parameters than this experiment could determine. The data suggested the conclusion that the infinite frequency resistance was not affected by excitation, which argued against major changes in either r_1 or r_2. Then, on the assumption of the single, frequency-dependent element z_3, its phase angle was not changed during excitation. The postulate of change in r_4 made the minimum value independent of frequency, which was reassuring, but for no certain reason. The decrease of resistance by perhaps a factor of over 100 made the 15 percent decrease of $|z_3|$ seem so relatively unimportant that it was ignored. Then we had the equivalent membrane circuit of figure 2:9b and the cable circuit of figure 2:10. Although a membrane resistance decrease in excitation was not the only possible explanation or necessarily the correct interpretation of the data, it was certainly the simplest and most attractive conclusion. So the principles of parsimony and prejudice were both strong supporting factors.

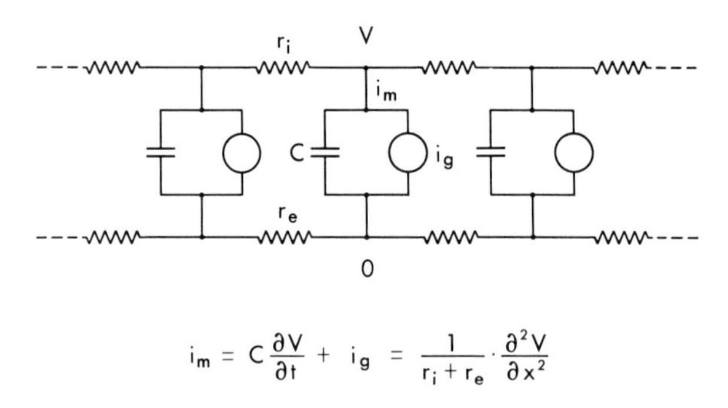

FIG. 2:10. Schematic and equation of general cable in which the nature of the membrane conductance current i_g is not specified.

In relation to the action potential, the membrane conductance change was found to be negligible until about the point of inflection of the rising phase of the potential. After corrections for finite effective electrode width, the conductance then rose very rapidly to reach its maximum somewhat before the maximum of the potential. It returned to its resting value more rapidly than the potential.

SQUID AXON

The investigation of excitation with the squid axon in the cell of figure 2:11 began the next spring. It started with only a barely detectable impedance change that was definitely a decrease. It soon progressed to the point where delightful oscilloscope records, figures 2:12, 2:13, and 2:16, were the reward for many and painful resets of the contacts on a Lucas spring rheotome in the dark (Cole and Curtis, 1938a, 1939). And the oscilloscope pattern was a greeting to Hodgkin on his first visit to Woods Hole—to which he responded with memorable enthusiasm. Although everything new had been done on *Nitella*, the squid conclusions were sharper.

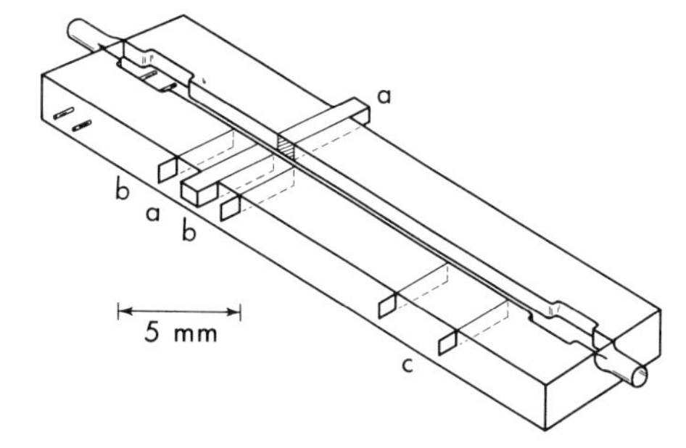

FIG. 2:11. Sketch of transverse measuring cell for a squid axon. The impedance was measured between the center *a* electrodes with the axon in the square trough. Stimulating electrodes are at the left and the various recorded potentials were between the *a*, *b*, and *c* electrodes.

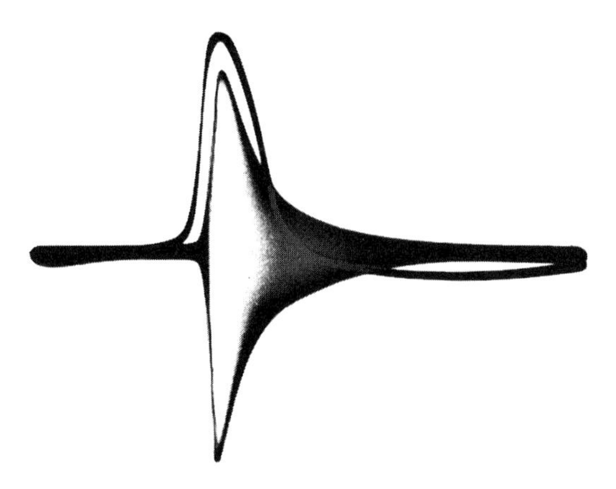

FIG. 2:12. Oscilloscope records of the action potential, line, and the membrane resistance decrease, band, from squid giant axon during the passage of an impulse. The time marks are at 1 msec intervals.

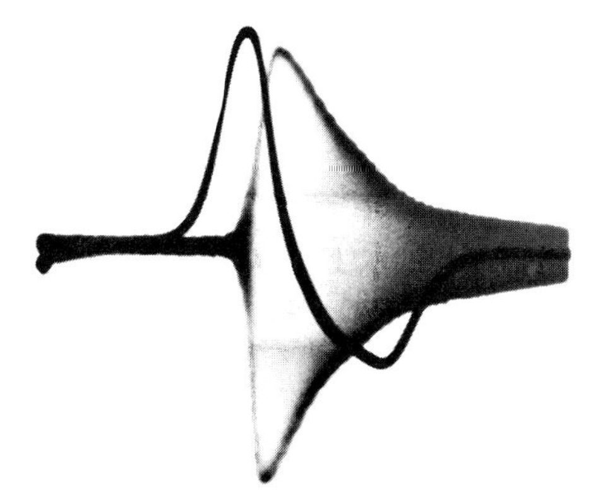

FIG. 2:13. Oscilloscope records of the first derivative of the action potential, line, and the membrane resistance decrease, band, from a squid giant axon during the passage of an impulse.

The consistent result that the peak values of the impedance change during the impulse lay on a circular arc as in figure 2:14 again permitted the interpretation of a membrane conductance increase which was constant over the frequency range. The apparent capacity change was only an average of 2 percent; the departure from the theoretical locus seen in figure 2:14 is the worst example that we found, while that of figure 2:15 is one of the best.

My claim that the time scales were not changed between the potential and the impedance records was challenged so sharply that we put one on each axis of the oscilloscope. It was difficult to disbelieve such Rorschach patterns as figure 2:16, but they contained

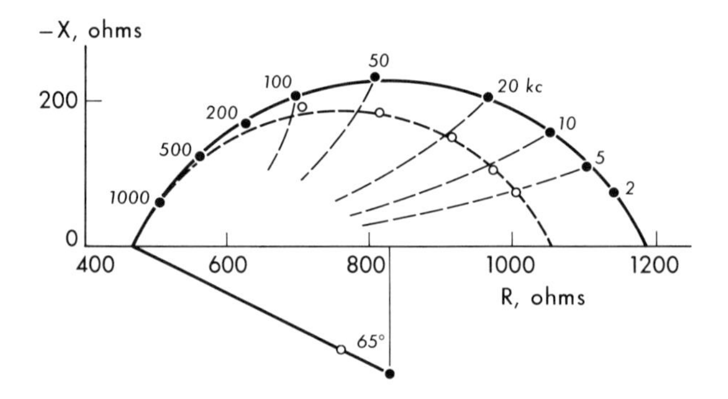

FIG. 2:14. Complex plane analysis of maximum impedance change of squid axon during impulse. The solid circles and circular arc are characteristics at rest and the open circles at peak values. The dashed circular arcs are the loci, at peak over the frequency range and for the peak at each frequency, calculated for a change of no parameter except the membrane resistance.

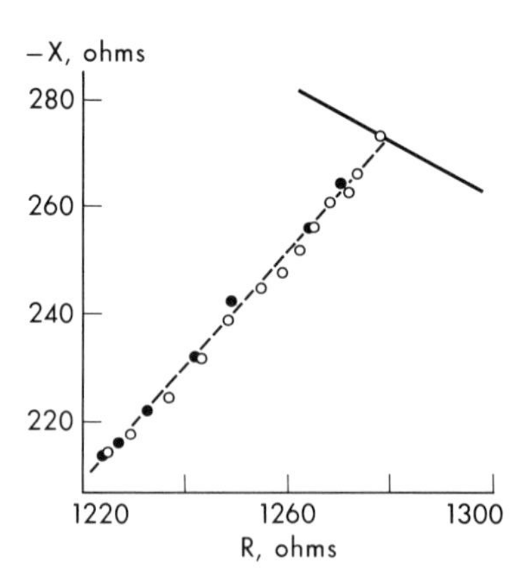

FIG. 2:15. Complex plane analysis of squid axon impedance during the impulse. The solid circles were measured at 10 kc during the rising phase and the open circles during the return to rest. The solid line is a portion of the resting locus and the dashed arc was calculated for a decrease of membrane resistance alone.

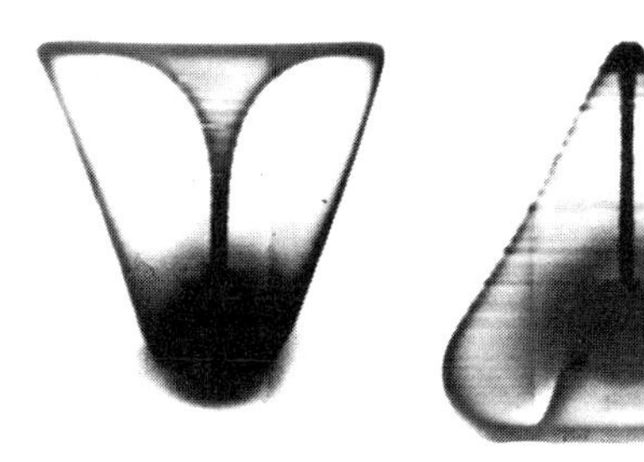

FIG. 2:16. Oscilloscope records showing membrane resistance decrease, horizontal, vs. action potential, left, and its derivative, right.

the distortions of amplifier speeds and electrode width. The problem of such corrections is important and widespread in analysis of experimental data. It is one of the solution—or an approximation—for the superposition integral equation of Duhamel,

$$F(x) = \int_{-\infty}^{x} G(g)H'(x - g)\, dg.$$

$F(x)$ is the observed result of the unknown cause, $G(x)$, and $H(x)$ is the result of a known cause—a step function in this case. The problem plagued A. V. Hill for decades in the analysis of his heat production records, and I do not know that there is even now a usefully simple and accurate method of solution without a computer. I tried an obvious piecewise approximation, only to get into hopeless complications. Later, Hill (1949) started anew at each step in the same

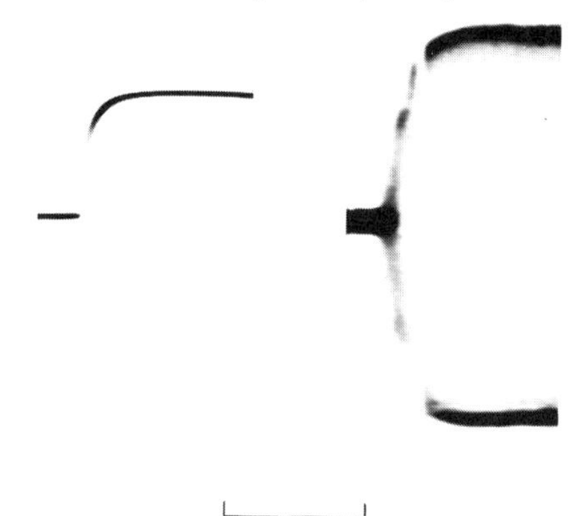

FIG. 2:17. Oscilloscope records of apparatus response to a step change of potential, left, and to a step change of resistance, right.

analysis—which made it practical. It could then be shown that the hidden residuals accumulated badly unless only a few, large pieces were chosen to approximate $F(x)$ and $H'(x)$. This Silberstein (1932) had emphasized in an analogous problem. The amplifier distortions of figure 2:17 were considerable, and overall corrections were made only with exponential approximations. These led to typical changes,

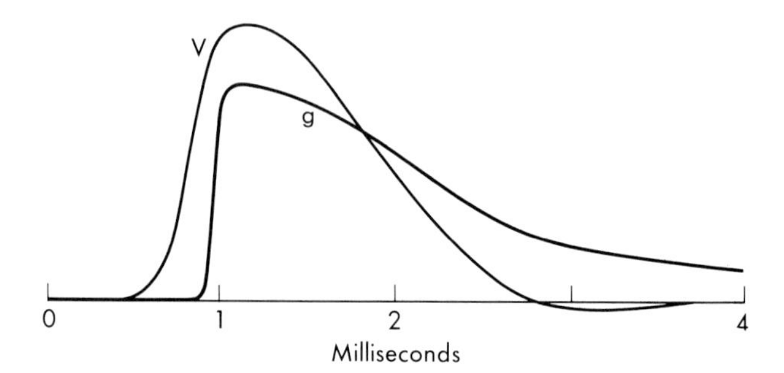

FIG. 2:18. Characteristic action potential V and membrane conductance increase g after corrections for amplifier delays and passage of impulse through the measuring region.

as in figure 2:18, in which the conductance increase is so rapid as to still be something of a puzzle. Probably our experiments should be repeated with faster equipment and with better corrections, rather than to rely on the best that could be done in 1938.

The average conductance increase by 28 mmho/cm^2 was substantial, and hopefully characteristic of animal membranes, whereas the longer duration than for *Nitella* was interesting. The constancy of the conductance change over the frequency range from 5 to 100 kc was quite reassuring. So this was thought of as a change from R_0 at rest to R_0' on excitation in figure 2:19, before the inductive reactance was discovered at lower frequencies.

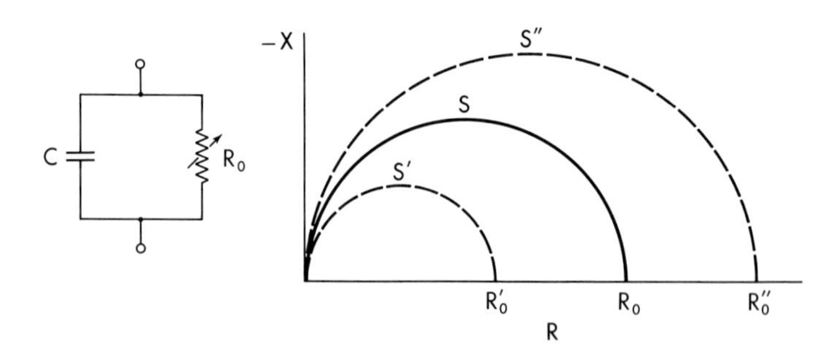

FIG. 2:19. Approximate equivalent membrane circuit and complex impedance locus for changes of the frequency independent resistance R_0.

There was now one additional conclusion to be drawn. It seemed reasonably certain that another and very important Bernstein hypothesis was essentially correct. The ion conduction did increase in excitation. But it was something of a surprise to me, and particularly to Ralph Lillie, that so little of the membrane, a few percent as represented by the capacity, was involved in the excitation process. Although the increase in ion permeability—as measured by conductance—was about fortyfold, the maximum figure corresponded, for a 100 Å membrane, to an average specific resistance of $2.5 \cdot 10\ \Omega\text{cm}^2$, or about 10^6 times that of sea water. So here again excitation could hardly be thought of as any extensive breakdown of the membrane.

These results particularly discouraged the explanation that the membrane impedance of the form $\bar{z}(j\omega)^{-\alpha}$ was the expression of an ion-conduction process similar to those found at metal electrodes. It now became much more reasonable to attribute this behavior to a rather solid dielectric with the equivalent resistance as an expression of internal dissipation without conduction. This explanation, too, must remain suspect until it has better foundation in theory.

Since these first indications that nerve activity did indeed involve an increase of membrane ion permeability, uncounted changes of cell and tissue impedances have been interpreted in this way. In only a few cases is there also a measure of the membrane capacity. The *Laminaria* data were reasonably well explained by conductance changes without appreciable corresponding variation of capacity. With chloroform, the frog sartorius behaved the same way, even as the membrane approached complete permeability and death, as found by Guttman (1939), figure 2:20.

Susumu Hōzawa, whom I had met doing beautiful Helmholtz pendulum experiments (1928) in Gildemeister's laboratory, arranged to follow the impedance changes of frog and human skin after nerve stimulation. He showed how to make separate recordings of the impedance components in a powerful application of the Schering bridge (1935). Forbes and Landis (1935) found an impedance change in the psychogalvanic reflex which could again be interpreted as an increase of membrane conductance.

It is certainly possible to question the interpretation of less exhaustive experiments but it seems expedient to expect that changes of membrane conductance are the usual cause of impedance changes, until proved otherwise. Among such earlier results are those of

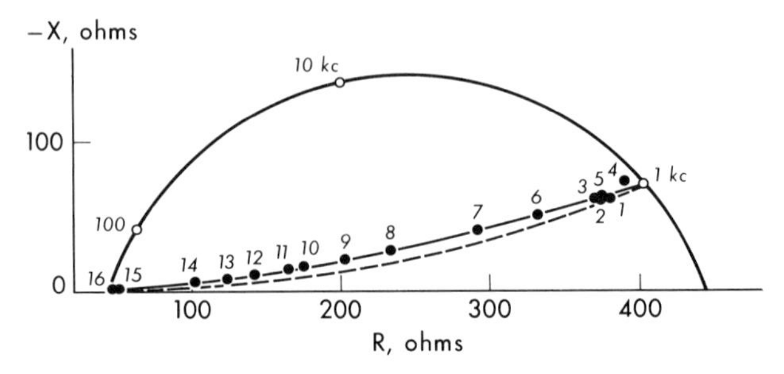

FIG. 2:20. Complex plane analysis of the action of chloroform on frog sartorius muscle. The open circles for three frequencies give the normal circular locus. The solid circles were obtained in the order indicated after chloroform was applied. The dashed line was computed for a decrease of membrane resistance without change of any of the other parameters.

Auger and Cotton (1937) on *Nitella*, Katz (1942) for frog skeletal muscle during activity, Fessard (1946) and Cox et al. (1946) on the electric organ, Tasaki and Mitzuguchi (1949) on the frog node, and Weidmann (1951) on Purkinje fibers.

Yet Rothschild confounded my prediction of a similar analysis for the rhythmic impedance changes in fish eggs (1940). On the same basis as for the other measurements, his data led (1947a,b) to the conclusion that the membrane capacity was oscillating without a detectible change of conductance!

CONDUCTION CURRENTS

There were now many new problems centering around the questions: What was the nature of the membrane conductance change? What did it have to do with excitation and with its propagation in an axon?

With the experimental results on *Nitella*—and a hint from Hodgkin—we had computed (Cole and Curtis, 1938b) some of the consequences of a partial membrane short circuit appearing at the point of inflection of a propagating action potential. This approach was investigated in detail by Offner, Weinberg, and Young (1940). They gave calculations for the velocity of propagation and discussion of the effect of such things as a delay in the onset of the conductance change.

There were experimental reasons to believe that, with all other parameters now known and linear, except for the nature of the membrane conductance, the Kelvin cable equation was an adequate description of the time and space phenomena in axons. And since there were no strong objections, this became—and has remained—a firm foundation for both theory and experiment, figure 2:10

$$i_m = C\frac{\partial V}{\partial t} + i_g = \frac{1}{r_e + r_i}\frac{\partial^2 V}{\partial x^2}. \quad (1:31)$$

For a uniformly propagating impulse with velocity θ,

$$\frac{\partial^2 V}{\partial x^2} = \frac{1}{\theta^2}\frac{\partial^2 V}{\partial t^2}$$

and so, corresponding to equation 1:49, figure 2:21

$$i_m = C\frac{dV}{dt} + i_g = \frac{1}{\theta^2(r_e + r_i)}\frac{d^2 V}{dt^2}. \quad (2:6)$$

Thus, with a knowledge of the fixed parameters and $V(t)$, i_g could be determined throughout the pulse. Unfortunately there was—and perhaps still is—a considerable degree of uncertainty as to the numerical values of the fixed parameters. And although the form of the action potential seemed quite consistent, we had no direct evidence of its peak amplitudes across the membrane, and there was some hesitancy in even assuming that this was equal to the resting potential which itself could only be estimated with some reservations (Cole and Curtis, 1938b).

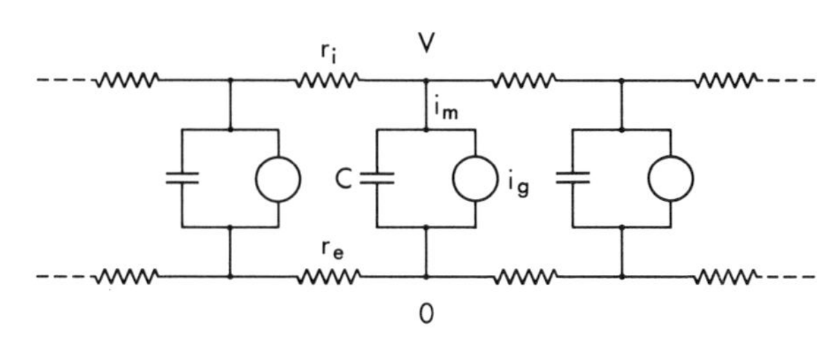

$$i_m = C\frac{dV}{dt} + i_g = \frac{1}{\theta^2(r_i + r_e)}\cdot\frac{d^2 V}{dt^2}$$

FIG. 2:21. Schematic and equation for a cable propagating at the constant speed θ.

However, the conductance change allowed a rather less cautious consideration of the conductance component of the membrane current. Something was changing the conductance after the inflection of the action potential, but not before. If the cause were an ion-current flow, we could assume it to be negligible up to the time t_f of the inflection—the "foot"—of the action potential and

$$i_g = -C\frac{dV}{dt} = \frac{1}{\theta^2(r_e + r_i)}\frac{d^2V}{dt^2}.$$

The action potential derivatives had been obtained in several ways depending on the resources and aptitudes of the interested investigators. We had first used numerical and graphical differentiation with its usual ragged result. Then spatial approximations from two or three electrodes along the axon proved highly satisfactory and without objectionable noise on even the second derivative, figure 2:22 (Cole and Curtis, 1939). In the other direction, Schmitt (1955) in particular left the neighborhood of the axon uncluttered and used electronic differentiation and integration.

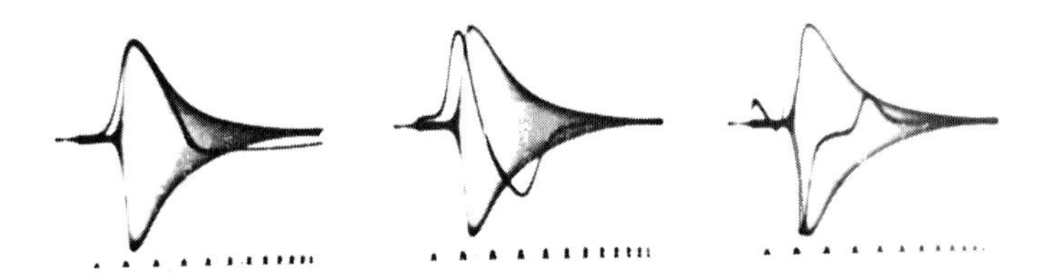

FIG. 2:22. Oscilloscope records of squid axon membrane resistance decrease, band, with lines for the membrane potential, left, potential first derivative, center, and second derivative, right. Time marks in msec.

It has been found in a variety of preparations, including *Nitella*, squid, and sartorius fiber (cf. Tasaki and Hagiwara, 1957; Taylor, 1963; Falk and Fatt, 1964), and it may be generally true that the foot of a propagating potential is rather exactly an exponential,

$$V = V_f \exp\left[(t - t_f)/\tau_f\right],\ t < t_f,$$

V_f is the potential at the inflection point, τ_f is the time constant. Then, to the extent that equation 2:6 is valid,

$$\tau_f = \frac{1}{C\theta^2(r_e + r_i)}$$

and

$$i_g = C\left[\tau_f \cdot \frac{d^2V}{dt^2} - \frac{dV}{dt}\right]. \qquad (2:7)$$

This could be calculated and was most interestingly plotted as a locus on the V, I plane figure 2:23. Since ohmic conductances lie in the first and third quadrants with positive slopes, the large excursion with a negative slope into the fourth quadrant—a wrong way current—was provocative to say the least. Although far from certain, values of $C = 0.14\ \mu\text{F/cm}$, $\tau_f = 0.04$ msec, and a peak value of $V_p = 110$ mV gave an inward maximum conductance current up to 1 mA/cm². This was an apparent negative conductance down to below $-100\ \Omega\text{cm}^2$ from the resting potential or only $-45\ \Omega\text{cm}^2$ at the maximum slope. The maximum positive slope of about $10\ \Omega\text{cm}^2$ just before the peak potential certainly deserved more attention than it got. The beginning of the conduction current somewhat before the potential inflection was also barely noticed.

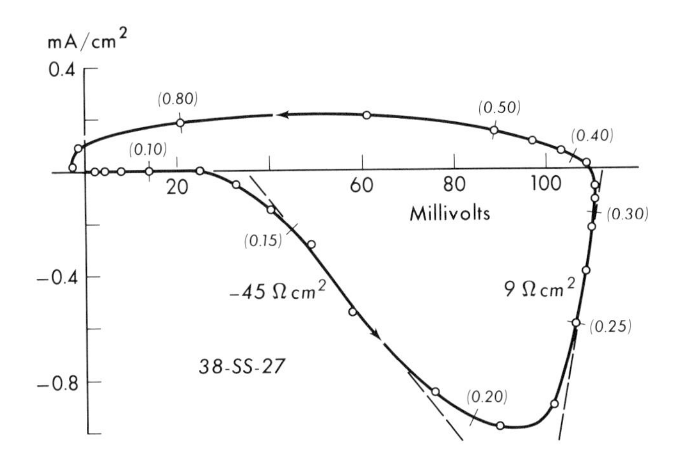

FIG. 2:23. Squid axon membrane ionic current-potential characteristic during an impulse. Computed on the assumption of a negligible conduction current during the exponential foot of the action potential. Times in msec in parentheses.

The experimental finding of a conductance change which was independent of frequency during an impulse was probably extremely good fortune. It certainly was a considerable source of confidence in the analysis and interpretation at a time when it was difficult to believe in them. There was no obvious cause or mechanism for a conductance change, and it seemed a little ominous that it should have the delightful characteristic of frequency-independence in addition to most obligingly allowing itself to appear directly as a leakage in parallel with the membrane capacity. But, no matter what the misgivings, such a conductance had to be free from reactance, represent a slope on the V, I plane, and extrapolate to an effective emf. With $g(t)$ and $V(t)$ given, I supposed that

$$I(V) = \int_0^V g\,dV \qquad (2:8)$$

and this produced a V, I trace lying entirely in the first quadrant. The disagreement of the two indirect measures of the conduction current led me to distrust both of them—so much so, in fact, that I looked for the possibility of a negative conductance in the impedance measurements. The other possibility, that both measures of the conduction current might be essentially correct, could not then have been understood, but it would have led to another extrapolated emf that has since become useful. So far as I can now find, in all of my search for relations between V, i_g, and g, I did not plot i_g vs. g. In this, the slopes have the dimension of a potential that might have proved helpful! However, without further information, this was then completely worthless as an objective means for analysis of the origin of either of the components or their combination.

Thévenin's theorem, the very useful and well-known tool of circuit analysis dating back to Helmholtz, completely defines the properties of any unknown network (Guillemin, 1935) for at least a momentarily steady state. This theorem states that the performance of a two terminal network of emf's and resistances is equivalent to a single emf and a single resistance (figure 2:24). This equivalent emf and resistance combination obviously changes as the elements of the network change, and in general, without further information, it cannot specify a unique network. It is, however, the simplest description of an unknown circuit, and as such may at least be a useful representation. Katz (1942) faced a similar problem and again I insisted that a resistance was $\partial V/\partial I$. However, with the background of two more decades, Jenerick (1963, 1964) was able to make good use of this approach.

The Ostwald-Bernstein postulate of a selective permeability to K^+ for the membrane at rest had been tacitly assumed in the absence of evidence for or against it. Then both the decrease of

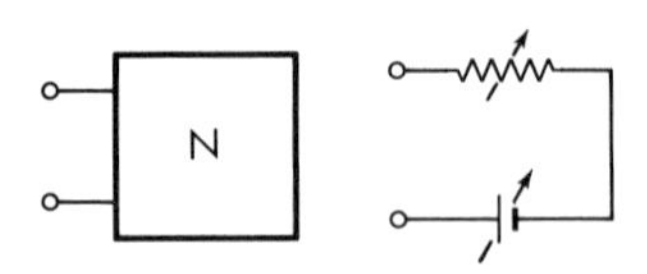

FIG. 2:24. Any circuit of any number of emfs and resistances in the two terminal box N, left, is electrically equivalent to the single emf and resistance, right, both of which may change as the original circuit components are changed.

membrane emf and the increase of conductance were easily attributed to an increase of permeability to another ion. The only thing I thought of was Cl^-, but I did not try to use it explicitly!

INSIDE AN AXON

OVERSHOOT

It now seems most interesting and probably quite characteristic that we had done only familiar things with the marvelous big squid axon—except when driven by desperation to the longitudinal measurements—and we had learned by bitter experience in dissection to handle it with great care. But after three years Curtis decided that he could run a long electrode from one end, through the axoplasm, into a region well removed from the injury at the point of entry. I was far from excited by the possibilities and even commented that an upside-down action potential would not be very interesting. But if it could be done it should. We knew the resting characteristic length λ of about 6 mm, so we had only to run a couple of cm down the inside of the half mm axon to get below a 5 percent effect of the injury on the measurement. Our action potentials (Curtis and Cole, 1940) were quite variable and inconclusive but we soon had word from Hodgkin and Huxley that they had done much the same thing, at about the same time and probably for much the same reason, but much better. In their preliminary note (Hodgkin and Huxley, 1939) with figure 2:25 they gave resting potentials about 50 mV which, even with a liquid junction correction, were far less than the average action potential of 90 mV. These results we fully confirmed (Curtis and Cole, 1942) except for the published action potential of 168 mV which I came to believe was probably the result of an overcorrection for the electrode and the amplifier input capacities. When Hodgkin and Huxley were able to publish their work in full after the war (1945), they most generously spoke of their confirmation of *our* work.

The distortion by the capillary electrode was a miserable problem for us. The electrolyte-filled glass capillary was a rather poor cable, with a high internal resistance and an appreciable capacity, but conventional, compensating networks were quite available. This, however, was not at all a conventional problem since the potential along the outside of the cable was our unknown and it was traveling with a wavelength comparable to the electrode length. So only an

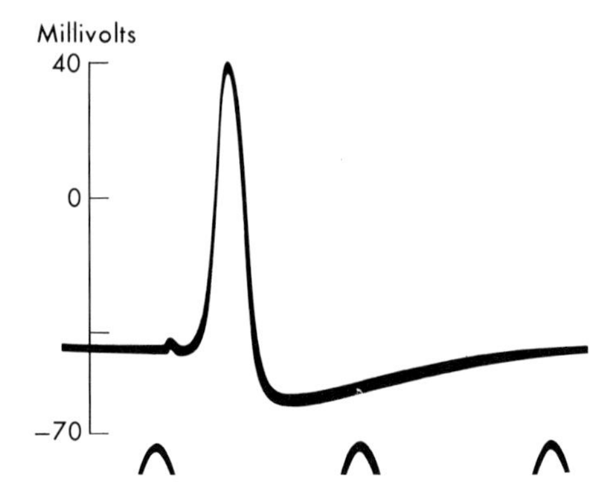

FIG. 2:25. Oscilloscope record for the potential of a capillary electrode inside the squid axon. The initial negative potential at rest is referred to the outside sea water. This potential changes sign during an impulse to overshoot the resting potential. Time, 2 msec intervals.

exponential compensation was used to give the fastest response to a step input. The inspired solution of an inside bare wire along the length of the capillary, as used by our Plymouth colleagues, was also the easy answer to that problem (Hodgkin and Huxley, 1945; Hodgkin and Katz, 1949). The wire became an effective shunt to the electrolyte for the fast changes of potential at which there was a significant capacity current across the capillary wall. But this same answer has been found essential in another recent and more controversial problem (p. 472; Chandler and Hodgkin, 1965).

The correction of the potentials, as measured in biological preparations, for liquid junctions in the electrode systems had long been ignored or approximated. However at this time it could not be ignored because if such corrections to the resting potential made it larger than the action potential, which was only a difference, there was no new problem. Approximations indicated that this was probably not the case and that these new observations were highly significant and startling. Not until much later was a more solid basis given for the liquid junction correction and for the validity of this conclusion (Cole and Moore, 1960b).

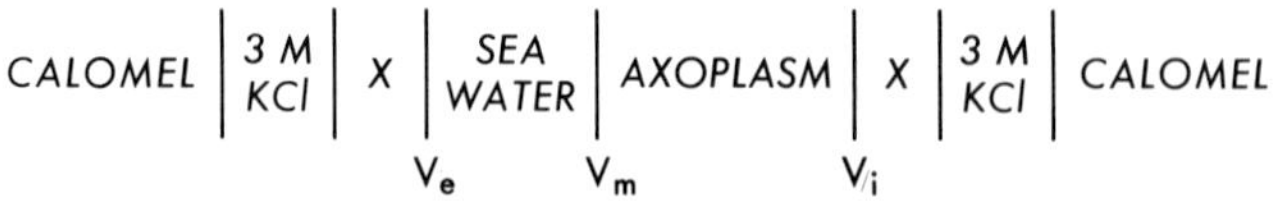

CALOMEL | 3 M KCl | X | SEA WATER | AXOPLASM | X | 3 M KCl | CALOMEL

V_e V_m V_i

FIG. 2:26. Equivalent electrochemical components for the potential of a micropipet electrode filled with electrolyte, X, and inserted into the axoplasm of a squid axon.

Membrane potential measurements were made on squid axons with a series of micropipets filled with KCl at concentrations from 0.03N to 3.0N or sea water. These observed values V were then represented by the chain of figure 2:26 in which the various pipet solutions X were the same on both sides. The liquid junction potentials V_e and V_i were to be calculated by the Henderson (1907; also MacInnes, 1939) equation

$$V_{12} = \frac{kT}{e} \cdot \frac{(\mathbf{U}_1 - \mathbf{V}_1) - (\mathbf{U}_2 - \mathbf{V}_2)}{(\mathbf{U}_1 + \mathbf{V}_1) - (\mathbf{U}_2 + \mathbf{V}_2)} \ln \frac{\mathbf{U}_1 + \mathbf{V}_1}{\mathbf{U}_2 + \mathbf{V}_2}$$

where **U**, **V** were the cation and anion conductances on the two sides of the junction as suggested by Lewis and Sargent (1909). The outside potential V_e could be calculated directly but V_i was less certain. The available analyses were used for the conductances of potassium, sodium, and chloride in axoplasm. Then with the apparently trivial, but necessary, assumption that the membrane potential V_m was independent of the electrolyte in the pipet, a reasonable value for

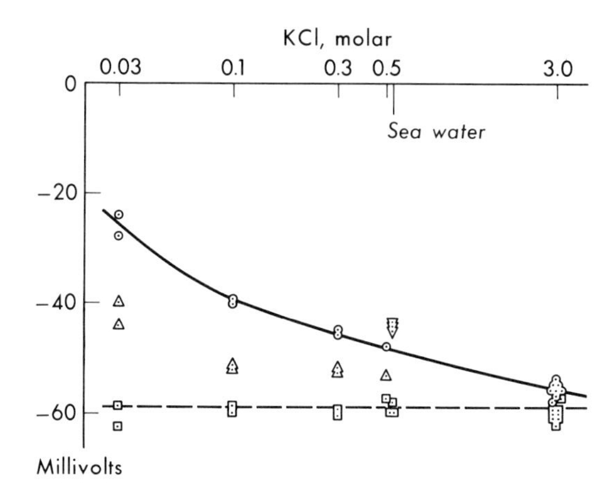

FIG. 2:27. Estimates of liquid junction and membrane potentials for squid axon. The circles and inverted triangles are for the potentials measured with the indicated KCl solutions and for sea water in the pipet. The intermediate triangles were obtained by adding the calculated external liquid junction potentials. The squares represent the fit to a constant membrane potential, dashed line, obtained by the best choice of a negative ion conductance for the axoplasm.

the remaining negative ion (presumably isethionate, Koechlin, 1955) conductance was found by a systematic cut and try process. On this basis the unknown axoplasm conductance **V** averaged 14.6 mmho/cm and the membrane potentials for the four axons of figure 2:27 averaged almost 60 mV. The junction correction for the squid axon in sea water gave an average of 4 mV for a 3N KCl pipet electrolyte that had now become standard. But also it was found that for iso-osmotic 0.5N KCl the correction was 10 mV and for sea water, 15 mV. These values were in good agreement with the early approximations and they certainly removed any lingering doubts that the action potentials in the range of 110 mV exceeded the resting potentials!

It now became possible and necessary to revise our earlier description (Cole and Curtis, 1938b) of the resting and action potentials to give figure 2:28. However, the local circuit current flows remained qualitatively unchanged because there was no reason then, as now, to question the use of cable theory.

There had been many earlier indications of such an "overshoot" (Grundfest, 1965) that had received but passing comment—if any. I had had no great misgivings in blaming unknown extraneous factors for such an observation in the attempts to get inside heart muscle culture cells, figure 2:29 (Hogg et al., 1934). But the many later measurements firmly established the overshoot as the trademark of all electrically excitable cells except apparently *Nitella*, and it, too, now seems to have joined the ranks (Kishimoto, personal communication). In connection with the more immediately profitable steady state effects of current flow to come, the records of the propagating action potential in a locally polarized region clearly indicated an absolute nature of the spike potential. But it, too, could not be recognized as an omen!

Then Bernstein was no longer adequate and the provocative evidence was in hand to require a spectacular major revision after World War II.

External Potassium

Another result of some importance and considerable interest was the decrease of the resting potential as the external potassium concentration K_e was raised, figure 2:30. The potential corresponded rather well with the Nernst-Boltzmann relation for high K_e, suggesting a simple K^+ permeability. This was very encouraging until

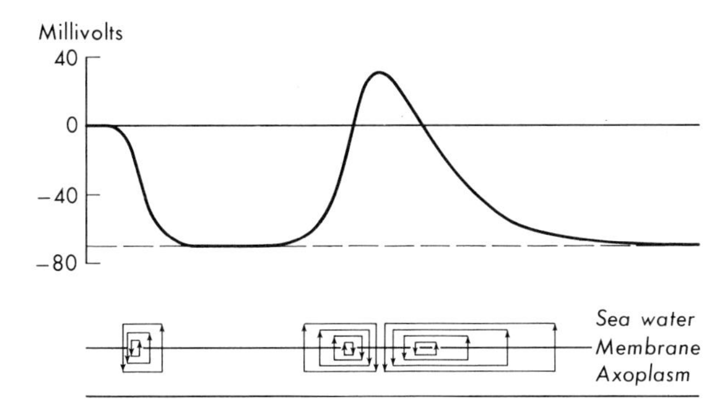

FIG. 2:28. Membrane potential, above, and local circuit current flows, below, at a killed end, left, and along an impulse, center.

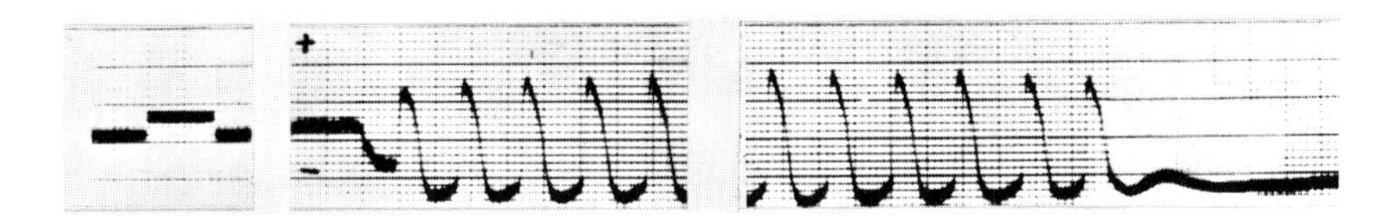

FIG. 2:29. Repetitive activity of a heart muscle culture after micropipet electrode was moved from base line potential, left, close to hot spot with resting potential, right, and overshoot in between. Calibration, 1 mV, 0.4 sec.

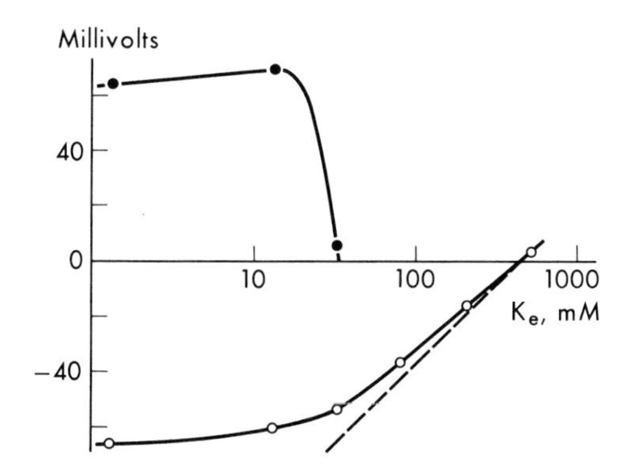

FIG. 2:30. Squid axon membrane potentials vs. external potassium ion concentration. Open and solid points, resting and action potentials, dashed line, equilibrium potentials.

it was realized that, in this region, the axon was not excitable and would not conduct an impulse. On the other hand, the relatively lesser dependence of potential on low K_e would then indicate a relatively higher permeability to another ion. But the axon was functional in this environment! It is very impressive that this functional relation between membrane potential and external potassium has been so widely confirmed, as summarized by Hodgkin (1951), and that there is not yet a completely satisfactory explanation.

Here, again, it is necessary to go beyond the classical idea of Bernstein.

RECTIFICATION

It seemed necessary to simplify the situation when the propagating impulse gave so little information. Obviously the membrane conductance depended on the current flow through it. The conductance increase during a conducted excitation, and the initiation of excitation by outward or depolarizing applied currents led to the expectation that outward currents would increase the conductance. But the outward total membrane current during the foot of the action potential gave little if any change of the membrane conductance. So it seemed highly advisable to establish a current–conductance relation before perhaps going further astray.

There had been a number of indications that the change of a membrane potential was not proportional to an applied current. Blinks had once again led the field with his work on large plant cells (1936b). Katz (1937), Hodgkin (1938), Arvanitaki (1939), and Pumphrey et al. (1940) had all shown subthreshold local responses near a cathode which could be interpreted as nonlinear membrane characteristics.

Impedance Change

In the further interests of simplicity our investigation was to be limited to the high-frequency transverse impedance as used for the impulse and to the steady state, except that the applied currents above threshold were to be blunted to see if the steady state effects depended upon an early excitation. There was a forlorn and unformulated hope that by a manipulation of a current flow it might be possible to maintain an excited state of the membrane long enough to get some idea as to what sort of a thing this state might be.

The experiments (Cole and Baker, 1941a) were highly satisfactory in that they gave the steady state conductance changes which were again independent of frequency. Also, the apparent effect on

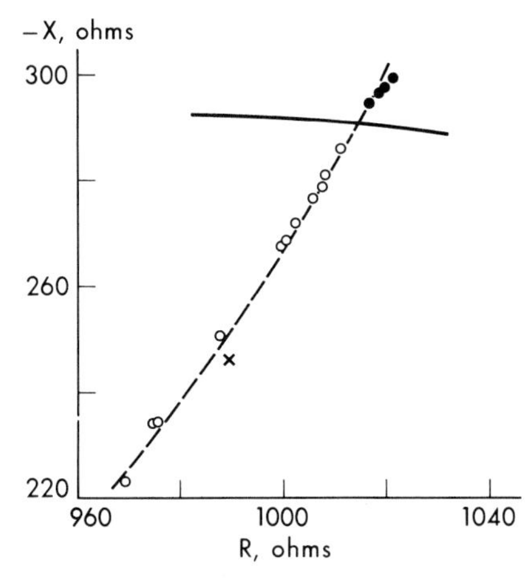

Fig. 2:31. Squid axon steady state impedance change at 20 kc with current flow on complex plane. Solid arc is part of normal frequency locus. Open circles are for outward current flow, cross is peak change during an impulse, and solid circles are for current inflow. Dashed arc is calculated locus for a change of membrane conductance alone.

the membrane capacity was even less for outward currents than during an impulse (figure 2:31). There were, however, definite indications of a capacity decrease for inward—or hyperpolarizing—currents. The difficulties of estimating local membrane currents in a cable with a nonlinear membrane were becoming apparent, and the conductance changes were not expressed in terms of the current density to which they should have been related. The conductance changes from the rest value saturated fairly well for both current directions—equal to about 20 Ωcm^2 decrease for outward and 200 Ωcm^2 increase for inward, figure 2:32. The former value is comparable to the change in an impulse, while the latter is considerably

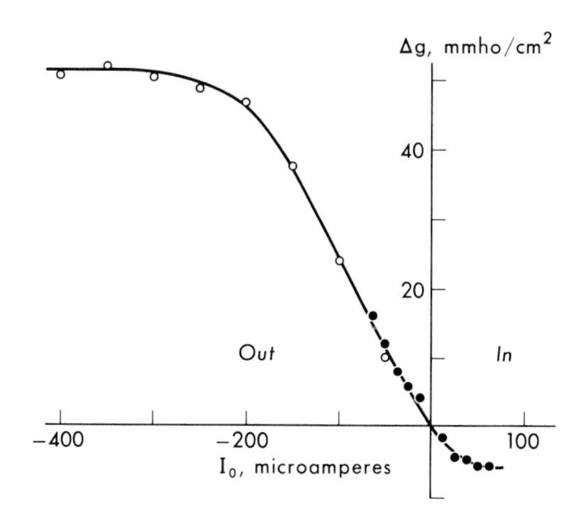

Fig. 2:32. Change of squid axon membrane conductance with impressed current.

lower than the direct current longitudinal measurements, which ranged from 400 to 1100 Ωcm². Thus these experiments clearly showed that the conductance change was closely related to a steady membrane current flow to make the conduction path a rather efficient rectifier. However, they still did not correlate directly with the current flow during an impulse.

Further, when it came to skeletal muscle, Katz (1942) found an increase of the impedance at 5 kc during current flow under the anode, and a decrease under the cathode. However, an "anomalous rectification" was obtained for the subthreshold steady state current-potential relation under the cathode (Katz, 1948). Another early experiment by Teorell (1949) showed an effect of an inward current flow on the impedance of frog skin over a range of frequencies. As for the squid axon this was interpreted to be the increase of a conductance in parallel with a constant capacity.

Current-Potential Relation

With a potential electrode inside the axon it became possible to investigate the current-potential relations of the membrane much more directly (Cole and Curtis, 1941), and perhaps also to find indications of any change of membrane potential. The direct current-potential records from the oscilloscope were pleasing, figure 2:33, and showed quite constant and reproducible relations such as in figure 2:34 for the several axons investigated. However, the measured current I_0 flowed between external electrodes, and was not obviously and directly related to the membrane current density i_m at the point where the membrane potential V_m was measured. The way out of this problem has been rather useful, although it still seems something of an analytical accident that it should have come out so simply. We found

$$i_m = \frac{r^2_e}{4(r_e + r_i)} \cdot I_0 \cdot \frac{dI_0}{dV_m}. \qquad (2:9)$$

This was called Cole's theorem by Hodgkin and it has even appeared in print as such (Taylor, 1963). It was quite frightening to find after the war that the original derivation included an assumption of linearity and it was still embarrassing even when a better derivation gave the same results.

There have been several subsequent derivations including my

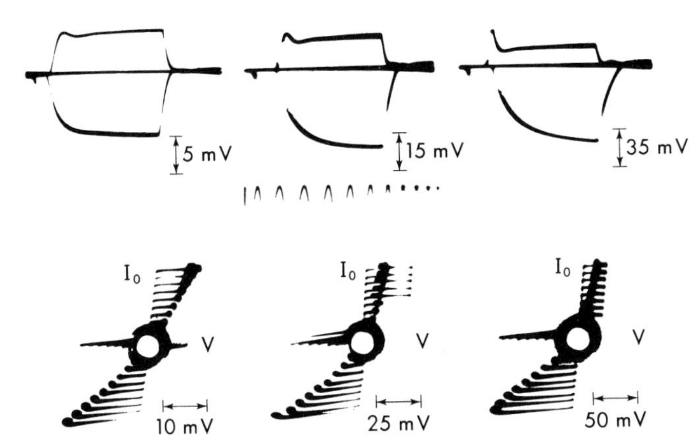

FIG. 2:33. Oscilloscope records of squid axon membrane potential change during current flow; outward, up; inward, down. Upper; left, 9.7 μA; center, 23.7 μA; right, 47.5 μA. Times at 2 msec intervals. Lower, maximum currents; left, 17.5 μA; center, 47.5 μA; right, 95 μA.

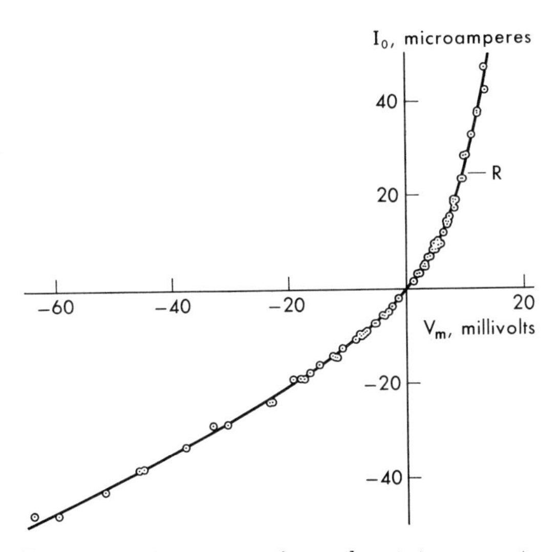

FIG. 2:34. Summary of steady state current-potential data for a squid axon. R shows the rheobase threshold.

own rather ponderous one (Cole, 1961). The key to the problem is to consider first the current-potential relations between an inside and an outside electrode. Considering only one side we have at the electrodes, $x = 0$,

$$\frac{dV_m}{dx} = -(r_e + r_i)I_0$$

and as before

$$\frac{dI_0}{dx} = -i_m$$

then

$$\left(\frac{dV_m}{dx}\right) \Big/ \left(\frac{dI_0}{dx}\right) = \frac{dV_m}{dI_0} = (r_i + r_e)\frac{I_0}{i_m}.$$

For both sides the total current and the increment are doubled to give

$$i_m = \frac{r_i + r_e}{4} \cdot I_0 \cdot \frac{dI_0}{dV_m}.$$

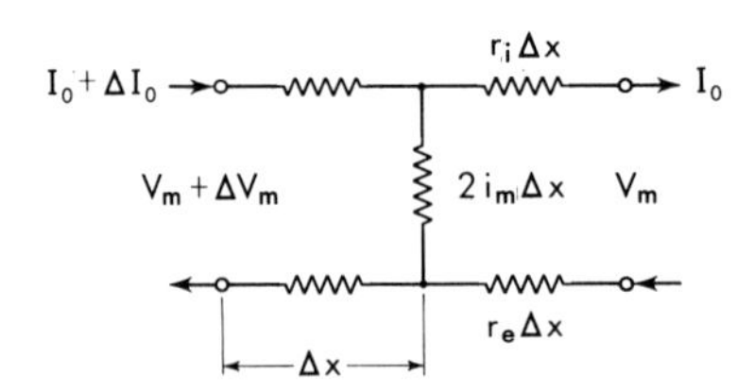

FIG. 2:35. An equivalent circuit element for derivation of the membrane characteristics of a nonlinear cable.

Another derivation that gives more insight into the physical basis is an extension of the cable impedance derivation given before. Here we start from the measurable relation between V_m and I_0 of figure 2:33. Then as we increase I_0 by ΔI_0, V_m will increase by ΔV_m and the point corresponding to the original V_m, I_0 will move away from the electrodes by a distance Δx. Over this distance we use the equivalent artificial cable section shown in figure 2:35 in which

$$\begin{aligned}\Delta V_m &= (r_i + r_e)(I_0 + \Delta I_0)\Delta x + (r_i + r_e)I_0\Delta x \\ &= 2(r_i + r_e)I_0\Delta x + (r_i + r_e)\Delta I_0\Delta x\end{aligned}$$

or

$$\frac{\Delta V_m}{\Delta I_0} = 2(r_i + r_e)\frac{I_0\Delta x}{\Delta I_0} + (r_i + r_e)\Delta x$$

and since

$$\Delta I_0 = 2i_m\Delta x$$

$$\frac{\Delta V_m}{\Delta I_0} \to \frac{dV_m}{dI_0} = (r_i + r_e)\frac{I_0}{i_m}$$

as $\Delta x \to 0$, to give the same result as above.

To go to the distant electrode case we have only to replace the real internal electrode by a hypothetical one and superpose the currents flowing directly from this and from the real outside electrode to the distant electrode and obtain the relation of equation 2:9.

The results of this analysis gave asymptotic conductances with a ratio of about 100 between the two directions of current flow, figures 2:36 and 2:37. This was at least as good as the early CuCuO rectifiers then available. However, the resting conductances were far higher than found before, and the action potential extremely

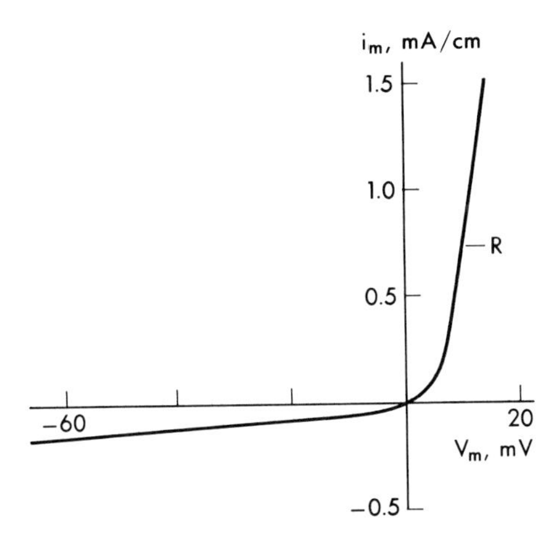

FIG. 2:36. The steady state current-potential characteristic for a squid axon membrane as calculated from longitudinal data for the axon.

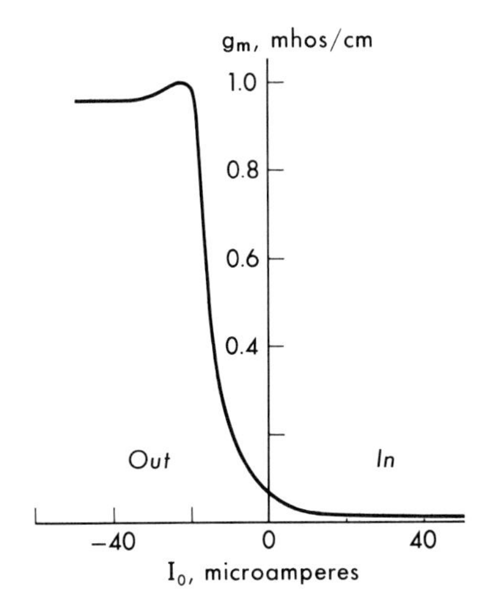

FIG. 2:37. The variational steady state slope conductance of a squid axon membrane as a function of the total current applied to the axon.

low for what still appear to be quite unknown reasons. This is still more puzzling when we note that the rheobase-stimulating potential was about 10 mV and in what has become the quite normal range. It may only be good luck that the results have been found at least qualitatively correct.

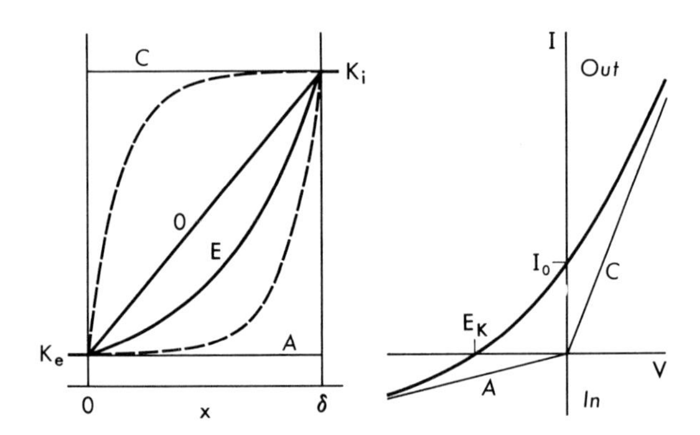

FIG. 2:38. A potassium theory of membrane rectification. The ion concentration distributions across the membrane thickness δ between K_e, external, and K_i, internal, are shown at the left. For equilibrium the effects of the concentration gradient and the electric field are balanced along E to give no current flow I, and the equilibrium potential E_K, right. With a membrane potential V of zero there is no field, the gradient of concentration is constant along O to give a short circuit current I_0. Extreme fields give the dashed profiles at the left and approaches to the limiting inward and outward conductances A and C.

Nonetheless it was easy to think of the rectification characteristic as an extension of the Bernstein model. If indeed only potassium ions were mobile in the membrane, then for the equilibrium potential E_K the ion concentration would vary across the membrane such as E in figure 2:38 with the electric field everywhere neutralizing the diffusion gradient. In the absence of the field the gradient would be constant and the ion concentration everywhere larger, line O. This would give the larger conductance for point I_0 according to equation 1:26. Large potentials $V \gg 0$ would tend to fill the membrane to the inside concentration K_i, and give a limiting high conductance C. In the opposite direction $V \ll E_K$ the ion concentration K_e and the limiting conductance A would be similarly reduced.

Many subsequent measurements have given an even higher ratio for steady state conductances in the two directions, figure 2:39 (Cole and Moore, 1960a, and others). If, then, the conductance were that of the potassium in the membrane and the partition coefficients remained constant, this ratio should not exceed that of the potassium ion concentrations on the two sides of the membrane. This simple idea was thus untenable for an axon in sea water and a K^+ ratio of about 20x.

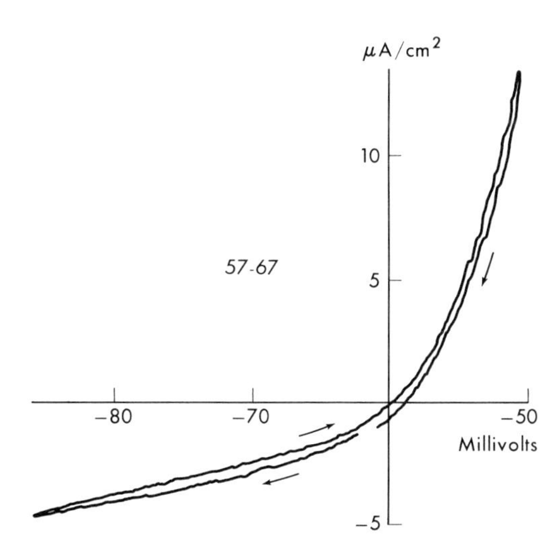

FIG. 2:39. Direct measurement of a squid axon membrane current-potential characteristic at near steady state with internal and external potential electrodes and extended current electrodes.

The differential conductance dI/dV was calculated, figure 2:37, for comparison with the high frequency results. The small peak at about rheobase was found in every experiment. It was suggestive of a negative conductance, but its significance was then only another puzzle to await the later explanation that is yet to come. I was very much impressed by the otherwise general similarity of these high-frequency and direct current conductances but the experimental evidence certainly did not require that they be either the same or different. At the time, both of these conductance changes by current flow were only thought of as changes at the zero frequency extrapolation R_0 of figure 2:19, to either R_0' or R_0''.

The nonlinear steady state characteristics of striated muscle fiber membranes became important but had been difficult to obtain until Adrian and Freygang made an interesting approach. Cremer had realized that a uniform current density would result from a parabolic external potential along a cable like structure and he showed how to obtain it (Taylor, 1963). But Adrian and Freygang (1962) injected current into a fiber to get an approximately parabolic internal potential V near an end. Treating the fiber as a cable with a potential V_0 and an open circuit at the end, $x = 0$, the solution of equation 1:37 is

$$V = V_0 \cosh x/\lambda.$$

Then for a difference ΔV between two micropipets near the end and a short distance l apart, and x/λ small,

$$V \doteq V_0 \text{ and } \Delta V/V \doteq l/\lambda^2 = r_i\, g_m\, \Delta x.$$

Physical Model Behavior

The many new experimental facts did much to replace the conjecture, hypothesis, and empiricism of the past by physical realities. It was true that the underlying mechanisms of the various phenomena were still unknown, or at best uncertain. But it was seen that the observed conductance behavior could produce a considerably advanced physical model for the subthreshold behavior and excitation of the membrane, and for impulse propagation by the axon.

The measurements of membrane capacity, inductive reactance, and resistance gave a model that was analytically identical with the general two-factor theory. This circuit model might then be expected to be at least as successful as the two-factor theory and, most importantly, to give a physical reality to the subthreshold and threshold phenomena. Although the analyses and comparisons had been limited to linear approximations, there was no doubt that the two-factor theory could be extended into nonlinearity by guessing at more and similarly uncertain parameters. On the other hand, the nonlinearity of the steady state membrane conductance was then another experimental fact and it could be included along with the facts of the capacity and inductive reactance as represented by the equivalent circuit, figure 2:40 (Cole, 1941), in calculating the membrane performance. The fact of a considerable conductance increase during impulse propagation gave first a simple explanation for the form of the impulse (Cole and Curtis, 1938b) and then led to a more comprehensive analysis and a prediction for the speed of the impulse (Offner et al., 1940).

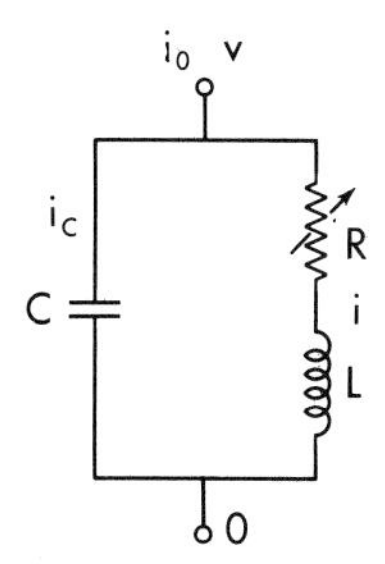

FIG. 2:40. An equivalent circuit for an approximate representation of the experimental squid axon impedance and rectification characteristics. The variable resistor R is given by a nonlinear current potential relation.

SUBTHRESHOLD

The steady state spread of membrane current density and potential was easily calculated, and the highly nonlinear membrane conductance required that these quantities be quite different near an anode from those under a cathode (figure 2:41, Cole, 1941), even considerably below rheobase. The long known difference between the electrotonic spreads at anode and cathode now had a simple explanation. It is worth noting that the curves of figure 2:41 were generated merely by sliding a single master curve for either the anode or the cathode along the distance axis. Furthermore, the sum of the effective anode and cathode resistances for a long stretch of axon turned out to be nearly constant as the current was increased. This accounted for the failure to find an overall rectification, although an analytical explanation remains to be found.

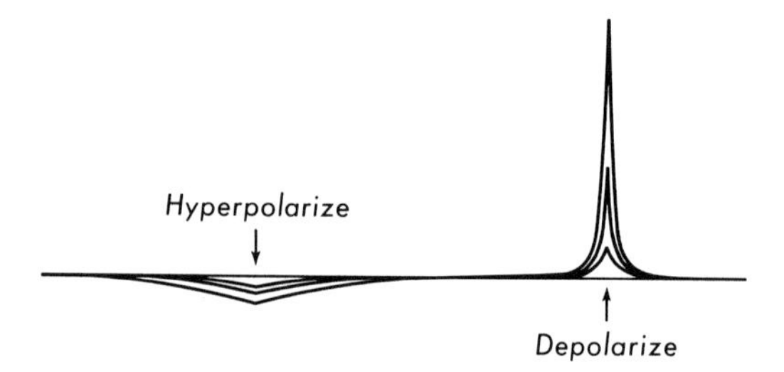

FIG. 2:41. Computed steady state membrane current densities at and near short anode and cathode electrodes applied to a squid axon. The applied currents are 0.25, 0.5, and 1.0 times the rheobase.

The transient axon behavior was more complicated. Looking first at the membrane equivalent circuit of figure 2:40 where the resistance R is now a function of the current $v_R/i = R(i)$,

$$LC \frac{\partial^2 v}{\partial t^2} + R(i)C \frac{\partial v}{\partial t} + v = L \frac{\partial i_m}{\partial t} + R(i)i_m \qquad (2:10)$$

and with the cable expression, equation 1:31,

$$\frac{\partial^2 v}{\partial x^2} = (r_e + r_i)i_m$$

we would then have a complete statement of the problem. A solution for this was not even attempted. Instead the potential across the membrane circuit alone was calculated for the application of a constant current i_0. Then, in place of equation 2:10,

$$i_C = C\frac{dv}{dt} = i_0 - i$$

and

$$v = L\frac{di}{dt} + v(i) \qquad (2:11)$$

or

$$\frac{dv}{dt} = \frac{i_0 - i}{C} \text{ and } \frac{di}{dt} = \frac{v - v(i)}{L}$$

giving

$$\frac{di}{dv} = \frac{C}{L} \cdot \frac{v - v(i)}{i_0 - i}. \qquad (2:12)$$

On the v, i plane the slope is then known everywhere and can be represented by short line segments to form a continuous path or trajectory such as the figures Von Kármán reproduced from van der Pol (1926) for several examples. In these the lines for constant slopes $di/dv = m$, called isoclines, were used as the principal guides for the trajectory. Of these the two orthoclines for $m = 0$ and $m = \infty$ are particularly important. In equation 2:12, $m = 0$ for $v = v_R$ so all trajectories cross the v_R vs. i characteristic parallel to the v axis. Similarly, $m = \infty$ for $i = i_m$ and the trajectories are all perpendicular to the i_m line.

Many problems such as this are easily, rapidly, and interestingly solved graphically. By a change of variable $x = v$ and $y = \sqrt{L/C}\, i$, this becomes

$$\frac{\Delta y}{\Delta x} = \frac{x - x_R}{y_0 - y_R}$$

and a step-by-step approximation on the x, y plane is again obtained with a transparent ruler or compass as shown in figure 2:42. However, as a concession to the obviously brutal approximation made in ignoring the difficult spatial distribution problem, the observed i_0 vs. v curve of figure 2:34 was used for calculations shown in figure 2:43. Not until much later could experiment produce data for direct comparison with the calculation on the basis of the i_m vs. v curve of figure 2:36.

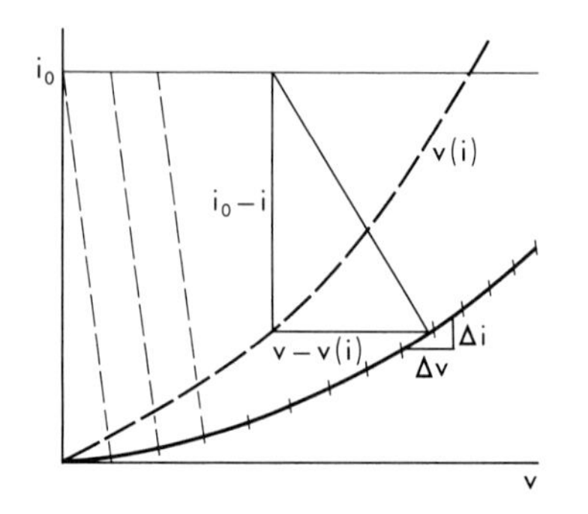

FIG. 2:42. Construction for the path on the current-potential plane after the application of a constant current to an equivalent squid axon circuit. The characteristic of R, the dissipative component, is $v(i)$ and the tangent to the path is given by the triangles. The dashed lines are for the conversion of the path into the potential-time relation.

Then by the more laborious integration of equation 2:11 indicated in figure 2:42 the v vs. t curves of figure 2:44 may be obtained. This step could often be avoided as it soon became possible to investigate and recognize the character of the solution on the phase plane. Nonetheless, here it was useful for comparison with our own experimental results (Cole and Curtis, 1941) and the more spectacular records of Arvanitaki (1939), which she considered from the original Poincaré point of view.

This combination of a highly nonlinear membrane conductance with the experimentally observed inductive reactance and dielectric capacity gave a simple equivalent circuit that had many properties interestingly similar to axon performance (Cole, 1941).

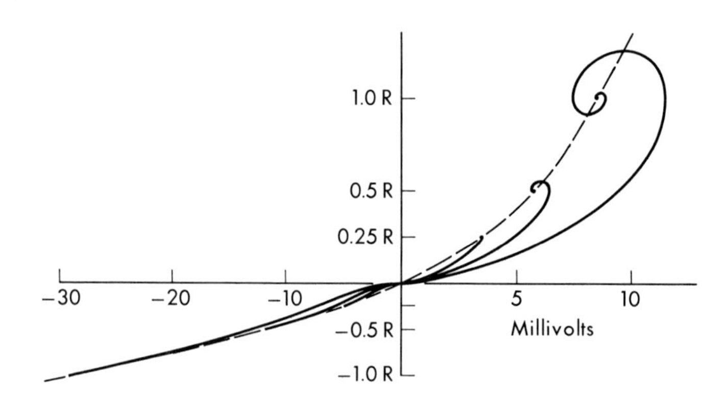

FIG. 2:43. The paths on a current-potential plane for a network approximately equivalent to the squid axon. To the right are the depolarizing, cathode, behaviors after currents of 0.25, 0.5 and 1.0 times rheobase. The hyperpolarizing, anode, paths are shown, on half scales, at the left.

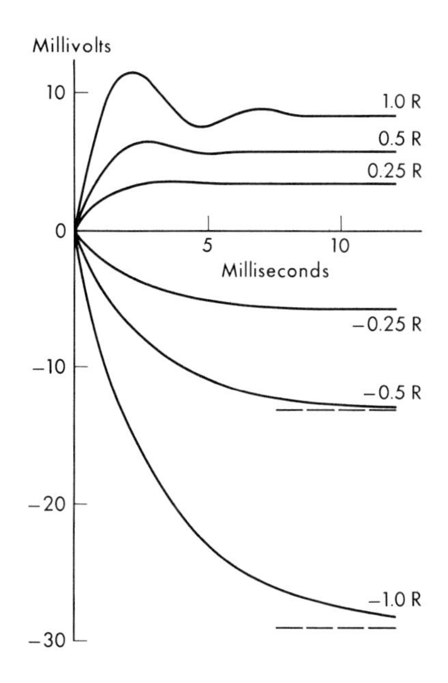

FIG. 2:44. The computed behaviors of a squid axon as presented on a current-potential plane were translated to the change of membrane potential in terms of the time after cathode, above, and anode, below, current steps given as indicated fractions of the rheobase.

EXCITATION

The analytical formulation of the circuit model was similar to the generalized Rashevsky-Monnier-Hill two-factor expression for excitation which, conversely, had been similarly represented on a K, L plane. The circuit then gave a basis for understanding the oscillations of threshold (Erlanger and Blair, 1936; Monnier and Coppée, 1939), as well as those of potential. The calculations of figures 2:43 and 2:44 were good representations of the many unpublished records Curtis and I had made of the membrane potential up to threshold after an applied step of current. A similar graphical analysis for $i_m = 0$ after a short pulse corresponded generally with the subthreshold observations of Katz (1937) and Hodgkin (1938) figure 2:45.

Of the obvious limitations of this model, the absence of a high frequency conductance is most easily rectified by the addition of the simple shunt resistor of figure 2:46. This was consistently ignored in the original work after it was found to contribute little but complication. The rectification was hopefully ascribed to potassium ion movement and reason was soon found to include the inductive reactance in this process. Both explanations were abundantly con-

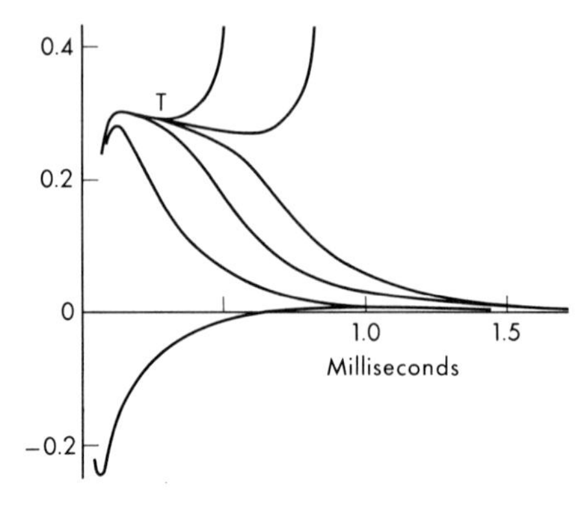

FIG. 2:45. Nerve potentials recorded near site of short current pulses, inward below, increasing outward above. At *T* the sub- and superthreshold potentials separate. Ordinates are fractions of spike heights.

firmed, but only much later (Hodgkin and Huxley, 1952). The disregard of a cable like spread of a stimulus was no less blatant than it was necessary—if only because an investigation of this contribution was not started until a quarter century later (Cooley et al., 1965; Cooley and Dodge, 1966; Noble and Stein, 1966). The assumption of a threshold potential had at least some experimental backing, but there was not even a vestige of an explanation for it. But it was simple and obvious and, to a first approximation, represented the bulk of the data available. Also, Curtis and I had seen it at a minimum slope condition for excitation with a considerably different time constant. Hill's case of complete accommodation could only be approximated as the resistance in the inductive branch approached zero. Furthermore, the model also ignored the experimental fact that many axons had not shown an inductive reactance and that, in at least some cases, these were quite superior axons.

These distressing and immediately important failures of the capacity, rectifier, and inductance model rather overshadowed its utility and reality. Nonetheless, it was the first description of subthreshold and excitation phenomena in terms of measurable physical properties of a membrane, and it is probably still the best approximation for their simple explanation in terms of potassium ion movements.

PROPAGATION

The fact of a high frequency conductance increase during a propagating impulse in *Nitella* (Cole and Curtis, 1938b) had consequences that were difficult to realize. Rushton's (1937b) theory appeared

Fig. 1:31. North Atlantic squid, *Loligo pealeii*, photographed in an aquarium at the Marine Biological Laboratory, Woods Hole, by Robert F. Sisson. These small specimens probably had a mantle length of about eight inches and giant axons almost a half millimeter in diameter. © National Geographic Magazine, March, 1967.

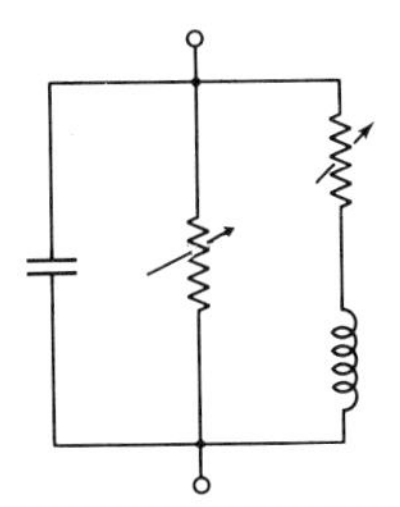

FIG. 2:46. A less approximate equivalent circuit to express the characteristics of a squid axon found by experiment. The shunt conductance which was added to represent the high frequency behavior of the axon was also a function of the current through it—or the potential across it.

during the experimental work but we felt that the conductance change far overshadowed any possible change of membrane emf. We were quite unwilling to recast his work into a more plausible model, so we started over again with the cable equation 1:32 in the form

$$\lambda^2 \frac{\partial^2 V}{\partial x^2} - \tau \frac{\partial V}{\partial t} - V = -E$$

with $\lambda = \sqrt{r_m/(r_i + r_e)}$, $\tau = r_m c_m$, and E for the membrane emf. On a coordinate s moving with the impulse speed θ, $s = x - \theta t$ and

$$\lambda^2 \frac{d^2 V}{ds^2} + \theta\tau \frac{dV}{ds} - V = -E. \tag{2:13}$$

This gave an exponential foot ahead of a conductance change and led to the conclusion that E was unchanged in this region. A reduction of either E or r_m or both in a short excited region then produced the peak of potential as a hyperbolic cosine followed by another exponential for the recovery tail.

Such a simple picture could give a very presentable action potential. But the change of E was arbitrary and could not be confirmed, so we went to the even simpler extreme of a localized short circuit, $E = 0$, $r_m = 0$. Outside the breakdown region this gave the characteristic lengths

$$\lambda_\theta = \lambda \Big/ \left(\frac{\theta\tau}{2\lambda} \pm \sqrt{1 + \frac{\theta^2\tau^2}{4\lambda^2}}\right)$$

ahead of or behind the short circuit. For large θ, this became $\lambda \to 1/\theta(r_i + r_e)c_m$ for the initial foot and $\lambda \to \theta\tau$ in the recovery tail.

About as we finished the *Nitella* work, Offner visited us with a proposal to do the same thing for his thesis under Gerard at Chicago. Not discouraged, he developed a theory (1939) of propagation based on membrane resistance change. Weinberg and Young did much the same on their own, but the three joined in the complete work (Offner et al., 1940). They also assumed uniform propagation but with the finite shunt conductance g_A, figure 2:47a, added for excitation. This happened at a threshold potential V_T to give the characteristic of figure 2:47b, and was removed at an arbitrary later time for recovery.

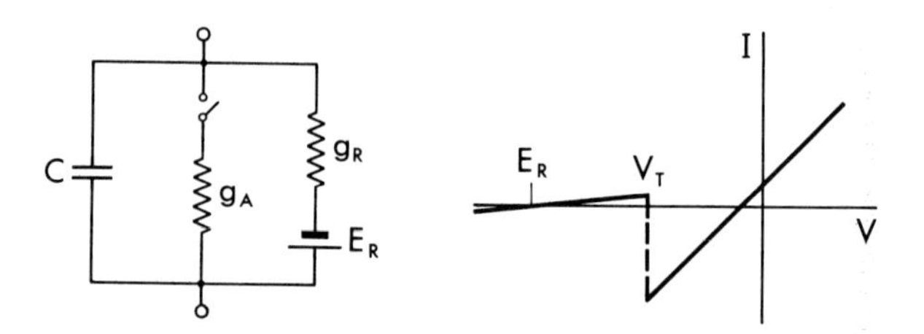

FIG. 2:47. A conductance theory of propagation. The membrane potential E_R and conductance g_R at rest are shunted by the active conductance g_A at an excitation threshold potential V_T.

The velocity was obtained from the three solutions of equation 2:13 and the effect of the recovery phase was found to be negligible. The two solutions of equation 2:13 for the rising phase alone are much simpler and lead to a velocity

$$\theta = \frac{\kappa}{\sqrt{1-\kappa}} \cdot \frac{1}{c_m} \cdot \sqrt{\frac{g_a}{r_e + r_i}}$$

where $\kappa = (V_T - E)/(V_P - V_T) < 0.5$ is the ratio of stimulus to response and it is assumed that $\kappa g_a \gg g_r$. An equivalent expression gave Offner et al. an impulse speed that was reasonable from our squid data but a hundred times too fast for *Nitella*.

CONCLUSIONS

The physical models of membrane behavior, which followed from the experimental facts expressed as equivalent circuits, were successful enough to focus attention on the mechanisms of ion conduction. These models were fragmentary and they were not consolidated into a consistent whole. They were difficult to believe, what with experimental difficulties and theoretical approximations. They certainly would have deserved oblivion except that they contained, in primitive form, the elements of what was to come.

Mechanisms

NONLINEARITY AND REACTANCE

The first attempt to explain an inductive reactance (Cole, 1941) was along the line then familiar for systems such as a telephone receiver (Kennelly, 1923) and a quartz crystal (Mason, 1948). In these an inductive reactance was a component of the motional impedance which resulted from the coupling of the electrical and mechanical elements. This system seemed so unlikely that it was easily deserted when a more attractive candidate appeared.

After we had shown that the squid axon inductive reactance was probably in the membrane, I told J. E. Becker and G. L. Pearson that we had 0.2 henry in a square centimeter of membrane 100 Å thick. They did not seem particularly surprised but agreed that they could not do quite as well—they had some 16 henries but it was in the considerably larger volume of a cubic millimeter. They then took the opportunity of satisfying my thoroughly aroused curiosity by showing me the manufacture and some properties and uses of the uranium oxide bead that was soon to appear as the Western Electric 1-A Thermistor (Pearson, 1940). In particular, they showed me an analysis of the linearized steady state impedance behavior of such a semiconductor with a negative temperature coefficient which was essentially duplicated recently by Mauro (1961).

Thermistor

The heat balance is given by

$$a\, dT/dt = P - b(T - T_0)$$

where P is the power supplied, T and T_0 the bead and ambient absolute temperatures, a the heat capacity, and b the thermal loss coefficient. In a steady state, the resistance of a semiconductor, uranium oxide in this case, is

$$R = V/I = AT^{1/2} \exp\,(B/T)$$

which is approximated by

$$R = \bar{R}[1 + k(T - T_0)] \qquad (2{:}14)$$

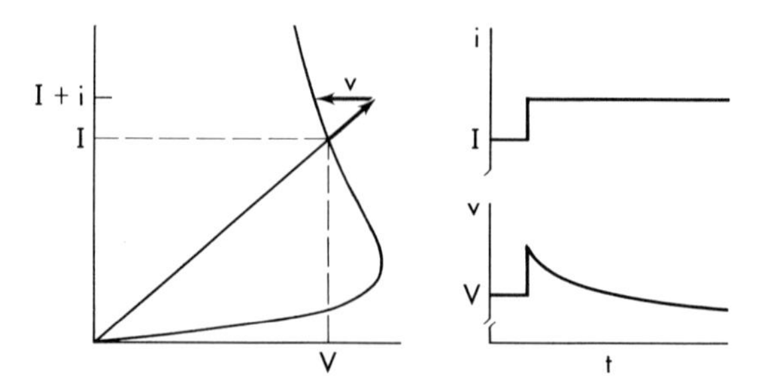

FIG. 2:48. An approximate representation for the steady state and transient behavior of a thermistor. At a sudden change of current from I to $I + i$ the potential immediately increases and then relaxes by an amount v as shown at the left on the current-potential plane and at the right as a function of time.

for the resistance temperature coefficient k, which is here negative.

The qualitative behavior of this system is well illustrated by the transient response of figure 2:48. For a sudden change from I to $I + i$ the change of $v = R_\infty i$ appears before the temperature has a chance to change. Then as the temperature increases by equation 2:14, v decreases as for an inductive circuit and eventually turns negative to give the negative steady state resistance characteristic.

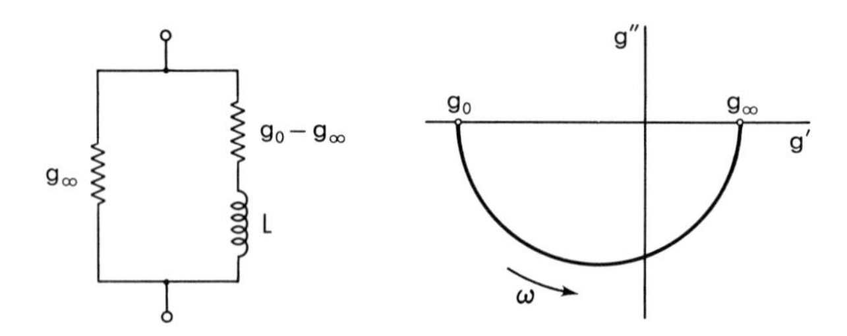

FIG. 2:49. Equivalent circuit, left, and complex conductance plane locus, right, for a thermistor polarized into the negative conductance region.

This performance is then that of the circuit of figure 2:49. Small variations of potential, current, and temperatures, v, i, and θ, near T give

$$(V + v)(I + i) = a\frac{d\theta}{dt} + b[(T - T_0) + \theta]$$

and so, since $VI = b(T - T_0)$

$$vI + Vi = (ap + b\theta)$$

for a sinusoidal input and $p = j\omega$.

Also, $$V + v = R(1 + k\theta)/(I + i),$$
and with $$V = RI$$
we have $$v = Rk\theta I + Ri.$$

Elimination of θ and V gives the conductance

$$g = \frac{i}{v} = \frac{1 - kRI^2/(b + ap)}{1 + kRI^2/(b + ap)} \cdot \frac{1}{R}.$$

For $p \to \infty$,

$$g \to g_\infty = 1/R$$

which is the normal isothermal resistance because there is no time for θ to change, shown in figure 2:48. For $p \to 0$,

$$g \to g_0 = \frac{1 - kRI^2/b}{1 + kRI^2/b} \cdot \frac{1}{R}$$

which is the variational conductance along the steady state characteristic also given in figure 2:48. Then

$$g - g_\infty = -\frac{2kI^2}{b + kRI^2} \cdot \frac{1}{1 + p\tau}$$

with $\tau = a/(b + kRI^2)$, to give the characteristic resistance-reactance network response. This response time, $\tau = L(g_\infty - g_0)$, gives $L = a/kI^2$, for the equivalent inductance, to complete the circuit and complex plane representations of figure 2:49. Thus the equivalent circuit consists of an inductance and two resistors. This was checked by bridge measurements on an ancient carbon filament lamp. But it was noted that the sign of the reactance is opposite to that of the temperature coefficient. This, too, was confirmed by measuring the impedance locus of a small tungsten filament lamp—such as was then being used as a high-pass filter and amplitude control in some electronic equipment.

At the same time I was very much impressed by the negative resistance characteristic of the thermistor and by the oscillator made of only a battery, a condenser, and a thermistor, operating both as a negative resistance and an inductance (Becker et al., 1947).

Formal Theory

When I was able to work at Princeton in the Fine Hall library I found that such anomalous reactances had long been well known in many fields. Probably the earliest was recognized in the electric arc (Busch, 1913), and, then relatively recently, the Dutch school had thoroughly explored the gas discharge phenomena (Seeliger, 1934). Over a wide range of pressure and current they had located several regions with either inductive or capacitive reactances. There were also several theories, each involving some time delay process. Then there was the reactance of vacuum tubes arising from the electron time of flight to limit seriously their use at very high frequencies. The same phenomenon appeared at near audio frequencies in the old gas detector tube. Strong electrolytes gave similar effects. Warburg's (1899) theory of diffusion polarization at a reversible electrode gave a capacitive reactance. The nonlinearity of the first Wien effect involved the time for formation of the ion cloud which was the basis of the Debye-Falkenhagen (1928) dispersion equivalent to a capacitive reactance.

It finally seemed quite obvious that all of the known processes responsible for nonlinear behavior required a definite time for operation. And since this time resulted in an anomalous reactance, such a reactance was to be expected in all nonlinear phenomena. Although this statement seems far too general to be true, I do not know of an exception. This could then be formalized (Cole, 1947; 1949) in a linearized domain

$$\frac{dv}{dt} = r_\infty \frac{di}{dt} + \frac{1}{\tau}(v - r_0 i) \tag{2:15}$$

in which r_∞, r_0 are the infinite and zero frequency resistances and τ is the time constant of the nonlinear process. This obviously gives a bilinear steady state impedance and an exponential transient to a step. In general the reactance is inductive or capacitive, figure 2:49 according as $r_0 - r_\infty$ is >0 or <0. This was certainly interesting and useful to the extent that it showed that a nonlinear process—such as had been well demonstrated for the squid axon—could be responsible for either the extra inductive or capacitive reactances that had been found.

Yet it has not been easy for some to realize that such an anomalous reactance is an entirely *linear* phenomenon and was discovered by *linear* measurements, even though it *arises* from a *nonlinear* process. Some have chosen to emphasize the origin of the reactance in a dissipative system (Mauro, 1961). A black box containing a thermistor must also conceal a battery to keep the thermistor warm if it is to show a reactance. Other systems and processes, such as the ion cloud relaxation and the diffusion polarization, may give a distinctly different reactance for small departures from equilibrium. These arise from a change in structure in a strictly linear domain.

FitzHugh developed this approach in a more general, but yet a more explicit, unpublished form. Let the potential of the system V be described by the current I, and a variable of state X,

$$V = f(I, X)$$

and this variable be given by

$$\dot{X} = g(I, X).$$

For small variations, v, i, x about a steady operating point

$$v = f_I i + f_X x; \quad dx/dt = g_I i + g_X x$$

where $f_I = \partial f/\partial I$, etc. Differentiation and elimination of x and dx/dt lead to

$$\frac{dv}{dt} = f_I \frac{di}{dt} + g_X \left(v - \frac{f_I g_X - f_X g_I}{g_X} i \right).$$

This shows the origin of r_0, r_∞, and τ.

All that was now needed was the membrane process which was described by the parameters r_∞, r_0, and τ as evaluated experimentally. But the danger of a serious logical mistake was also realized. Because of the difficulty of fitting a conventional electromagnetic inductance into a thin membrane, a serious and successful search was made for another type of mechanism to produce the same effect. Why then should we be so complacent with the membrane capacity of 1 $\mu F/cm^2$ just because it seemed to be explained by an orthodox electrostatic polarization? As this question was first recognized what is still the best line of reasoning was applied. If the capacity is evidence of a change of resistance with current flow, why should it remain so constant with large increases of resistance under an anode and with large decreases, both under a cathode and with deterioration to inexcitability and approaching death? A yet better question was to appear later: why should such a capacity appear with *no* change of resistance!

However, none of the nonlinear and reactive models, such as those mentioned, seemed to have any place in a living cell membrane on the one hand, and the passage of potassium ions through the squid axon membrane was an extremely attractive—even probable—

explanation for its nonlinearity on the other hand. On this basis the r_0 and r_∞ of equation 2:15 and in figure 2:38 were at least approximately known, but the details of the process would have to be explored in order to have any basis for an estimate of τ. In some way, that I cannot now remember or reconstruct, it became obvious that the nonlinearity not only had to produce a reactance but also that the potassium ion might be expected to give the membrane an inductive reactance. This entirely qualitative explanation was merely that as the membrane potential was suddenly changed, from say the equilibrium potential E_K to a more highly conducting potential, the ion concentration at every point in the membrane could be expected to increase as in figure 2:38; that this increase of the ion content of the membrane could only come across the boundaries of the membrane; and that it would thus take time to establish a new steady state. All that remained was to evaluate τ; to do this it was necessary to go to the theory of the motion of ions in an electric field and by diffusion in concentration gradients.

ELECTRODIFFUSION MODELS

It had come to be highly probable that the membrane capacity of 1 $\mu F/cm^2$ represented a passive, not to say inert, component of cell membranes while such an unbelievable thing as an inductive reactance could be expected as a part of a nonlinear system. The axon membrane did indeed have a nonlinear conductance which was attributed to a potassium ion flow, and this could easily result in an inductive reactance.

For a large inward field the steady state ion concentration in the membrane would be low, corresponding to the outside concentration K_e, while for a large outward field the membrane concentration would increase everywhere to correspond to K_i as indicated in figure 2:38 (Cole, 1941). Then the transition from a low to a high conductance would be determined by the time required to bring in the additional ions and distribute them across the membrane, as shown in figure 2:50. This process was obviously of the kind formally described by equation 2:15 and an inductive equivalent circuit. But a quantitative description of such a model was not available, and it was necessary to turn to the theory of the motion of ions in concentration gradients and electric fields as it had been developed for strong electrolytes.

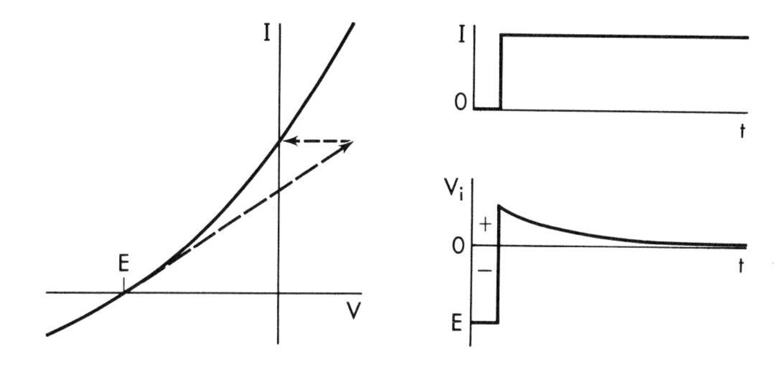

FIG. 2:50. A potassium theory of squid axon membrane steady state rectification and inductive transient behavior. As the inside potential V is made less negative than the equilibrium potential E the increased ion content of the membrane gives the rising steady state current-potential characteristic, solid line left. At the start of a sudden outward current, top right, the membrane potential changes along the initial dashed line, left, given by the equilibrium ion concentration distribution. As ions enter at the inside of the membrane faster than they leave at the outside the overall resistance decreases and the potential falls along the second dashed line, left, to give an inductive response as shown, lower right.

Classical Theory

The fundamentals of the theory of strong electrolytes had been firmly established by Nernst (1888, 1889) and heavily relied upon by Bernstein (1902) as a basis for his membrane theory. In addition to the equilibrium potential, such as for potassium ion, equation 1:23

$$E_{\mathrm{K}} = \frac{kT}{e} \ln \frac{\mathrm{K}_e}{\mathrm{K}_i}$$

Nernst had also contributed the diffusion coefficient of an electrolyte and the diffusion potential between two concentrations of an electrolyte. Equation 1:25 is extended for a symmetrical, univalent electrolyte with concentrations n and m of ions with charges $+e$ and $-e$ and mobilities u and v, the current flow I, in concentration gradients and an electric field X,

$$I = uekT \frac{dn}{dx} - une^2X$$

$$- vekT \frac{dm}{dx} + vme^2X.$$

For approximate electroneutrality $n \approx m$ and for a zero current flow

$$X = \frac{u - v}{u + v} \cdot \frac{kT}{e} \cdot \frac{1}{n} \cdot \frac{dn}{dx}$$

and the diffusion potential is

$$V = \frac{u - v}{u + v} \cdot \frac{kT}{e} \ln \frac{n_e}{n_i}$$

between the concentrations n_e and n_i.

Planck (1890a) examined the general potential theory problem, as stated by Nernst, with power and elegance, and proved the existence of a unique potential within an electrolyte bounded by current sources and sinks in a nonconducting surface. He then restricted himself first to the general analysis of the junction between two parallel planes over which the ion concentrations and potentials were fixed, and later (1890b) to the detailed classical calculation for the potential of such a boundary with no net current flow (MacInnes, 1939).

The Nernst and Planck results have been most extensively used, modified, improved, and approximated for the interpretation of biological and artificial membrane problems and as a guide to the design of experiments. So much work has been done on these electrodiffusion systems, and for so long, as to produce a bibliography almost incomparably larger than that for any other membrane model.

Although there seems to be no conclusive evidence that either an electrodiffusion process is or is not important for biological membranes, the system deserves attention as the most highly developed, the best known, and the most used model available.

Much of the work has been done on steady state resting potentials without net current flow, considerably less on the steady state relations between currents and potentials, with very little on the kinetics of change from one steady state to another.

As we follow Planck and repeat the procedure in deriving the Laplace equation 1:5, we have the flow of ion i across the element $\Delta y \Delta z$ at x given by

$$\Delta i(x) = \left[-D \frac{\partial n}{\partial x} - neu \frac{\partial V}{\partial x} \right] \Delta y \Delta z.$$

D, the diffusion coefficient, and u, the mechanical mobility, were related by Einstein (1905) as $D = ukT$. Then

$$\Delta i(x) = -u[kT\partial n/\partial x + ne\partial V/\partial x]\,\Delta y\Delta z$$

and at $x + \Delta x$,

$$\Delta i(x + \Delta x) = \Delta i(x) + \frac{\partial}{\partial x}\Delta i(x)\Delta x$$
$$= \Delta i(x) - \frac{\partial}{\partial x}\left[ukT \cdot \frac{\partial n}{\partial x} + neu\frac{\partial V}{\partial x}\right]\Delta x\Delta y\Delta z.$$

Similarly, for the y and z directions, to give the net accumulation in $\Delta x\Delta y\Delta z$ for each ion,

$$\frac{\partial n_i}{\partial t} = -\text{div}\,[u_ikT\,\text{grad}\,n_i + n_ie_iu_i\,\text{grad}\,V] \qquad (2{:}16)$$

which is the continuity equation for electrodiffusion.

With a free charge q we modify the Laplace equation to satisfy the Gauss law

$$\epsilon\int_s \Delta i \cdot d\sigma = 4\pi q.$$

As before, the integral over an element of volume $\Delta x\Delta y\Delta z$ is given by

$$\int_s X\,ds = \left(\frac{\partial X}{\partial x} + \frac{\partial Y}{\partial y} + \frac{\partial Z}{\partial z}\right)\Delta x\Delta y\Delta z$$

so

$$\text{div}\,\Delta i = -\text{div grad}\,V = \frac{4\pi}{\epsilon}\cdot\frac{q}{\Delta x\Delta y\Delta z}$$

to give the Poisson equation 1:4, $\nabla^2 V = -4\pi q/\epsilon$, and here

$$\nabla^2 V = -\frac{4\pi}{\epsilon}\Sigma n_ie_i. \qquad (2{:}17)$$

We now have a complete statement of the laws upon which the problems of classical electrodiffusion depend. There is no general solution or even a general approach to the solution of this system of equations. It is usually found that the general answers require a knowledge of each of the ion concentrations at every point. Every one of the useful results of these equations—and there are many—has come as the result of good approximations. And these approximations have come from good physical intuition and experience as often as from shrewd analysis.

Liquid junction

Restricting himself to a single dimension x in which n and V vary, Planck showed that the field at each point will change with a time constant $\tau_S = \epsilon/4\pi g$ after a change of current where g is the conductivity. This "charging process" is the dielectric relaxation of Maxwell (1873) with $\tau_S = RC$ for the capacity C and resistance R of an element, and as Planck pointed out, it is very fast for 1.0N HCl with $\tau_S = 2 \cdot 10^{-11}$ sec. This charging process is completed exponentially according to

$$i = gX + c\, dX/dt$$

and arrives at the quasi steady state, $r_\infty = 1/g$ as shown in figure 2:51, or as shown in figure 2:52 by the equivalent sinusoidal impedance locus.

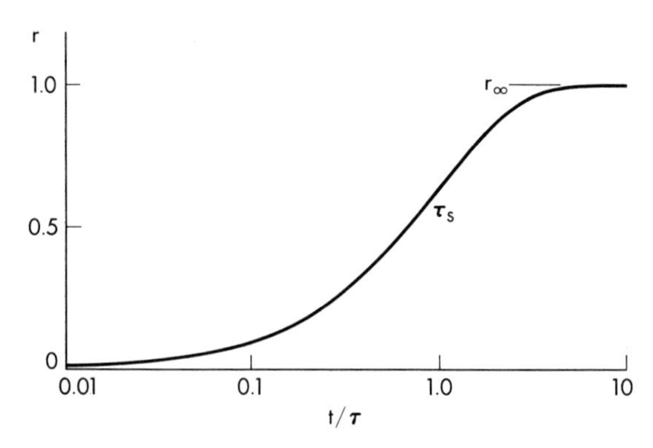

Fig. 2:51. The simple form of the Maxwell-Planck charging process after the application of a constant current to a uniform electrolyte. The effective resistance $r = v(t)/i$ is shown as a function of t/τ, on logarithmic scale, as it starts from zero to approach the steady state value r_∞ at long times.

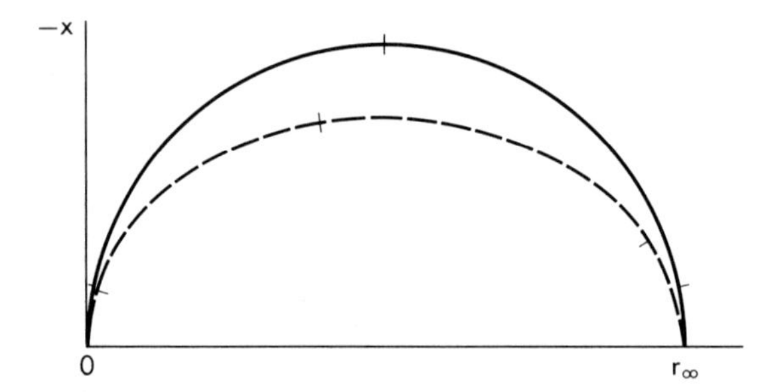

Fig. 2:52. The complex impedance locus for a uniform electrolyte, solid arc, and for a linear concentration profile across a membrane, dashed locus. The central frequencies are for $\omega\tau = 1$ and on the two sides the points for $\omega\tau = 10$ and 0.1 are shown.

Such a statement of the charging process is only true, however, in a region of uniform conductance, whereas most of our electrodiffusion problems are interesting because the conductance varies across the membrane. Although the problem can be stated in transient terms it is, as usual, much simpler to consider the complex impedance formulation

$$z(\omega) = \int_0^\delta \frac{dx}{g(x) + j\omega c}.$$

Apparently this can only be evaluated in the unusual cases where the ion distributions are known. As an illustration, the impedance locus for the simple case of a linear twenty to one variation of g across a boundary is given in figure 2:52.

After the charging transient is over, an additional field

$$\Delta X = \Delta I / g$$

will have appeared and this will remain as a quasi steady state indicated by r_∞ in figures 2:51 and 2:52 as long as the ions have not moved enough to cause an appreciable change of potential across the membrane. As the ions begin to move to carry the new current, the concentrations and the field interact and change toward their new steady state values in what Planck terms the slower principal nonlinear process. This is given in one dimension from equations 2:16 and 2:17 as

$$\begin{aligned} \frac{\partial n_i}{\partial t} &= u_i \left[kT \frac{\partial^2 n_i}{\partial x^2} - e \frac{\partial}{\partial x} (n_i X) \right] \\ \frac{\partial m_i}{\partial t} &= v_i \left[kT \frac{\partial^2 m_i}{\partial x^2} + e \frac{\partial}{\partial x} (m_i X) \right] \\ \frac{\partial X}{\partial x} &= \frac{4\pi e}{\epsilon} \Sigma (n_i - m_i) \end{aligned} \tag{2:18}$$

with concentrations, mobilities, and charges of n_i and m_i, u_i and v_i, and e and $-e$ respectively for the positive and the negative ions. In general, the process is quite complicated and neither Planck nor anyone else, until long after, gave an expression for its speed. We will, however, approximate it later in some special cases. For the present we may represent the redistribution by the formal equation 2:15 as another exponential process with the time constant τ_D reaching one or the other of the two steady states of figure 2:53 according as r_0 is larger or smaller than r_∞. Similarly, these two alternate processes are presented as loci D' and D'' on the complex impedance plane (figure 2:54) and as equivalent circuits (figure 2:55).

Then with approximate electroneutrality, $\Sigma n_i = \Sigma m_i$, Planck

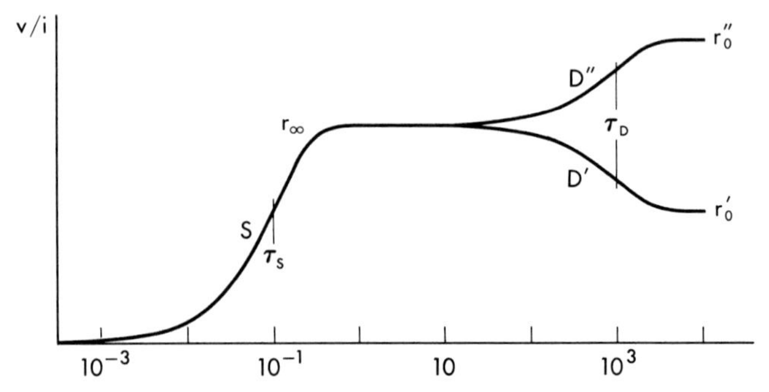

FIG. 2:53. Curve S is the charging process with a time constant $\tau_S = 0.1$ reaching a quasi steady state r_∞. The subsequent curves D', D'' are simplified forms of a redistribution process with a time constant $\tau_D = 10^3$. One or the other of the nonlinear steady states r_0' or r_0'' is reached depending upon whether the ion content of the membrane is increased or decreased.

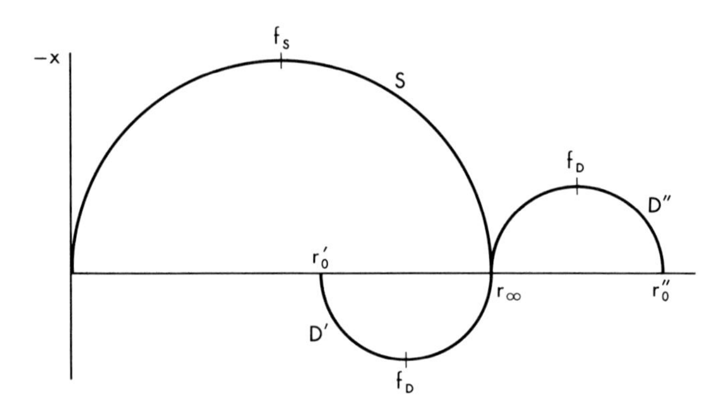

FIG. 2:54. Complex impedance locus for charging process S with characteristic frequency f_S, and redistribution processes with frequency f_D. The latter locus D' or D'' is inductive or capacitive according as $r_0 = r_0' < r_\infty$ or $r_0 = r_0'' > r_\infty$.

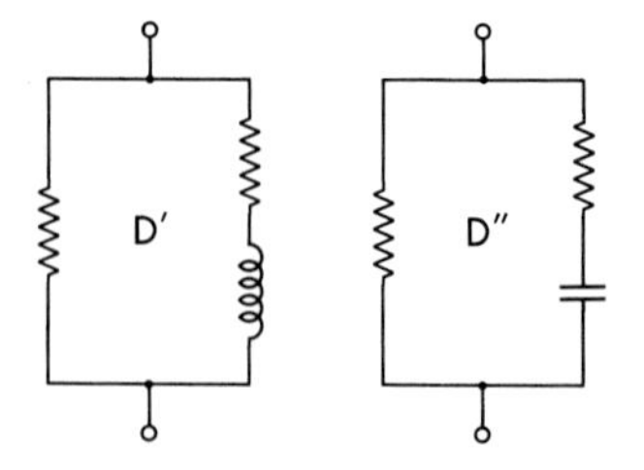

FIG. 2:55. Approximate equivalent circuits for the redistribution processes D' and D'' with the characteristics shown in figures 2:53 and 2:54.

generalized equations 1:26 and 1:27. He eliminated time by summing the equations 2:18 to give

$$\frac{\partial}{\partial x}\left[kT\frac{\partial}{\partial x}(\mathbf{U}-\mathbf{V})+eX(\mathbf{U}+\mathbf{V})\right]=0$$

where $\mathbf{U} = \Sigma u_i n_i$ and $\mathbf{V} = \Sigma v_i m_i$. Integration gave

$$X = -\frac{I}{e^2(\mathbf{U}+\mathbf{V})} - \frac{kT}{e}\cdot\frac{1}{\mathbf{U}+\mathbf{V}}\cdot\frac{\partial}{\partial x}(\mathbf{U}-\mathbf{V})$$

for a current density I and the potential across a region $0 \leq x \leq \delta$,

$$V = -\int_0^\delta X\,dx = \frac{I}{e^2}\int_0^\delta \frac{dx}{\mathbf{U}+\mathbf{V}} + \frac{kT}{e}\int_0^\delta \frac{d(\mathbf{U}-\mathbf{V})}{\mathbf{U}+\mathbf{V}}. \quad (2{:}19)$$

Planck identified the second integral as an instantaneous diffusion potential E_∞ to be found as the current is abruptly stopped, and the first integral, implicitly, as the corresponding instantaneous resistance R_∞ of the region. Here "instantaneous" is used to describe the time interval between the completion of the charging process and an appreciable beginning of the electrodiffusion. Each point on a V, I diagram, figure 2:56 is then located and described by different values of these two parameters which are also the elements of the instantaneous Thévenin circuit.

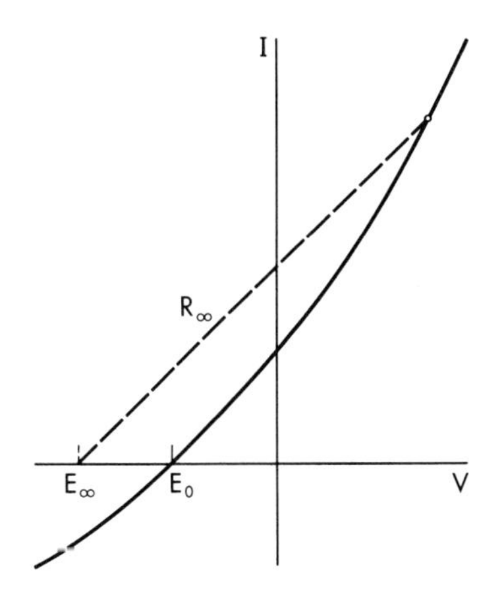

FIG. 2:56. Schematic potential vs. current diagram for a general electrodiffusion membrane. The solid line is the steady state characteristic. The dashed line shows the instantaneous resistance and emf as given by Planck.

The equations have been manipulated in many ways but $n_i(x)$ and $m_i(x)$ cannot be determined, except for approximate electroneutrality, even in a steady state. All but a few of the solutions for special cases are quite cumbersome.

Following Planck (1890b), in a steady state the individual currents I_i^+ for positive ions of concentration n_i and mobility u_i are

$$kTe\frac{dn_i}{dx} - n_i e^2 X = I_i^+/u_i$$

and similarly for negative ions,

$$-kTe\frac{dm_i}{dx} - m_i e^2 X = I_i^-/v_i.$$

With approximate electric neutrality

$$\Sigma n_i = \Sigma m_i = c$$

and

$$kTe\, dc/dx + ce^2X = \Sigma I_i^+/u_i \equiv I^+/u$$
$$-kTe\, dc/dx + ce^2X = \Sigma I_i^-/u_i \equiv I^-/v.$$

By subtraction,

$$2kTe\, dc/dx = I^+/v - I^-/u$$

a constant, so

$$c = Ax + B$$

and the total concentration varies linearly from one boundary $c(0)$ to the other $c(\delta)$. By addition,

$$-2ce^2X = I^+/u + I^-/v$$

a constant, so

$$X = C/(Ax + B).$$

The field is thus constant for $c(0) = c(\delta)$ and $V = -X\delta$. Otherwise the field is hyperbolic and $V = D \log c(\delta)/c(0)$. The constants A, B, C, and D are to be determined and may be complicated—as are the resulting expressions for $n(x)$ and $m(x)$.

Single ion currents

In a prophetic but long-ignored paper, Behn (1897) considered the net steady state flux of an ion, following Planck's treatment of the general constrained liquid junction. With the advent of tracer isotopes the problem attained a new importance and Behn's result was rediscovered by Ussing (1949) and it has been derived and discussed by Teorell (1951, 1953). Starting with the single ion equation 1:25,

$$I = une\left[\frac{kT}{n}\frac{dn}{dx} + e\frac{dV}{dx}\right]$$

direct substitutions of Planck's expressions for $n(x)$ and $dV(x)/dx$ led to a positive ion current,

$$I_n^+ = u \cdot \frac{ekT}{\delta} [\xi n(\delta) - n(0)]B^+$$

and negative ion current,

$$I_m^- = -v \cdot \frac{ekT}{\delta} [m(\delta) - \xi m(0)]B^-$$

where the Behn coefficients

$$B^+ = \frac{N(\delta) - N(0)}{\ln [N(\delta)/N(0)]} \cdot \frac{\ln [\xi N(\delta)/N(0)]}{\xi N(\delta) - N(0)}$$

$$B^- = \frac{M(\delta) - M(0)}{\ln [M(\delta)/M(0)]} \cdot \frac{\ln [M(\delta)/\xi M(0)]}{M(\delta) - \xi M(0)}$$

with $\xi = \exp eV/kT$, $N = \Sigma n$, and $M = \Sigma m$. The currents for all positive ions are then in the proportions

$$I_1^+ : I_2^+ : \ldots = u_1[\xi n_1(\delta) - n_1(0)] : u_2[\xi n_1(\delta) - n_2(0)] : \ldots$$

as presented by Behn.

Diffusion polarization

The principal contributions to the kinetics of electrodiffusion processes were begun by Warburg (1899) for a reversible electrode. Here the entire current was carried by one ion at the electrode surface. If $n \approx m \approx c$, this current for a negative ion

$$I = -2v\, ekT\, dc/dx$$

and with a negligible field X in the body of the electrolyte

$$\frac{\partial c}{\partial t} = D \frac{\partial^2 c}{\partial x^2}$$

where the diffusion coefficient $D = (u + v)kT$. For an alternating current and small changes of concentration, γ, from an initial c_0

$$c = \gamma \exp pt + c_0$$

$$\frac{d^2\gamma}{dx^2} - p\gamma/D = 0$$

$$\gamma = A \exp(-x\sqrt{p/D}) + B \exp(x\sqrt{p/D})$$

as for a Kelvin cable without leakage from equation 1:35.

This becomes

$$\gamma = \bar{\gamma} \exp(-x\sqrt{p\tau})$$

for $\gamma \to 0$ as $x \to \infty$, $\gamma = \bar{\gamma}$ at the electrode $x = 0$, $\tau = 1/D$, and so

$$I = 2v\,ekT\bar{\gamma}\sqrt{p\tau}$$

while the change of electrode potential

$$\Delta E = kT\bar{\gamma}/ec_0.$$

The impedance at the electrode surface

$$Z = \Delta E/I = 1/2ve^2c_0\sqrt{p\tau}$$

is analogous to that of the simple cable. The corresponding transient potentials after a unit step of current are

$$\Delta E(t) = \sqrt{\pi v e^2 c_0} \cdot \sqrt{t/\tau}$$

by the Fourier transform, Campbell and Foster (1931) No. 520.

These results were the basis of Nernst's (1908) theory of nerve excitation. Hill (1910) promptly improved the theory with two reversible electrodes—semipermeable membranes—a distance δ apart. Labes and Lullies (1932) computed the impedance of this "Hill layer" and used it in their extensive calculations of the impedance of nerve. Lullies (1937) showed their results on the complex plane.

Such a membrane will be symmetrical about the midpoint, $x = 0$ and $B = -A$. With an electrode at $\delta/2$ this is analogous to the simple cable short-circuited at the corresponding length, equation 1:37, and we find for both electrodes, with $\tau = \delta^2/4D$,

$$Z = \frac{\delta}{2ve^2c_0} \cdot \frac{\tanh\sqrt{p\tau}}{\sqrt{p\tau}}.$$

The frequency-dependent part of this impedance was tabulated by Kennelly (1921) and goes from 1 to 0 as $p = j\omega$ goes from 0 to ∞ as shown in figure 2:57. Then $R_0 - R_\infty = \delta/2ve^2c_0$ with $R_\infty = \delta/(u + v)e^2c_0$ for the bulk of the electrolyte between the electrodes. As happens so often, the transient solution for $E(t)$ after a step of current is not available in closed form. Series solutions are given in Hill (1910), Carslaw and Jaeger (1947), and Walker et al. (1968). An interesting feature of this dispersion is that, like the Debye-Falkenhagen dispersion for strong electrolytes, it appears in a system which is linear as it is perturbed from equilibrium.

Further Theory

Soon after my first impedance work on *Arbacia* eggs I had the singular good fortune to spend the year 1928–29 with Professor

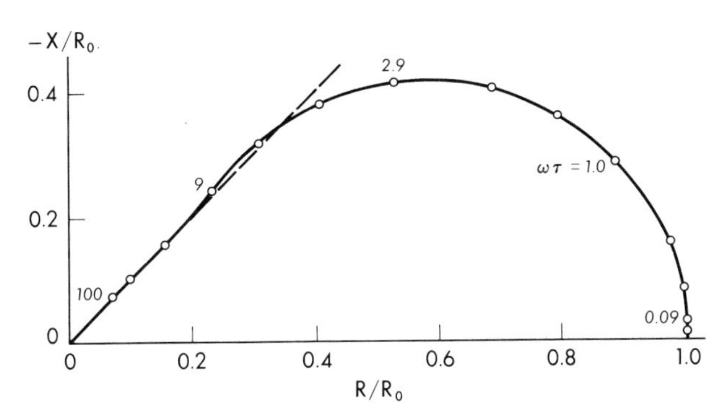

FIG. 2:57. The calculated complex impedance locus for a finite electrolyte between two reversible electrodes. Diffusion polarization near the electrodes dominates at high frequencies and is replaced by a constant gradient at low frequencies approaching steady state.

Peter Debye at the Physikalisches Institut in Leipzig. The Debye-Hückel electrolyte theory had achieved full stature, Debye had the dipole theory of dielectrics translated and I read "Polar Molecules" in proof, the Debye-Falkenhagen high-frequency dispersion in electrolytes had just appeared, the x-ray scattering of gases was in process, and the Falkenhagen-Dole theory of viscosity was done while I was there.

At that time I was vaguely hopeful that the membrane phase angle was some part of membrane function. Although the analogy to dielectrics became interesting and impressive, there was also the similar characteristic of polarizable electrodes in which ion movements were a necessary part. But Debye guided me firmly into problems of strong electrolytes that might represent membrane situations—without regard to the problems of dielectrics and electrodes that are still mostly unexplained.

I rapidly became acquainted with Kohlrausch's theory of conductance and Nernst's theory of electrodiffusion in strong electrolytes, with Planck's statement of it and its application to liquid junctions, and with Einstein's Brownian motion theory, the tools of strong electrolyte theory with which Debye worked. Many theoretical membrane models were set up and numerous attempts were made to obtain solutions. Those that were successful were mostly trivial and known but not obviously useful. More of the problems remained as interesting challenges in spite of my efforts to follow

Debye's emphatic advice—"approximate and approximate until you can get a solution, and then go back to find how bad your approximations are and correct for them." It is hardly surprising that I came to feel I would find out more about cell membranes by experiment than by analysis.

Single ion model

After a decade, many additional and undigested experimental results were available and badly in need of explanation. Electrodiffusion was still the best known approach, and it had become increasingly attractive and promising. But neither Planck's work nor the various efforts I had made in Leipzig seemed to apply or to give any clue as to the time constant for a transition from one steady state to another.

Mott (1939) had just given his theory of semiconductor rectification, and Debye suggested that I look at an analogous model for ion conduction across a membrane. For a single positive ion, equation 2:16,

$$\frac{\partial n}{\partial t} = u \frac{\partial}{\partial x}\left[kT \frac{\partial n}{\partial x} - neX\right]$$

and in steady state,

$$I = kT\, eu\, dn/dx - ne^2uX.$$

Then for so low a concentration that X is approximately constant and $n = n(0)$ at $x = 0$, $n = n(\delta)$ at $x = \delta$,

$$I = e^2uX \frac{n(\delta) - n(0) \exp \delta/\lambda}{1 - \exp \delta/\lambda}$$

where $\lambda = kT/eX$. From this

$$I = 0 \text{ at } E = -X\delta = (kT/e) \ln [n(0)/n(\delta)]$$
$$I \rightarrow n(\delta)e^2uV/\delta \text{ for } V \gg E$$
$$I \rightarrow n(0)e^2uV/\delta \text{ for } V \ll E$$

and

$$n = \frac{[n(0) - n(\delta)] \exp x/\lambda + n(\delta) - n(0) \exp \delta/\lambda}{1 - \exp \delta/\lambda}. \qquad (2{:}20)$$

These are the results as obtained and used by Mott.

We proceed further by noting that the variational resistance r_0 for slow small changes of I is

$$\frac{1}{r_0} = \frac{dI}{dV} = \frac{I}{V} + ue^2 \cdot \frac{eV}{kT} \cdot \frac{\exp \delta/\lambda}{(1 - \exp \delta/\lambda)^2} \cdot \frac{n(\delta) - n(0)}{\delta}, \qquad (2{:}21)$$

which corresponds to equation 2:8 as in figure 2:58. However for a change of current without change of the ion profile we have the instantaneous resistance, equation 1:26,

$$r_\infty = \frac{1}{ue^2} \int_0^\delta \frac{dx}{n(x)} \tag{2:22}$$

where $n(x)$ is given by equation 2:19, and

$$V = E + r_\infty I$$

with equation 1:23,

$$E = (kT/e) \ln [n(0)/n(\delta)].$$

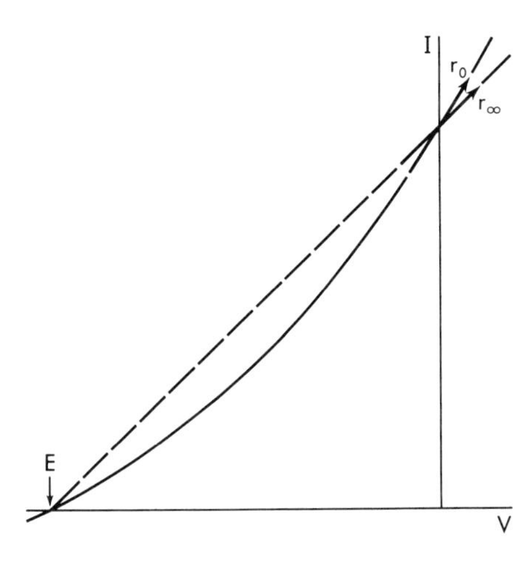

FIG. 2:58. Potential-current relation for a single ion membrane theory with the chord and slope resistances r_∞ and r_0 approached at infinite and zero frequencies.

With r_0 and r_∞ of figure 2:58 given by equations 2:21 and 2:22, we must concern ourselves with the transition from r_∞ to r_0. Returning to equation 2:16 and by a change of variables,

$$\frac{\partial n}{\partial \theta} = \frac{\partial^2 n}{\partial s^2} - \frac{\partial n}{\partial s}$$

where

$$s = x/\lambda, \ \lambda = kT/eX$$
$$\theta = t/\tau, \ \ \tau = kT/ue^2X^2.$$

We now expect a transition time of the order of

$$\tau_D = \frac{kT}{ue^2X^2} = \frac{\delta^2}{ukT}\left(\frac{kT}{e}\right)^2 \cdot \frac{1}{V^2} = \frac{\delta^2}{4D}\left(\frac{2kT}{eV}\right)^2 \tag{2:23}$$

where $D = ukT$ is the diffusion coefficient. From dimensions it was to be expected that for a pure diffusion process $\tau = \delta^2/4D$; with a constant field this is only modified by $(2kT/eV)^2$, which is of the order of unity for usual biological processes.

Another and quite different check was made on this estimate of a redistribution time along the lines of the qualitative reasoning first used. In the neighborhood of $X = 0$ we may approximate from equation 2:20

$$n(x) = n(0) + [n(\delta) - n(0)]\left[1 - \frac{1}{2} \cdot \frac{Xe}{kT}(\delta - x)\right]x/\delta \quad (2:24)$$

and for $X = 0$,

$$n(x) = n(0) + [n(\delta) - n(0)]x/\delta.$$

So during the transient after a reduction of X to zero, $n(x)$ will increase by an amount

$$\Delta n(x) = \frac{1}{2}\frac{Xe}{kT}(\delta - x)x/\delta$$

and the ion content of the membrane will increase by

$$q = \int_0^\delta \Delta n(x)\, dx = \frac{1}{12} \cdot \frac{Xe}{kT}[n(\delta) - n(0)]\,\delta^2.$$

This initial deficit will have to be filled by a net inflow across both boundaries. This process is pure diffusion; the initial rate of inflow is

$$\frac{dq}{dt} = ukT\left[\left(\frac{d\Delta n}{dx}\right)_\delta - \left(\frac{d\Delta n}{dx}\right)_0\right] = ueX[n(\delta) - n(0)].$$

If the process is exponential in time, then at $t = 0$

$$dq/dt = q/\tau_D$$

and we find

$$\tau_D = \frac{1}{12} \cdot \frac{\delta^2}{ukT}.$$

Since this and the dimensional estimate were of the same order of magnitude there was reason to expect both to have some validity.

It was then possible to speculate. The rectification steady state characteristic of the squid membrane was most easily attributable to potassium, at least qualitatively. Then for a time constant

$$\tau_D = kT/X^2e^2u$$

a mean gradient of 50 mV per 100 Å and $\tau_D = 2$ msec corresponding to the maximum inductive reactance at 150 c, we find $ue = 2 \cdot 10^{-8}$ cm/sec per volt/cm. This is about 10^{-5} of the nominal value for potassium in water of about $8 \cdot 10^{-4}$ cm/sec per volt/cm. With an average specific resistance for a 100 Å membrane of 1 kΩcm²,

$$r = 10^3\ \Omega\text{cm}^2/10^{-6}\ \text{cm} = 10^9\ \Omega\text{cm}$$

and by $1/r = ne^2u$ the equivalent ion concentration comes to about 10^{-3} molar. But with a figure of 25 Ωcm² during excitation or steady state depolarization, we have this effective concentration approaching 0.1 molar, which was indeed a startling figure.

As a result I arrived at the conclusion that the high resistance of such a living membrane was largely to be ascribed to an extremely low mobility of an ion in the membrane structure, with the amazing corollary that the ion concentration might not be far below that of the usual biological electrolytes.

The basis in theory for these results was not particularly solid, there was no experimental check, and the work was not published for nearly a decade (Cole, 1947, 1949); but after more than twenty years, right or wrong, the results remained essentially unchanged as the only estimates for an ion mobility or an ion concentration in a living membrane.

The subsequent work on the simple electrodiffusion theory has only recently been reviewed and compared with various data on the squid axon membrane (Cole, 1965a). Partly because of this review and partly because the theoretical developments are inaccurate, incomplete, inconsistent, and, even worse, of rather uncertain utility, they will be presented here in even less detail than before.

The first estimates of a redistribution time for the single ion electrodiffusion model were very interesting. They gave the first and, until recently, the only kinetic model of ion movement through a living membrane that provided any explanation for the provocative new measurements made on the squid giant axon membrane. These calculations were, however, very crude and unsatisfactory. Furthermore, they did not provide a hopeful basis or even suggest a promising approach for the simple Planck problems of more than one mobile ion in a membrane. With the subsequent concentration on increasingly complicated steady state membrane problems with fixed charges (Teorell, 1951, 1953) and mixed valence types (Schlögl, 1954), even the simplest kinetic problems seemed to deserve more and better consideration than I had given them.

In the course of some numerical transient calculations, I realized that the constant field single ion model was the analogue of a classical heat flow problem in which heat flows without loss along a wire moving between a source and a sink. I had also been impressed by the crude approximation of the calculated ion profiles during current flow to a sinusoid half wave and a linear term across the membrane. Carslaw and Jaeger (1947) had given an explicit solution for the temperature distribution after a sudden change in the source temperature. Then Richard FitzHugh gave me an explicit solution, which corresponded to a sudden change of wire velocity in the heat problem.

Returning to the differential equation 2:16

$$\frac{\partial n}{\partial t} = ukT\frac{\partial^2 n}{\partial x^2} - ueX\frac{\partial n}{\partial x}$$

the transition from an initial steady state for X_0 to another value X, we have

$$n(x,t) - n(x,\infty) = \exp\left[\frac{x}{2\lambda} - \frac{t}{4\tau}\right] \cdot \sum_{p=1}^{\infty} b_p \exp\left[-\frac{p^2\pi^2 t}{\delta^2}\right] \sin\left[\frac{p\pi x}{\delta}\right]$$

and for this the difference between the initial and final steady states has been expressed as a Fourier sine series

$$n(x,0) - n(x,\infty) = \exp(x/2\lambda) \cdot \sum_{p=1}^{\infty} b_p \sin(p\pi x/\delta).$$

The redistribution transient is then described by the series of time constants,

$$\tau_p = \frac{\delta^2}{\pi^2 ukT} \cdot \frac{1}{p^2 + (eV/2\pi kT)^2} \tag{2:25}$$

into which both the electric and the diffusion forces enter. The first estimates had considered the two components separately but arrived at surprisingly good values for the coefficients. Here for the leading term $p = 1$, we find that the time constant is halved for a change of V from zero to 150 mV, while for higher terms the corresponding potential changes go much farther beyond the usual biological range. Thus we will not make serious errors if we ignore the electric terms and consider only the diffusion aspects of the ion redistribution kinetics. We do, however, need to have an estimate of the amplitudes b_p of the more important terms; here again it is simplest to relax a small field to zero. By equation 2:24 the change of ion distribution is parabolic across the membrane and this may be expanded,

$$(\delta - x)x = \frac{8\delta^2}{\pi^3}\left[\sin \pi x/\delta + \frac{\sin 3\pi x/\delta}{3^2} + \frac{\sin 5\pi x/\delta}{5^2} + \cdots\right].$$

In the symmetrical case the terms for p even are absent, while the third and fifth harmonics, with time constants $\tau_1/3$ and $\tau_1/5$, have the respective relative amplitudes of $\frac{1}{9}$ and $\frac{1}{25}$. Thus, as is apparent by inspection, a parabola and a half sine wave are good approximations for each other. A simple exponential transient or a semicircular impedance locus with the leading time constant,

$$\tau_D = \delta^2/\pi^2 ukT$$

is then a good approximation to the redistribution process. As the field is moved away from zero, the linear perturbation dn/dX becomes considerably more complicated and has not been investigated. The direct calculation of the impedance that was an earlier stumbling block also has not been carried through satisfactorily. But, as mentioned, constant field calculations for $n(x)$ could be approximated by $\sin \pi x/\delta$ and perhaps even better by $\exp (x/2\lambda) \sin (\pi x/\delta)$, so that there are as yet no impressive danger signals attached to the use of the single redistribution time.

The use and the approximation of constant field assumptions have been given widespread discussion from many points of view. After a sudden change of current, ΔI, and after the charging transient, these fields will everywhere be changed by an amount

$$\Delta X = \Delta I/e^2 \sum (n_i u_i + m_i v_i).$$

In general, this field can only be calculated with a knowledge of each n_i, m_i across the membrane. Also, in general, it will not be constant; certainly in some cases a constant field approximation may be grossly in error, especially during the early part of a redistribution transient.

Consequently the field is constant only in a limited class of steady state cases, and usually not at all in transient conditions. So when the constant field assumption is made the consequences should be examined.

The constant field assumption is particularly questionable for the single ion theory in which there is no basis to assume even an ap proximate electroneutrality. Here we may change equations 2:18 to dimensionless variables

$$N = n/\bar{n};\ S = \kappa x;\ \theta = t/ukT\kappa^2;$$

$$F = \frac{e}{\kappa kT} X;\ I = \frac{i}{u\kappa\bar{n}ekT};\ \kappa^2 = \frac{4\pi\bar{n}e^2}{\epsilon kT}$$

where $\bar{n}$ is a central concentration and $1/\kappa$ is the Debye ion cloud thickness. Then

$$\frac{\partial N}{\partial \theta} = \frac{\partial}{\partial S}\left[\frac{\partial N}{\partial S} - NF\right]$$

$$\frac{\partial F}{\partial S} = N. \qquad (2:26)$$

I have not found a useful way to deal with the complete transient, even in this form.

Several ion models

Returning to the problems of more than one mobile ion, we are able to draw upon the experience with the single ion to give at least a qualitative estimate of their kinetic behavior.

A very popular—and certainly interesting—membrane model is the relatively simple one of two salt solutions having a common negative ion and the same total concentration at both boundaries such as 0.5M KCl, 0.05M NaCl inside and 0.5M NaCl, 0.05M KCl outside. The problem is further simplified, and perhaps made more realistic, by the assumption of a uniform fixed concentration, $\bar{m}$, of immobile ions throughout the membrane. The positive ions have concentrations n_1 and n_2, and mobilities u_1 and u_2,

$$n_1(0) + n_2(0) = n_1(\delta) + n_2(\delta) = \bar{m}$$

and for approximate steady state electroneutrality,

$$n_1(x) + n_2(x) \simeq \bar{m}.$$

The steady state currents

$$I_1 = -u_1 ekT\frac{dn_1}{dx} + u_1 n_1 e^2 X$$

$$I_2 = -u_2 ekT\frac{dn_2}{dx} + u_2 n_2 e^2 X.$$

By addition and rearrangement with $\mathbf{U} = u_1 n_1 + u_2 n_2$ and $I = I_1 + I_2$,

$$X = \frac{kT}{e} \cdot \frac{1}{\mathbf{U}} \cdot \frac{d\mathbf{U}}{dx} - \frac{I}{e^2 u}.$$

Since X is constant and $n_1(x)$ and $n_2(x)$ are known in this case and similar to the single ion distribution of equation 2:20, this may be integrated to give explicit, if somewhat ponderous, results. For the transients after a small change of current we start again with the Nernst-Planck-Einstein equations 2:18 in the form,

$$\frac{\partial n_1}{\partial t} = u_1 \frac{\partial}{\partial x}\left[kT \frac{\partial n_1}{\partial x} - n_1 eX\right]$$

$$\frac{\partial n_2}{\partial t} = u_2 \frac{\partial}{\partial x}\left[kT \frac{\partial n_2}{\partial x} - n_2 eX\right]$$

$$\frac{\partial X}{\partial x} = \frac{4\pi e}{2}(n_1 + n_2 - \overline{m}).$$

By addition of the first two,

$$\frac{\partial}{\partial t}(n_1 + n_2) = kT \frac{\partial^2 \mathbf{U}}{\partial x^2} - eX \frac{\partial \mathbf{U}}{\partial x} - e\mathbf{U}\frac{\partial X}{\partial x}$$

and from substitution of the third with $q = n_1 + n_2 - \overline{m}$,

$$\frac{\partial q}{\partial t} = kT \frac{\partial^2 \mathbf{U}}{\partial x^2} - eX \frac{\partial \mathbf{U}}{\partial x} - \frac{4\pi e^2 \mathbf{U}}{\epsilon} \cdot q.$$

The last term describes a local exponential charging process, and with Planck we again assume that the time constant for it, $\tau_S = \epsilon/4\pi e^2\mathbf{U}$, is everywhere short as compared to that for the redistribution process. After this process is over we have

$$\frac{\partial}{\partial t}(n_1 + n_2 - \overline{m}) = kT \frac{\partial^2}{\partial x^2}(u_1 n_1 + u_2 n_2) - eX \frac{\partial}{\partial x}(u_1 n_1 + u_2 n_2)$$

and for small changes $n_1 = \bar{n}_1 + v_1$, $n_2 = \bar{n}_2 + v_2$

$$\frac{\partial}{\partial t}(v_1 + v_2) = kT \frac{\partial^2}{\partial x^2}(u_1 v_1 + u_2 v_2) - eX \frac{\partial}{\partial x}(u_1 v_1 + u_2 v_2).$$

This may be rewritten as

$$\frac{\partial q}{\partial t} = \frac{\partial^2}{\partial x^2}\bar{u}kTq - eX \frac{\partial}{\partial x}\bar{u}q \tag{2:27}$$

in which

$$\bar{u} = \frac{u_1 v_1 + u_2 v_2}{v_1 + v_2}$$

is a weighted mean mobility lying between u_1 and u_2 which can be expected to be a function of both position and time. This may be integrated over small intervals Δx and Δt for which $\bar{u}$ may be considered as constant, as indeed Planck (1890a) described the calculation of the redistribution process. On the other hand, we may consider that there will be a most frequent value of $\bar{u}$ in space and time which can be taken as constant for an approximate solution of equation 2:27. This is of the same form as equation 2:16 for the single ion and has a similar solution, equation 2:25. Again we may approximate, by ignoring the effect of the field in the physiological range and assuming $q(x)$ is sinusoidal to obtain the principal time constant of the redistribution

$$\tau_D = \delta^2/\pi^2\bar{u}kT.$$

The appearance of a mean mobility is analogous to, although less rigorous than for, the diffusion coefficient of an electrolyte as given by Nernst on the basis of approximate electroneutrality. In this case, however, we must be prepared to find, and perhaps expect, a spectrum for τ_D centered on the one calculated to produce aberrations of the single time constant transient and impedance phenomena.

With this result it first became possible (Cole, 1965a) to estimate a redistribution time to compare with the charging time given by Planck. Instead of his example of 1.0N HCl we may take different mixtures of HCl and KCl at the boundaries with a uniform total concentration of 2.0N and the same concentration of fixed negative ions within the membrane. At a point where $n_H = n_K = 1.0$N the resistance is

$$r_\infty = 1/[n_H u_H + n_K u_K]e^2 = 3\ \Omega\text{cm}$$

and the capacity,

$$c = \epsilon/4\pi = 7.1 \cdot 10^{-12}\ \text{F/cm}$$

to give

$$\tau_S = r_\infty c = 2.2 \cdot 10^{-11}\ \text{sec or } f_S = 1/2\pi\tau_S = 7.2 \cdot 10^9\ \text{c.}$$

For the redistribution we take an average value

$$\bar{u} = (u_H + u_K)/2 = 3.4 \cdot 10^{-3}\ \text{cm/sec per volt/cm}$$

and for a thickness $\delta = 100$ Å, we have

$$\tau_D = 2.4 \cdot 10^{-9}\ \text{sec or } f_D = 6.6 \cdot 10^7\ \text{c.}$$

This redistribution is also quite fast; although it will overlap the charging appreciably, Planck must certainly have thought of junctions so much thicker than 100 Å that his separation of the two dispersions was amply justified.

The general steady state case for any number of univalent mobile ions was developed by Planck and has been reproduced many times.

The pattern of development is much the same as for the single and two ion problems. The results are similar except that they are expressed in terms of the two variables, $\mathbf{U} = \Sigma u_i n_i$ for positive ions and $\mathbf{V} = \Sigma v_i m_i$ for negative ions, and that analytical solutions are not usually found for $n_i(x)$ and $m_i(x)$. Consequently the expressions from equation 2:19

$$r_\infty = \frac{1}{e^2}\int \frac{dx}{\mathbf{U}+\mathbf{V}}, \quad E_\infty = \frac{kT}{e}\int \frac{d(\mathbf{U}-\mathbf{V})}{\mathbf{U}+\mathbf{V}}, \quad r_0 = dV/dI$$

must be considered in this form.

The redistribution time, only mentioned qualitatively by Planck as being long compared to the charging process, has been developed in a manner analogous to that given above for the two mobile ions. With the same assumptions, approximations, and uncertainties, I produced (Cole, 1965a) the time constant

$$\tau_D = \delta^2/\pi^2\, \overline{u+v}\, kT$$

where $\overline{u+v}$ is a principal effective mobility

$$\overline{u+v} = \frac{\Sigma u\nu + \Sigma v\mu}{\Sigma \nu + \Sigma \mu}$$

and ν and μ are perturbations of n and m from their steady state values. The spectrum of $\overline{u+v}$ can be expected to lie between the extremes of the mobilities of the fastest and of the slowest mobilities

$$u_{\text{max}} + v_{\text{max}} \geq \overline{u+v} \geq u_{\text{min}} + v_{\text{min}}$$

and again to produce distortions of the simple consequences of a single time constant process, equation 2:15.

Here again it probably is not worth while to attempt to do much more with analysis. Cohen and Cooley (1965) have given computer solutions for a few problems of large transients. These show some complications that do not have obvious physical explanations and the application of these results to the interpretation of biological measurements is yet to be made. Yet Offner (1967 and in manuscript) found reason to expect both anomalous capacitive and inductive reactance in an electrodiffusion membrane.

Permeabilities and Conductances

The concept of permeability is obvious. At least qualitatively, it is a flow of a species across a membrane as a function of an applied driving force on each individual—the "go" for a given "push." The measure of a permeability can also be simple where the flow is proportional to the force—such as for a slow flow of water through a pipe and the flow of ions through a uniform electrolyte. Here the flow, I, may be proportional to the force, F, and constant in the direction of flow, $I = GF$. This permeability G may also remain constant, independent of F and independent of time, to represent a linear, steady state system. Such permeabilities have been measured for many membranes, materials, and forces and they have been expressed in a wide variety of units. Each of these permeabilities has an operational significance but any relationships between them may be difficult to find. We can restrict ourselves to simple electro-diffusion, where some unification is possible, with the full awareness that it may not apply to living or other membranes. But we may also hope that added factors and other processes, such as hydrostatic pressure, narrow channels, and potential barriers, may be expressed in similar terms.

Starting again with the Nernst-Planck-Einstein equation for the flow of a single species in a steady state,

$$I = u(kT\, dn/dx - neX) \tag{1:25}$$

consider first the case of an uncharged molecule or an ion in the absence of a field. Here

$$I \int_0^\delta dx/ukT = n(\delta) - n(0)$$

in general, or if u is constant

$$I = \frac{D}{\delta}\,[n(\delta) - n(0)].$$

This is Fick's law in which the concentration difference is usually considered as the driving force and the permeability, $G = D/\delta$, is in units of cm/sec. It is, however, more generally useful to express the force as the gradient of a potential—in this case the chemical potential, $d\mu = kT\, d\,(\ln n)$. Then

$$I \int_0^\delta dx/nu = \mu(\delta) - \mu(0)$$

which shows that the permeability is of the nature of a conductance un and for u again a constant,

$$I = \frac{u[n(\delta) - n(0)]}{\ln\,[n(\delta)/n(0)]} \cdot \frac{1}{\delta}\,[\mu(\delta) - \mu(0)]$$

the permeability is given explicitly. Further, the total flow of any number of uncharged species is given by the sum of such individual terms.

Next, in the absence of a concentration gradient, an individual ion current

$$I/ue^2 = dV/dx$$

and for u constant

$$I = \frac{ue^2}{\delta} V(\delta)$$

with $V(0) = 0$. The permeability for each ion is here its conductance and the total current is given by the sum of the currents and so by the sum of the individual conductances—as discovered by Kohlrausch.

Returning to the more general steady state ion flow, the current

$$I/une = \frac{d}{dx} [\mu + eV] = \frac{d\bar{\mu}}{dx}$$

and

$$I \int_0^\delta dx/une^2 = [\bar{\mu}(\delta) - \bar{\mu}(0)]/e$$

where $\bar{\mu}$ is the electrochemical potential. The permeability is now the electrical conductance

$$G = e^2 \Big/ \int_0^\delta dx/un$$

and then, for each and every ion,

$$I = G(V - E)$$

where $E = \dfrac{kT}{e} \ln \dfrac{n(0)}{n(\delta)}$ is the Nernst equilibrium potential at which the particular ion current flow is zero. Then, to the extent that u and n are known, this ion conductance can be calculated and the behavior for all ions is given by

$$I = \Sigma I_i - \Sigma G_i(V - E_i).$$

This may be put in the form

$$V = \frac{I}{\Sigma G_i} + \frac{\Sigma G_i E_i}{\Sigma G_i}$$

which is the equivalent of Planck's expression, equation 2:19. Then two conductances can be determined for two known values of E.

However, the measurement of all such conductances must be instantaneous. They must be completed before any changes of concentrations become appreciable or in a time of much less than the redistribution times.

Consequently we are led to an ion conductance as a useful expression of a membrane ion permeability, and any prediction of an ion mobility and concentration across a membrane permits a prediction of its conductance.

Goldman constant field theory

In cases where ion concentrations are so low that the Poisson equation, $dX/dx \sim 0$, as assumed by Mott, or where the total concentrations are equal at the two boundaries, as Planck showed, the field may be approximately uniform. Here the individual steady state ion currents were summed by Goldman (1943) to give

$$I = V \frac{\Lambda^+ \xi - \Lambda^-}{\xi - 1} \tag{2:28}$$

where

$$\Lambda^+ = \frac{e^2}{\delta} [\mathbf{U}(0) + \mathbf{V}(\delta)],\ \Lambda^- = \frac{e^2}{\delta} [\mathbf{U}(\delta) + \mathbf{V}(0)]$$

are the limiting conductances for $V \to \pm\infty$ and $\mathbf{U} = \Sigma un$, $\mathbf{V} = \Sigma vm$, and $\xi = \exp eV/kT$ are the Planck variables. The diffusion or resting potential, with $I = 0$, is given by

$$V_0 = \frac{kT}{e} \ln \frac{\Lambda^-}{\Lambda^+}.$$

As it has come to be written (Hodgkin and Katz, 1949),

$$V_0 = \frac{kT}{e} \ln \frac{\Sigma un(\delta) + \Sigma vm(0)}{\Sigma un(0) + \Sigma vm(\delta)} \tag{2:29}$$

in which u and v are referred to as permeabilities, another definition of permeability appears. This is also true for the equivalent expression in terms of transference numbers (Dray and Sollner, 1956).

The Goldman theory might have been extended beyond a constant field by including Behn's result. The current,

$$I = \frac{ekT}{\delta} [B^+ \xi \Sigma un(\delta) - B^+ \Sigma un(0)$$
$$- B^- \Sigma vm(\delta) + B^- \xi \Sigma vm(0)]$$

and the resting potential,

$$V = \frac{kT}{e} \ln \frac{B^{+}\Sigma un(0) + B^{-}\Sigma vm(\delta)}{B^{+}\Sigma un(\delta) + B^{-}\Sigma vm(0)}.$$

If we ignore the fact that B depends on potential to give the complication of Planck's transcendental equation, the permeabilities may be written $P^{+} = B^{+}u$ and $P^{-} = B^{-}v$ to give

$$V = \frac{kT}{e} \ln \frac{\Sigma P^{+}n(0) + \Sigma P^{-}m(\delta)}{\Sigma P^{+}n(\delta) + \Sigma P^{-}m(0)}.$$

The identity of this form with the original Goldman equation may help justify their use as at least a mode of expression for experimental results!

The Goldman theory is a foundation that has been much more widely exploited than either Goldman or I could have predicted when he completed it for his thesis. Nor can we even now believe that it is a solid foundation. The tremendous use of the theory seems to me evidence for the need of such a tool. At the same time it demonstrates that the more complete formulations cannot now be used with a comparable effectiveness.

Constant field approximation

The use of a constant field has been widespread and fruitful and a better understanding of its limitations may be helpful. For a single ion in steady state, equations 2:26 become

$$\frac{dN}{dS} = NF - I$$

$$\frac{dF}{dS} = N.$$

Here there is more basis to suspect an analytical solution for constant I but without it we turn first to $I = 0$. By substitution and integration

$$N = F^2/2 + A$$

to give the parabolas on the F vs. N plane displaced vertically according to the constant as shown in figure 2:59. Furthermore, N and F are easily obtained as one or another function of S depending on whether $A >$ or < 0, although it is a considerable nuisance to determine the constants. For $A = 0$, it is found that

$$\bar{S} = \kappa\delta = \sqrt{2}[1/\sqrt{N(0)} - 1/\sqrt{N(\delta)}]$$

to classify membranes roughly as thin or thick according as $A < 0$ or $A > 0$ or as $\delta < 1/\kappa$ or $\delta > 1/\kappa$. For the thin membrane $\kappa\delta = 0.1$, F is an excellent approximation to a constant and N is given directly by equation 2:25, and for $\kappa\delta = 1.0$ the variation of F from the constant field is not alarming. However, as $\kappa\delta$ increases further, F eventually reverses at a minimum of N. This situation could have been anticipated as one in which there is a Debye cloud extending in from each boundary of the membrane.

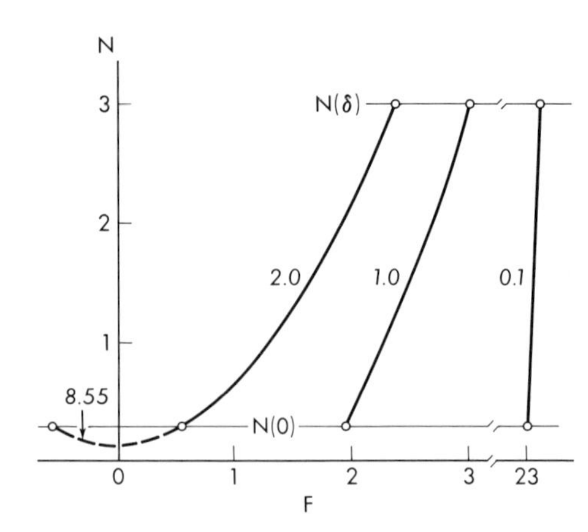

FIG. 2:59. Ion concentration N vs. field F across a single ion membrane at zero current and for four thicknesses. The field is essentially constant for the thin membrane 0.1, shows increasing variation at 1.0 and 2.0, follows the latter path but continues to actually reverse for 8.55.

Unless and until a solution is carried out for $I \neq 0$ we must rely upon point by point numerical solutions for each specific problem. A few solutions have been obtained graphically on the F vs. N plane where

$$\frac{\Delta N}{\Delta F} = F - I/N$$

as shown in figure 2:60a. Although it may be tedious to match the end points, the constant field has been found a good approximation for the thin membrane cases that were worked out. This is an interesting elementary example for analogue calculation (figure 2:60b). It is a trivial problem for almost any digital computer and it is easily scheduled for hand calculation.

A more complete analytical solution with space charge and current flow was obtained (Bass, 1964) in terms of Airy functions for different concentrations of a single symmetrical electrolyte at the boundaries of the junction. Even here the problem of equation 2:18 was

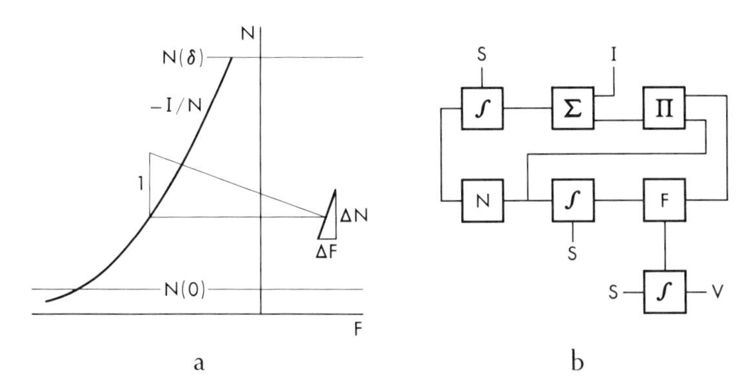

FIG. 2:60. (a.) A graphical construction for the ion concentration N, vs. field F, across a single ion membrane with current flow I. (b.) A block diagram for computation of behavior of a single ion membrane.

simplified by neglecting the energy density of the electric field, $\epsilon X^2/8\pi$, in comparison to the thermal, $2nkT$. The ratio becomes

$$\epsilon X^2/16\pi nkT = 10^{-7}V^2/M\delta = 10^{-3}/M$$

for a potential $V = 100$ mV across a membrane of thickness $\delta = 100$ Å. So the molar concentration M was assumed to be much larger than one millimole in this membrane. Even with an asymptotic approximation, this solution is not easy to use. Another explicit analytical solution for equations 2:18 appeared (Sinharay and Meltzer, 1964) also in terms of Airy functions. This does not indicate any possibility of the appearance of a negative resistance as predicted by an approximate analysis (Skinner, 1955). However, a digital program has been also developed as more practical!

There is consequently good reason to rely almost entirely on automatic computation for which this and some more general problems are not at all difficult. Bruner (1965a) developed the complete steady state theory extensively for this purpose and gave examples (1965b) of potential vs. current for a single ion membrane. There was unfortunately no discussion of the mechanisms primarily responsible for the characteristic features of these results.

We may expect constant field results to be reasonably good for some steady state situations but open to question in other steady states and in all transient situations. But when we can go ahead on this basis and have no simple alternative, we will proceed to do so!

Fixed Charge and Other Modifications

Since Goldman, there have been extensive developments of the electrodiffusion theory that are almost entirely without an anchor in experiments on artificial or living membranes.

Teorell (1935) and Meyer and Sievers (1936) independently began to consider a uniform concentration of immobile, or fixed, charges within the membrane. These problems were extended by Teorell (1951, 1953), taking into account the Donnan potentials at the boundaries and extending the Planck treatment to include the effects of current flow across the membrane. The solutions were restricted to local electroneutrality, symmetrical salts, and steady state, but a constant field was not assumed. Individual ion fluxes, currents, and conductances were calculated, at least in principle, and many special cases were worked out explicitly. However, the rectification ratios were limited much as before and the intercepts of the limiting tangents of V, I curves were again at the origin. The equilibrium ion distribution across the Donnan boundaries was derived explicitly by Bartlett and Kromhout (1952). Schlögl (1954) treated the general problem of a fixed charge membrane without the restriction to symmetrical salt components, and gave an extended treatment of the allowed and forbidden ranges of parameters that had plagued Teorell. In its generality, this theory permits the two high-field conductances to extrapolate back to zero current and intersect at a potential other than zero. I found this a great relief because such behavior corresponds to observations on the squid membrane, figure 2:36, that the restricted theories such as equation 2:28 could not produce.

Conti and Eisenman (1965b) continued the development for the steady state characteristics of a membrane with an arbitrary distribution of fixed charges and a single mobile ion of opposite charge. They investigated the nonsteady state problem (1965a) without net current flow across the membrane. Here they find that the membrane potential is so largely dominated by the ion distributions at the boundaries that it comes near to the steady state value while the internal concentrations are still far from their final values. This is particularly useful for such things as glass electrodes of various kinds. Conti and Eisenman (1966) revived the classical electrodiffusion theory as a model for a semipermeable membrane with mobile sites trapped within the membrane, such as a liquid ion exchanger. Eisenman and Conti (1965) reviewed some of this work and its possible application to biological membrane problems.

EXPERIMENTS

Further experiments, figures 1:47, 1:48, 1:49, with the squid axon (Cole and Marmont, 1942) had usually shown the added low-frequency capacity anomalies that Baker and I (1941b) had suspected. The change to an inductive reactance for increased external potassium might then correspond to a change of the resting potential from a value more negative than E_K to one more positive. The increased capacitive reactance for added external calcium then corresponded to a hyperpolarization, while the inductive reactance and decreased steady state resistance, and the consequent dramatic resonance corresponded to a depolarization.

Subsequent experiments (Cole and Marmont, 1942) had shown a small added low-frequency capacitive reactance as a more probable normal resting characteristic which became much larger in high calcium. This, then, was to be qualitatively explained as the behavior to be expected on the poorly conducting limb of a rectification characteristic which was at least self-consistent with the considerable increase of the dc membrane resistance found in high-calcium media.

On the other hand, a marked increase of capacitive reactance over that corresponding to the 1 $\mu F/cm^2$ of the membrane was often found at frequencies approaching 1 kc and, in addition to the lower frequency inductive reactance, was a strong indication of yet another and faster component. Subsequent dissection of the data by subtraction of the constant phase angle passive capacitive admittance produced an ionic admittance locus, figure 1:51, much as would be predicted for two reactive components—one inductive, seen at low frequencies, and the other capacitive, appearing at somewhat higher frequencies. This complication carried the problem considerably beyond the simple nonlinear potassium conductance picture, which was as far as I had been able to go.

Much of the work on artificial membranes has stemmed from that of Loeb (1920) and Michaelis (1926) on collodion. It has grown prodigiously and in many directions as exemplified by the extensive bibliography of Sollner et al. (1954). Quite recently the new synthetic ion exchange resins have received attention because of their strength and variety.

By far the most effort has been expended on the steady state potentials without current flow, and Planck's explanation of them, and the electrochemical mechanisms involved. Much less attention has been paid to the relations of current flow and potential to the measurement of internal concentrations. But I know of almost no measurements by transient or impedance approaches on any of these systems to which electrodiffusion might apply.

Nonetheless, the rectification by ion replacement was recognized by Blinks (1930). He showed a record that seems certainly to be an anomalous capacitive reactance in a thin collodion membrane between KCl and HCl. Although obviously not exponential, the time constant is about one second. Assuming a thickness of 0.5 μ, the mean mobility becomes 10^{-8} cm/sec per volt/cm on the basis of simple diffusion equation 2:23. However, the applied potential was 0.5 volt so the electric term should predominate according to equation 2:25 to give the mean mobility 10^{-9} cm/sec per volt/cm. Further assuming a 4000 Ω resistance and a 1 cm^2 membrane, its specific resistance is $8 \cdot 10^7$ Ωcm and the concentration, corresponding to a mobility of 10^{-9} cm/sec per volt/cm, is then in the neighborhood of 0.1N. Unfortunately I know of no way to check such a figure.

In his extension of our impedance measurements (Dean, Curtis and Cole, 1940) on the Dean and Gatty films, Goldman (1943) showed not only clear rectifications but also definite indications below 100 c of what were probably anomalous capacitive reactances. The membrane thickness is also a problem here, but on the hopeful assumption of 100 Å, mobilities and concentrations much the same as for the squid axon appear, as well as constant phase angle capacities ranging from 0.005 to 0.5 $\mu F/cm^2$.

Teorell (1948) independently proposed an electrodiffusion explanation for the squid membrane inductive reactance. He demonstrated this by the transient potential after the start of a current across a porous glass membrane from LiCl to HCl solutions and by the Fourier analysis to give the impedance. The time constant was about six minutes and, if the mean mobility were 10^{-3} cm/sec per volt/cm, the glass membrane would have been 3 mm thick, which may be reasonable. However, this inductive behavior was preceded by a capacitive effect which was ten times faster, but hardly a dielectric relaxation. Later Teorell (1949) mentioned a separate moving boundary explanation which may have complicated the simple electrodiffusion in such a thick membrane.

I have long admired the polarization electrode impedance theory but I never was entirely convinced that it had been adequately confirmed even for a reversible electrode. The most complete and satisfactory experimental investigation of electrodiffusion theory that I know was given by Walker and Eisenman (1966) as an ideal example of an ion exchanger with mobile sites. They used HCl between AgCl electrodes. They checked the linear concentration and exponential potential profiles, the overall V vs. I and transient characteristics. They further found g_∞ linear and extrapolated E_∞ as given by Planck for sudden changes of current. Further Walker et al. (1968) have compared the transient behavior very satisfactorily with the corresponding theory and expect a similar agreement for the sinusoidal characteristics. Segal (1967) measured the impedances of both cation and anion ion exchangers from 0.01 to 0.5 c. He found them to conform generally to the Warburg equation and showed that most of the impedance was in the external unstirred solution. An extension of the theory allowed calculations of small flows of co-ions.

Summary

The electrodiffusion theory of Planck anticipated—at least in principle—its application to explain the squid axon membrane phenomena that appeared about 1940. In particular, it was simplified for the single ion, potassium, and extended to estimate the kinetic characteristics. The model was eminently satisfactory in general—so much so that the discrepancies seemed minor.

The steady state rectification was in the right direction and of the right kind, although the ratio of the limiting slopes was too large and the intersection of the asymptotes was not given by simple theory.

The ion redistribution process explained either an inductive or capacitive anomalous reactance very nicely and gave a very low ion mobility, 10^{-8} cm/sec per volt/cm, in the membrane. But this led to a seemingly high ion concentration and there was no basis to expect the two such different reactances that were in one membrane at the same time. However, there were a few experimental models for which the theory gave a possible explanation.

Stability of Systems

Before Von Kármán's (1940) review I had paid but scant attention to the problems and complexities of instability and maintained oscillations. Van der Pol's (1926) article had been interesting, I had liked Volterra's (1931) theory of competing species, Wegel's (1930) theory of the larynx, and the thermistor oscillator (Becker, 1947) was a novelty and impressive. Then all these things took on a new importance.

Several different new measurements suggested a capacity, rectifier, inductance equivalent circuit for the squid axon membrane. The behavior of this circuit could be made at least reasonably consistent with a wide variety of subthreshold and threshold data. The rectification and the inductance of the circuit were attributed to potassium, and they had an attractive and at least possible explanation in a simple electrodiffusion of that ion. No part of this otherwise happy situation showed any hint of a threshold, an action potential, of even the sustained sinusoidal oscillations that Arvanitaki (1939) had demonstrated, figure 2:61, and others were to find.

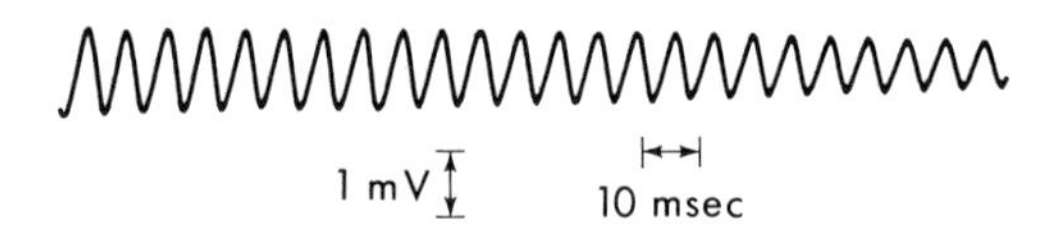

FIG. 2:61. Slowly decrementing oscillations of potential found for *Sepia* axon at the end of a train of repetitive action potentials.

But Von Kármán opened a new and exciting approach and led me back through Liénard (1928) and van der Pol to Poincaré (1881, 1882), whom Arvanitaki had invoked. This and the considerable Japanese and Russian literature were mostly concerned with relaxation oscillations and the closed trajectories, called limit cycles, encountered in vacuum tube circuits. At the same time the negative slopes and "wrong way" currents seen in the propagating impulse (figure 2:23) had made the extension of the nerve membrane nonlinearity into a negative resistance region seem a reasonable and vastly intriguing possibility.

The difficulties also seemed to be considerable. The theory was scattered, seldom elementary, and dealt with few systems which would go through a wild excursion upon slight provocation—such as an inverted pendulum or a nerve membrane. The analysis always started with known parameters to arrive at performance, and nowhere was there even a chapter on "how to determine a nonlinear component from the behavior of a system." On the experimental side there was no direct measure of a negative resistance, and a nerve fiber had ignored my feeble effort to make excitation stand still in space and time for measurement.

It was necessary to guess even the kind of negative resistance that might be involved. The distinction between the two types shown on a V vs. I diagram (figure 2:62) was then well recognized. The difference had been pointed out by Barkhausen (1926) and vacuum tube circuitry for both types had been developed (Herold, 1935). The one characteristic, similar to a letter N, gave three values of V over a range of I and for the other, S, characteristic I was multiple valued for a range of V but I found it particularly difficult to understand the nature of the difference between the two. Given only the central limb AB of figure 2:62, I could find no basis to decide whether it would be stable for constant current or constant potential. Since the stability seemed to involve topological relations to the axes, I asked Solomon Lefschetz how to tell one side of such a line from the other. I still do not know the answer and I am also far from certain that it is even a good question, but it was a gambit that won enthusiastic encouragement and help. Another result of this question was, as Lefschetz said recently, that it started the work he had carried on for more than twenty years.

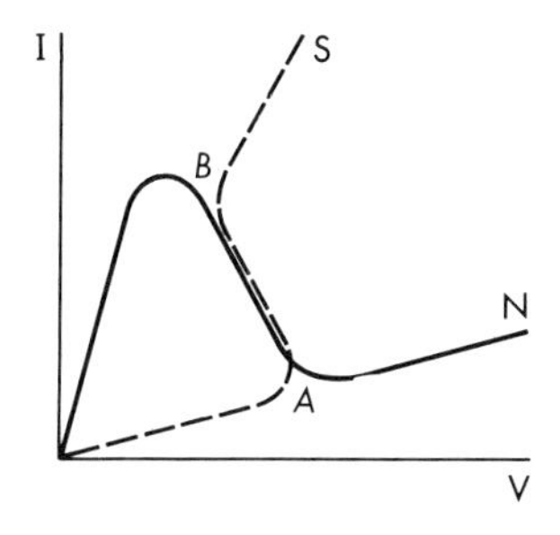

FIG. 2:62. Schematic potential vs. current characteristics for two negative conductance systems. The dashed line represents the S type which may be stable for a constant current and the solid line the N type with stability under a fixed potential.

OSCILLATIONS

The literature gave ample basis to expect that an explanation of the Arvanitaki subthreshold oscillations would appear just by assuming a mechanism with a hidden N negative resistance characteristic in the membrane. More than enough energy to supply the losses would be provided by the negative resistance region and would tend to increase the amplitude of the oscillation. As the oscillation went beyond the negative region this extra energy, carried in the capacity and inductance of the equivalent circuit, would be dissipated. The energy then would be reduced to its original value as the system returned to the negative resistance region.

The modification of the equations for the circuit of figure 2:40 to include a negative resistance in the steady state rectification was suggested by the slight irregularities found experimentally (figure 2:37), but the calculation of the performance was neither obvious nor simple. I turned to the van der Pol arrangement in which the passive RLC circuit was activated by a vacuum tube $f(V)$ in figure 2:63 that was equivalent to an N type negative resistance. My circuit, however, led to an equation that was more general than van der Pol's and was soon treated by Levinson and Smith (1942).

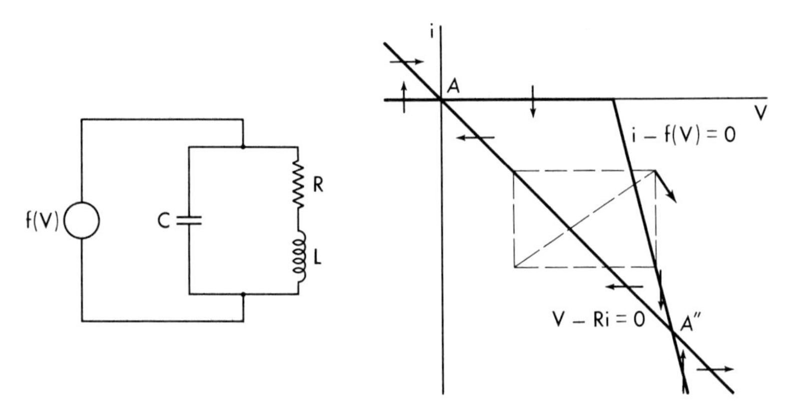

FIG. 2:63. (a.) Equivalent circuit for the calculation of membrane oscillations sustained by a negative conductance element $f(V)$. (b.) Graphical construction for tracing trajectories of an oscillatory membrane on the potential vs. current plane. The horizontal orthoclines are the resistance line, the vertical orthoclines are the piecewise linear negative conductance $f(V)$ with the stable (A) and unstable (A'') singular points.

The convenient transformation of Liénard (1928) did not produce an easy graphical solution, so the problem was attacked directly.

On rewriting equations 2:11,

$$\frac{dV}{dt} = -\frac{i + f(V)}{C}, \qquad \frac{di}{dt} = \frac{V - Ri}{L}. \tag{2:30}$$

A cubic N characteristic had been used by van der Pol for $f(V)$, and the potassium rectification data were available for R. But it was convenient at this stage to work with three straight line segments for $f(V)$ and to make R constant and now $-i$ is plotted upward on the V, i plane to look more like subsequent developments. By change of notation equations 2:30 become

$$\frac{dV}{dt} = -[i + f(V)], \qquad \frac{di}{dt} = V - Ri \tag{2:31}$$

so $di/dt = 0$ on the orthocline $i = V/R$ and $dV/dt = 0$ along $i = -f(V)$, as shown in figures 2:63 and 2:64.

The general pattern of the solutions could now be seen but some

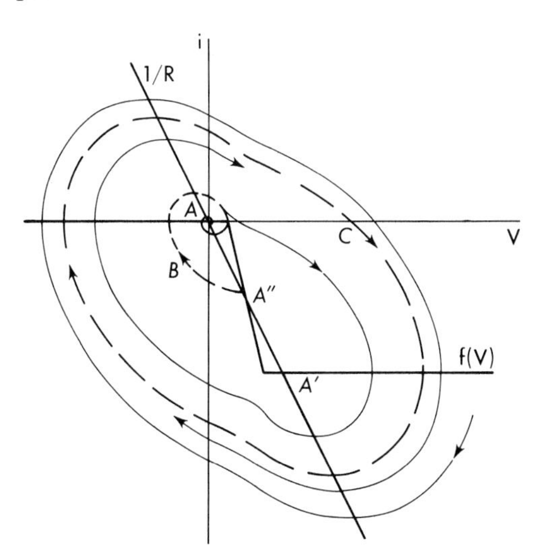

FIG. 2:64. Graphical calculation of trajectories for oscillatory membrane with resistance R and negative conductance $f(V)$. Initial points inside the dashed line B will reach the stable point A and those outside will spiral out to reach the stable oscillatory limit cycle, dashed trajectory C. Points outside C will spiral in to it.

important details were not obvious so it was necessary to trace some trajectories. The quotient of equation 2:31 is

$$\frac{di}{dV} = -\frac{V - Ri}{i + f(V)}$$

and as figure 2:64 shows, this slope too is found directly by similar triangles on the V, i plane. The example of figure 2:64 showed:

Two points A, A' which were stable, approached by spiral trajectories or damped oscillations after small displacements from the unstable point A'';

Two closed trajectories or limit cycles, B around A and a similar one about A', which were unstable because any nearby point can only move farther away;

Another limit cycle, C, enclosing both the stable points and the two unstable limit cycles. This limit cycle however is stable because it is ultimately reached from all exterior points and from all interior points not inside the unstable limit cycles.

Once again it would have been an additional chore to obtain the V, t relations. But without them it was apparent that a small displacement from the resting potential would give continuous, near sinusoidal oscillations with such a negative resistance. And probably with the additional postulate of only a slight spontaneous shift of some parameter this partly hypothetical membrane could initiate oscillations almost as beautiful as those Arvanitaki had persuaded a real membrane to produce (figure 2:61).

ACTIVITY

The measurements of the steady state rectification and its time constant, attributable to potassium ion flow, had adequately expressed many steady state and subthreshold observations. With a single ad hoc hypothesis of a threshold potential, they had contained, considerably better than qualitatively, most of the successes of the formal two-factor theories of excitation as well as some oscillatory phenomena which required a more general theory. The experimental facts plus the entirely hopeful assumption of an N type negative resistance component in the membrane gave a reasonable representation of the Arvanitaki subthreshold oscillations and might possibly be the explanation for them.

There was still no physical explanation of a threshold or of an action potential, but the hints seemed quite clear. There could be no doubt that the threshold was the expression of some kind of an instability of the membrane system. The membrane potential records of Katz and Hodgkin with pulse stimuli, figure 2:45, and of our own with current steps, made it seem quite probable that this instability appeared directly in the membrane potential. The turn of these records just above threshold along with those of a propagating impulse on the V, I plane were of the kind that might result from an N negative resistance.

Van der Pol had used a cubic N characteristic on the phase plane that should apply to an excitable membrane. Passing the rheobase, the subthreshold oscillation maximum, of figure 2:33, opened up into a tremendous excursion into the fourth quadrant as seen in the established, propagating impulse (figure 2:23). This seemed clearly to require an N characteristic on the V, I plane such as was then found only in gas discharge, the dynatron, and the passive iron wire, in the midst of many other systems which were all in the S category. The search for a satisfactory representation became particularly difficult when the accommodation phenomena were included. Excitation had appeared at a rather constant change of membrane potential during rapidly rising applied currents. Below a lower, liminal, gradient of current rise, the potential followed the current nearly along the steady state rectification curve without excitation. Otherwise, a piecewise linear approximation to a cubic characteristic appeared to be a possibility as well as a highly probable means to obtain an absolute all-or-none threshold.

I was unable to cope with the complexities of the circuit of figure 2:63 beyond the subthreshold oscillations, so I worked on only the factual membrane capacity and a hypothetical N resistance—as shown in figure 2:65a—ignoring the factual rectification and its long time constant. Here we have only

$$I = C\,dV/dt + f(V)$$

and we may again use a straight line symbolic representation of a cubic for $f(V)$ as shown in figure 2:65b. The time course of V is easily obtained analytically as one or two exponentials with appropriate constants, but we will first consider the general behavior.

The application of a constant current $I < I_0 = f(V_1)$ as in figure 2:65b will give three equilibrium potentials, of which the middle one is unstable. For $I = I_0$ the equilibrium $V = V_1$ is stable for

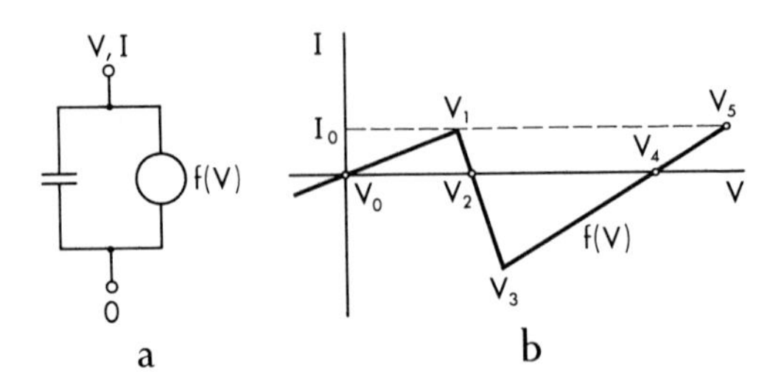

FIG. 2:65. (a.) Behavior of a membrane with a hypothetical negative conductance element $f(V)$. After a constant current is applied the potential will change from V_0 towards V_1. Above the threshold I_0, the potential will abruptly increase from V_1 to V_5. (b.) Without steady external current flow, a short pulse is subthreshold if it produces a potential $V < V_2$ which returns to V_0, but above threshold for $V > V_2$ which goes to V_4.

displacements to the left and unstable to the right. After the application of a current but slightly larger than I_0 there remains only the single stable potential at V_5. There is then a minimum of dV/dt as the potential passes the threshold V_1. This was the typical behavior with or without the point of inflection, as we had found in so many records after a near threshold constant current.

For $I = 0$ after a short shock there are again three points of equilibrium V_0, V_2, and V_4, where $dV/dt = 0$. At $V < V_0$ the conductive, ionic current is inward and the capacitive current outward so $dV/dt > 0$ and $V \to V_0$. Similarly for $V_0 < V < V_2$, $dV/dt < 0$ and $V \to V_0$ to make V_0 a stable point; the same procedure shows V_4 stable for $V > V_2$. Looking now at V_2, we have already found that V moves to the left away from V_2 for $V < V_2$ and to the right for $V > V_2$, so V_2 is a point of unstable equilibrium—a threshold. And indeed the analytical solutions near this point were quite good representations of Hodgkin's records, leaving a near threshold potential slowly at first and then more rapidly, as given by a positive exponential.

Hodgkin and I had both noticed the difference of threshold, V_1 for a steady current and V_2 for a short pulse, predicted by this hypothetical N conduction characteristic, but I know of no evidence either for or against the prediction. Above the threshold the calculated potential rose by two exponentials to the steady state points

V_5 or V_4, from which there was no recovery. Here again there was no specific attempt to detect an effect of a steady current on the spike amplitude, but there were no changes large enough to attract particular attention either at the site of stimulation or during the passage of an impulse.

Suggestive and helpful as the N negative resistance was, it was also very confusing, to me at least. I was fully convinced that a resistance as measured by a Wheatstone bridge was $\Delta V/\Delta I$ and independent of frequency. The impedance change during an impulse had been reasonably interpreted in these terms. The rectification and anomalous reactance had been satisfactorily correlated and attributed to potassium flow. The thermistor, which had provoked the general reactance concept, and its negative resistance were well understood intuitively and by elementary analysis. But I did not then seriously consider the invisible, hypothetical N characteristic in these terms and thought of it rather as a frequency independent phenomenon. Then, in going from a positive slope and positive resistance to the negative slope and resistance, the measured resistance would have to increase toward plus infinity and return from minus infinity to the final negative value. There was no experimental support for this except an occasional slight bump at the start of the impedance decrease during an impulse after a questionable correction for apparatus distortion. I also suspect that on the same basis I may have persuaded Bernhard Katz (1942) to desert some perfectly good interpretations of impedance data on muscle excitation.

One important step was made in spite of the conceptual difficul-

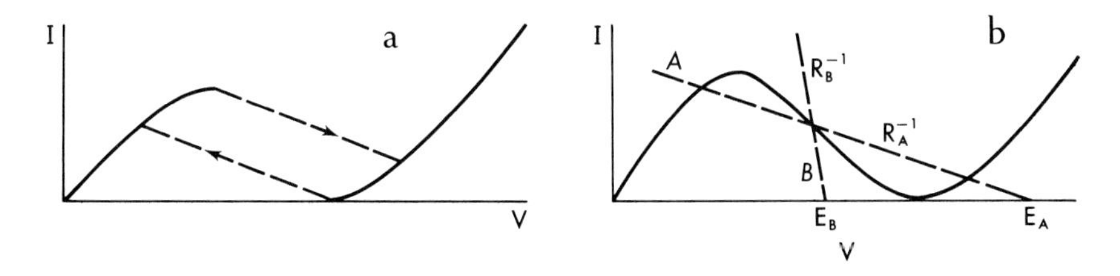

FIG. 2:66. Behavior of a passive iron wire. (a.) As the potential was increased the current first rose along the left solid line, then moved to the right spontaneously along the upper dashed line, and finally followed the right solid line. Returning, the lower dashed line was the spontaneous path. (b.) If the surface had the characteristic given by the solid line, an emf E_A applied through a resistance R_A to give the load line A would be unstable at the center intersection. At this same point, the load line B with a smaller R_B would give stability.

ties. In the midst of them James Bartlett came to me with some old passivated iron data, such as figure 2:66a. I can remember being impressed and puzzled by the transient current-potential lines between the passive and active states. They were absolutely straight and parallel. I finally realized that they were the load lines of the external circuit and that this analogue of a nerve fiber must have an N negative resistance characteristic, such as $I = f(V)$.

In this equation we now have $I = (E - V)/R$, where E and R are the external emf and resistance of figure 2:66b. For the three equilibria as shown by A the center one is again unstable, but with a low external resistance represented by B the single equilibrium will be stable. It was not then difficult to persuade Bartlett to investigate this inaccessible region for a small area, to discourage propagation, and a low resistance source of potential, to allow stability. These experiments gave discouragingly complicated information (Bartlett, 1945), but the approach has been widely used. The results on passivated metal surfaces are only now being interpreted as the interplay of a variety of physical chemical processes (Franck and FitzHugh, 1961; Ord and Bartlett, 1965), but the concept of stable potential control for an irritable system had been evolved and applied.

This was undoubtedly an old concept but Steimel (1930) has been credited with the first clear and conclusive demonstration of the universal validity of the current or potential control principle. The circuitry must also be quite old for it is only the dual of the conventional potentiometer. However, there has been much less need to measure or maintain a current with a fixed or no potential difference.

Polin (1935, 1941) was able to patent and use the arrangement for prevention of ship corrosion and Ussing (1952) applied it to begin his long investigation of the transport of ions across frog skin in the absence of an electrochemical potential difference.

PROPAGATION

We had considered the propagation of a localized short circuit as a simplified model of the data on *Nitella* (Cole and Curtis, 1938) and the model was similarly applicable to the squid axon. Offner, Weinberg, and Young (1940) then showed that a sufficient decrease from the resting resistance, such as found for *Nitella* and squid, at a critical potential, which was an assumption, would support a propagating impulse. They examined some extensions, including a delay in the resistance change, and some consequences—of which a reasonable impulse speed was most important. It seemed obvious that an N type resistance with its negative region could also support propagation and produce the ionic current as a function of the membrane potential similar to what was required by the calculations during the rising phase of the propagating action potential. But so far as I know this analysis was not made until quite recently, and then only at the end of a long, tortuous and almost circuitous path of experiment and analysis that demonstrated both the necessity and the reality of an N type negative resistance characteristic. A calculation of the rising phase action potential that could have been done before World War II is illustrative of the problems of the propagating impulse as well as being elementary and rather realistic.

The general cable equation 1:31 is

$$\frac{1}{r_e + r_i} \cdot \frac{\partial^2 V}{\partial x^2} = c_m \frac{\partial V}{\partial t} + i_g.$$

For a large external medium $r_e \to 0$, for a constant speed of propagation θ,

$$\partial^2 V/\partial x^2 = (\partial^2 V/\partial t^2)/\theta^2$$

and for an axon radius a

$$\frac{d^2 V}{dt^2} - K \frac{dV}{dt} = K \frac{f(V)}{C}$$

where $K = 2\theta^2 r_2 C/a$, r_2 is the axoplasm resistivity, C and $f(V)$ are the capacity and conduction current for a unit membrane area. This equation has now been solved for a cubic $f(V)$ (FitzHugh, 1968), and it can be solved for linear segments of $f(V)$, giving exponentials for positive slopes and increasing sinusoids for negative slopes. The solution for a three segment characteristic is then completed by matching up the boundary conditions at the junctions of the segments. This will determine the speed θ, although the process may be much more difficult than was carried out by Offner et al.

We may also state the problem for the phase plane V vs. y, where $y = dV/dt$, as

$$\frac{dy}{dV} = K \frac{y + f(V)/C}{y}. \tag{2:32}$$

This may be solved step by step, numerically or graphically, for a given $f(V)/C$ and with various values of K until one is found to arrive at $y = 0$ for $f(V) = 0$ at the final value of V for this rising phase. As an example I have chosen the characteristic of figure 2:67. This illustration is for $r_2 = 30\ \Omega\text{cm}$, $C = 1\ \mu\text{F/cm}^2$, and $2a = 480\ \mu$, $\theta = 20$ m/sec to give $K = 10^4\ \text{sec}^{-1}$. The orthoclines are $dy/dt = 0$ for $y = -f(V)/C$ and $dV/dt = 0$ for $y = 0$ as indicated.

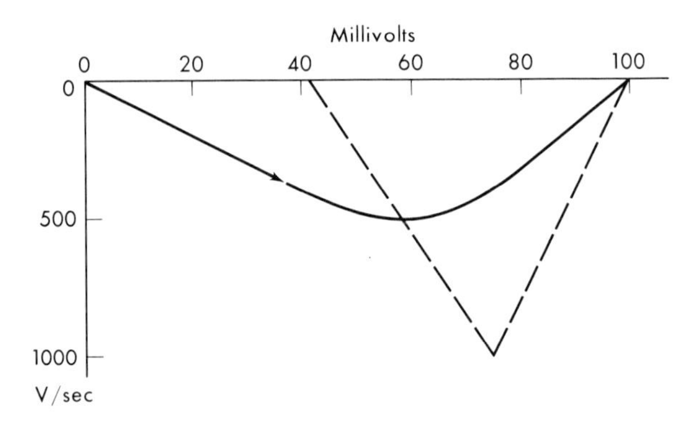

FIG. 2:67. A calculation of the rising phase of a propagating action potential. The membrane was assumed to have the piecewise linear characteristic given by the dashed line. On this phase plane, dV/dt vs. V, the exponential foot appears as a straight line, the hyperbolic excitation in the negative conductance region is an elliptic arc, and the final, active, exponential is again linear.

The resting conductance is made zero because the effect of a resistance of 1 kΩcm^2 is only slight but somewhat more difficult to take into account. In this region, with $f(V) = 0$, $dy/dV = K$ and, since the trajectory starts at $t = -\infty$, $y = 0$ for $V = V_3$, but we notice that this is an unstable point so that a trajectory must be aimed quite precisely if it is to arrive at this point. The necessary slope may be found by the conventional solution for the negative rate constant of equation 2:32. However since the slope is constant,

$$\frac{dy}{dV} = \frac{y}{V - V_3} = K\,\frac{y + g_3\,(V - V_3)/C}{y}$$

so

$$y^2 - K(V - V_3)y - K(V - V_3)^2 g_3/C = 0$$

and

$$y = -\tfrac{1}{2}\,K(V_3 - V)\,[1 - \sqrt{1 + 4g_3/KC}].$$

Here I chose $g_3 = 0.040$ mho/cm^2, corresponding to the average

25 Ωcm² found at the last part of the rising phase in the propagating squid action potential to give

$$y = 15.5(V - V_3).$$

The peak potential was taken as 100 mV and the maximum inward current as 1 mA/cm² at 75 mV. The trajectory should be nearly a segment of an ellipse in the negative resistance region, and it became apparent that it was not far from a circle. Consequently V_1 was changed until a circle tangent to the upper line at V_2 was also tangent to the lower one at V_1, to give the solution shown.

This showed that an axon with such an N type negative resistance characteristic can propagate at least a rising phase of an action potential. Upon a second integration the potential vs. time curve should have a reasonable appearance, and a conductance measured as $g(V) = f(V)/(V_3 - V)$ should closely resemble the corrected experimental curves.

RECOVERY

The existence of a threshold and some of its characteristics had been obtained on the assumption of an N type negative resistance and the neglect of the slowly rising rectification current attributed to potassium. On this same basis a reasonable facsimile of the rising phase of an action potential could be produced, and the current assumption that such an action could be propagated at a reasonable speed has only recently been confirmed.

There was, however, no return to the initial resting potential, or recovery phase of these action potentials; this was the next problem. The neglected outward current flow would appear with a time constant, L/R, of the order of a millisecond. It seemed certain that this could more than neutralize the inward negative resistance current and bring the potential back to the starting point again. The same problem had been set up formally, in equations 2:31 and figure 2:64, to investigate the possibility of subthreshold oscillations. This had shown a satisfactory all-or-none threshold, and I fully expected to find an N characteristic that would not produce a stable limit cycle with the other and known normal axon properties.

Eliminating i between equations 2:31,

$$C\frac{d^2V}{dt^2} + \frac{df}{dt} + \frac{V}{L} + \frac{R}{L}\left[C\frac{dV}{dt} + f\right] = 0$$

changing the time scale by the factor $\sqrt{LC}$, writing

$$\frac{df}{dt} = \frac{df}{dV} \cdot \frac{dV}{dt} = f' \frac{dV}{dt}$$

and

$$\frac{d^2V}{dt^2} = \frac{d}{dt}\left(\frac{dV}{dt}\right) = \frac{d}{dV}\left(\frac{dV}{dt}\right) \cdot \frac{dV}{dt} = y \frac{dy}{dV}$$

where

$$y = \frac{dV}{dt}$$

gives

$$y \frac{dy}{dV} + \left[\sqrt{\frac{L}{C}} f' + \sqrt{\frac{C}{L}} R\right] y + V + Rf = 0.$$

Now using the Liénard transformation to eliminate f', which is in fact a return to the variable i,

$$z = y + \sqrt{L/C}\, f \text{ and } \frac{dz}{dV} = \frac{dy}{dV} + \sqrt{\frac{L}{C}}\, f'$$

gives

$$\frac{dz}{dV} = \frac{R\sqrt{C/L}\, z + V}{\sqrt{L/C}\, f + z}$$

and

$$\frac{dz}{dt} = -[R\sqrt{C/L},\, z + V],\ \frac{dV}{dt} = -[\sqrt{L/C}\, f - z].$$

On the V, z plane the orthocline $dz/dt = 0$ is $R\sqrt{C/L}\, \bar{z} + \bar{V} = 0$ and for $dV/dt = 0$, $\sqrt{L/C}\, \bar{\bar{f}} - \bar{\bar{z}} = 0$. We have again at the point V, z for $\bar{\bar{V}} = V$,

$$\frac{dV}{dt} = -[\sqrt{L/C}\, \bar{\bar{f}} - z] = z - \bar{\bar{z}}$$

and for $\bar{z} = z$

$$\frac{dz}{dt} = -[R\sqrt{C/L}\, z + V] + [R\sqrt{C/L}\, z + \bar{V}] = \bar{V} - V.$$

By similar triangles dz/dV can be constructed as in figure 2:63, which is more involved than the simple Liénard procedure.

The choice of the constants to use in the integrations was somewhat arbitrary, but considerably more important reservations had to be remembered. No one of the constants had been measured at all directly; each had been evolved as a result of some, and sometimes questionable, calculations on the best, but still none too good, experimental data. The potassium resistance was taken to be $R = 200\ \Omega\text{cm}^2$ as a compromise between a resting resistance of $1{,}000\ \Omega\text{cm}^2$ and a limiting depolarization resistance of the order of $100\ \Omega\text{cm}^2$. The inductance used was the first value found, $L = 0.2$ henry cm^2. The capacity, $C = 1\ \mu\text{F/cm}^2$, also had not been confirmed for the squid axon and rested, as did similar values for other membranes, on a considerable structure of measurement and analysis. The subthreshold calculations were better than qualitatively successful without any use of an inward current component or a negative resistance, so none was assumed below threshold. The negative resistance region was begun at a nominal threshold of 20 mV above the resting potential. A maximum inward current of $1\ \text{mA/cm}^2$ was placed at a depolarization of 60 mV. This current was a little more than was abstracted from the propagating impulse potential, and the potential was somewhat less. The inward current then turned outward at 120 mV for a rather large action potential. The unit of time $\sqrt{LC} = 0.45$ msec and in this unit

$$\frac{dV}{dt} = -[450\,f - z], \qquad \frac{dz}{dt} = -[.45\,z + V].$$

For $f = 0.5\ \text{mA/cm}^2$ at $V = 0$ mV, $\bar{\bar{z}} = -0.225\ V/\text{sec}$ of the V orthocline $dV/dt = 0$ and similarly for $dz/dt = 0$ at 60 mV, $\bar{z} = 0.135$ to give the orthocline loci of figure 2:68. No great effort was

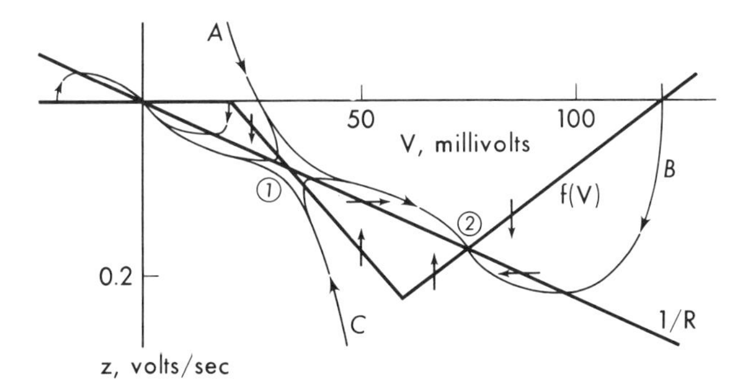

FIG. 2:68. Calculations of recovery for an oscillatory membrane with a resistance R and a piecewise linear negative conductance $f(V)$. Near threshold, path A may return to rest or go on to stable active point 2. A simple return from the peak potential along B will also stop at 2. The end of a far-flung excursion such as C might pass the unstable point 1 to return to rest.

made to match all possible characteristics, and better values might have been chosen. The value of dV/dt at 60 mV is 200 V/sec, which may be too low. The effective resistance above 60 mV is 46 Ωcm^2, considerably above the 25 Ωcm^2 from impedance or 9 Ωcm^2 from the potentials during a propagating impulse. Furthermore, the steady state total ionic current was inward at 60 mV and nearby, whereas only a slight hint of such behavior had been seen experimentally at a lower potential.

The resting potential is seen to be stable for small displacements. The threshold is strictly all-or-none, a trajectory such as A going on a large excursion if it does not return quite promptly to rest. But trouble came at the third singular point which is also stable and forms a trap that only a superthreshold, a far-flung trajectory such as C, can avoid. I was determined—to my later regret—to have an all-or-none threshold, and this type of instability was not only a nice way to achieve it but, as far as I could see, it was the only way. However, none of the possibly sensible or obviously ridiculous modifications of this N type negative resistance could produce a decent action potential recovery. Since this was all guesswork, without tangible experimental evidence, I gave it up, bewildered and disgusted.

NONLINEAR MECHANICS

From restricted channels the first two parts of Minorsky's "Introduction to Non-linear Mechanics" (Minorsky, 1947) became available near the end of World War II. In this were assembled, in orderly form and for the first time, the considerable but widely scattered literature on the analysis and behavior of nonlinear systems. After some definitions and the introduction of a few practical examples, Minorsky showed that the problems could usually be reduced by series expansion to the form

$$\frac{dx}{dt} = ax + by$$

$$\frac{dy}{dt} = cx + dy.$$

Then the lines $dx/dt = 0$, $dy/dt = 0$ are what we may call orthoclines as particular cases of the isoclines used by van der Pol. Then on the x, y plane the slope is given by

$$\frac{dy}{dx} = \frac{cx + dy}{ax + by}$$

except at the intersections of the orthoclines, called singular points, at which $dy/dx = 0/0$. The behavior of the system in the neighborhood of these points was, however, completely determined by the coefficients, a, b, c, and d, or specifically by the roots of the characteristic equation

$$S^2 - (a + d)\,S + (ad - bc) = S^2 - pS + q = 0.$$

There were then five cases with five different behaviors on the x, y plane, figure 2:69, depending on the roots, S_1 and S_2, at each singular point:

1. If S_1 and S_2 are real and of opposite sign, $q < 0$, one has a

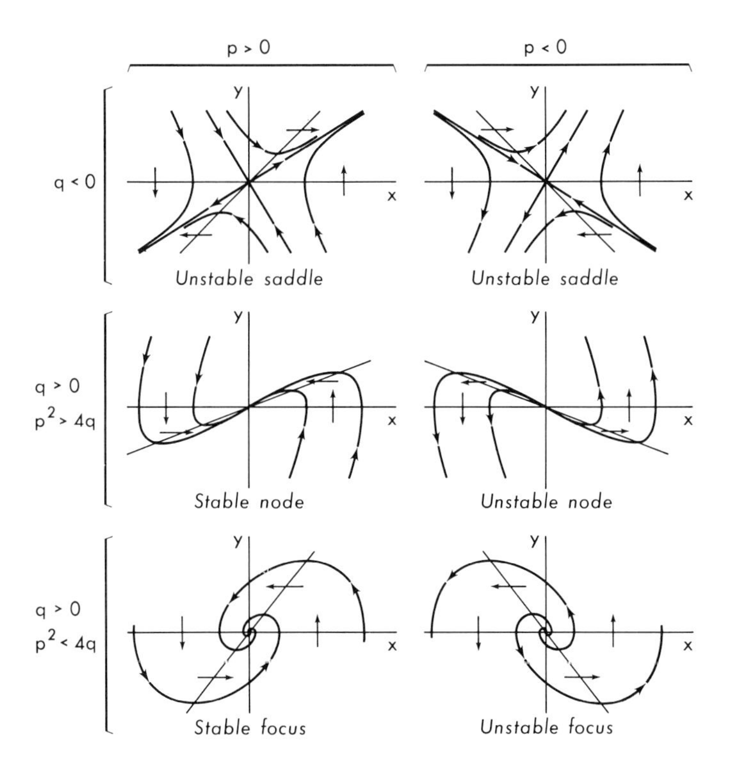

FIG. 2:69. A summary of the trajectories in the neighborhood of a singular point. The parameters p and q are coefficients of the characteristic equation. The vertical and horizontal arrows show the path direction across the orthoclines.

saddle point. The trajectories usually approach and then leave the vicinity of such a point as in figure 2:69 top.

2. If S_1 and S_2 are real and negative, $p > 0$, $q > 0$, $p^2 > 4q$, one has a stable nodal point, or

3. If S_1 and S_2 are real and positive, $p < 0$, $q > 0$, $p^2 > 4q$, one has an unstable nodal point. Each trajectory approaches or leaves such points with a definite slope as shown in figure 2:69, center, corresponding to an exponential on the phase plane of $y = dx/dt$.

4. If S_1 and S_2 are conjugate complex with a real part >0, $p > 0$, $q > 0$, $p^2 < 4q$, one has a stable focal point or

5. If S_1 and S_2 are conjugate complex with a real part <0, $p < 0$, $q > 0$, $p^2 < 4q$, one has an unstable focal point. Each of these trajectories approaches or leaves the focal points without a definite slope, as given in figure 2:69 bottom, corresponding to decreasing or increasing sinusoidal oscillations on a phase plane.

This made it quite obvious that the stable resting potential must be a stable nodal or focal point, and there was experimental evidence to show that it might be either. Then the well established all-or-none law and my many personal observations of its validity could only be represented by a saddle point, clearly dividing the subthreshold from the superthreshold trajectories. But the superthreshold behavior was an insuperable problem. Clearly a stable nodal or focal point in the positive membrane potential region gave no possibility for the experimental fact of recovery.

An exception was the possibility of an unstable limit cycle around such a point as described by Levinson and Smith (1942). But if this described the membrane, someone should at some time have maneuvered it inside the limit cycle to go to the point inside. Another saddle point in this potential region would only transfer the problems to a higher potential, and furthermore there was no experimental reason to worry about such an additional complication. So I was left with only the possibility of an unstable nodal or focal point for a third singular point. It seemed entirely reasonable that all of the trajectories emerging from such a point should reach the stable resting singular point and permit an excitation trajectory to go around them all. But on the phase plane of $x = V$, $y = dV/dt$ I could produce no plausible membrane characteristics, $I = f(V)$, to give such a nice and reasonable representation of nerve excitation.

Models and Analogues

The words model and analogue have been used often and in many connections and connotations. There have been extensive, and not always definitive, attempts to distinguish between the meanings of the words in some fields. This has not been particularly true for representations of the impressive array of phenomena that living nerves have been found to produce. Probably the things that may be called models or analogues are themselves of more importance than their classification in one category or the other. But if there are obvious differences the use of the proper word may contribute to clarity and brevity.

The concept of an analogue is clear. It has some properties that are similar to those of the original without regard to any overall likeness between the two. Thus a passive iron wire is an analogue of a nerve or an electromagnetic inductance may be an analogue of a thermistor. Although the analogues are of vastly different composition and structure they perform much the same as the original in some ways. With the analogue as more illustrative than descriptive, it is useful to look upon a model—theoretical or experimental—as an attempt to express and interpret the properties of the original as completely as possible. Thus the model should be based on physical reality.

Although an equivalent circuit of capacity, resistance and inductance is itself an analogue for a squid membrane, the measurements on which it was based are physical characteristics of the membrane. So a description of excitation in these terms may be a model. The application of physical laws to a potassium ion permeable membrane between different concentrations of that ion gives the characteristics of capacity, rectification, and inductance. This is more certainly a model of the membrane. It is also a more fragile model because more facts from the living membrane show it to be inadequate; so it became an analogue! The analytical correlation and presentation of experimental facts without any basis in physical reality, such as the two-factor formulations of excitation, may be called a theory but they do not come into the category of a theoretical model.

I have not always used such a distinction between models, analogues, and theories. Sometimes a term, particularly model or theory, has become too firmly established and in other situations I may not be sure which is most appropriate.

PASSIVE IRON WIRE

An iron wire in concentrated nitric acid becomes passive and quiescent. It can be activated into a local, energetic reaction by the touch of a piece of zinc or at the local cathode of an external battery. Once initiated this activation will move along the wire at a constant speed. Then, if the stimulus has been removed, the iron will revert to its initial passive state and this recovery will terminate the activity as they both propagate. Wilhelm Ostwald (1900) was struck by the resemblance between this behavior and that of nerve. The iron wire model has since been extensively investigated and has become the best and most widely used nerve analogue. As reviewed by Bonhoeffer (1948) and Franck (1956), Ostwald started Heathcote to work on it in 1900. The next and by far the most complete details of the analogies were published by R. S. Lillie (1936) between 1920 and 1935. Bonhoeffer and then Franck carried these further and both turned to the underlying physical chemistry of the analogue as had Bartlett (1945). I have been sufficiently impressed by the complexities of even the simpler passivity phenomena (Franck and FitzHugh, 1961; Ord and Bartlett, 1965) to venture the opinion that an understanding of the behavior may come sooner and be simpler for a real nerve membrane than for this analogue.

BONHOEFFER MODEL

Late in World War II, Leo Szilard pointed out an article in a copy of *Naturwissenschaften* that had come to the Metallurgical Laboratory through unexplained channels. In it, Bonhoeffer (1943) had begun to study the passive iron wire and similar systems both as models for nerve and as interesting problems of physical chemistry.

He set out to show that chemical kinetics could lead to relaxation oscillations, and evolved a rather complete description of the passive iron behavior from qualitative characteristics and in neurophysiological terms (Bonhoeffer, 1948, 1953). He used thermal ignition and a tuned circuit with a negative resistance as illustrations of instability. Following van der Pol, he showed the threshold and repetitive responses for increasing stimuli as trajectories on an "activation," x, vs. "refractoriness," y plane, (figures 2:70a, b, and c). Here $\dot{x} = E(x, y)$ and $\dot{y} = R(x, y)$ as indicated. This put many phenomena into a compact, if formal, representation that could be made into quite a satisfactory extension of the Rashevsky-Monnier-Hill model beyond excitation and through the complexities of response and recovery. I found, quite by accident at the start of our computation work at the National Bureau of Standards, that Alt (1948, 1954) had worked out the steady state propagation of burning in a fuse. In this he used the Arrhenius expression for the reaction rate and represented many of the calculations on the phase plane.

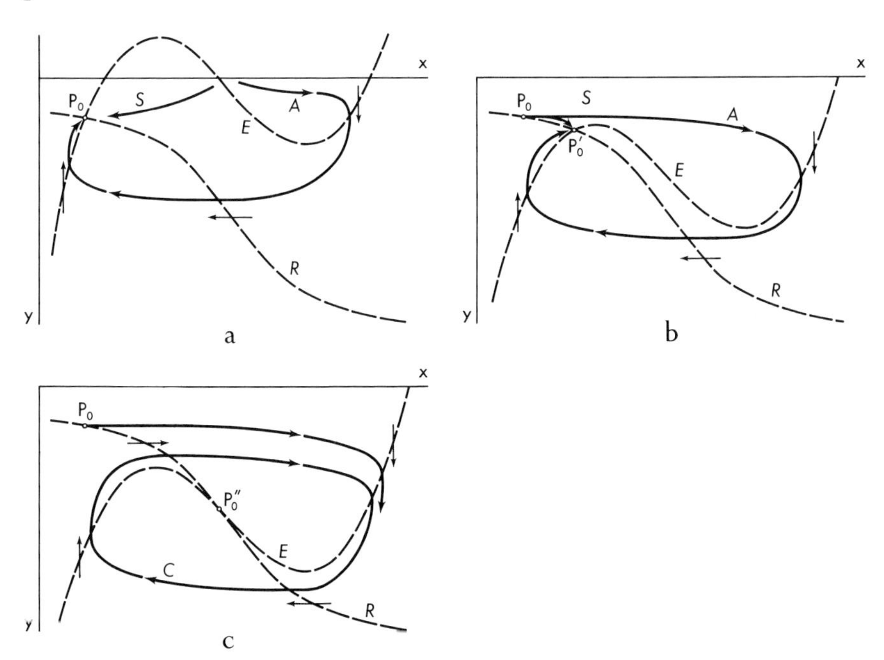

FIG. 2:70. A presentation of excitation and response by Bonhoeffer on an activation x vs. refractoriness y plane in terms of an excitation characteristic E and a refractory process R. (a.) After a short stimulus the solid line S is the subthreshold return to rest P_0 and the path A is the active response. (b.) After a constant stimulus the subthreshold path goes directly to the new stable point P_0' and above threshold the response is given by A. (c.) For a larger constant stimulus, the trajectory will spiral in to the closed path C representing a repetitive response.

Although Bonhoeffer was more interested in repetitive responses such as figure 2:70c, this model could also produce single responses to single stimuli. However, the size of the response varied continuously with the size of the stimulus as shown in figure 2:70a and b, and I was so deeply committed to an all-or-none threshold that I protested both to Bonhoeffer and to Max Delbrück who had translated the 1948 article. I argued that I had seen probably a few thousand impulses but not one had been half size, and so they put the orthoclines close enough together and for far enough to give a negligible probability for my finding a violation of the all-or-none threshold. And when I invoked a dozen friends who had each seen hundreds of thousands of all-or-none impulses, they only pushed the orthoclines even closer together. I insisted upon arguing this point with Bonhoeffer before he died, and I regret that I could not tell him when I was later convinced that I was wrong on two counts.

BONHOEFFER, VAN DER POL (BVP) MODEL

I have been very interested—and only a little chagrined—to watch a return to this topological technique of analysis and method of representation. In the intervening years the new and better data that became available for the squid membrane made it so probable as to be practically certain that this membrane is not strictly all-or-none. And then in a determined—not to say ruthless—simplification and approximation to these squid data, FitzHugh (1960, 1961) evolved what he most appropriately calls the Bonhoeffer-van der Pol, BVP, model. In it the two variables, x and y, are now identified physically as the fast excitatory membrane potential and a slower recovery process, respectively. They are represented by the equations

$$\dot{x} = c(x - x^3/3 + y + z)$$
$$\dot{y} = -(x - a + by)/c \qquad (2:33)$$

in which a, b, c are constants and z is the stimulus and particular examples are shown on the x, y plane of figures 2:71 and 4:60. Here we note that there is but a single stable singular point at the intersection of the orthoclines, $\dot{x} = 0$, $\dot{y} = 0$. The threshold is then not absolute but may be made as sharp as desired by adjustment of the constants. In figure 2:71 the subthreshold S and active A

responses are for current steps given by $z = -0.124$ and -0.128 while a rather fast train of impulses appears at $z = -0.4$. Responses for near threshold pulses are in figure 4:60.

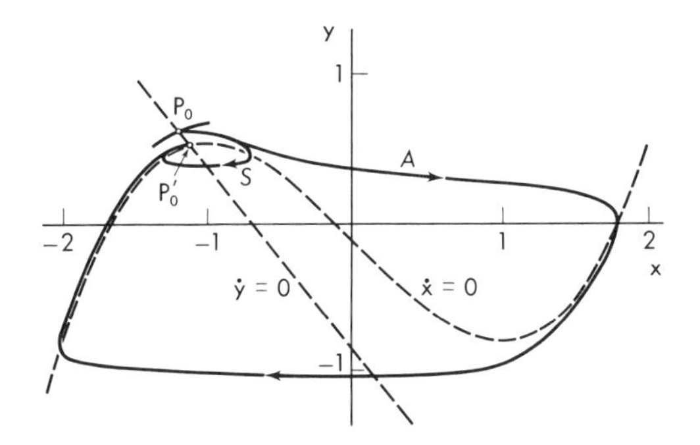

FIG. 2:71. The *BVP* model in which x is the fast, excitatory potential and y is a slow recovery process. The dashed orthoclines are given by the straight line for $dy/dt = 0$ and the cubic for $dx/dt = 0$. After a constant current the path S is subthreshold and A is the active response.

Soon after this work appeared, Nagumo et al. (1962) analyzed the growth of a local stimulus to a full fledged propagating impulse and the loss of a subthreshold response in this system and simulated it with a train of membrane analogue circuits of figure 2:72 in which tunnel diodes gave the cubic characteristic.

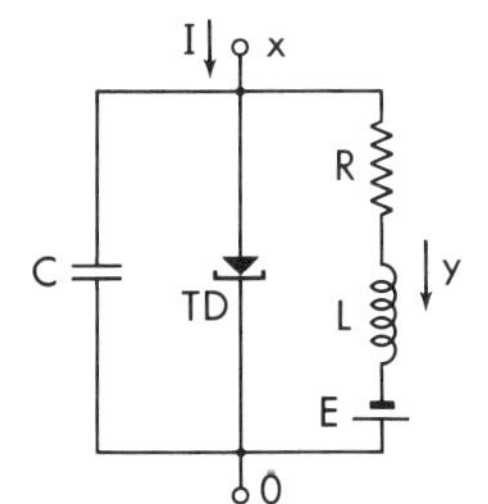

FIG. 2:72. An equivalent circuit for the *BVP* model. A tunnel diode *TD* provides the cubic characteristic. The x parameter is the potential across the circuit and y is the current through the resistance-inductance arm.

I can only conclude by repeating that Bonhoeffer was qualitatively correct about the threshold. I am not pleased to have been wrong and I am very disgusted to have had such complete faith in the all-or-none law. Nonetheless it was interesting as well as good fun.

IRREVERSIBLE PROCESS THEORY

As a student of Bonhoeffer, Uhlrich Franck too became interested in the passive iron wire model as a problem of electrochemistry and as an analogue for nerve behavior. Beyond the iron wire and the fuse he looked at the general properties of unstable systems from the irreversible process theory of Onsager (1931a,b) and particularly Prigogine (1947), and illustrated them by the hydraulic analogy used by Hill (1936) and many others.

The simple case of two forces, X_1, X_2, and two fluxes, J_1, J_2, is stated as

$$J_1 = L_{11}X_1 + L_{12}X_2$$
$$J_2 = L_{21}X_1 + L_{22}X_2$$

where the coefficients L are of a conductance character. Energy is then dissipated at the rate

$$\phi = X_1J_1 + X_2J_2 = T\dot{S}$$

where $\dot{S}$ is the rate of entropy production. It is found that for a steady state $\dot{S}$ is either a minimum or a maximum and that the system is correspondingly stable or unstable. The relations between the coefficients in these two situations are analogous to those found in the nonlinear systems analysis.

Franck (1956) then applied this analysis to the hydraulic system to arrive at the W.C. analogue of nerve activity. He further tabulated some examples of physical and chemical systems and the coefficients which may be negative to give instability. In particular he pointed out the two electrical cases of the dynatron and arc types, and N and S on V, I coordinates, as being duals of each other and so requiring dual circuitry.

OTHER ANALOGUES

It is certainly a matter of common experience that many people can enjoy and appreciate a gadget much more than a differential equation. Further than that, the simpler the gadget and the fewer the hidden parts, the more impressed and interested they are. And I am no exception. Analogies are fun—and useful unless they convey a wrong impression. I am sure that I have only seen or heard of a small fraction of the analogues for nerve that many and very bright people have thought of over a couple of centuries.

One of the simplest, best known, and most used examples of the nature of the nerve impulse is the row of dominoes stood on end, close together. When one is pushed enough a wave of "going down" proceeds to the end of the stack. A fuse is about the same but in Stockholm my very sophisticated audience found a row of matches (Swedish) much more amusing. The very old Chinese flip-flop—or Jacobs ladder—has the advantage of an apparent recovery but don't try to make it too long. The first thermistor—which was such a revelation to me—could just as well illustrate what nerve is all about except that it apparently had an *S* characteristic. Now, however, there are thermistors with positive resistance temperature coefficients which have the more appropriate *N* characteristic.

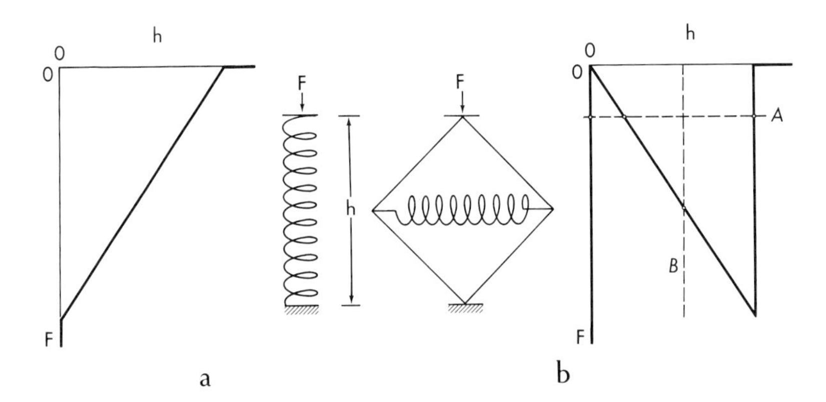

FIG. 2:73. (a.) The usual positive spring requires an increasing force as it is compressed. (b.) The negative spring linkage requires a decreasing force as it is compressed. A constant force, such as weight *A*, results in an unstable equilibrium but a very stiff spring *B* can provide a stable equilibrium anywhere in the negative spring region.

My particular favorite analogy over the past decade has come to be the negative spring (Cole, 1965b). It is so ingenious that I wish I could claim that I had had such an idea—but I suspect that it has long been well known. The ordinary, usual, positive spring is shown with its force-length behavior in figure 2:73a. This particular gadget—which my friends on the West Coast know and perhaps appreciate more than those farther to the East—is shown with its characteristic in figure 2:73b. If the behavior of the usual spring of figure 2:73a is to be called positive there is then no alternative but to label the characteristic of figure 2:73b as that of a negative spring.

If this is indeed an analogue of an axon membrane, what it does depends upon what we ask it to do. Suppose for example that I apply a weight $F = mg$ as given by A. There are obviously three equilibrium positions where the downward force of the weight equals the upward force of the gadget. It is easily—and rather spectacularly—shown that the low and the high positions are indeed "two stable states" while the in-between equilibrium is easily shown to be unstable. Then it is to be expected that for an increase of h less the critical value this analogue will return to the resting state—as after an inadequate stimulus. And, for a displacement about the unstable equilibrium position the device goes rapidly—and noisily—to the active state that is propagated as a quasi stable state in a nerve impulse. But a very strong spring, such as B, analogous to that used on the iron wire, will hold this system in a stable equilibrium.

General Summary

The linear characteristics of the squid axon had given its specifications as a submarine cable. Such a passive cable was quite inadequate by comparison with the axon. This model could not account for excitation or its propagation as a nerve impulse—which required a local energy release for its superior performance. There were a number of formal theories of excitation and propagation. Although these could describe many of the phenomena, they did so in terms of variables and parameters which usually were not physically defined and usually not measured or measurable.

First for *Nitella* and then in the squid axon an increase of membrane conductance was found during an impulse. This supported Bernstein. Then direct measures of the membrane potential confounded this hypothesis by reversing during the impulse.

Venturing beyond the linear region, the membrane conductance was found to be extremely nonlinear and this behavior was closely correlated with the physiological activity below threshold. The membrane rectification and inductive reactance were attributed to the potassium ion. They gave a useful model for both the electrical and physiological behavior of axons below and at threshold. These two phenomena could be correlated in general and, specifically on the basis of electrodiffusion theory, they gave estimates for the mobility and concentration of the potassium ion in the squid membrane.

Although experiments were able to describe excitation data in terms of physical measurements and advance explanations for it in terms of potassium ion movement, the action potential active response still had to be described in terms of variables that were without definition as physical quantities and the overshoot was a complete enigma.

Nonetheless, cable theory considerations and the known parameters of the squid axon required that the membrane perform as a negative resistance. Formal topological investigations of the stability of such systems were at first successful only to the extent that they were removed from experimental physical reality. But eventually the two were to meet.

Over many years many analogies for a nerve impulse had been produced. Although these did not suggest plausible mechanisms for axon membrane performance, they served as more or less interesting demonstrations of nerve-like behavior, and most could be shown to have formal topological characteristics that were similar to those of a living axon.

References

Adrian, E. D., 1914. The all-or-none principle in nerve. J. Physiol., *47:* 460–474.

Adrian, E. D., and Bronk, D. W., 1928. The discharge of impulses in motor nerve fibres. Part I. Impulses in single fibres of the phrenic nerve. J. Physiol., *66:* 81–101.

Adrian, R. H., and Freygang, W. H., 1962. Potassium and chloride permeability of frog muscle membrane. J. Physiol., *163:* 61–103.

Alt, Franz L., 1948. Steady state solutions of the equation of burning. Ballistic Research Laboratories Report #682, Aberdeen Proving Ground, Md.

Alt, Franz L., 1954. Recent results on the equations of burning. Proc. Intl. Math. Cong., Amsterdam, Sept.

Arvanitaki, A., 1939. Recherches sur la réponse oscillatoire locale de l'axone géant isolé de "Sepia." Arch. Intl. Physiol., *49:* 209–256.

Auger, D., and Cotton, E., 1937. Etude de capacité et de resistance accompagnant le développement de la variation négative chez les Characées. C. R. Soc. Biol., *124:* 1211–1214.

Barkhausen, H., 1926. Warum kehren sich die fürden Lichtbogen gültigen Stabilitatsbedingungen bei Elektronenröhren um? Phys. Zeit., *27:* 43–46.

Bartlett, J. H., 1945. Transient anode phenomena. Trans. Electrochem. Soc., *87:* 521–545.

Bartlett, J. H., and Kromhout, R. A., 1952. The Donnan equilibrium. Bull. Math. Biophys., *14:* 385–391.

Bass, L., 1964. Potential of liquid junctions. Trans. Farad. Soc., *60:* 1914–1919.

Becker, J. A., 1947. Properties and uses of thermistors—Thermally sensitive resistors. Bell Syst. Tech. J., *26:* 170–212.

Behn, U. A. R., 1897. Ueber wechselseitige Diffusion von Electrolyten in verdünnten wässerigen Lösungen, insbesondere über Diffusion gegen das Concentrationsgefälle. Ann. Physik u. Chem., *62:* 54–67.

Bernstein, J., 1902. Untersuchungen zur Thermodynamik der bioelektrischen Ströme. Erster Theil. Arch. ges. Physiol., *92:* 521–562.

Bishop, G. H., 1928. The effect of nerve reactance on the threshold of nerve during galvanic current flow. Am. J. Physiol., *85:* 417–431.

Blair, H. A., 1932. On the intensity-time relations for stimulation by electric currents. I. J. Gen. Physiol., *15:* 709–729, II. ibid., 731–755.

Blinks, L. R., 1930. The variation of electrical resistance with applied potential. II. Thin collodion films. J. Gen. Physiol., *14:* 127–138.

Blinks, L. R., 1936a. The polarization capacity and resistance of Valonia. I. Alternating current measurements. J. Gen. Physiol., *19:* 673–691.

Blinks, L. R., 1936b. The effects of current flow in large plant cells. Cold Spring Harbor Symp. Quant. Biol., *4:* 34–42.

Bonhoeffer, K. F., 1943. Zur Theorie des elektrischen Reizes. Naturwissen., *31:* 270–275.
Bonhoeffer, K. F., 1948. Activation of passive iron as a model for the excitation of nerve. J. Gen. Physiol., *32:* 69–91.
Bonhoeffer, K. F., 1953. Modelle der Nervenerregung. Naturwissen., *40:* 301–311.
Brink, F., Jr., 1951. Excitation and conduction in the neuron. In Stevens, S. S. (ed.). Handbook of experimental psychology, New York, Wiley, p. 50–93.
Brink, F., Bronk, D. W., and Larrabee, M. G., 1941. Chemical excitation of nerve. Ann. N. Y. Acad. Sci., *47:* 457–485.
Bruner, L. J., 1965a. The electrical conductance of semipermeable membranes. I. A formal analysis. Biophys. J., *5:* 867–886.
Bruner, L. J., 1965b. The electrical conductance of semipermeable membranes. II. Unipolar flow, symmetric electrolytes. Biophys. J., *5:* 887–908.
Bullock, T. H., 1959. The neurone doctrine and electrophysiology. Science, *129:* 997–1002.
Busch, V., 1913. Stabilität, Labilität und Pendelungen in der Elektrotecknik. Leipzig, Hirzel.
Campbell, G. A., and Foster, R. M., 1931. Fourier integrals for practical applications. Bell Telephone System Monograph B584. (Reprinted New York, Van Nostrand, 1948).
Carslaw, H. S., and Jaeger, J. C., 1947. Conduction of heat in solids. (2nd ed., 1959) Oxford, Clarendon Press.
Chandler, W. K., and Hodgkin, A. L., 1965. The effect of internal sodium on the action potential in the presence of different internal anions. J. Physiol., *181:* 594–611.
Cohen, J., and Cooley, J. W., 1965. The numerical solution of the time-dependent Nernst-Planck equations. Biophys. J., *5:* 145–162.
Cole, K. S., 1933. Electric excitation in nerve. Cold Spring Harbor Symp. Quant. Biol., *1:* 131–138.
Cole, K. S., 1934. Alternating current conductance and direct current excitation of nerve. Science, *79:* 164–165.
Cole, K. S., 1941. Rectification and inductance in the squid giant axon. J. Gen. Physiol., *25:* 29–51.
Cole, K. S., 1947. Four lectures on biophysics. Rio de Janeiro, Institute of Biophysics, Univ. of Brazil.
Cole, K. S., 1949. Some physical aspects of bioelectric phenomena. Proc. Nat. Acad. Sci. U.S.A., *35:* 558–566.
Cole, K. S., 1955. Ions, potentials and the nerve impulse. In Shedlovsky, T. (ed.), Electrochemistry in biology and medicine, New York, Wiley, p. 120–140.
Cole, K. S., 1961. Non-linear current-potential relations in an axon membrane. J. Gen. Physiol., *44:* 1055–1057.
Cole, K. S., 1965a. Electrodiffusion models for the membrane of squid giant axon. Physiol. Rev., *45:* 340–370.
Cole, K. S., 1965b. Theory, experiment and the nerve impulse. In Waterman, T. H. and Morowitz, H. J. (eds.), Theoretical and mathematical biology. New York, Blaisdell, p. 136–171.
Cole, K. S., and Baker, R. F., 1941a. Transverse impedance of the squid giant axon during current flow. J. Gen. Physiol., *24:* 535–549.
Cole, K. S., and Baker, R. F., 1941b. Longitudinal impedance of the squid giant axon. J. Gen. Physiol., *24:* 771–788.
Cole, K. S., and Curtis, H. J., 1936. Electric impedance of nerve and muscle. Cold Spring Harbor Symp. Quant. Biol., *4:* 73–89.

Cole, K. S., and Curtis, H. J., 1938a. Electrical impedance of nerve during activity. Nature, *142:* 209.

Cole, K. S., and Curtis, H. J., 1938b. Electric impedance of *Nitella* during activity. J. Gen. Physiol., *22:* 37–64.

Cole, K. S., and Curtis, H. J., 1939. Electric impedance of the squid giant axon during activity. J. Gen. Physiol., *22:* 649–670.

Cole, K. S., and Curtis, H. J., 1941. Membrane potential of the squid giant axon during current flow. J. Gen. Physiol., *24:* 551–563.

Cole, K. S., and Marmont, G., 1942. The effect of ionic environment upon the longitudinal impedance of the squid giant axon. Fed. Proc., *1:* 15–16.

Cole, K. S., and Moore, J. W., 1960a. Ionic current measurements in the squid giant axon. J. Gen. Physiol., *44:* 123–167.

Cole, K. S., and Moore, J. W., 1960b. Liquid junction and membrane potentials of the squid giant axon. J. Gen. Physiol., *43:* 971–980.

Conti, F., and Eisenman, G., 1965a. The non-steady state membrane potential of ion exchangers with fixed sites. Biophys. J., *5:* 247–256.

Conti, F., and Eisenman, G., 1965b. The steady state properties of ion exchange membranes with fixed sites. Biophys. J., *5:* 511–530.

Conti, F., and Eisenman, G., 1966. The steady-state properties of an ion exchange membrane with mobile sites. Biophys. J., *6:* 227–246.

Cooley, J. W., and Dodge, F. A., Jr., 1966. Digital computer solutions for excitation and propagation of the nerve impulse. Biophys. J., *6:* 583–599.

Cooley, J., Dodge, F., and Cohen, H., 1965. Digital computer solutions for excitable membrane models. J. Cell. Comp. Physiol., *66*, Suppl. *2:* 99–109.

Cox, R. T., Coates, C. W., and Brown, M. V., 1946. Electrical characteristics of electric tissue. Ann. N. Y. Acad. Sci., *47:* 487–500.

Cremer, M., 1900. Ueber Wellen und Pseudowellen. Zeit. für Biol., *40:* 393–418.

Curtis, H. J., and Cole, K. S., 1940. Membrane action potentials from the squid giant axon. J. Cell. Comp. Physiol., *15:* 147–157.

Curtis, H. J., and Cole, K. S., 1942. Membrane resting and action potentials from the squid giant axon. J. Cell. Comp. Physiol., *19:* 135–144.

Davis, H. T., 1936. The Theory of Linear Operators. Bloomington, Ind., Principia Press.

Dean, R. B., Curtis, H. J., and Cole, K. S., 1940. Impedance of biomolecular films. Science, *91:* 50–51.

Debye, P., and Falkenhagen, H., 1928. Dispersion von Leitfähigkeit und Dielektrizitätskonstante bei starken Elektrolyten. Physik. Zeits., *29:* 121–132.

Dray, S., and Sollner, K., 1956. A theory of dynamic polyionic potentials across membranes of ideal ionic selectivity. Biochem. Biophys. Acta, *21:* 126–136.

Einstein, A., 1905. Über die von der molekularkinetischen Theorie der Wärme geforderte Bewegung von in ruhenden Flüssigkeiten suspendierten Teilchen. Ann. der Physik, *17:* 549–560.

Eisenman, G., and Conti, F., 1965. Some implications for biology of recent theoretical and experimental studies of ion permeation in model membranes. J. Gen. Physiol., *48:* (5, part 2) 65–73.

Erlanger, J., and Blair, E. A., 1936. Observations on repetitive responses in axons. Am. J. Physiol., *114:* 328–361.

Falk, G., and Fatt, P., 1964. Linear electrical properties of striated muscle fibres observed with intracellular electrodes. Proc. Roy. Soc. (London) B., *160:* 69–123.

Fessard, A., 1946. Some basic aspects of the behavior of electric plates. Ann. N. Y. Acad. Sci., *47:* 501–514.

FitzHugh, R., 1955. Mathematical models of threshold phenomena in the nerve membrane. Bull. Math. Biophys., *17:* 257–278.

FitzHugh, R., 1960. Thresholds and plateaus in the Hodgkin-Huxley nerve equations. J. Gen. Physiol., *43:* 867–896.

FitzHugh, R., 1961. Impulses and physiological states in theoretical models of nerve membrane. Biophys. J., *1:* 445–466.

FitzHugh, R., 1966. Theoretical effect of temperature on threshold in the Hodgkin-Huxley model. J. Gen. Physiol. *49:* 989–1005.

FitzHugh, R., 1968. Mathematical models of excitation and propagation in nerve. In Schwan, H. (ed.), Bioelectronics, New York, McGraw-Hill.

Forbes, T. W., and Landis, C., 1935. The limiting A.C. frequency for the exhibition of the galvanic skin ("psychogalvanic") response. J. Gen. Psychol., *13:* 188–193.

Franck, U. F., 1956. Models for biological excitation processes. Prog. Biophys. Biophys. Chem., *6:* 171–206.

Franck, U. F., and FitzHugh, R., 1961. Periodische Elektrodenprozesse und ihre Beschreibung durch ein mathematisches Modell. Zeit. für Elektrochem., *65:* 156–168.

Goldman, D. E., 1943. Potential, impedance, and rectification in membranes. J. Gen. Physiol., *27:* 37–60.

Grundfest, H., 1965. Julius Bernstein, Ludimar Hermann, and the discovery of the axon spike overshoot. Arch. ital. Biol., *103:* 483–490.

Guillemin, E. A., 1935. Communication networks. Vol. II, New York, Wiley, p. 181.

Guttman, R., 1939. The electrical impedance of muscle during the action of narcotics and other agents. J. Gen. Physiol., *22:* 567–591.

Henderson, P., 1907. Zur Thermodynamik der Flüssigkeitsketten. Zeit. physik. Chem., *59:* 118–127.

Hermann, L., 1899. Zur Theorie der Erregungsleitung und der elektrischen Erregung. Arch. ges. Physiol., *75:* 574–590.

Herold, E. W., 1935. Negative resistance and devices for obtaining it. Proc. Inst. Radio Eng., *23:* 1201–1223.

Hill, A. V., 1910. A new mathematical treatment of changes of ionic concentration in muscle and nerve under the action of electric currents, with a theory as to their mode of excitation. J. Physiol., *40:* 190–224.

Hill, A. V., 1936. Excitation and accommodation in nerve. Proc. Roy. Soc. (London) B., *119:* 305–355.

Hill, A. V., 1949. The numerical analysis of records to eliminate time-lag. J. Sci. Instr., *26:* 56–57.

Hodgkin, A. L., 1938. The subthreshold potentials in a crustacean nerve fibre. Proc. Roy. Soc. (London), B., *126:* 87–121.

Hodgkin, A. L., 1951. The ionic basis of electrical activity in nerve and muscle. Biol. Rev., *26:* 339 409.

Hodgkin, A. L., 1954. A note on conduction velocity. J. Physiol., *125:* 221–224.

Hodgkin, A. L., and Huxley, A. F., 1939. Action potentials recorded from inside a nerve fibre. Nature, *144:* 710–711.

Hodgkin, A. L., and Huxley, A. F., 1945. Resting and action potentials in single nerve fibres. J. Physiol., *104:* 176–195.

Hodgkin, A. L., and Huxley, A. F., 1952. Currents carried by sodium and potassium ions through the membrane of the giant axon of *Loligo*. J. Physiol., *116:* 449–472.

Hodgkin, A. L., and Katz, B., 1949. The effect of sodium ions on the electrical activity of the giant axon of the squid. J. Physiol., *108:* 37–77.

Hogg, B. M., Goss, C. M., and Cole, K. S., 1934. Potentials in embryo rat heart muscle cultures. Proc. Soc. Exp. Biol. and Med., *32:* 304–307.

Hōzawa, S., 1928. Studien über die Polarisation der Haut. I. Arch. ges. Physiol., *219:* 111–140.

Hōzawa, S., 1935. Die "Lichtbandmethode." Zeit. für Biol., *96:* 586–607.

Jenerick, H., 1963. Phase plane trajectories of the muscle spike potential. Biophys. J., *3:* 363–377.

Jenerick, H., 1964. An analysis of the striated muscle fiber action current. Biophys. J., *4:* 77–91.

Katz, B., 1937. Experimental evidence for a non-conducted response of nerve to subthreshold stimulation. Proc. Roy. Soc. (London) B., *124:* 244–276.

Katz, B., 1939. Electric excitation of nerve , London, Oxford University Press.

Katz, B., 1942. Impedance changes in frog's muscle associated with electrotonic and "endplate" potentials. J. Neurophysiol., *5:* 169–184.

Katz, B., 1948. The electrical properties of the muscle fibre membrane. Proc. Roy. Soc. (London) B., *135:* 506–534.

Kennelly, A. E., 1921. Tables of complex hyperbolic and circular functions. Cambridge, Harvard University Press.

Kennelly, A. E., 1923. Electrical vibration instruments. New York, Macmillan.

Koechlin, B. A., 1955. On the chemical composition of the axoplasm of squid giant nerve fibers with particular reference to its ion pattern. J. Biophys. Biochem. Cytol., *1:* 511–529.

Labes, R., and Lullies, H., 1932. Analyses der Nerveneigenschaften durch Wechselstrommessungen mit Hilfe der Membrankernleitertheorie. Arch. ges. Physiol., *230:* 738–770.

Lapicque, L., 1907. Recherches quantitatives sur l'excitation électrique des nerfs traitée comme une polarisation. J. Physiol. Path. gen., *9:* 620–635.

Le Fevre, P. G., 1950. Excitation characteristics of the squid giant axon: A test of excitation theory in a case of rapid accommodation. J. Gen. Physiol., *34:* 19–36.

Levinson, N., and Smith, O. K., 1942. A general equation for relaxation oscillations. Duke Math. J., *9:* 382–403.

Lewis, G. N., and Sargent, L. W., 1909. Potentials between liquids. J. Am. Chem. Soc., *31:* 363–367.

Liénard, A., 1928. Etude des oscillations entretenue. Revue Générale de l'Electricité, *23:* 901–912, 946–954.

Lillie, R. S., 1936. The passive iron wire model of protoplasmic and nervous transmission and its physiological analogues. Biol. Rev., *11:* 181–209.

Loeb, J., 1920. Ionic radius and ionic efficiency. J. Gen. Physiol., *2:* 673–687.

Lorente de Nó, R., 1947. A study of nerve physiology. N. Y., Studies from the Rockefeller Institute, Vols. *131*, *132*.

Lucas, K., 1909. The "all or none" contraction of the amphibian skeletal muscle fibre. J. Physiol., *38:* 113–133.

Lullies, H., 1937. Die Messung und Bedeutung der elektrolytischen Polarization im Nerven. Biol. Rev., *12:* 338–356.

MacInnes, D. A., 1939. The principles of electrochemistry. New York, Reinhold, Ch. 13.

Mason, W. P., 1948. Electromechanical transducers and wave filters. 2nd edition. New York, Van Nostrand.

Mauro, A., 1961. Anomalous impedance, a phenomenological property of time-variant resistance. Biophys. J., *1:* 353–372.

Maxwell, J. C., 1873. A treatise on electricity and magnetism. 2nd edition 1881, 3rd edition 1892. Oxford, Clarendon Press.

Meyer, K. H., and Sievers, J. F., 1936. La perméabilité des membranes. I. Théorie de la perméabilité ionique. Helv. Chim. Acta., *19:* 649–664.

Michaelis, L., 1926. Die Permeabilität von Membranen. Naturwissen., *14:* 33–42.

Minorsky, N., 1947. Introduction to non-linear mechanics. Ann Arbor, Michigan, J. W. Edwards.

Monnier, A. M., 1934. L'excitation electrique des tissus. Paris, Hermann.

Monnier, A. M., and Coppée, G., 1939. Nouvelles recherches sur la résonance des tissus excitables. Arch. Intl. Physiol., *48:* 129–180.

Mott, N. F., 1939. The theory of crystal rectifiers. Proc. Roy. Soc. A., *171:* 27–38.

Nagumo, J., Arimoto, S., and Yoshizawa, S., 1962. An active pulse transmission line simulating nerve axon. Proc. Inst. Radio Eng., *50:* 2061–2070.

Nernst, W., 1888. Zur Kinetik der in Lösung befindlichen Körper: Theorie der Diffusion. Zeit. physik. Chem., *2:* 613–637.

Nernst, W., 1889. Die elektromotorische Wirksamkeit der Ionen. Zeit. physik. Chem., *4:* 129–181.

Nernst, W., 1899. Zur Theorie der elektrischen Reizung. Nach. K. Gesell. (Göttingen), *1899:* 104–108.

Nernst, W., 1908. Zur Theorie der elektrischen Reizung. Arch. ges. Physiol., *122:* 275–314.

Noble, D., and Stein, R. B., 1966. The threshold conditions for initiation of action potentials in excitable cells. J. Physiol., *187:* 129–142.

Offner, F., 1939. Circuit theory of nervous conduction. Am. J. Physiol., *126:* 594.

Offner, F. F., 1967. Transient forces in membrane relaxation. Biophys. Soc. Abst. TD14.

Offner, F., Weinberg, A., and Young, G., 1940. Nerve conduction theory: Some mathematical consequences of Bernstein's model. Bull. Math. Biophys., *2:* 89–103.

Onsager, L., 1931a. Reciprocal relations in irreversible processes. I. Physic. Rev., *37:* 405–426.

Onsager, L., 1931b. Reciprocal relations in irreversible processes. II. Physic. Rev., *38:* 2265–2279.

Ord, J. L., and Bartlett, J. H., 1965. Electrical behavior of passive iron. J. Electrochem. Soc., *112:* 160–166.

Ostwald, W., 1900. Periodische Erscheinungen bei der Auflösung des Chrom in Säuren. (Parts I and II). Zeit. physik Chem., *35:* 33–76 and 204–256.

Parrack, H. O., 1940. Excitability of the excised and circulated frog's sciatic nerve. Am. J. Physiol., *130:* 481–495.

Pearson, G. L., 1940. Thermistors, their characteristics and uses. Bell Lab. Record, *19:* 106–111.

Planck, M., 1890a. Ueber die Erregung von Elektricität und Wärme in Elektrolyten. Ann. Physik u. Chem., Neue folge, *39:* 161–186.

Planck, M., 1890b. Ueber die Potentialdifferenz zwischen zwei verdünnten Lösungen binärer Elektrolyte. Ann. Physik u. Chem., Neue folge, *40:* 561–576.

Poincaré, M. H., 1881. Memoire sur les courbes définies par une équation differentielle. J. de Math., 3e série, *7:* 375–422.

Poincaré, M. H., 1882. Memoire sur les courbes définies par une équation différentielle. J. de Math., 3e série, *8:* 251–296.

Pol, B., van der, 1926. On relaxation oscillations. Phil. Mag., 7th series, *2:* 978–992.

Polin, H. S., 1935. Corrosion Preventative. U. S. Patent #2021,519. Date of issue 11/19/1935.

Polin, H. S., 1941. Corrosion Preventative System. U. S. Patent #2221,997. Date of issue 11/19/1940.

Pregogine, I., 1947. Etude thermodynamique des phénoménes irréversibles. Paris, Dunod.

Pumphrey, R. J., and Young, J. Z., 1938. The rates of conduction of nerve fibres of various diameters in cephalopods. J. Exp. Biol., *15:* 453–466.

Pumphrey, R. J., Schmitt, O. H., and Young, J. Z., 1940. Correlation of local excitability with local physiological response in the giant axon of the squid (*Loligo*). J. Physiol., *98:* 47–72.

Rashevsky, N., 1933a. Outline of a physico-mathematical theory of excitation and inhibition. Protoplasma, *20:* 42–56.

Rashevsky, N., 1933b. Some physico-mathematical aspects of nerve. Physics, *4:* 341–349.

Rashevsky, N., 1935. Some physico-mathematical aspects of nerve conduction. II. Physics, *6:* 308–314.

Rosenblueth, A., Wiener, N., Pitts, W., and Garcia Ramos, J., 1948. An account of the spike potential of axons. J. Cell. Comp. Physiol., *32:* 275–317.

Rothschild, Lord, 1940. Rhythmical impedance changes in the trout's egg. Nature, *145:* 744.

Rothschild, Lord, 1947a. Rhythmical impedance changes in salmon eggs. Nature, *159:* 134–135.

Rothschild, Lord, 1947b. Spontaneous rhythmical impedance changes in the egg of the trout. II. J. Exp. Biol., *23:* 267–276.

Rushton, W. A. H., 1937a. A graphical solution of a differential equation with application to Hill's treatment of nerve excitation. Proc. Roy. Soc. (London) B., *123:* 382–395.

Rushton, W. A. H., 1937b. Initiation of the propagated disturbance. Proc. Roy. Soc. (London) B., *124:* 210–243.

Schlögl, R., 1954. Elektrodiffusion in freier Lösung und geladenen Membranen. Zeit. physik. Chem. (Neue folge), *1:* 305–339.

Schmitt, O. H., 1937. An electrical theory of nerve impulse propagation. Am. J. Physiol., *119:* 399.

Schmitt, O. H., 1955. Dynamic negative admittance components in statically stable membranes. In Shedlovsky, T. (ed.) Electrochemistry in biology and medicine. New York, John Wiley, p. 91–120.

Seeliger, R., 1927. Einfuhrung in die Physik der Gasentladungen. Leipzig, Barth (Reprint Ann Arbor, Edwards, 1944).

Segal, J. R., 1967. Electrical capacitance of ion-exchange membranes. J. Theoret. Biol., *14:* 11–34.

Silberstein, L., 1932. Determination of the spectral composition of X-ray radiation from filtration data. J. Opt. Soc. Am., *22:* 265–280.

Sinharay, H., and Meltzer, B., 1964. Characteristics of insulator diodes determined by space-charge and diffusion. Solid State Electronics, *7:* 125–136.

Skinner, S. M., 1955. Diffusion, static charges and the conduction of electricity in nonmetallic solids by a single charge carrier. II. J. Appl. Physics, *26:* 509–518.

Sollner, K., Dray, S., Grim, E., and Neihof, R., 1954. Electrochemical studies with model membranes. In Clarke, H. T. (ed.). Ion transport across membranes. New York, Academic Press, p. 144–188.

Steimel, K., 1930. Die Stabilität und die Selbserregung elektrischer Kreise mit organen fallender Charakteristik. Zeit. f. Hochfrequenz, *36:* 161–172.
Tasaki, I., 1949. Collision of two nerve impulses in the nerve fibre. Biochim. Biophys. Acta., *3:* 494–497.
Tasaki, I., 1959. Conduction of the nerve impulse. In Field, J. (ed.) Handbook of physiology, 1. Washington, D. C., American Physiological Soc., p. 75–121.
Tasaki, I., and Hagiwara, S., 1957. Capacity of muscle fiber membrane. Am. J. Physiol., *188:* 423–429.
Tasaki, I., and Mizuguchi, K., 1949. The changes in the electric impedance during activity and the effect of alkaloids and polarization upon the bioelectric processes in the myelinated nerve fibre. Biochim. Biophys. Acta, *3:* 484–493.
Taylor, R. E., 1963. Cable theory. In Nastuk, W. L. (ed.) Physical techniques in biological research. Vol. *6*. Ch. 4. New York, Academic Press, p. 219–262.
Teorell, T., 1935. An attempt to formulate a quantitative theory of membrane permeability. Proc. Soc. Exp. Biol. and Med., *33:* 282–285.
Teorell, T., 1948. Membrane electrophoresis in relation to bio-electrical polarization effects. Nature, *162:* 961.
Teorell, T., 1949. Membrane electrophoresis in relation to bio-electrical polarization effects. Arch. Sci. Physiol., *3:* 205–219.
Teorell, T., 1951. Zur quantitativen Behandlung der Membranpermeabilität. Zeit. Elektrochem., *55:* 460–469.
Teorell, T., 1953. Transport processes and electrical phenomena in ionic membranes. In Butler, J. A. V. and Randall, J. T. (eds.) Progress in biophysics and biophysical chemistry. Vol. 3, London, Academic Press, p. 305–369.
Ussing, H. H., 1949. The distinction by means of tracers between active transport and diffusion. Acta Physiol. Scand., *19:* 43–56.
Ussing, H. H., 1952. Some aspects of the application of tracers in permeability studies. Advanc. Enzymol., *13:* 21–65.
Volterra, V., 1931. Lecons sur la theorie mathematique de la lutte pour la vie. Paris, Gauthier-Villars.
Von Kármán, Th., 1940. The engineer grapples with nonlinear problems. Bull. Am. Math. Soc., *46:* 615–683.
Walker, J. L., Jr., and Eiesnman, G., 1966. A test of the theory of the steady-state properties of an ion exchange membrane with mobile sites and dissociated counterions. Biophys. J., *6:* 513–533.
Walker, J. L., Jr., Sandblom, J., and Eisenman, G., 1968. Theory and test of the transient properties of an ion exchange membrane with mobile sites and dissociated counterions. (In manuscript.)
Warburg, E., 1899. Ueber das Verhalten sogenannter unpolarisibarer Elektroden gegen Wechselstrom. Ann. physik. Chem., *67:* 493–499.
Wegel, R. L., 1930. Theory of vibration of the larynx. Bell Sys. Tech. J., *6:* 207–227.
Weidmann, S., 1951. Effect of current flow on the membrane potential of cardiac muscle. J. Physiol., *115:* 227–236.
Young, G., 1937. Note on excitation theories. Psychometrika, *2:* 103–106.

PART III

Ion Conductances and an Impulse

Contents to Part III

Taming the Axon

CABLE FRUSTRATIONS AND LIMITATIONS

In spite of the considerable qualitative success in describing linear and subthreshold squid axon characteristics there were two rather obvious deficiencies. First was that there was no readily available way to compute the time course of the membrane potential after the application of a constant current from outside the axon. Even if the complete time and current relations of the membrane conductance were known, the solution of the cable equation was a laborious task. This John Tukey proposed to do in 1941 with the then available IBM card equipment. It should have worked had we known what characteristics to put in, but I could not make guesses that seemed to be worth the effort. Conversely, it was not found possible to so manipulate the cable equation that it could give the transient conductance characteristics in somewhat the same manner as the steady state properties had been extracted.

A rather frustrating limitation of the steady state impedance measurements was that it was not possible to get enough current flow through the membrane for a measurement except by the longitudinal axon electrode arrangement. We had been limited to experiments at the resting potential and I very much hoped that the effects of current flow on the steady state impedance would provide the link between the rectification curve for potential vs. current and the effect of current flow on the frequency-independent change of conductance with current flow. It was obvious that the spatial distribution of both alternating and direct currents in the electrode regions would make the analysis difficult and quite likely impossible. But having been lucky on the simple steady state rectification, I had a blind hope that some sense might be made of the data. Marmont and I worked for the better part of a summer on a wide variety of polarizing schemes, none of which had much potentiality and none of which gave useful data—so there was no problem of interpretation.

Even worse was the persistent question, "How does an impulse start?" Curtis and I took many records of the potential near the threshold, above and below, and almost as many, probably ill-advised and certainly futile, attempts were made to see something through the complexity of the cable equation. One thing that was consistently true was the simple and rather absurd observation that if the potential did not show an inflection it was subthreshold whereas it never returned from an inflection until after a full-blown response.

A considerable variety of experimental and analytical approaches had produced at least as much variety of ideas and information. In fact, I felt that there must be enough available in the background to provide a simple concept of the workings of a nerve membrane if one but had that touch of genius that could ignore everything but the important. It had become rather obvious—although I am not sure it was so clearly stated—that variation along the length of the distributed system was an ever present, an apparently almost insurmountable, obstacle. The axon and the cable equation must correspond to each other. But the equation could only rarely be used to interpret the axon performance into a useful form. Conversely the partial differential equation was so difficult to manage that practically no experimentation could be done on it to attempt to match the axon behavior particularly under the rather complicated and difficult conditions that had been tried. So there was no real beginning of an understanding of excitation and its propagation. Along with the many facets that had been seen clearly was the somewhat intuitive conviction that somehow, somewhere, an *N* type of negative resistance played the major role. The situation was obviously ripe for new ideas.

SPACE AND CURRENT CONTROL

During an inquisition on our aims and failures, L. J. Savage asked with characteristic shrewdness why we did not put a current-carrying electrode inside the axon and measure directly some of the things we were failing to get indirectly. I protested that electrode impedance would predominate because we could not make electrodes good enough to allow them to have so small an area. Quite undaunted, Marmont soon came up with the proposal for a long current-carrying electrode on the axis of the axon, a coaxial external measuring electrode with equipotential guards at each end, and an electronic feedback membrane current control with direct measurement of the membrane potential. This arrangement would have the great advantage that the current density and potential would be the same at each time and at all points of the known measuring area of the membrane, while the guards would dispose of the complicated time and distance phenomena at the ends. This would allow for the first time a direct measure of absolute membrane current densities and their relation to the membrane potential—only a few of which had been rather indirectly calculated. And I pointed out that potential as well as current could and should be controlled.

This new approach involved considerable experimental complications—in the axon mounting, the internal and external electrodes, circulation of external medium, and the speed and stability of the control electronics (Marmont, 1949). But the advantage from the point of view of theory and analysis was difficult to overestimate.

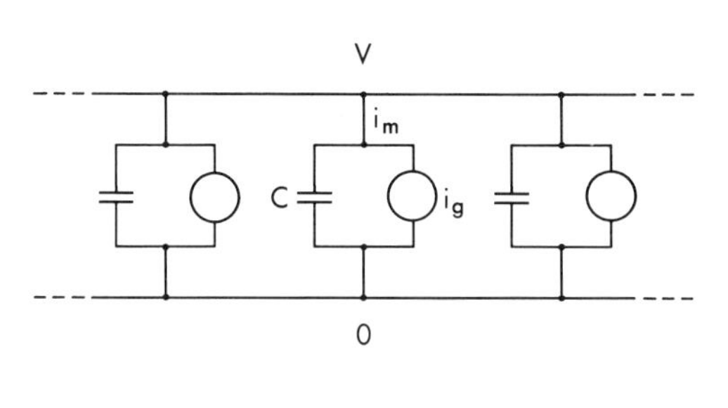

$$i_m = C \frac{dV}{dt} + i_g$$

FIG. 3:1. An axon with extended internal and external electrodes is represented by the equivalent cable, above, with the inside electrode at a potential V relative to the outside electrode and by the simplified cable equation, below, for the membrane current.

In the Kelvin (1885) equation 1:31 and figure 3:1, we now had the membrane potential independent of the space coordinate along the measuring chamber. With $\partial^2 V/\partial x^2 = 0$, we were left with

$$i_m = C \frac{dV}{dt} + i_g(V, t)$$

an ordinary differential equation. Although certainly nonlinear, to an unknown but most probably violent degree, this equation was vastly closer to the unknowns of the problem than before. As a phenomenological description, it could be said that the axon had been robbed of its ancient right to propagate an impulse by eliminating the local circuit currents, $\partial^2 V/\partial x^2$, by which an active region normally reached ahead to move itself along the axon. Or it could be maintained that the internal and external electrodes enabled an impulse approaching one end to reach over the entire electrode length simultaneously to give a near infinite velocity.

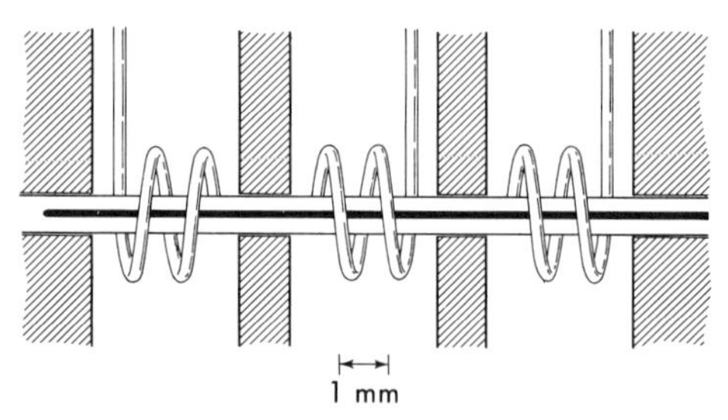

FIG. 3:2. Diagram of squid axon and electrode arrangement for membrane controls. The axon passed through holes in the shaded insulating partitions. The central outside spiral electrode, for current and potential measurement, was between similar guard electrodes. The internal electrode, solid line, was inserted from the right to lie in the center of the axon.

The guarded electrode measurement of figure 3:2 had been introduced by Kelvin and was so well known that its importance was obvious—after it had been suggested. It required that the gap between the measuring and guard electrodes be small enough to allow only a negligible variation of membrane current density and potential in this region. This has been a potential theory problem of enough complexity that it has usually been solved only approximately and by a resistance network analogue. The other obvious requirement for the guards was that they be long enough to attenuate anything happening at the outer ends and prevent any effect in the central region. This also has not been an easy problem either to state or to solve as it has appeared in many different forms and been approximated in many different ways.

Only much later were critical evaluations of the effectiveness of such a system made. Measurements of membrane potential during impulse propagation along an axon without and with the free axial

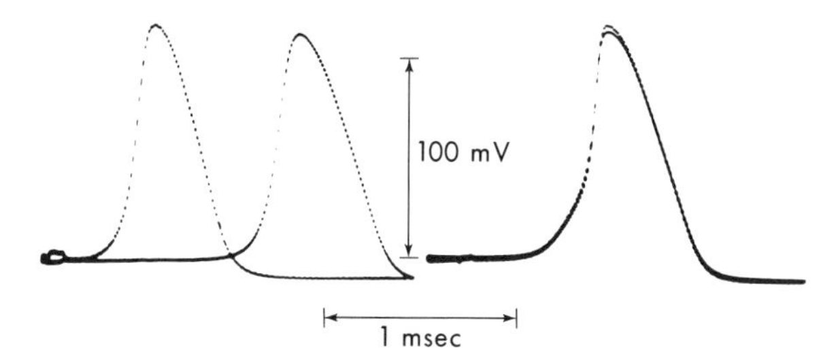

FIG. 3:3. An illustration of the effect of an internal, axial electrode on the response of a squid axon. Before electrode insertion, left, the action potential propagated normally between two microelectrodes, 15 mm apart. After electrode insertion, right, the action potentials appeared practically identical and simultaneously at the recording electrodes.

electrode, figure 3:3, by del Castillo and Moore (1959) show the near simultaneity in the latter case. On the other hand, with a nearly ideal internal electrode at a fixed potential, figure 3:4 (Cole and Moore, 1960a), the central region is virtually immune to extensive and normal potential changes beyond the electrode regions.

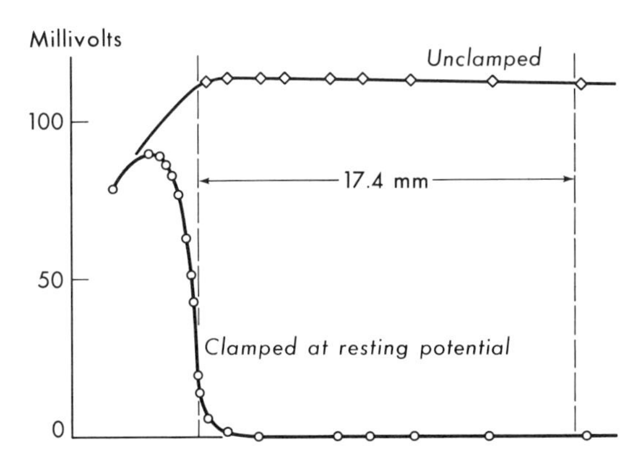

FIG. 3:4. An illustration of internal electrode control of the squid axon. The axon was stimulated at the left, the axial electrode was between the vertical dashed lines, and potentials were measured with a microelectrode. The action potential remained constant, above, with the axial electrode floating, but was quickly extinguished, below, as it passed the end of the electrode fixed at the rest potential.

Quite aside from the immediate possibilities, this approach has seemed to me to be a beautiful example of the interplay of experiment and theory. Much of the previous progress had been made possible by varied and usually rather elementary theoretical considerations. But a limit had been reached practically at which the appropriate theory was far too difficult—for us at least—to be useful. Since no way could be found to appreciably simplify the theory for existing experiments the impasse could only be broken by very considerably complicating the experiment; placing the axon in a highly nonphysiological situation, for which the theory was almost immeasurably simplified. This is no isolated example of a valuable strategy—if in experimental trouble look at theory and if in theoretical trouble look at experiment.

The control concept had been highly developed during World War II, principally with feedback electronics. It was widely applied afterward and promoted to a considerable extent under the name of cybernetics by Norbert Wiener (1948). In general the difference between the actual and the desired position of a system is used to control the power to reduce this error. The control of the membrane current, which was certainly an early application of the concept to biological experiment, is illustrated in figures 3:5 and 3:6. The actual current flow I through the membrane m between the inner and outer electrodes, i and e, produces the potential $V_R = RI$ across the standard resistor R. The desired current I_C is expressed as a command potential $V_C = RI_C$. The difference between V_R and V_C, the error signal, is then applied to the amplifier, of transconductance δ, to produce the current I. Then

$$I = \delta(V_C - V_R) = \delta R(I_C - I)$$

and

$$I = I_C/(1 + 1/\delta R)$$

to show that the error in I is less than 1 percent for $\delta R > 100$.

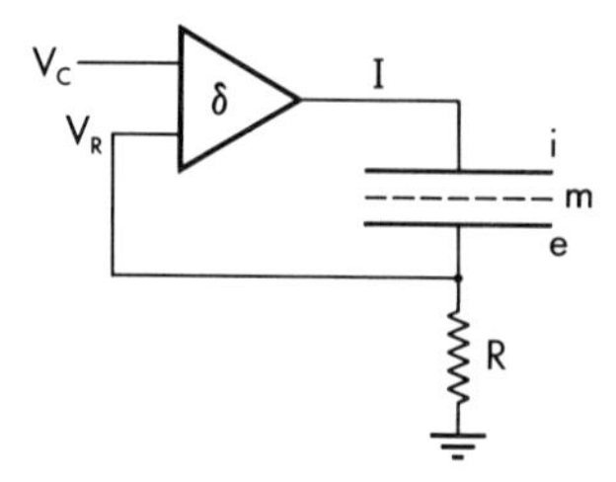

FIG. 3:5. Diagram for control of current I through axon membrane m. The difference between the potential V_R across the resistor R and the command potential V_C controls the output current I of amplifier δ to internal electrode.

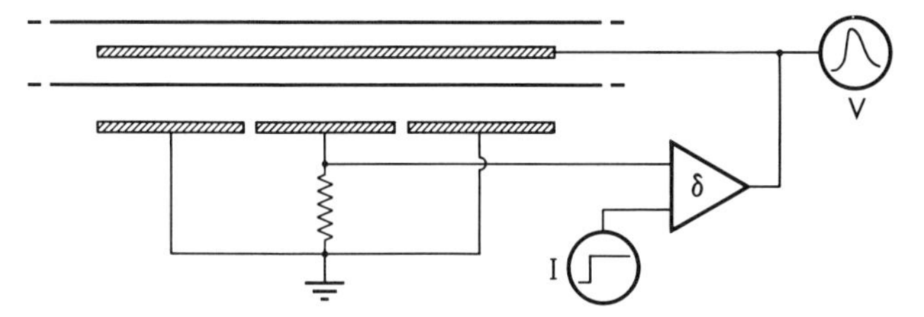

FIG. 3:6. The current through the membrane and the resistor to outside central measuring electrode is maintained equal to the command I by amplifier δ and the resultant internal potential is given at V.

This indeed relegated our previous constant current arrangement of many megohms, many batteries and the mechanical contacts of the Lucas spring rheotome to the horse and buggy category. However, as must be expected for any instrument of great power it has had to be used with a commensurate understanding and care that have given rise to a truly vast literature on the theory and practice of control.

It is an axiom of progress that the difficulty and uncertainty of any step into the unknown is replaced by ease and elegance as the power and certainty of both technique and concept close over the old to go on to the new frontiers.

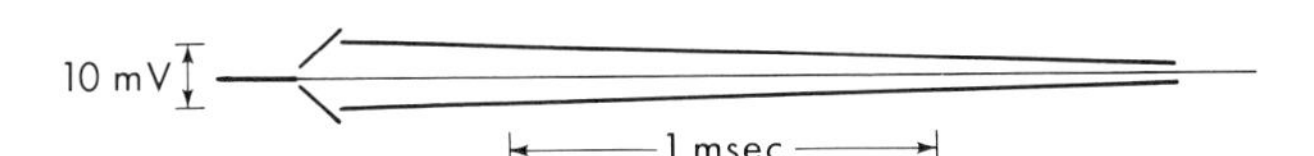

Fig. 3:7. The squid axon membrane potential during short small in and out pulses of current, left, and the subsequent returns without external current flow, right.

Application of a short current pulse to the squid axon membrane, figure 3:7, gave a sudden jump of potential as the result of the resistances between the internal and external electrodes. The potential then began to rise nearly linearly to give a measure of the membrane capacity. As the current was returned to zero on command the reverse of the initial jump was followed by a nearly exponential return to the rest potential. This time constant then gave the resting membrane resistance since $\tau = RC$. For changes of potential somewhat less than 10 mV, the amplitudes and time courses were quite impressively independent of the direction of current flow, to demonstrate linearity over this range. Under other conditions, the linear response might be a damped oscillation, figure 3:37, which seemed to correspond to some depolarization.

As a consequence, some inductance had to be included in the equivalent circuit such as had been given by a theoretical K channel. These records which could be taken in a few seconds then gave quite directly the membrane capacity and resistance values that originally had taken many weeks of experiment and analysis or, in some cases, such as the inductive reactance, that had taken years.

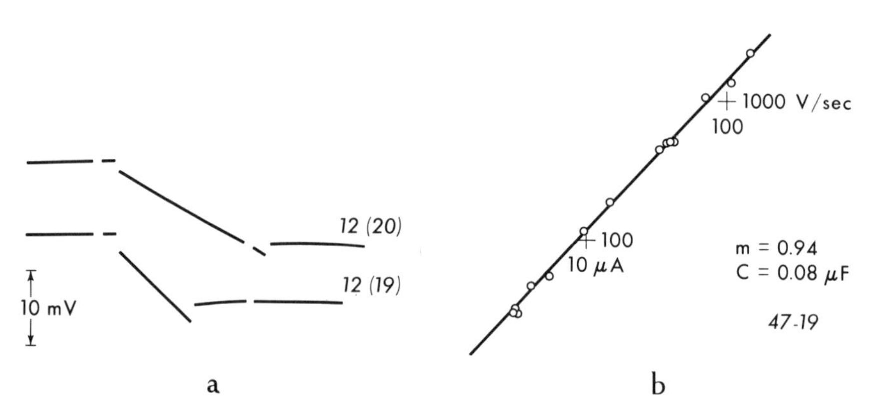

Fig. 3:8. (a.) Initial changes of membrane potential for short pulses of current showing the on and off jumps corresponding to the series resistance, and the intermediate charging of the axon membrane capacity. (b.) The initial charging of an axon membrane capacity as a function of the applied current, on logarithmic scales.

The early rate of rise of the potential, figure 3:8a, gave an average for the membrane capacity of 1.2 $\mu F/cm^2$. Although this was somewhat higher than usually found before or since, it was again a reassuring as well as a simple and direct measurement.

Another consequence of these experiments with current control was that the early rate of rise of potential was not constantly proportional to the current. It varied as a small power of the durations of the pulse, figure 3:8b, which were far below the time constant of the membrane. The variation of the dielectric current after application of a potential had been rather completely calculated in series, even in the difficult range $t \approx \tau_0$. The calculation for the potential after application of a current to such a dielectric with a series resistance was rather more unpleasant. The Heaviside operational calculus gives it in incomplete gamma functions. In a crude way it seemed possible to correlate the observation with the dielectric parameter α although a detailed analysis of the individual curves was not undertaken. Hodgkin et al. (1952) found a variation from an exponential in the transient current after a membrane potential step that was qualitatively to be expected from a dielectric —initially too rapid and later too slow.

On analysis the jumps of potential at the on and off of the current, such as figure 3:8a, were found to be surprisingly large. The direct calculation gives

$$\begin{aligned} R &= R_0 + R_1 + R_2 \\ &= R_0 + \frac{1}{2\pi l}\left[r_1 \int_{a_0}^{a_1} d\rho/\rho + r_2 \int_{a_2}^{a_0} d\rho/\rho\right] \\ &= R_0 + \frac{1}{2\pi l}\left[r_1 \ln\,(a_1/a_0) + r_2 \ln\,(a_0/a_2)\right] \end{aligned}$$

where R_0 is the resistance of the electrodes over the length l of the measuring section, r_1 and r_2 are the specific resistances of sea water and axoplasm, and a_0, a_1, and a_2 are the radii of the axon membrane and of the sea water and axial electrodes. The resistance of some electrodes had been measured as a function of frequency or by pulses in sea water in the cell and were reasonably low. The specific resistance of the axoplasm had shown considerable variability in earlier experiments, although the ranges did overlap slightly. However, the calculation of r_2 from these potential measurements gave excessive values averaging about 100 Ωcm. If we refused to accept them, the only alternative seemed to be that there was considerable resistance in series with the capacity in the membrane structure. This would be from 5 to 3 Ωcm^2 according as the axoplasm was once or twice the resistivity of sea water.

Such a resistance was not a complete surprise because the internal resistivities from suspension measurements had shown a variability that was unexpected for so well-controlled a biological system. It was realized that a thin resistive layer close to the capacitive structure might have to be included as part of the equivalent internal resistivity,

$$\hat{r}_2 = r_2 + r_3/a. \qquad (1{:}12)$$

This more direct appearance of complexity in the membrane was confirmed by Hodgkin et al. (1952). They went on to infer, from sea water-choline mixtures for the external medium, that such a membrane resistance might be 80 percent outside and 20 percent inside.

As such a current is continued or increased we can predict symmetry only so long as the membrane is linear with the parameters independent of current. For simplicity—perhaps near the equilibrium potential for potassium—we may assume exponential potentials under these conditions of about 1 msec time constant for both

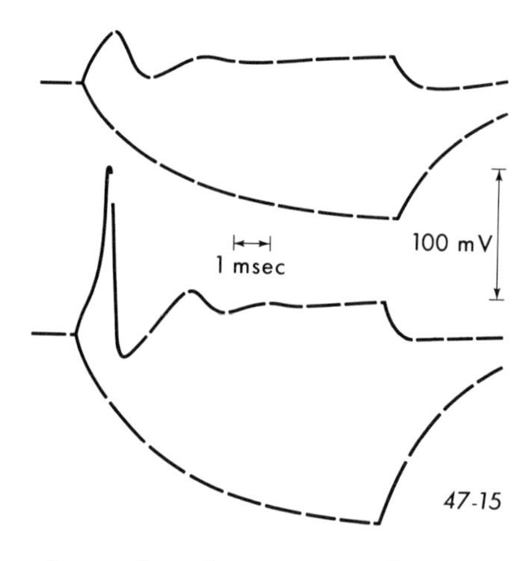

FIG. 3:9. The squid axon membrane potential during and after long inward and outward pulses of current. Subthreshold response shown by upper records and superthreshold behavior shown in lower curves.

an inward and an outward current. In figure 3:9 (Cole, 1949) we have gone considerably beyond the linear range while still below threshold. There is a prolonged transient with an increased steady state potential for the inward current, but a definitely oscillatory transient and lower steady state for the outward current. These responses are quite satisfactorily similar to those calculated for the simple rectifier and inductance model, figure 2:44. For the outward current oscillatory potential there could be little doubt of the need for an inductive reactance but for the hyperpolarization we might expect a capacitive reactance from the poorly conducting side of a potassium rectifier. Although there is considerable analytical difference between a circuit with two capacities and an overdamped inductance-capacity circuit, the two can be nearly enough equivalent so that an easy distinction cannot be made. For example, the impedances Z of the two circuits, figure 3:10 are similar:

$$Z = \frac{R_1(Lp + R_2)}{R_1LC_1p^2 + (L + R_1R_2C_1)p + R_1 + R_2}$$

and

$$Z = \frac{R_1(R_2C_2p + 1)}{R_1R_2C_1C_2p^2 + [R_1C_1 + (R_1 + R_2)C_2]p + 1}.$$

Although an analysis was not attempted, the two capacitor circuit seems the more probable. In some ways the similarity of these tran-

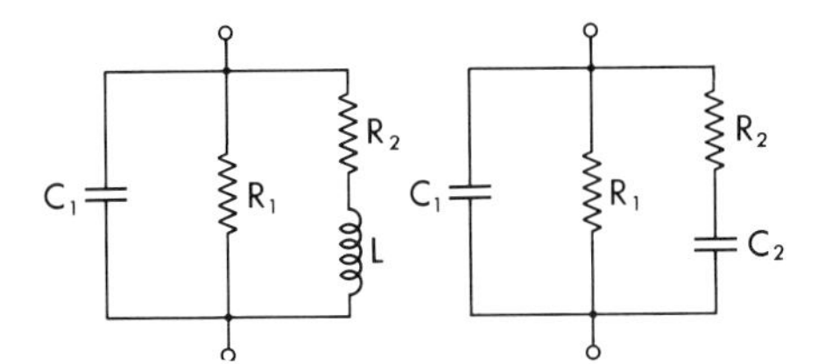

FIG. 3:10. Two circuits which may give responses similar to those of an axon membrane with inward current.

sients to those which had been found with distributed parameters (Curtis and Cole, 1940) was disturbing. They made one wonder if it ought not to have been possible to make some useful approximation to the cable geometry. Perhaps this could have been done earlier and more easily than the ultimate experimental reduction of the problem to one of lumped parameters.

THRESHOLD

A slight further increase of the current step gave again the full-blown action potential response as the threshold was passed to give rheobase I_0, figures 3:9 and 3:11a. At the other extreme an effectively instantaneous 10 μsec pulse, figure 3:11b, and an intermediate 0.1 msec pulse, figure 3:11c, also showed a latent period before it was obvious whether the stimulus was above or below threshold. It was now possible to repeat the classical stimulus strength-duration threshold experiment under conditions of uniform current density and membrane potential. This appeared to be quite important because in the midst of cable analysis problems it had seemed entirely possible that empirical formulations such as those of Rashevsky, Monnier, and Hill might reflect to a considerable extent the distance-time characteristics of the nerve in the experimental situation. Most of the experiments were presented on a normalized basis in figure 3:12 (Cole, 1955). It is clear that the constant quantity $It = Q_0$ relation is quite well followed at short times—as Hill had rather strongly emphasized. Since this corresponded to a constant change of potential across the membrane capacity alone, it was a simple thing to postulate this as the criterion of excitation. The dielectric loss characteristic of the membrane should then modify the constant quantity relation. Whether or not this might be detectable has not been checked.

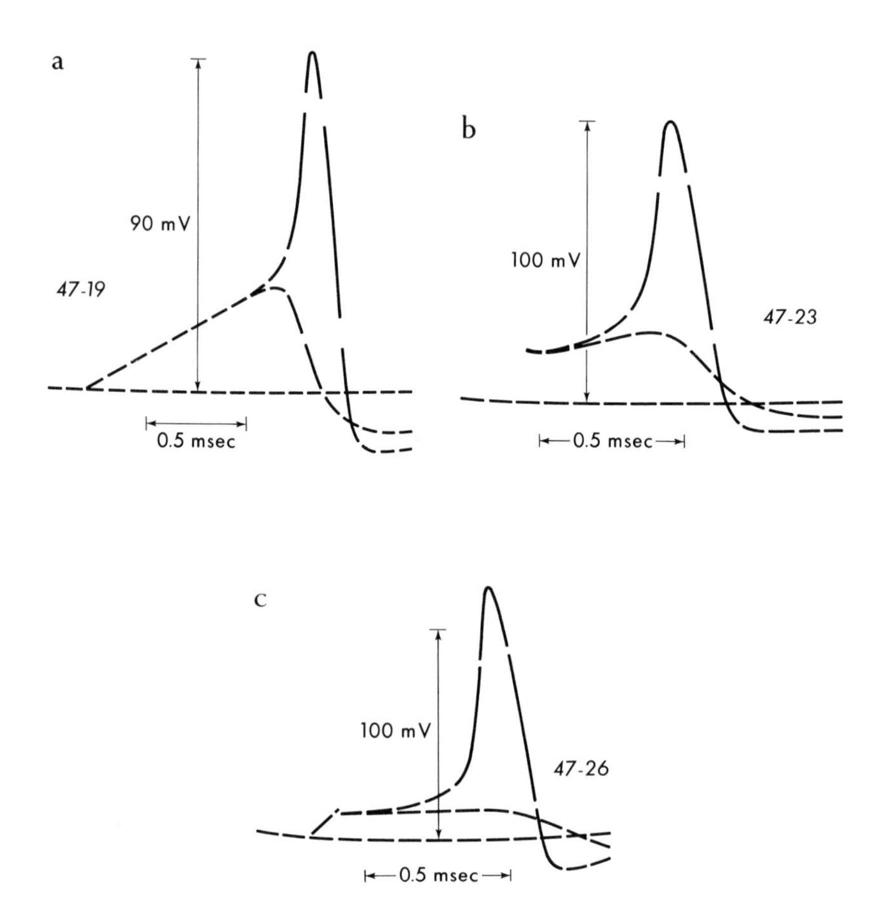

FIG. 3:11. Subthreshold and threshold responses of squid axon membranes to outward current of *a*, long, *b*, short, and *c*, intermediate duration.

The sharpness of the transition from the constant quantity threshold at short times to the rheobase at long times was rather surprising. This was to be measured (FitzHugh, 1966) as

$$\sigma = I(\tau)/I_0$$

where $\tau = Q_0/I_0$. The assumptions of a constant capacity and resistance and a threshold potential change for the membrane are equivalent to the Rashevsky-Monnier-Hill formulation without accommodation, giving the simple exponential of figure 3:12 with

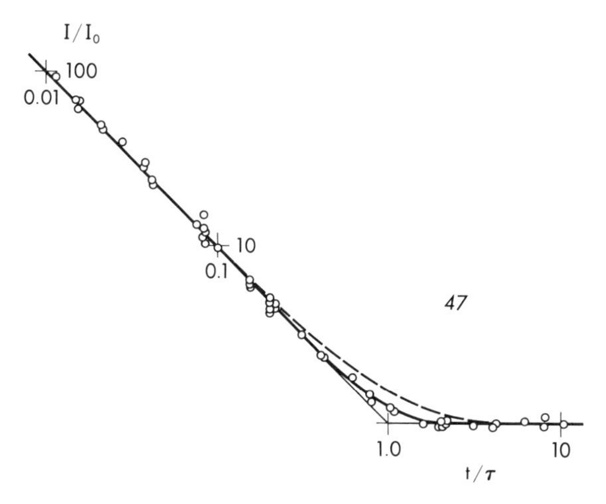

Fig. 3:12. Threshold current strength I vs. duration t expressed in terms of rheobase I_0 and a time constant τ, and on logarithmic scales. The dashed curve is for a single exponential function.

$\sigma = 1.582$. The introduction of accommodation makes the transition more abrupt. Carried to an extreme, Hill's expression—as modified to give the apparent rather than the "true" rheobase which is not observable—leads to the indeterminate 0/0 as $\lambda \to k$. This had been evaluated long before and was extended by Le Fevre (1950).

In Hill's terminology the solution of equations 2:2 gives the threshold stimulus I for a duration t as

$$I = I_0(1 - k/\lambda)/[\exp(-t/\lambda) - \exp(-t/k)]$$

where I_0 is the "true" rheobase and k and λ are the time constants of excitation and accommodation. At short times $It \to kI_0 = Q_0$ to give the constant quantity relation. The threshold is a minimum for

$$t^* = \left[1 \Big/ \left(\frac{1}{\lambda} - \frac{1}{k}\right)\right] \ln k/\lambda$$

to give the "observed" rheobase

$$I^* = I_0(\lambda/k)^{1/(\lambda/k-1)}.$$

But for $\lambda = k(1 + \delta)$, where δ is small,

$$I \to I_0 k/t \exp(-t/k)$$

and

$$I^* \approx I_0\, e \text{ at } t = k$$

or

$$I/I^* \approx \frac{1}{e} \cdot \frac{k}{t} \exp\left(\frac{t}{k}\right)$$

for $t \leq k$ (Cole et al., 1955).

This strength-duration relation shows a transition from constant quantity to rheobase, $\sigma = 1.455$, only somewhat sharper for $\lambda \to k$ than for $\lambda = \infty$, $\sigma = 1.582$. It still was not adequate to express the data of figure 3:12 with $\sigma = 1.26$. Thus, as was discussed earlier (Cole, 1941), the restriction to real time constants—as made implicitly by Hill, explicitly by Monnier and Coppée (1939) and emphatically by Katz (1939)—was not only unnecessary but also ill-advised in view of the oscillatory nature of the subthreshold membrane potential. However, a detailed comparison of the more general formulation of the two-time constant formulation of excitation—or the equivalent calculation for an *RLC* membrane—was only made much later (FitzHugh, 1968).

The absolute values of the thresholds, which I had been unwilling even to estimate before, were now directly available. Certainly the most interesting was the constant quantity Q_0 at short times. This averaged $2 \cdot 10^{-8}$ coulomb/cm², to give about one ion pair separated for a 300 Å square. This further gave an initial threshold change of membrane potential $V_0 = Q_0/C = 16$ mV which was not unexpected. The rheobase I_0 was more variable but gave 35 μA/cm² for the average at room temperatures of 26–29°C from the considerable electronics without compensating air conditioning. This then gave a time constant $\tau = Q_0/I_0 = 0.6$ msec. If the membrane resistance remained at its nominal rest value of 1000 Ωcm², we calculated a steady state threshold potential change of 35 mV which was probably as good as the assumption. In the next summer, 1948, with mostly new and redesigned equipment (Marmont, 1949) we found a much higher rheobase, 28 μA/cm² at 15°C with an average Q_{10} of 2.0 from 5° to 25°C. Subsequently Hagiwara and Oomura (1958) reported a threshold potential change independent of stimulus duration. Then Sjodin and Mullins (1958) found a threshold decrease at 1 msec with increase of temperature. More recently, Guttman (1962, 1966) has shown Q_0 to be nearly independent of temperature at $1.4 \cdot 10^{-8}$ coulomb/cm². In spite of a difference in technique, she also reports a rheobase of 17 μA/cm² at 20°C with a Q_{10} of 1.9, to give more confidence in the earlier results.

A considerable amount of other work was carried through; one unpublished result was the correlation by Marmont of the constant quantity threshold with the external sodium. The currents required to block the response were found to be very large from the inflection through the peak and some attempts were made to determine effective resistances through a spike. These were quite difficult to measure because of the spontaneous variation of the latent period.

In the course of all threshold experiments, except for rheobase, the stimulating membrane current was returned to its original zero value after the pulse. At no time were there any additional currents such as had been present because of the spatial derivative $\partial^2 V/\partial x^2$ at the site of stimulation in an unskewered axon. Yet the time courses of the various potentials were remarkably little changed in form from those in which these local circuit currents had been supposed to have contributed, such as figures 2:3 and 3:3.

I had definitely hoped, without any particular reason, that interesting and useful differences would appear between the behaviors of the axons with and without the local circuit currents. But the linear and subthreshold behaviors were quite nearly and rather reasonably approximate. The threshold and subsequent responses were without significant differences. Although I was quite disappointed, I came to realize that there was a considerable clarification of the current vs. potential relations. The earlier representation of figure 2:23 had been based on the hypothesis that there had been little ionic membrane current flow before the conductance increase had started. Here, however, there was no need to balance off the capacitive and the local circuit current as in equation 2:7 to obtain the appropriate scale factors for the rest of the analysis.

The total membrane current was required to be zero with a considerable accuracy. There was no good reason to expect that the membrane capacity current was much different from $i_c = C\, dV/dt$. Consequently, the ionic or "leakage" current must be

$$i_g = -C\, dV/dt$$

and this was quite significant—even without scale factors which were rather easy to obtain. Then the V_m vs. i_g relation, which was actually plotted as a phase plane locus (figure 3:13), showed clearly the nature of the excitation process. After a reduction of the membrane potential by a threshold amount, current started to flow through the ionic path. But, as this further reduced the capacity

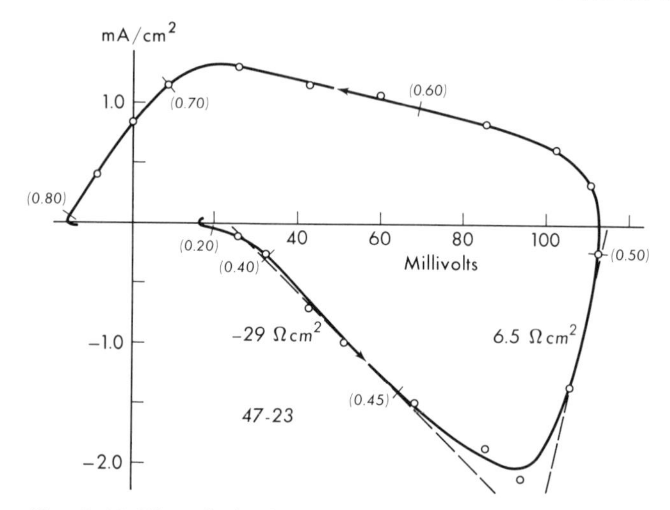

FIG. 3:13. The relation between the squid axon membrane potential and the ionic current density during an excitation. The figures in parentheses are times in msec after the stimulus.

charge and the membrane potential, the ionic current did not tend to decrease as it would for a circuit with an emf equal to the resting potential and constant resistances. Quite the contrary, as the membrane potential moved farther from the rest value, the inward ionic current increased to give a response of an autocatalytic nature. After the membrane potential had come near to zero this unstable action began to slow up, and after the spike of the action potential the ionic current reversed to follow a rather reasonable path back to zero as the membrane potential returned during recovery toward its rest value.

The initial excitation process with a zero external net current seemed to involve two important interacting processes: first, the membrane capacity was supplying the energy for the process, and second, the ionic path was acting as a negative resistance in that for a major part of excitation the current through it was increasing as the potential decreased. The numerical value of about $-30\ \Omega cm^2$ seemed to have no particular significance except that it was reasonably close to the $-45\ \Omega cm^2$ obtained so painfully and uncertainly before the war. Again the positive resistances before the peak of the potential agreed in the neighborhood of $8\ \Omega cm^2$. This was considerably lower than the average of $25\ \Omega cm^2$ from the first and far less direct measurement.

But most puzzling was the fact that in this active process the membrane potential was moving farther from the only known emf in the system, the potassium potential E_K, near the resting potential.

POTENTIAL CONTROL

In remembering the constant potential as it had been used to stabilize the iron wire (Bartlett, 1945) it seemed that the application of this concept was particularly promising for this axon situation. The application of a potential change from the resting potential would require a transient discharge of the membrane capacity. But after this the capacity current would remain zero to the extent that dV/dt approximated zero and the capacity could neither deliver nor absorb energy. This then appeared to be the ultimate in the simplification of the cable equation since now only an ionic current could flow and it could be directly measured, figure 3:14.

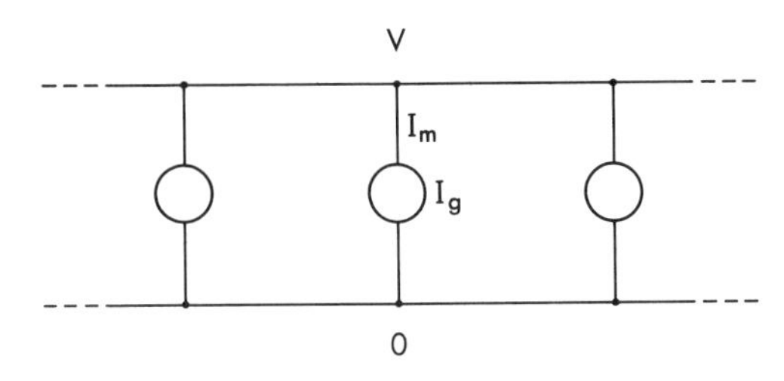

$I_m = I_g$

FIG. 3:14. With long internal and external electrodes the internal potential V is uniform and constant in time, and the membrane current flow is confined to the ionic path.

Probably at least as important was the long accumulating conviction that the membrane was a sort of an N characteristic negative resistance. Since the axon had demonstrated its instability for constant current in producing an all-or-none action potential there was good reason to hope that the membrane might be stable under a constant potential and without the threshold and all-or-none behavior required in an unstable situation. Still another factor was sheer curiosity. No matter if all of the reasons were wrong, it would

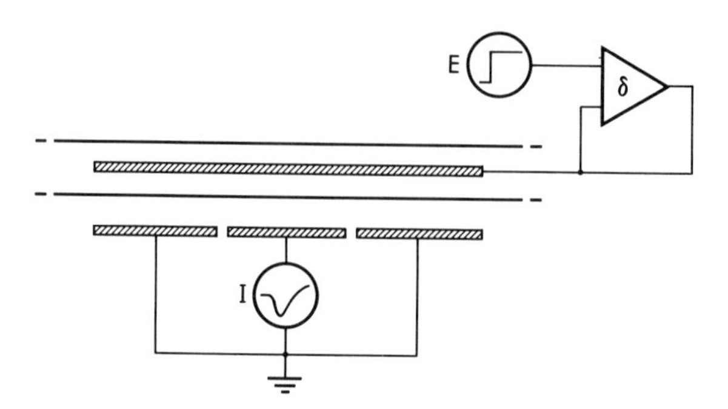

FIG. 3:15. Control of membrane potential by comparison with the command E. The amplifier δ supplies the necessary current and the central electrode portion is measured, I.

be interesting to see the answer of the membrane to the new question "What current do you have to produce for a change of potential?" And, finally, it was a question that the membrane could be made to answer—the experiment could now be done.

The rearrangements were rather minor and the schematic is shown in figure 3:15. The electrodes were silver-silver chloride, and the membrane potential was measured between the axial electrode and the outside central electrode. The difference between this potential V and the command E was now corrected by the current I supplied from the amplifier to be approximately equivalent to an emf E with negligible resistance applied directly to the electrodes.

There were difficulties, and I could obtain only rather sparse records on a few axons. These were entirely self-consistent; the results shown for one experiment in 1947, figure 3:16, are representative. The first and most immediately important conclusion was that there was no trace of a threshold or unstable behavior as a function of either time or potential. The early inward current was small and slow as it first appeared above 18 mV of depolarization and increased smoothly in amplitude and speed up to 38 mV. As it came in faster at larger potential changes it decreased in size to disappear between 64 and 128 mV. Then the current later turned smoothly at all potentials to flow in the outward direction; this came earlier and the final current became larger as the potential step was further increased. It thus seemed quite certain that this was yet another step in taming a nerve membrane by the further applications of rather simple, straightforward physical principles, concepts, and techniques. We had now succeeded not only in preventing the propagation of an impulse, so that the excitation stood still, but had also so constrained the membrane that it no longer responded in its usual all-or-none fashion.

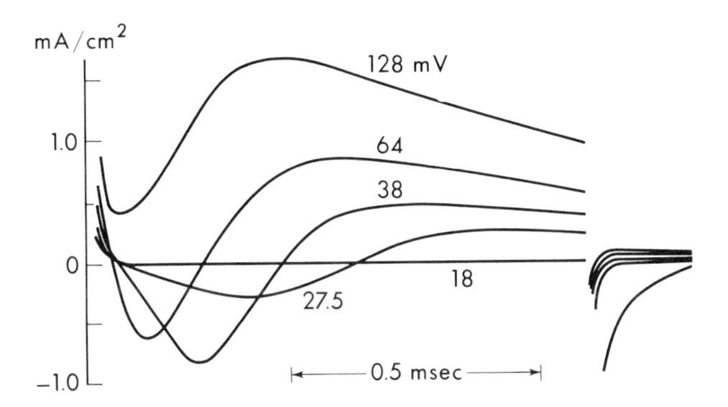

FIG. 3:16. Squid axon membrane current densities after changes of potential from the resting potential as shown.

The early inward current flowing against the resting potential had come to be expected, but again I was greatly disappointed not to find a steady state negative resistance. Even though the extent of my ignorance and confusion was more clearly revealed, I was very pleased by the direct records of the amplitude and form of the currents. They gave good basis for at least a qualitative explanation of the initiation, rise, and recovery of the action potential and its propagation (Cole, 1949).

PHYSIOLOGICAL IMPLICATIONS

As the membrane potential was reduced rapidly enough and by perhaps 20 mV from its rest value of −60 mV, a small amount of inward current would flow. Before this current was over it would have reduced the potential still more by the charge drawn from the membrane capacity. This cumulative process would continue with increasing rapidity, both because of the increasing amount of inward current flow and because it started more promptly as the potential increased. The rate of potential increase would probably reach a maximum approaching 800 V/sec somewhere in the neighborhood of a 50 mV depolarization—a little beyond the potential for the maximum inward current. Then the rate should fall off because of the decreasing inward current and the potential would have to reach a maximum where there was no longer an inward current, somewhere between 64 and 128 mV from these data. After the maximum, the later outward current at all potentials would steadily recharge the capacity back to the original resting potential.

It was not noted at the time, but it could have been pointed out that the limited duration, the transient characteristic, of the inward early current at all potentials clearly implied a liminal gradient of current rise for excitation. Although a fast pulse or step can produce the exciting inward current, a sufficiently slow rise will only approximate the steady state V, I relation, which is stable under all conditions. It seemed certain that there was a sufficient early rate of rise of the potential to support conduction at a nominal 20 m/sec. Even more spectacular is the fact that these data—quite irrespective of their interpretations—have subsequently been shown to contain most of the important facts of classical axonology.

Analysis and calculations

There were a number of questions as to the adequacy of the records. The first was the interpretation of the beginning transient. In part to understand this and in part to look at the whole phenomenon from another point of view, the records, such as figure 3:17, were plotted on the current phase plane, I vs. $\dot{I}$ (figure 3:18). The initial transient was clearly nearly exponential, as expected, but with a time constant of 8 μsec it was not entirely over before the current turned to become inward. It was, however, easily possible to extrapolate the charging current to infinite time and from this to extrapolate the ion current to zero time. As shown by the dash lines, this gave a good basis to expect that the ion current started with a zero slope as a function of time—which it should do if for no other reason than that the potential took a finite time to change.

The falling off of the outward current with time was not to be expected on the basis of earlier experiments. The steady state rectification had been found without a long creep of potential corresponding to this rather steady decline. Measurements of the electrodes alone showed that they did indeed polarize by an amount and at a rate comparable to that found with the axon. An approximate correction could then be made as indicated by the solid line

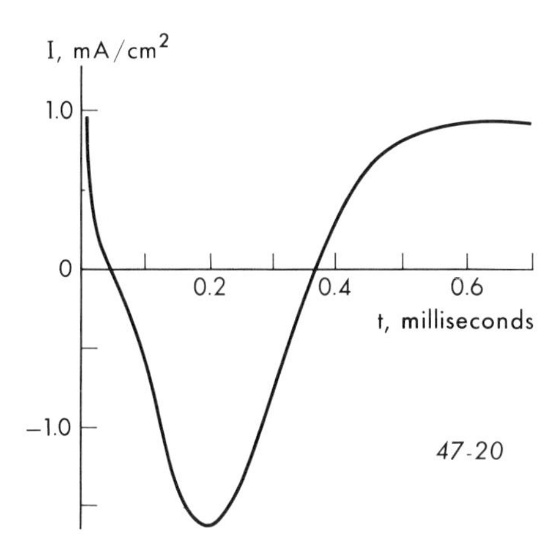

Fig. 3:17. Squid axon membrane current density after a change of potential of 38 mV from the resting potential.

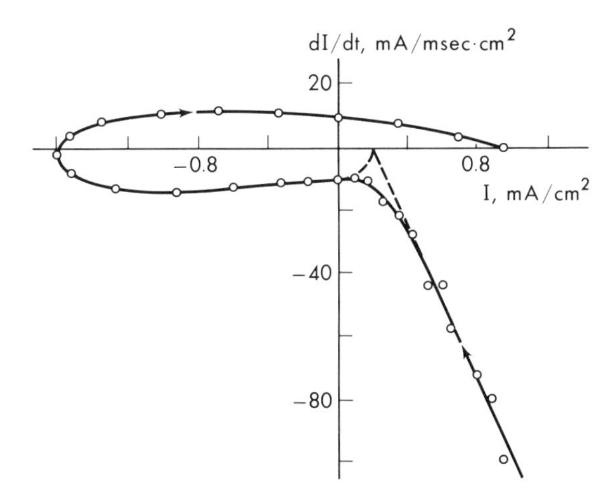

Fig. 3:18. Phase diagram of membrane current I vs. dI/dt after a potential change of 38 mV. The initial exponential, lower right, is the charging of the capacity and it is separated from the beginning of the ionic current by the dashed lines.

at the long time end of figure 3:18 although for this purpose only the maximum of the outward current was used.

The axial electrode supplied current to the guards with fringing, and usually excitation and propagation beyond their ends, and it was not known how much this current might alter the potential of the electrode in the center section. It was more apparent that the membrane potential change in the center section was indeed not constant during the applied step across the electrodes. The resistance R of the electrodes, axoplasm, sea water, and membrane gave a changing total current because of the membrane capacity component. These combined to give the potential and current corrections,

$$\Delta V = RI; \qquad \Delta I = RC\, dI/dt. \tag{3:1}$$

The resistive error was considerable, but the capacitive component was found to be small and negligible except during the beginning transient. This defect in the experiment could not be corrected to estimate what the current course would have been if a true step of potential had been applied across the membrane capacity, and no attempt was then made to better the experiment. The currents as functions of the membrane potentials at the time were as shown in figure 3:19.

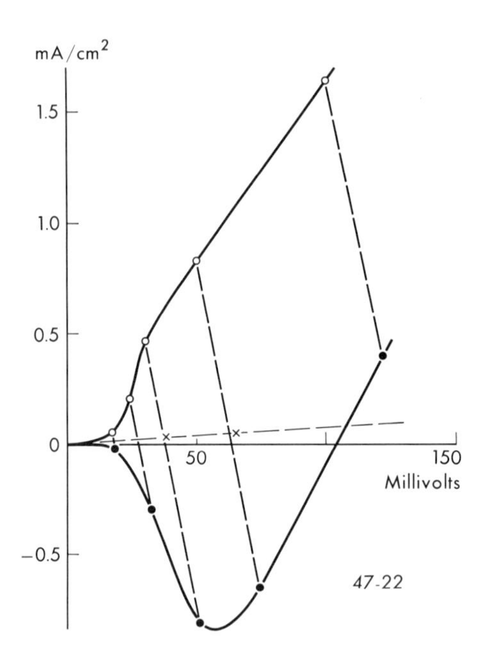

Fig. 3:19. Squid axon membrane current densities after step changes of potential. The initial currents, after the capacitive transients, were shown by crosses and the dashed line through them. The peaks of the early currents, solid circles, and the near steady state late currents, open circles, were given for the membrane potentials after correction for the series resistance, along the diagonal dashed lines.

It was indeed interesting that the initial transients agreed remarkably well in producing a constant, zero time resistance independent of the applied step. A complication in the steady state relation was quite apparent and seemed correlated, as in the first rectification experiments, with the rheobase of excitation. The effective negative resistance of the rapidly increasing inward current was about $-35\ \Omega\text{cm}^2$, a principal item of interest. The near parallelism of the early peak and steady state lines to give nearly equal conductances at larger potential steps was another curiosity. It was noted, too, that the early peak current reversed at the maximum potential changes!

These were, however, quite minor observations and worries. I had no theory to explain the early inward current. The steady state current continued to be the sort of thing that might be attributed to potassium, and I tried in many different ways to modify such a system to account for the transient behavior. Nor could I find a satisfactory and useful way to present the data.

These data gave a highly satisfactory qualitative description of excitation, response, and recovery and some interesting and useful numerical values. But it seemed certain that there must be a way to calculate any and all of the usual physiological results from these membrane data obtained at constant potential. The only attempt that I carried through gave a reasonable facsimile of an action potential on the basis of assumptions and procedures that I could not justify.

In my qualitative estimates of the consequences of the constant potential characteristics I had merely assumed that dI/dt at each I was a function of V only. This meant that after a change of potential from $V = V_1$ to $V = V_2$, where $V_2 > V_1$, the current pattern was the same is if the change from $V = 0$ to $V = V_2$ had been made at some particular time after the change had been made from $V = 0$ to $V = V_1$. In other words, the behavior at a particular value of the current was entirely independent of what had been done to achieve such a current.

The most obvious approach to the direct use of the new data was to consider that the membrane current was a function of the potential and time, $I(V, t)$. Then a small change of current ΔI would be given by

$$\Delta I = \left(\frac{\partial I}{\partial V}\right)_t \Delta V + \left(\frac{\partial I}{\partial t}\right)_V \Delta t$$

or

$$\frac{\Delta I}{\Delta t} = \left(\frac{\partial I}{\partial V}\right)_t \frac{\Delta V}{\Delta t} + \left(\frac{\partial I}{\partial t}\right)_V.$$

For $y = -C\,dV/dt = I$,

$$\frac{dy}{dt} = -\frac{1}{C}\left(\frac{\partial I}{\partial V}\right)_t \cdot y + \left(\frac{\partial I}{\partial t}\right)_V$$

and

$$\frac{dy}{dV} = \left(\frac{\partial I}{\partial V}\right)_t - \frac{C}{y}\left(\frac{\partial I}{\partial t}\right)_V. \tag{3:2}$$

$(\partial I/\partial t)_V$ was easily determined from the experimental curves and could be interpolated reasonably well not only to give $I(V, t)$ but also to give $\dot{I}(V, I)$, the form needed for a graphical integration of equation 3:2. There was however no obvious measure of $(\partial I/\partial V)_t$ which gave r_∞ as I had been using it.

I had ample evidence not only that r_∞ was a significant quantity but also that it had a wide range of variation—as during excitation, propagation, current flow, and for changes of external ions. But the only thing to do was to ignore this term, although it was not necessary to do so explicitly. As before, for a zero external current we write

$$dy/dV = \dot{I}/I \text{ where } y = dV/dt.$$

Curves of I vs. t at constant V were interpolated between the experimental results and quite complete representations of $\dot{I}$ as a function of I and V could be assembled. As a first approximation, the inward currents from figure 3:16 before the peak could be expressed as

$$I = -k(V - V_0)t,$$

where V_0 is the potential at which no inward current appears. $\frac{dI}{dV} = \frac{CkV}{I}$ or $I = \pm\sqrt{Ck(V - V_0)}$, and similarly $V = A \exp(\pm t/\tau) + B$. Thus to the extent that the approximation is valid, the rising part of a response was a straight line on the V, I plane and an exponential in time.

The graphical solution of equation 3:2 was again obtained by the construction shown in figure 3:20 in which the length $\overline{V}$ is given as $\dot{I}(V, I)$. For an initial depolarization of 22.5 mV the solution of figure 3:21 was obtained. The initial trajectory was surprisingly near to the straight line given by the approximation and to an exponential foot for the action potential. Another difficulty appears as this trajectory arrives at the orthocline $\dot{I} = 0$ because there is no place for it to go. But the boundary $\dot{I} = 0$ is only the crests of the I vs. t curves, so it was highly practical to assume that the trajectory rode down them. Also there was neither a critical point at which to stop nor evidence of any other path.

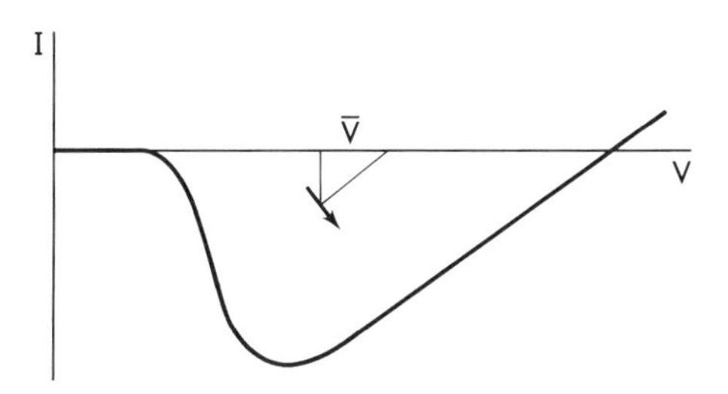

FIG. 3:20. Graphical construction for an action potential trajectory in which $\bar{V}$ was given by potential control data.

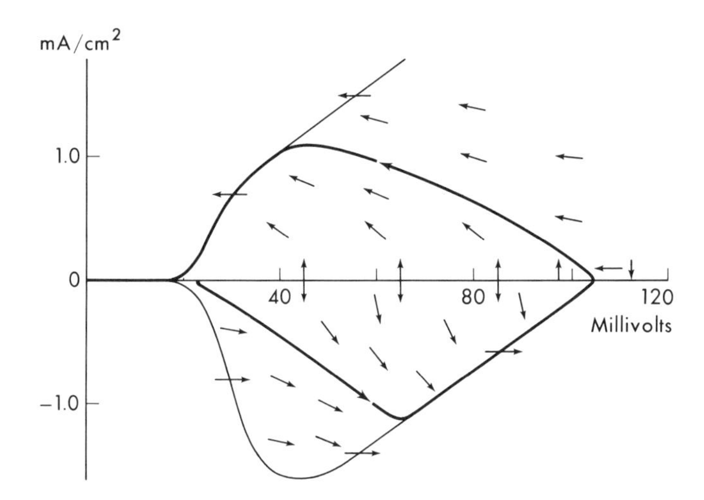

FIG. 3:21. Action potential trajectory, heavy line, computed from potential control data. The vertical orthoclines are on the potential axis and the horizontal orthoclines are the light lines, the peaks of the early currents, below, and the steady state currents, above.

The singular point shows evidence of being a stable point so this system would approach it exponentially and remain indefinitely. There was again the arbitrary procedure of a small displacement to allow $I = 0$ for $V < V_0$ and so to cross the orthocline $I = 0$ in a finite time. Thereafter, the solution proceeded from the now unstable point much as before until the final orthocline $\dot{I} = 0$ was encountered; it again had to be assumed that this was the terminal leg of the course. This procedure gave a plausible action potential but it was so studded with unsupported assumptions that it could not justify much confidence. Even the last part of Minorsky, which dealt with transitions from one phase plane to another, was not much help when it became available (1947).

SUMMARY

Once again the hints and frustrations of the past, the new possibilities of material and method, and a willingness to do the undone had indeed led to a taming of the axon. The elimination of impulse propagation and the complexities of the cable equation had not even been dreams before. Yet only an internal axial electrode and an outside electrode with its guards had given a uniform current density over the measured membrane and made the impulse stand still. This gave us directly a body of membrane facts, adding to and replacing vague conclusions from approximate calculations based on uncertain assumptions.

The simplicity of the situation, the power of attack, and the clarity of the membrane performance focused attention on excitation itself and led to a direct test of hopeful ideas as to the nature of normal instability. The simple, primitive command of the membrane potential did indeed eliminate the threshold behavior in the observed ionic current. As functions of potential and time, these currents accounted for, and sometimes could describe quantitatively, much of classical electrophysiology—threshold and accommodation, rate of rise and amplitude of spike potential, and recovery.

Atlantic Crossings

THE SODIUM HYPOTHESIS

Hodgkin had written to me briefly of the work with Katz in the summer of 1947, showing that sodium was responsible for the excitation and the overshoot of the squid axon action potential beyond the resting potential, but I was far from realizing either the implications or the consequences. He, however, was very interested in my account of the summer's work and came to Chicago in early 1948 to talk about it, to tell us in detail of the sodium work (Hodgkin and Katz, 1949), and to learn more of the Ling and Gerard (1949) microelectrode work on muscle. Hodgkin told how after the worst of the Battle of Britain he had convinced himself that the overshoot of the action potential, as it more than neutralized the resting potential, could only be the expression of an increased permeability to sodium. As soon as it was possible to get to Plymouth again, he and Katz had shown that the action potential was indeed somewhat proportional to the log of the external sodium ion concentration Na_e for even hyper-osmotic solutions. This they interpreted by the Goldman (1943) equation 2:29 as a transient increase of the rest permeability for sodium relative to potassium from $P_K:P_{Na}:P_{Cl} = 1.0:0.04:0.45$, which could express our K_e data. These ratios became $P_K:P_{Na}:P_{Cl} = 1.0:20:0.45$ at the peak of the action and then reversed back again to give a large P_K with $P_K:P_{Na}:P_{Cl} = 1.8:0:0.45$ in recovery. This in turn subsided as the membrane and its potential returned to rest. Thus the effects of sodium and potassium shown in figures 3:22a and 2:28 were at least described.

We repeated the experiments the following summer with choline instead of sucrose replacement of sodium. We also used space and current control in place of the freely propagating impulse to confirm and somewhat extend the results (figure 3:22b) (Cole, 1955). Later, with Nastuk, Hodgkin (1950) further showed this sodium theory to apply very nicely to the sartorius muscle fiber membrane. Huxley and Stämpfli (1951 a,b) then did the same for the frog node.

The original results and these additional items certainly made

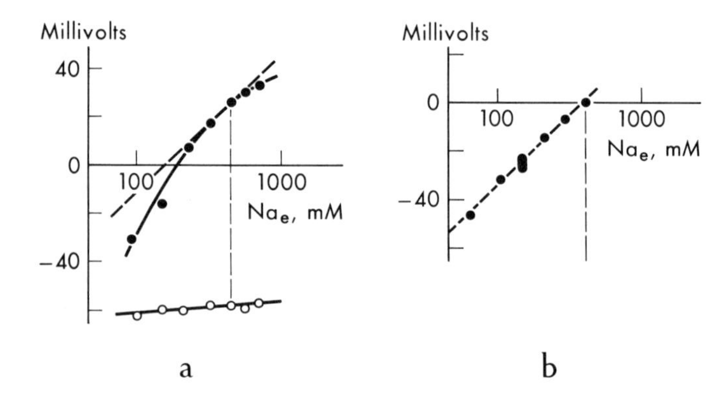

FIG. 3:22. (a.) Squid axon resting potentials, open circles, and action potentials, solid circles, for external sodium concentrations on logarithmic scale. The vertical dashed line is normal sea water. The slant dashed line is the theoretical sodium equilibrium potential. (b.) Squid axon action potential without propagation for normal external sodium, vertical dashed line, and change as replaced by choline. The diagonal dashed line is theoretical sodium equilibrium potential.

the sodium explanation very attractive (Hodgkin, 1951) and, furthermore, there was no obvious alternative. Yet this explanation was hardly beyond the hypothesis stage. It added to the evidence and suggestion of sodium participation given by Overton (1902)—yet to be recalled. But the Goldman permeabilities were only convenient numerical devices. There was no indication that they were at all applicable in this case and, as Goldman himself had shown, the theory was but a poor approximation to data on current flow in the squid membrane. More and more powerful support was needed—so Hodgkin came to Chicago for a tutorial on what was to become known as the voltage clamp.

THE VOLTAGE CLAMP

But before he came to Chicago, Hodgkin had evolved with Huxley and Katz (1949) a carrier system for sodium which required that an inward sodium current appear immediately upon depolarization of the membrane potential and then decay somewhat exponentially as it was inactivated, perhaps by calcium. Although he could not convince me that my data had any other interpretation than an inward current arising from a linear small outward current,

Hodgkin insisted that the obvious electrode complications and the obvious negative resistance behavior might conceal a large initial inward current. He could not be convinced otherwise until my results had been confirmed at Plymouth the next season with considerably improved electrode and electronics systems, and he and Huxley had ultimately to abandon the theory (Hodgkin and Huxley, 1952d).

As Hodgkin and Huxley went ahead with amazing speed (Hodgkin, 1964) and as I became involved in national defense, I had occasional reports from them. But again I did not appreciate the beautiful simplicity of the fundamental concepts and the spectacular detail and successes of the analyses they were able to make of potential control data. It was only after Hodgkin had sent me drafts of their manuscripts before publication and before the Cold Spring Harbor Symposium of 1952 that I began to understand what had grown from my simple idea to tame the squid axon.

This speed of application has seemed to me a very powerful example of the benefits of rapid and free communication. Hodgkin and Huxley had not progressed to internal current electrodes, to guards, or to control systems. They had not developed or used the membrane potential control concept. I could not publish my work immediately or in appropriate detail but, by free exchange of methods and results, they were able within a year to repeat all my work with very considerable improvements (Hodgkin and Huxley, 1952e, Discussion). This may well have saved several years at the least!

What Makes a Nerve Impulse Go

It was now possible to describe a nerve impulse, both with the simple beauty of the ideas as they first appeared and with a confidence made possible by the extensive investigation and detailed elaborate analysis that gave quantitative support to these ideas.

The cable theory had increased steadily in power and was unchallenged. It gave an impressive description and interpretation of the internal, external, and membrane current flows during propagation. The membrane capacity clearly played an important, but passive, role while activity was confined to the membrane conductance. As modified for the newer squid axon facts, the analysis for *Nitella* illustrated in figure 2:28 (Cole and Curtis, 1938b) still applied except for a better understanding of the conductance. It is however far simpler to make the complete presentation of the squid impulse propagation on the basis of the new facts and interpretations.

Most nerves have a high internal concentration of K_i^+ and a low Na_i^+, separated by a membrane from a usual surrounding of more or less dilute sea water in which the Na_e^+ far exceeds the K_e^+. At rest the nerve membranes have a slight permeability to K^+ ions; some of these ions run out to leave an excess of negative ions inside. The resulting electric field finally sends K^+ ions back as fast as they come out, to produce an equilibrium, with the interior some 70 mV negative in potential, E_K of figure 3:23. As an impulse approaches —at a speed of 20 m/sec or 40 mph—this potential increases toward zero and a membrane path for Na^+ begins to open. As a few of these ions move under the drive of both the electric field and the concentration gradient they begin to neutralize the excess internal negative ion concentration and further raise the internal potential toward zero. This potential change further and more rapidly increases the Na^+ inward flow, in a runaway cycle, past a zero membrane potential. The inward electric drive is thus lost and later replaced by an opposition as an inside excess of these positive charges reverses the potential. This positive potential slows down

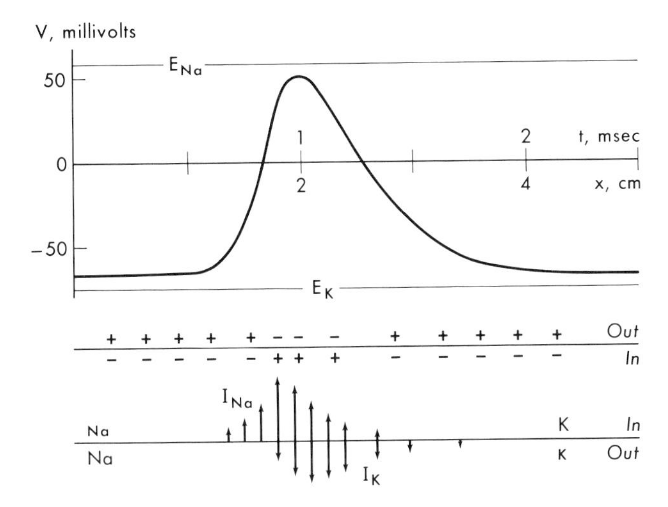

FIG. 3:23. Descriptions of an impulse in a squid axon propagating to the left as a function of time t at one point or distance x at one time. At rest, left, the potential, top, is near the potassium potential E_K, there is a deficit of positive charges inside the axon, center, and the ion currents, bottom, are negligible. The increase of potential as the impulse approaches, allows an inward sodium current I_{Na} which further increases V and I_{Na}. As the potential approaches E_{Na}, I_{Na} decreases, I_K increases until, after the peak of the spike, the outward I_K dominates to return V to the rest potential.

the inward flow as it now approaches a new equilibrium of more than +50 mV, E_{Na}. Also, during the later stages of this process, and more slowly, the sodium paths begin to present more difficulty. But now the K^+ is being allowed to cross more freely to give a maximum to the potential and a turn back toward the starting point. As the Na^+ flow is stopped, the more sedate recovery is executed by the K^+ ion outflow that reaches a maximum rate again at about the zero potential. The later declining rate comes in part from the opposing electric drive, when again there is a negative ion excess inside, and then from the reduction of the K^+ ease of movement to the near minimal value from which it started.

So the impulse is only a rapid gain of Na^+ followed by a slower loss of K^+. This obviously cannot continue indefinitely without another, apparently quite different and much slower, metabolic process to pump out the extra Na^+ and replace it by K^+. On this there is a vast literature beginning with Dean (1941).

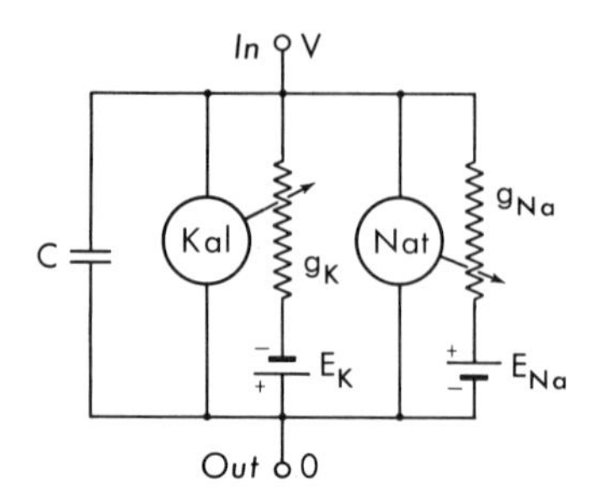

FIG. 3:24. A circuit which represents the electrical behavior of an element of a squid axon membrane in terms of two imaginary characters, Kal and Nat, which control the potassium and sodium conductances, g_K and g_{Na}, under orders from the membrane potential V.

Still without going into the exotic intricacies of the mathematical detail, it is possible to present the essential results of the voltage clamp analysis and to apply them to the excitation process. As shown in an equivalent electrical circuit, figure 3:24, the emf E_K is about -70 mV, the value at which there can be no net flow of potassium ions in or out. Similarly, E_{Na} is over $+50$ mV. At rest the conductance g_K plus a small fixed conductance (not shown) is about 1 mmho/cm^2. g_{Na} is vanishingly small so the resting potential is near E_K. After a 50 mV change of potential, figure 3:25, to about -20 mV, the control system dubbed Nat increases the conductance g_{Na}, to give an inward sodium current, and then returns more slowly toward zero. The system called Kal increases g_K rather slowly to maintain it steadily. For lesser commands of potential, such as to -45 mV, neither system reacts so much or so fast—rather like small boys to a whispered suggestion. But then,

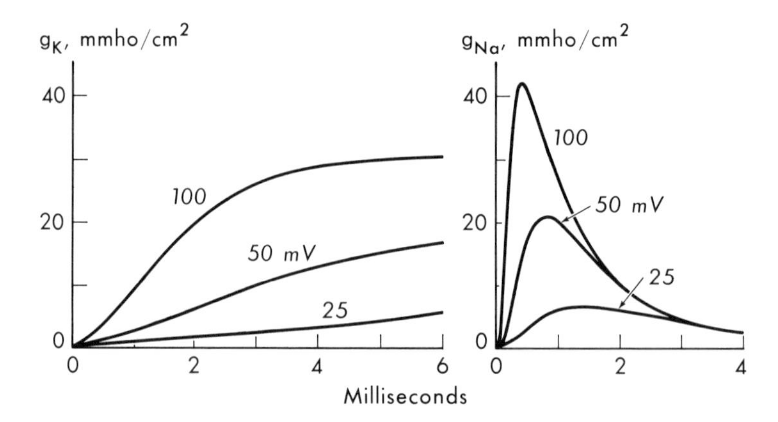

FIG. 3:25. The membrane ion conductances, g_K, g_{Na}, after 25, 50, and 100 mV increases of membrane potential from rest.

again like small boys, they perform with more enthusiasm at the shouted command of +30 mV to reach higher levels of conductance more promptly.

For a small short pulse of outward current applied at one end of an axon, the less negative charge on the capacity spreads along the axon and then decays at all points. After a larger pulse Nat opens up a little to tend to hold the potential with some inward flow of Na^+; this spreads only a short distance along the axon before it is brought back by Kal's slow response. A yet larger pulse increasing the end potential by some 20 mV will persuade Nat to produce a net inward current quite soon. Not only will this current increase the end potential, but it will spread along the axon by the internal and external current paths to bring them closer to the potential at which a net inward current will appear. Then, as at the site of stimulation, Nat responds to give a faster and larger g_{Na} and the full excitation course is under way. As it goes towards the full response it also sends an increasing amount of inward current farther and farther along the axon to bring first near and then far points into activity. The process asymptotically becomes the constant velocity, self-sustaining, self-determining impulse that neither knows nor cares what started it.

Simple and obvious though this process appears and well demonstrated as it was by Hodgkin (1938) with figure 2:45, it has only been much later that computation in any detail was begun (Cooley et al. 1965; Cooley and Dodge, 1966; Noble and Stein, 1966).

The Hodgkin-Huxley Axon

The series of papers by Hodgkin, Huxley and Katz (1952) and by Hodgkin and Huxley (1952a–d) are remarkable in many ways. From the—perhaps near—view of fifteen years it is hard to believe that this collection will not remain an obvious turning point in electrophysiology and membrane biophysics. It assembled within its grasp much of everything that had gone before and molded it all into a self-contained, coherent structure. But this structure, which I have called the "Hodgkin-Huxley axon," is far more than a synthesis of the past. It is built entirely on the new information provided by the voltage clamp and, up to now at least, is not available by any other single approach or combination of experiments.

This new knowledge, which contains so much of the old, is then analyzed and stated in terms of ion conductances, a kind of ion permeabilities. The delightfully simple concept of ion conductance is most attractive and very useful—so much so that it can be quite deceptive. It can be deceptive because known systems with analogous conductances, such as a thermistor, are not of an elementary simplicity. It is also highly deceptive if we forget for a moment that the conductances are but expressions of the entirely unknown process by which ions get through membranes. So these conductances are the center both of information and of ignorance. As this ignorance is replaced by understanding it seems most likely that the conductances will play a crucial part. Even if not, they must go down in history as the first clear, general, and succinct statement of the problems of ion permeability.

Other notable features of the Hodgkin-Huxley axon may be somewhat a matter of personal opinion. I have been very impressed by the fact that in the last analysis the ionic conductances are completely described in terms of an initial condition, a time constant and a steady state, and that each of these is a function only of the membrane potential. This description is certainly extraordinarily successful and—to the extent that it is successful—it seems to me to argue that the ionic characteristics of the membrane are uniquely determined by the potential difference across it.

If indeed the ionic properties are entirely determined by the membrane potential, they should be most simply and directly observed and analyzed by experiments in which the membrane potential is the independent variable. This might then help us to understand why the voltage clamp is so much more powerful than was to be expected on the basis of the simple stability concept from which it was developed. And it would give yet another example—if any were needed—of the credo that things should be tried if only because they can be.

Certainly no one can deny that the Hodgkin-Huxley axon is a highly specific, detailed, and quantitative structure. Even with the numerous experiments from which it was constructed it is axiomatic that other, and certainly different, experiments cannot produce the same numerical values for the specification of the many functions and coefficients. The corollary of this is that the Hodgkin-Huxley axon can, has been, and will continue to be improved upon to both minor and major extents. A vague target would have been much more difficult for sharpshooters, but even a highly visible, clearly defined bull's-eye is not easily or extensively damaged by bird shot.

After one has talked with the authors and has at least a superficial knowledge and understanding of the 126 page series, he may come to think of the prodigious intellectual and physical effort and the accomplishment that these pages represent. Obviously, before the beginning of 1948, Hodgkin, Huxley, and Katz all had a full appreciation of the significance of the transient sodium permeability during an impulse, and rather clear ideas of its rise and fall. But I have no indications that they had any particular plans as to how to go further at the time when I made my voltage clamp concept and 1947 data available to them. Yet in the proceedings of the 1949 Paris symposium, they gave full details of their experimental equipment and presented enough results to show that they were rather sure of where they were going and how they would get there. At the Copenhagen Physiological Congress in the summer of 1950, Hodgkin and Huxley showed me some of their data analyzed into sodium and potassium conductances as functions of time and potential. Then the *Journal of Physiology* reports that the manuscripts of the first four papers were received on 24 October 1951, and the fifth on 10 March 1952. This achievement, on the basis of only two squid seasons at Plymouth and four years elapsed time, seems as much of a record for speed as it is for progress.

We may also have impressions about the presentation of the material in the papers. In some ways the amount of detail is surprising and, at times, even oppressive. On the other hand, the conclusions are clearly drawn, and candid estimates of reliability are available. One wonders if the order of making the connections between the final calculations and physiology was fully intended or somewhat subjective. I, at least, have been curious as to the basis on which the more complex and difficult physiological phenomena were given precedence over the simpler, more elementary, and more obviously physical behaviors that came later!

But one can hardly avoid the feeling of a relentless drive toward a well-defined objective and utterly complete concentration upon it. The whys, wherefores, and most of the oblique consequences are ignored in the splendid isolation of the authors in the midst of the developments as they are unfolded. But if they had undertaken to reflect upon such a thing—even a little, as I now do—500 pages might not have been enough.

As we turn to a more detailed concern with the work which culminated in the summary paper by Hodgkin and Huxley (1952d), it is certainly not necessary, and probably not appropriate, to give any extended account of the procedures of experiment and analysis that have been so widely and variously presented. Instead, I devote myself more to comment of a descriptive and editorial nature, although I have been far from dispassionate in my viewpoints.

EXPERIMENTAL BASIS

The underlying concept and technique of control of the membrane potential remained the same as in the original work. Current was passed through the membrane between long internal and external electrodes. This current was supplied electronically as needed to produce the required potentials and changes of potential across the membrane. Several important modifications were made which largely eliminated the previous sources of error that had been pointed out to Hodgkin. Separate potential and current electrodes were used both inside and outside the axon, figure 3:26, and so avoided the error in potential caused by polarization of the current electrodes that Huxley (1964) emphasized. The internal potential electrode was effective only over the central, guarded section of the axon, and its potential could not be as much affected as that of a

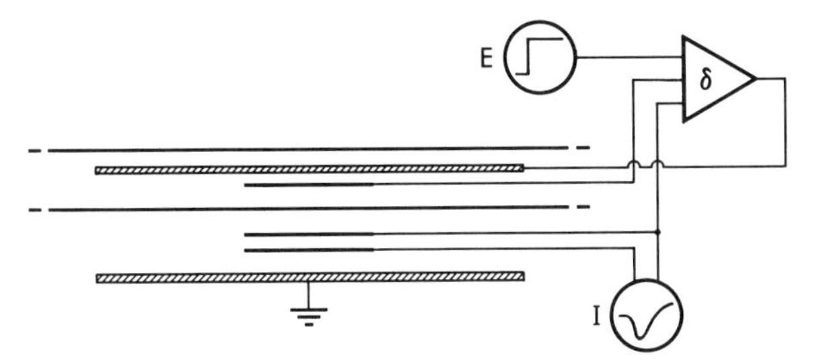

FIG. 3:26. Improved voltage clamp of squid axon membrane. The potential difference between control electrodes inside and outside the axon was compared with the command E by the control amplifier δ. The output current was supplied to the long inside and outside electrodes and the current flow in the central section I was measured between short external electrodes.

single long current electrode by the uncontrolled activity near the ends of the guards.

The electronic control system incorporated a compensation, that was new, for the variations of membrane potential produced by current flow through the axoplasm, sea water, and the effective series resistance in the membrane. This compensation was used when the need for accuracy justified the hazard of instability. The average membrane series resistance appeared to be 5.7 Ωcm^2 and 80 percent of this was correlated with the resistance of the external solution. The process of charging the membrane capacity after a change of potential was a nearly exponential one with a time constant of about 6 μsec without compensation, and the average membrane capacity was 0.9 $\mu F/cm^2$.

The time courses for the ionic currents were shown to be very dependent on temperature, increasing in speed by a factor of 3 for 10°C increase of temperature. The less certain effect of temperature on the amplitudes was an increase of about 1.3 times per 10°C rise.

The authors first identified the early current as sodium. In sea water it reversed from inward to outward at absolute potentials of +45 to +50 mV (figure 3:27) to correspond to the sodium ion equilibrium potential

$$E_{\mathrm{Na}} = \frac{kT}{e} \log \mathrm{Na}_e/\mathrm{Na}_i \qquad (1{:}23)$$

as calculated for the sea water and axoplasm concentration, Na_e and Na_i. As the external sodium was replaced by choline the reversal

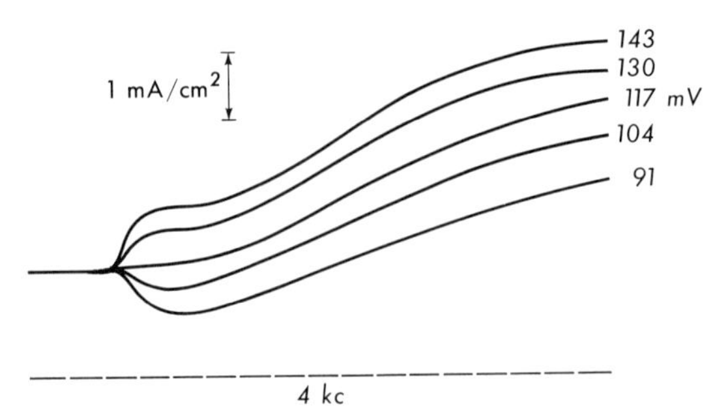

FIG. 3:27. The early component of membrane ionic current reversed in direction at about an increase of 117 mV for the membrane potential step from rest.

potential was decreased as expected from equation 1:23, until in sodium free sea water only outward early current was found (figure 3:28). So this current appeared to be sodium moving under concentration and potential differences across the membrane.

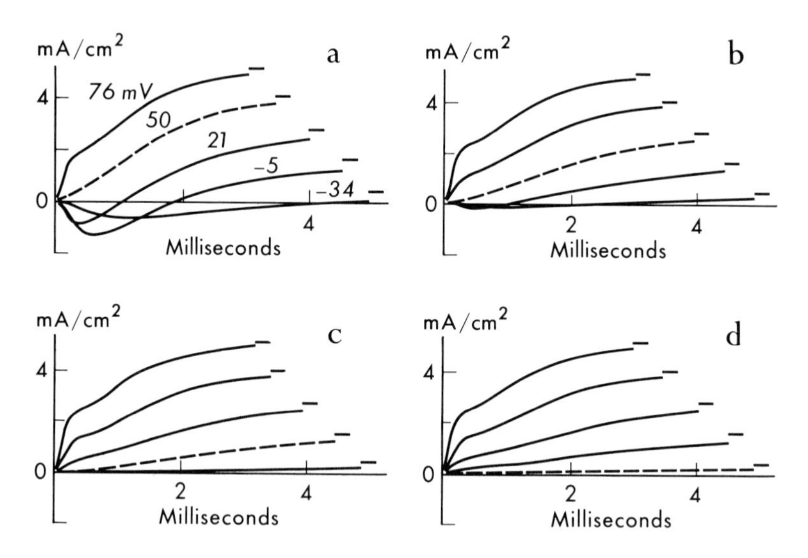

FIG. 3:28. Theoretical experiment to show effect of external sodium concentration on membrane currents in voltage clamps as shown in *a*. *a*, normal, 100 percent, sodium; *b*, 30 percent; *c*, 10 percent; *d*, 3 percent. The rest potential is -65 mV. The steady state current is shown at the end of each curve. The current at E_{Na} for each concentration is given by the dashed line in which there is no early, sodium, current.

The remaining current was unaffected by the external sodium and was identified as potassium. It could then be recorded directly at each $V = E_{Na}$ as E_{Na} was changed by changes of Na_e. I_K could also be interpolated for V between two values of E_{Na} because it did not change during the early part of I_{Na} and because the time course of I_{Na} was not affected by Na_e. The separated currents, as shown in figure 3:29, were then expressed as the conductances

$$g_{Na} = I_{Na}/(V - E_{Na}), \qquad g_K = I_K/(V - E_K).$$

The maximum values of these conductances were equivalent to about 30 Ωcm^2 for $V > 0$, but fell about as $\exp(-V/5)$ for $V < -20$ mV. The maximum rates of change, dg/dt, decreased steadily in the range from $V = +60$ to -60 mV.

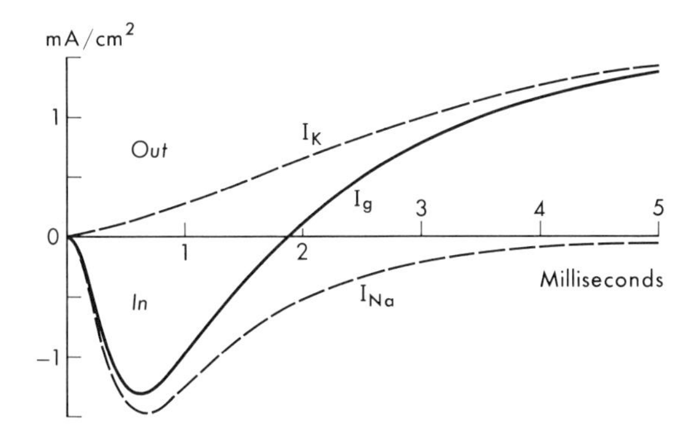

FIG. 3:29. Solid line is the ionic current I_g after a voltage clamp to 56 mV above the rest potential. This was analyzed into the early transient inward sodium current I_{Na} and the later outward potassium current I_K.

The conductances were found to be continuous through sudden changes of V, as demonstrated by the linearity of the instantaneous currents of figure 3:30. During an early sodium current the conductance line intersected the third, leakage, component at the expected E_{Na} with a slope $g_{Na} = 33$ mmho/cm² (figure 3:30a). This confirmed the result of the current separation procedure and, most significantly, further identified g_{Na} as an instantaneous—or infinite frequency—conductance. The similar result after the end of the sodium current gives E_K and g_K with the same interpretations (figure 3:30b).

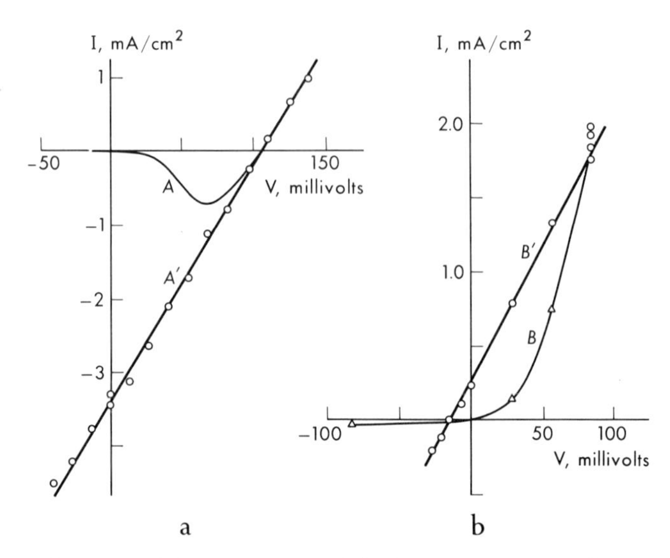

FIG. 3:30. Direct measures of the instantaneous membrane conductances. In *a* the early currents at 0.28 msec after an initial potential change are given by *A*. At this time after an initial clamp of 110 mV the potential was again changed to give the circles represented by an instantaneous sodium conductance and potential of line *A′*. The curve *B* of *b* is the later current, at 0.63 msec after the initial potential change. The change at this time from 84 mV to other potentials gave the circles and the instantaneous potassium conductance and potential of line *B′*.

After the potassium current I_K had been found by elimination between the voltage clamp current I_g and I_g' at the different external sodium ion concentrations Na_e and Na_e', the sodium currents I_{Na} and I_{Na}' at both of these concentrations were also available. Such dependence of an ion current I across a membrane was to be related to the ion concentrations at the boundaries by the "independence principle" to be presented later.

This procedure could then be used to calculate the sodium and potassium currents in altered external and internal ionic environments from the values obtained under the normal conditions. The peak sodium currents in 30 percent and 10 percent sodium calculated on this basis from corresponding values in 100 percent normal sodium sea water agreed well with the experimental results when account was taken of the change of resting potential and the corresponding change of the sodium inactivation factor.

Finally the sodium conductance was dissected into two factors. The initial rapid rise, the sodium "on" process, reached its steady state in a msec or less. Then an inactivation took over and this "off" process was found to reduce the sodium conductance by increasing amounts over the range from −90 to −20 mV and with increasing speed from −60 to −20 mV.

As converted back from conductances, these ionic current components are shown in figure 3:31 after a voltage clamp at one potential.

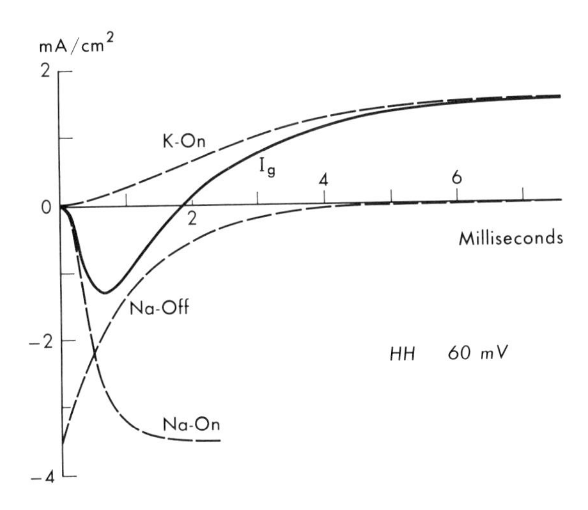

FIG. 3:31. Complete analysis of the ionic membrane current I_g after a voltage clamp to 60 mV from rest. The sodium current is turned on rapidly and turned off more slowly by inactivation. The potassium current is turned on slowly to reach a steady state.

MATHEMATICAL DESCRIPTION

Hodgkin and Huxley express the results of their voltage clamp experiments in terms of three dimensionless parameters m, h, n, which vary between zero and one to describe the "on" and "off" of the sodium conductance and the "on" of the potassium conductance. Each of these parameters is assumed to obey first order kinetics:

$$\frac{dn}{dt} = \alpha_n(\underline{n} - n) - \beta_n n, \text{ etc.} \qquad (3:3)$$

α_n is a forward and β_n a backward rate constant, and in near normal situations they are functions of only the membrane potential V. In order to fit the experimental data Hodgkin and Huxley expressed the conductances as

$$g_K = \bar{g}_K n^4 \text{ and } g_{Na} = \bar{g}_{Na} m^3 h. \tag{3:4}$$

The analytical expressions for the α's and β's which best fit the data were chosen to conform, as far as possible, to the pattern of the Goldman analysis. In the original form, for depolarization from the rest potential given by $V < 0$, they are:

$$\begin{aligned}
\alpha_n &= \frac{0.01(V+10)}{\exp\frac{V+10}{10} - 1}; & \beta_n &= 0.125 \exp(V/80) \\
\alpha_m &= \frac{0.1(V+25)}{\exp\frac{V+25}{10} - 1}, & \beta_m &= 4 \exp(V/18) \\
\alpha_h &= 0.7 \exp(V/20); & \beta_h &= \frac{1}{\exp\frac{V+30}{10} + 1}.
\end{aligned} \tag{3:5}$$

Then there were the constants

$$\begin{aligned}
\bar{g}_K &= 36 \text{ mmho/cm}^2; & E_K &= 12 \text{ mV} \\
\bar{g}_{Na} &= 120 \text{ mmho/cm}^2; & E_{Na} &= -115 \text{ mV}
\end{aligned}$$

and also the linear leakage with

$$g_L = 0.3 \text{ mmho/cm}^2, \qquad E_L = -10.6 \text{ mV}.$$

I have found it convenient to use the alternative expressions

$$\frac{dn}{dt} = (\underline{n} - n)/\tau_n, \text{ etc.}$$

where $\underline{n} = \alpha_n/(\alpha_n + \beta_n)$ is the steady state value of n at V, and $\tau_n = 1/(\alpha_n + \beta_n)$ is the corresponding time constant. This formulation has been presented in figure 3:32 (Cole, 1955) partly in graphical form with the numerical constants on an absolute membrane potential scale.

DISCUSSION

There are many rather obvious comments that can be made on this analytical model of the squid axon membrane. Some have been expressed in print formally, others in printed discussions, far more

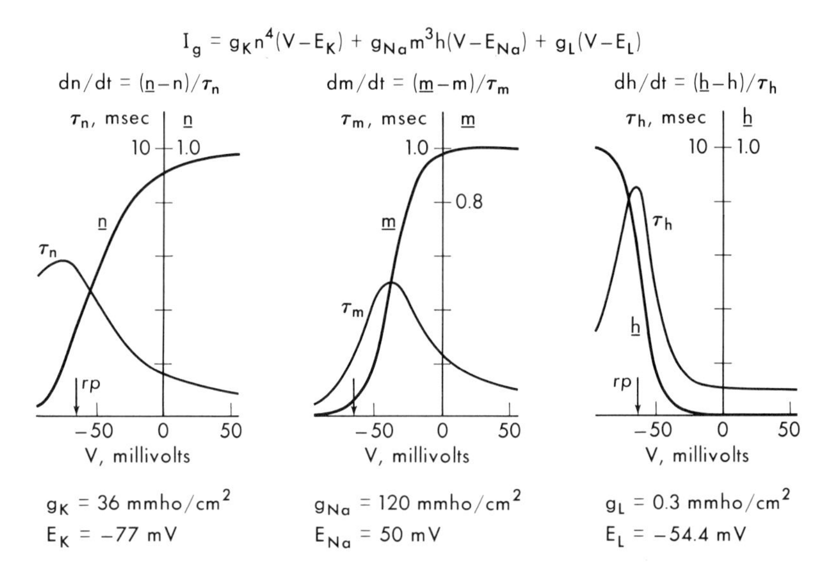

FIG. 3:32. Representation of the Hodgkin-Huxley equations for a normal squid axon. The membrane current I_g is given at the top in terms of the constants, bottom, and the parameters n, m, and h. The constants in the differential equations for these parameters are shown graphically as functions of the membrane potential; *rp* is the potential at rest.

in meeting discussions, while the uninhibited and often uninformed written and oral person-to-person arguments all over the world are quite beyond estimate. I have made a few comments myself, usually in a quite limited context and for a particular audience. Among the more recent and more general reviews, Noble (1966) succeeded well in his aim to give an explanatory account of the theory.

In the first place, it should be completely clear that at the very least the equations are empirical expressions of the voltage clamp data. Hodgkin and Huxley do show that they are a reasonably good representation of these data. These were simple calculations and a few results are shown in figure 3:33. Although there were definite and consistent differences, the agreement was reasonably good and attention was turned to broader and far more difficult tests.

With such a description of the membrane ion current as a function of membrane potential—as acquired from axons in which excitation and propagation were prevented—it is now possible to retrace the painful steps back to the physiology of a normally functioning axon because the additional factors are rather simple and have long been known.

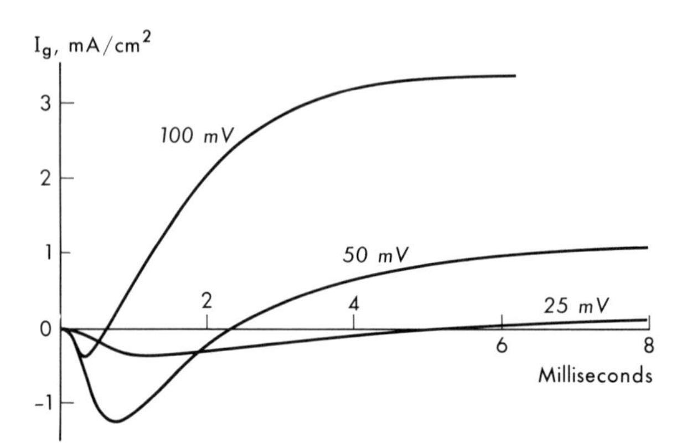

FIG. 3:33. The ionic membrane currents after the indicated potential increases as calculated from the Hodgkin-Huxley equations.

The addition of the membrane capacity completes the description of the membrane current

$$I_m = C\, dV/dt + I_g.$$

These were termed "membrane" currents and potentials; the less ambiguous term "space-clamped" has also been used. In the free axon without internal or external clamping electrodes we come at long last back to the cable and the Kelvin equation 1:31 of figure 3:34.

$$\frac{1}{r_e + r_i} \cdot \frac{\partial^2 V}{\partial x^2} = C\frac{\partial V}{\partial t} + I_g.$$

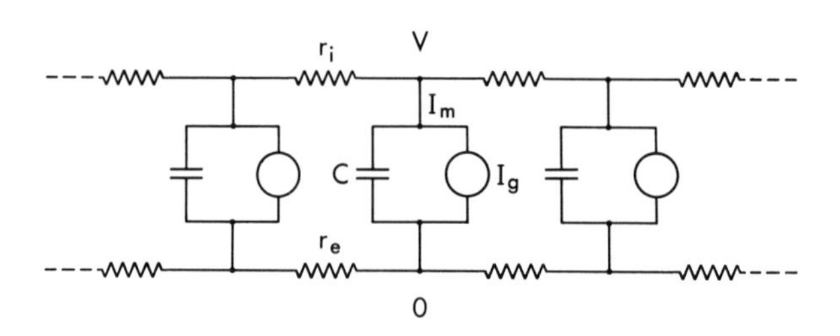

$$I_m = C\frac{\partial V}{\partial t} + I_g = \frac{1}{r_i + r_e} \cdot \frac{\partial^2 V}{\partial x^2}$$

FIG. 3:34. The cable equivalent of the squid axon in which I_g of the general, partial differential, cable equation may be given by the Hodgkin-Huxley equations.

In principle, there are solutions of this parabolic partial differential equation that can be computed—such as for the initiation or blockade of a propagating impulse. Although this completely general description of an axon was first too formidable for use except under the auspices of IBM, the restriction to a uniformly propagating impulse has been reasonably practical and useful. For $V = f(t - x/\theta)$, where θ is the speed of the impulse, equation 2:6 gave

$$\frac{1}{\theta^2(r_e + r_i)} \cdot \frac{d^2V}{dt^2} = C\frac{dV}{dt} + I_g.$$

This is again an ordinary differential equation for the cable, figure 3:49, but now containing the unknown parameter θ.

Far above and beyond their primitive use as expressions of voltage clamp currents, Hodgkin and Huxley have interpreted these data as sodium and potassium ion currents. The equations are thus a formal and complete presentation of the "Sodium Theory"—essentially replacing the earlier, comparatively qualitative, description of it. Although it should not be necessary, perhaps a more complete descriptive title should be on record such as:

> "A Theory that a Sodium Ion Permeability, Controlled by the Membrane Potential, is Responsible for the Impulse Excitation and Propagation of the Squid Axon in Its Normal Environment and Under Normal Conditions, with the Hope and Expectation that the Theory Will Apply to Some Other Excitable Cells and Under Some Other Conditions."

There have been interpretations of the sodium theory to mean that sodium *and only sodium* ions could have this importance. Although this interpretation could be shown not to be true, it hardly seems possible that this could condemn the theory. The mistakes are both of premise and of logic.

The authors are quite restrained and cautious in their speculation as to the mechanisms which might underlie their description of the ion permeabilities. They have obviously devoted considerable attention to models with multivalent membrane-soluble carriers. They felt forced to withdraw one pre-clamp attempt and have had to so restrict the carrier concentration and parameters as to make the model a difficult one to support.

The choice of enzyme kinetics in equations 3:3 may have been one of convenience or one based on some hope and belief that it might correspond to a physical mechanism. At least conceptually, this is a very attractive and useful approach. They can then present the parameters n, m, and h as the probabilities that sites on or in the membrane may be activated to permit the reception and passage of an ion. If the simultaneous activation of four such sites—in series or in parallel—is necessary for the passage of potassium ions, the probability is n^4 and proportional to the conductance g_K. Similarly, if three sites are necessary to pass sodium we have $g_{Na} \sim m^3$. If the inactivation of but a single site can block sodium and the probability of this is $1 - h$, then $g_{Na} \sim m^3h$.

The complexity of the equations makes them far too formidable to be useful for casual discussions and conclusions—without a pocket computer that is not yet available, except perhaps in classified missiles or satellites.

The equations are so specific as to invite alterations and to require modifications from axon to axon and tissue to tissue. They have also been constructed to express the performance of an average Plymouth axon over a limited range of variables. They seem quite satisfactory for Na_e variation from normal to zero, but there is no evidence here of the effect of K_e or of other ions, except choline, or of molecules. The potential range is restricted, too, the data not going much beyond the normal E_{Na} on the positive side and usually not more than tens of millivolts more negative than E_K. Here again, there is no hint of what might happen as this range is exceeded, and the equations are perhaps fortunate choices to the extent that they apply so widely. The time span of the experiments is impressive—from about 10 μsec to a few tens of milliseconds. Although significant things may happen sooner they do not seem likely—particularly in retrospect—to be of great importance except perhaps as a key. There are many interesting and very important nerve functions that extend far beyond this range of investigation. It should be expected that apparently constant parameters would change in hundreds of milliseconds and more or that new factors would appear at such times. Once again, the success of any such extrapolations should be as cherished as a four-leaf clover, and a failure should not condemn the whole structure.

It seems singularly unfortunate to me that the HH axon has been rather persistently thought of and referred to as a theory of ion permeability. In particular, it seems most inappropriate to attack it as the "equivalent circuit model" of ion permeability and to propose vague concepts as alternatives.

Insofar as the HH axon separates the sodium and potassium components of the ion current it is indeed postulating separate membrane mechanisms for these components under relatively normal conditions and, to that extent, it is certainly a membrane model. Beyond this, the equations and the equivalent circuit are purely empirical expressions of experimental fact. They are only a description of these two black boxes as seen from the outside of the membrane and they rely upon no models or hypotheses of the internal membrane mechanisms. These expressions of membrane behavior are mostly accurately to be thought of as only giving performances that are reasonably similar to those of the membrane. They certainly cannot be considered as possible models, nor more than interesting and useful analogues for the ion permeability mechanisms of the membrane.

The use of electrical conductances in the equations and the equivalent circuit of the HH axon has apparently aroused reactions that seem more emotional than rational. A conductance in general is a very old concept and well established fact. It has been possible to so extend and generalize it as to become the basis of a highly sophisticated group of technologies upon which much of our present civilization depends. These mathematical structures are both powerful and awesome. They are to be feared, but they are also useful. Insofar as a current and a potential difference are linear functions of each other they are most simply represented by a conductance—no matter what the process by which the current is carried. It is an experimental fact that for the squid axon membrane the ionic current is proportional to a change of potential difference—for as much as tens of millivolts, for times up to a hundred microseconds and for frequencies down to ten kilocycles. So, to the extent that this is true it is only a wasteful perversity that will refuse to express the behavior of the membrane as a conductance or will refuse to accept the challenge to explain such measurements in terms of a permeability mechanism. As a consequence any permeability model is not competing with the HH axon. These equations can be judged only as to the extent to which they agree with experimental fact. And, to the extent that the axon is an adequate representation of fact, any permeability theory or model must conform to the performance of the analogue—as well as to explain that performance. The HH axon is a goal not a competitor of theory.

The definitions of the conductances were presented bluntly and only as a means to represent the ion current components,

$$I_K = g_K(V - E_K), \; I_{Na} = g_{Na}(V - E_{Na}).$$

But then it was shown that such a formal conductance was continuous throughout the capacitive transient after a step change of potential, whereas the current was not. This agreed with our earlier high frequency results. Then also this conductance at short times was shown to be a constant parameter, independent of V and I as shown in figure 3:30, to further support and extend the linear definition. However, from their recounting of these results I have thought that Hodgkin and Huxley had been somewhat surprised —as well as pleased—that their formal definition had turned out to be a real, operationally defined, experimentally observable, conductance. Rather unfortunately, they did not elaborate, except by inference, upon the significance of such a conductance in a nonlinear system. They were undoubtedly wise to avoid any interpretation as specific as the potassium electrodiffusion model, although it would have given a helpful concept, while the formal expression for any nonlinear system was probably too abstract. Nonetheless, it would have helped appreciably for them to have emphasized more the purely physical reality of this instantaneous, infinite frequency, chord conductance, g_∞. In particular they might have contrasted it with the steady state, zero frequency, slope conductance,

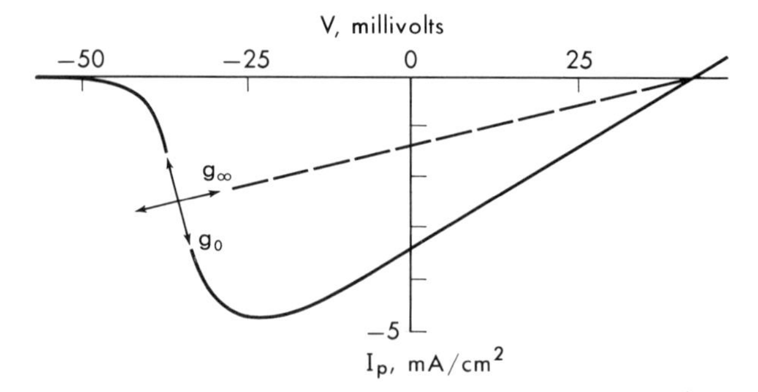

FIG. 3:35. The limiting conductances for the squid membrane sodium current. At short times or at high frequencies the conductance is along the dashed line g_∞. In the quasi steady state, given by the solid line, at low frequencies or long times, the conductance g_0 is negative.

g_0 as an alternative, entirely equivalent expression of the steady state membrane behavior, as shown in figures 2:58 or 3:35. Thus

$$g_0 = dI/dV = g_\infty(V) + \frac{dg_\infty(V)}{dt}(V - E)$$

and the basis for Hodgkin's explanation of their use of g_∞ is obvious—it is far more practical to manipulate analytically. On the other hand g_∞ is deceptive in that under all physically realizable conditions it is positive even when a system may have a negative value of g_0 representing the important potential instability. The corresponding condition for $g_0 < 0$,

$$g_\infty(V) + \frac{dg_\infty(V)}{dV}(V - E) < 0$$

is not obvious or even easily recognized. It is well worth noting that the HH steady state g_K and the peak g_{Na} increased rapidly over a limited region of V, so the corresponding g_0 can be negative if, and only if, $V < E$ in this region, as shown in figure 3:36 (Cole, 1965).

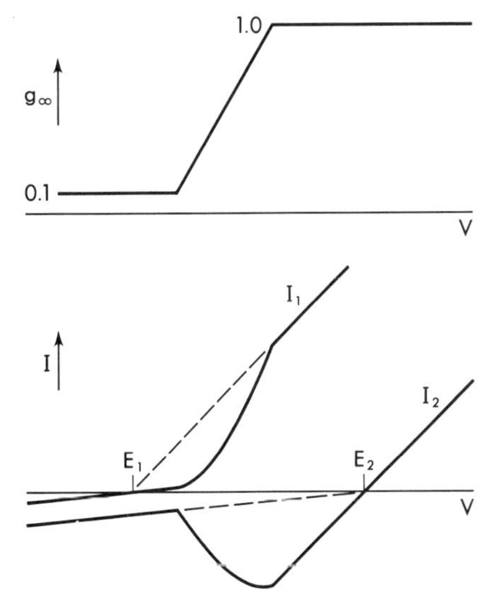

FIG. 3:36. The origin of a negative steady state membrane conductance. Above is an assumed high frequency conductance as it depends on potential. In a system with a low equilibrium potential E_1 the current I_1 vs. potential V characteristic approximates an ideal rectification. When the equilibrium potential is high, E_2, the current I_2 may show a negative steady state conductance in the region of decreasing high frequency conductance as given above.

It is further worth emphasizing—as Hodgkin and Huxley did not—the unique consequences of the application of the voltage clamp concept:

First and foremost, it permitted detailed investigation of the N type negative resistance characteristic which is inaccessible under the normal conditions of instability and all-or-none excitation.

Next, the clamp made a short-time analysis possible. Under constant current the passive time constant of the resting membrane RC is about 1 msec, which considerably confounds the analysis of the conductance changes with similar time characteristics, even with the power and precision of impedance measurements. However, with a potential source of low effective resistance R_S the membrane capacity is charged with the time, $CRR_S/(R + R_S) \approx R_SC$, of the order of 10 μsec, to permit direct observation of the entire subsequent time course of the conductance changes.

Finally, after its prompt establishment, the clamp maintains a new and constant driving potential, whether it be the electrochemical potential or the total electrical potential, on each and every ion across its conduction or permeability path. This vastly simplified the analysis and permitted the use of the membrane potential as the one and the only independent variable of the experiment.

Significance for mechanism

The complete and spectacular dependence of the entire structure of the HH formulation on the membrane potential—and the membrane potential alone—makes it seem highly probable that this variable occupies a position of predominant importance in control of the ion conduction process in the membrane. As a consequence, it seems quite reasonable to expect that a theory of ion permeability in which the membrane potential is the independent parameter has an initially high a priori probability of being a successful theory. A corrollary, perhaps even additional support for this view, is the evidence of the entirely unanticipated and unique power of the voltage clamp.

Stated in more classical and widely endorsed terms, this structure is already more than a superficial successor to Gasser's dictum to electrophysiologists, "Never infer a process from a potential." Quite the contrary, it is now possible to describe the ion conduction processes completely in terms of the membrane potential—and the membrane potential alone. The fact that the mechanisms of the processes are still unknown should and must humble us somewhat, and there may even be those who would now welcome the dual of Gasser's prohibition—"Never infer a mechanism from a current."

The time and potential behaviors of the conductances are in some ways comparable to a primitive electrodiffusion model, and suggest that changes of ion concentrations and electric field through the thickness of the membrane may give a theory that will be useful. Such problems can be so highly specific and so difficult that it may be some time before either a profitable approach can be found or the apparently reasonable possibilities eliminated (Cole, 1965). On the other hand, the past popularity and the many difficult accomplishments in this field of theory can in no way guarantee their applicability, and may indeed argue against it. So the conductances have not yet been found to be, and are not yet to be considered as, indicative of mechanism.

The temperature characteristics are so marked as to require that a model include them. The amplitude coefficient for $\bar{g}_K$ and $\bar{g}_{Na}$ of 1.3 over a 10°C interval (Hodgkin et al., 1952; Moore, 1958) corresponds to 2.7 percent per degree and is only a little high for average ion mobilities in aqueous and nonaqueous solutions. However, the membrane implications are far from clear. The temperature effect on the times of conductance changes τ_n, τ_m, τ_h, with a $Q_{10} = 3$, may be no more demanding than other considerations, but it does point emphatically to the presence of a high-energy process—almost 20 kcal/mole or 0.8 eV.

Another indication along this same line is the abrupt reduction of the conductances as the membrane potential is made more negative in the region $-20 \text{ mV} > V_m > -80 \text{ mV}$. Although $\exp(-V/5)$ may be an indication of a multiplicity of processes or of valence, it too may be evidence of a similar high-energy process.

We can only conclude that this first full harvest from the voltage clamp data and analyses has not only given us challenging questions in a compact form; it has also provided us with many hints for answers of high importance and wide application in living cell membranes.

Tests of the Hodgkin-Huxley Axon

The achievements of Hodgkin and Huxley were rather awe-inspiring but they were even more challenging. It was obviously important to test and confirm or modify their experimental and analytical work and some of its implications. It was necessary to find what limits there were, if any, to the physiological content of the voltage clamp data as given in their expressions for them. It was necessary, as far as possible, to interpret the many ramifications of relatively normal physiological behavior in terms of sodium and potassium conductances. It was similarly important to rewrite much of neuropharmacology in these same terms, and particularly to look for structural and kinetic implications of altered function.

There was no guarantee that the squid axon membrane was in any way representative of anything but itself. And indeed, in 1952, by saying at Cold Spring Harbor that he was "not a unitarian," Hodgkin made it quite clear that he and Huxley were then in no way implying any more general applications of their work. But the similarities of so many cells and tissues, that Hodgkin (1951) had himself just emphasized, made this view seem unduly cautious, and it seemed far more promising to go ahead on the assumption that these new approaches were not limited to the squid axon. It is, of course, true that almost every excitable tissue has some rather unique characteristic that should appear in voltage clamp data, unless the great value of the voltage clamp were limited to the squid axon membrane alone. It was obvious that new techniques would have to be developed for application of the voltage clamp concept to other cells.

The greatest challenge of all was—and still is—the explanation of the only unknowns in the Hodgkin-Huxley formulation, the sodium and potassium conductances. These were described so precisely and were such adequate expressions of so much physiology that any theory which could explain them quantitatively would be sufficient—for physiology at least. Conversely, a theory which failed to account for these conductances seemed destined to be incomplete or wrong in some, probably important, aspect. Theory thus has a clear and shining goal. Even if one is determined not to believe the Hodgkin-Huxley interpretations of the voltage clamp data, these data themselves are important facts of membrane behavior—to the extent that the experimental conditions are known. Thus a representation or an explanation of the voltage clamp data is a first and more primitive necessary condition on any model or theory—although it may turn out not to be a sufficient condition. The HH axon presents the ion current across the squid axon membrane in all of its complicated dependence upon membrane potential. In combination with the passive linear elements of the axon, the membrane capacity, and the internal and external electrolytic conductances, we have what appears to be reasonably complete and reasonably accurate information on all important components of the axon. Consequently, we should expect this model to answer all of the questions that have been asked of real axons. Many of these questions should be asked and for many reasons.

The ion equations are based entirely on voltage clamp data. There is no a priori guarantee that they have any correspondence with the behavior of the squid axon under any other conditions. To the extent that the equations do correspond to real axon performance, we have reason to believe that there remains no important mistake in the long series of many and variegated logical and theoretical or experimental events from Bernstein on. As experience and confidence increase, the model may first be used as a compact representation of the experimental facts and then for prediction and test. Perhaps finally it becomes an authoritative statement, not to be denied or questioned except by the higher authority of experiment, of what the axon is and does. The representation has undoubted limitations. These need to be explored and understood and the equations modified, extended, or replaced as may be necessary and possible.

As the model becomes reasonably well established it is all the more important to ask it questions that a real axon cannot answer—perhaps because elements are not accessible, because of thermal fluctuations, because it cannot be kept sufficiently constant, or because it may not even survive.

The additional elements of curiosity and challenge are far from minor. Many things should be and often are tried just because they can be.

COMPUTING THE MEMBRANE AND THE AXON

It seems highly probable from the mathematical point of view that there are solutions for every physically reasonable problem. This has so often been true that it could be accepted as a plausible assumption, but the existence of a unique solution has been demonstrated (Cole et al., 1955). It is however usually quite another matter to find the solutions.

Analytical solutions are practically impossible because of the extreme nonlinearity of the experimental relations between the ionic current and electric potential of the membrane. For small variations the equations can easily be linearized to give constant coefficients in the differential equations. These equations may still be of an order as high as four, and although much analysis is available on the properties of the solutions, the calculation of numerical information may be tedious. On the other hand, it may be possible to approximate on the basis of physical considerations or for special purposes to achieve simpler and significant results.

Similarly the well-developed topological methods, that have become available under the designation of nonlinear mechanics, have not been carried far enough into three or four space to be useful for the HH axon but can be very valuable in simplified situations (FitzHugh, 1960).

The explosive growth of the entire automatic computer field since the early 1950's has now made almost any computational power available anywhere—so much so that it can be assumed that almost no axon computations can be done without a computer and that a computer can be available for almost any problem. There seems little reason not to believe that the "almosts" will practically vanish in a few years, just as the novelty of space men and electric can openers has already worn off. So it seems appropriate to again stress the prodigious feats of only a decade and a half ago that seem now but small, fast, and easy steps toward the frontiers of today and perhaps of tomorrow. The experiments of Hodgkin, Huxley, and Katz and Hodgkin and Huxley were no mean feats, but it is even more impressive to remember that all calculations in the final paper were done on a desk calculator. For this reason alone these results should be listed first and together. But there are important items at so many levels of complexity that these first calculations will be dispersed among them.

The account of the calculations will be arranged somewhat in the order of apparent complexity. In working back toward a description of a functioning nerve in an animal, this rather reverses much of the order of the experimental development. But it is far from obvious that the most difficult problems are the most important. It may be found that an understanding of even very simple problems is so difficult that the rest are only a little worse.

Up to the present, most of the emphasis has had to be placed on space clamp calculations in which current density and potential difference are uniform over an area, figure 3:1. Here we have first the linear approximations which are valid for a few millivolts, then the nonlinear subthreshold and the threshold excitation phenomena in the range of 10–20 mV, and finally, the stationary action potentials. As the space variable is again considered, calculation of uniform propagation has become a readily available routine, figure 3:49; doubtless the starting and stopping of an impulse will become increasingly easy to work out for the free, normal axon of figure 3:34.

Methods

The simplest approach to the HH equations, 3:3, 3:4, 3:5, is that of an approximation from the Taylor series (Hodgkin and Huxley, 1952d; Chandler et al., 1962):

$$f(x + \delta x) = f(x) + \left.\frac{\partial f}{\partial x}\right|_x \delta x + \left.\frac{\partial^2 f}{\partial x^2}\right|_x \frac{(\delta x)^2}{2!} + \left.\frac{\partial^3 f}{\partial x^3}\right|_x \frac{(\delta x)^3}{3!} + \cdots$$

For a sufficiently small value of δx this approaches:

$$\delta f = f(x + \delta x) - f(x) = \left.\frac{\partial f}{\partial x}\right|_x \delta x.$$

The equation:

$$I = C_m \frac{dV}{dt} + \bar{g}_{\mathrm{K}} n^4 (V - E_{\mathrm{K}}) + \bar{g}_{\mathrm{Na}} m^3 h (V - E_{\mathrm{Na}}) + \bar{g}_L (V - E_L)$$

then becomes:

$$\delta I = C_m \frac{d\delta V}{dt} + \bar{g}_{\mathrm{K}} \underline{n}^4 \delta V + \bar{g}_{\mathrm{Na}} \underline{m}^3 \underline{h} \delta V + \bar{g}_L \delta V + 4\bar{g}_{\mathrm{K}} \underline{n}^3 (\bar{V} - E_{\mathrm{K}}) \delta n + 3\bar{g}_{\mathrm{Na}} \underline{m}^2 \underline{h} (\bar{V} - E_{\mathrm{Na}}) \delta m + \bar{g}_{\mathrm{Na}} \underline{m}^3 (\bar{V} - E_{\mathrm{Na}}) \delta h$$

where δI, δV, δn, δm, δh are the small linear variations from the steady state values $\bar{I}$, $\bar{V}$, $\underline{n}$, $\underline{m}$, $\underline{h}$. In the same way,

$$\frac{d}{dt}(\delta n) = \frac{\partial \alpha_n}{\partial V}\,\delta V - (\alpha_n + \beta_n)\delta n - \underline{n}\,\frac{\partial(\alpha_n + \beta_n)}{\partial V} \cdot \delta V \quad (3{:}6)$$

with similar expressions for δm and δh. Hodgkin and Huxley then eliminated δn, δm, and δh between the four equations to obtain a fourth order linear differential equation with constant coefficients for δV. This may be written in terms of the operator p in the form

$$g(p) = pC + g_\infty + \frac{g_m}{1 + p\tau_m} + \frac{g_n}{1 + p\tau_n} + \frac{g_h}{1 + p\tau_h} \quad (3{:}7)$$

and corresponds to the circuit shown in figure 3:39.

Upon leaving the range of a few millivolts in which a linear approximation is useful, it becomes necessary to rely entirely on numerical methods. Hodgkin and Huxley tried to use the then current automatic computer at Cambridge in their original work, but unfortunate mistakes and delays forced them to rely entirely on a desk calculator. For these integrations they employed the successive difference process of Hartree (1932–1933) which they present in outline. Although this requires judgment and is tedious, it has the compensation that the computer is in touch with all stages of the process. He is able to see trends, detect errors, modify intervals, and make approximations and interpolations, and must derive a considerable satisfaction in seeing what is going on as the solutions progress (Huxley, 1964). Yet Huxley has admitted that it took about a day to compute a single space-clamped response and a week for a free propagating impulse, while Hodgkin has often emphasized his colleague's amazing speed and accuracy with a desk calculator.

An analogue computer has many of the same advantages as a desk calculator although it has limited accuracy and is quite inadequate for some problems. Nonetheless, FitzHugh (unpublished) has been able to supplement the direct solution by a process of successive approximations to give satisfactory impulse speeds and action potentials in free propagation problems. An analogue computer can present a solution graphically in a few minutes, but it is moderately complicated and usually needs tender, loving maintenance.

For automatic digital computation, the Runge-Kutta single-step process was first used on the Standards Eastern Automatic Computer (SEAC) at the National Bureau of Standards as outlined by Cole et al. (1955). It has since been used on other computers—IBM 704, 650, Honeywell 800, EDSAC I, II—and by other investigators —FitzHugh and Antosiewicz (1959) and Huxley (1959a).

As a general technique of computation, the effects of temperature have been dealt with as originally outlined by Hodgkin and Huxley. Rather than to change all of the α's and β's or all of the τ's, it is far simpler to change only C and the time scale correspondingly. Thus at a temperature T, C becomes ϕC, the unit of time becomes $1/\phi$ where, for $Q_{10} = 3$, $\phi = 3^{(T-6.3)/10}$.

Space clamp

Linear approximation

Hodgkin and Huxley solved the fourth order linear differential equation by Horner's root reduction method (Householder, 1953) to obtain two real roots and a complex conjugate pair, all with negative real parts, after the application of a small current step at their resting potential. This gave a slightly underdamped response very similar to experimental records (figure 3:37). They also used this result to compare with the step-by-step numerical calculation that was necessary where the linear approximation was no longer adequate.

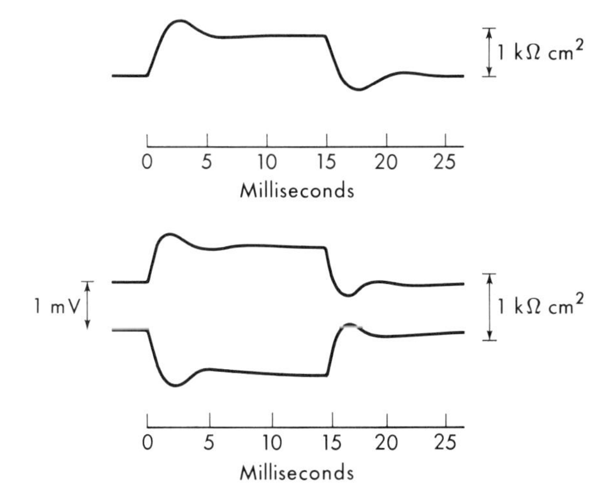

FIG. 3:37. Linear responses of membranes to applied current pulses; above as calculated from Hodgkin-Huxley equations; below as recorded for a real axon.

Hodgkin and Huxley pointed out that the n and h processes produced inductive reactances. The larger potassium process was equivalent at 25°C to the circuit of figure 3:38a. The inductance of 0.4 henry cm² became larger by a factor of 3 for each 10°C decrease of temperature, and increased as $V \rightarrow E_K$ to be replaced by a capacity for $V < E_K$. This inductance was noted as reasonably close to the experimental value of 0.2 henry cm². Cole and Baker (1941) and Hodgkin and Huxley (1952d) concurred with the earlier conjecture in ascribing it primarily to potassium. Although use of the probably preferable equivalent circuit of figure 3:38b would have widened the divergence, this could quite well be explained by a higher resting potential in the earlier experiments. The HH axon produces a potassium inductive reactance that varies from infinity to 0.08 henry cm² over the membrane potential range from 12 mV of hyperpolarization to 120 mV of depolarization. This is combined with a steady state potassium and leakage conductance from 0.3 to 36 mmho/cm² over the same range. All in all, these results rather well confirm my model, its use, and interpretation of twenty-odd years before where excitation was not involved!

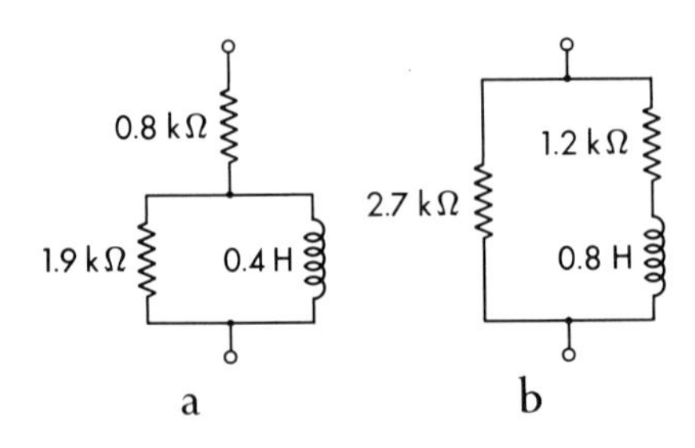

FIG. 3:38. Composite equivalent circuits for the ion conductance of the Hodgkin-Huxley axon at the resting potential.

Chandler et al. (1962) have considered the solutions of equations 3:6 more generally and in somewhat more detail. This was done, quite belatedly, for the highly practical purpose of determining the nature and the limits of stability of an axon membrane in an imperfect voltage clamp; the numerical calculations were carried through on an IBM 650.

The stability of dynamical systems was of great mathematical interest at the beginning of this century. The results gave the foundations of the theory that has become so important with electronic and other new powerful technologies. In particular, the stability of an equilibrium can be investigated by a theorem of Liapunov

(1907; Minorsky, 1947; Lefschetz, 1957). This shows that a nonlinear system is almost always stable or unstable according as the linearized system is stable or unstable.

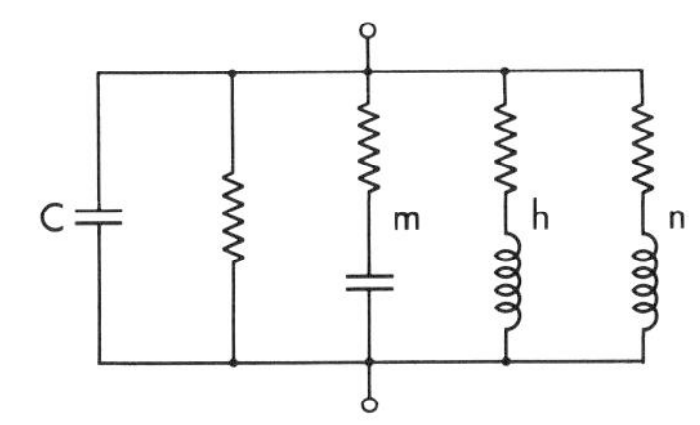

FIG. 3:39. Complete equivalent circuit for linear responses of the Hodgkin-Huxley axon. The m, h, and n branches are for the frequency dependent characteristics of the m, h, and n parameters of the equations. Every element of the circuit, except the membrane capacity C, changes with the potential difference across the membrane.

The linearized membrane system of figure 3:39 in series with a conductance g has solutions

$$\delta x_i = a_i \exp \lambda t$$

where λ satisfies

$$g + F(\lambda) = 0$$

and

$$F(\lambda) = g_\infty + \lambda C + \frac{g_m}{1 + \lambda\tau_m} + \frac{g_n}{1 + \lambda\tau_n} + \frac{g_h}{1 + \lambda\tau_h}. \quad (3:8)$$

The system is stable or unstable as the real parts of every $\lambda < 0$ or > 0. $F(\lambda)$ for real values of λ is shown for the resting potential in figure 3:40. There are simple poles at $-1/\tau_m$, $-1/\tau_n$, and $-1/\tau_h$ with the latter two quite close together. For $g > 0.812$ mmho/cm^2, the four roots are all real and negative. Thus, the system is stable for a series resistance less than about 1230 Ωcm^2. In the other direction, at least one of the real roots becomes negative if g is more negative than -0.9 mmho/cm^2. This corresponds to a series negative resistance less than -1100 Ωcm^2 and is about what is to be expected from the steady state V, I characteristic at the resting potential. In the region $-0.9 < g < 0.81$ there are two negative real roots and a pair of conjugate complex roots which can be represented on the complex plane, $\lambda^* = \lambda' \pm \lambda''$. On the root locus, indicated schematically in figure 3:41, two roots become pure imaginaries as g increases through a critical value g_c to pass from instability to stability. This locus has not been calculated carefully nor has a simple procedure been found to evaluate g_c. However, Hodgkin and Huxley found stability at $g = 0$ so at least $0 > g_c > -0.897$, and from figure 3:41, $g_c \approx -0.5$ mmho/cm^2.

Hearon (1964) showed by a more general analysis from linear kinetics that $\Sigma\bar{g} = g + g_\infty + g_m + g_n + g_h > 0$ is a necessary condition for stability, while $\Sigma\bar{g} < 0$ is a sufficient condition for instability. There is however a $Q > 0$ defining a region $Q > 2\bar{g} > 0$ which may be stable or unstable.

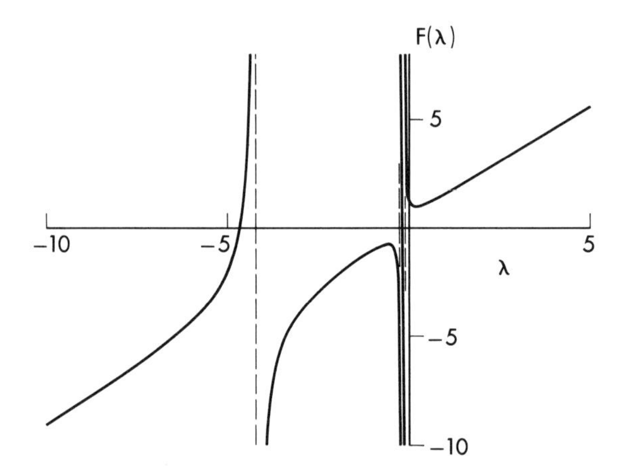

FIG. 3:40. The behavior of $F(\lambda)$ of the characteristic equation at the resting potential and for real values of λ. The three simple poles are indicated by the vertical dashed lines. As the conductance g across the membrane is made increasingly negative, the entire pattern moves to the right. The system reaches instability when the first real root turns positive.

When I asked if they had calculated any impedances directly, Hodgkin said they had not thought of it and it was too bad because it would have been so easy. So I may outline it here. Returning to equations 3:6,

$$\frac{dn}{dt} = \frac{\underline{n} - n}{\tau_n}, \text{ etc.}$$

where $\underline{n} = \alpha_n/(\alpha_n + \beta_n)$, $\tau_n = 1/(\alpha_n + \beta_n)$, etc., and applying a sinusoidal potential of frequency $\omega/2\pi$ and replacing d/dt by p, we have $n = \underline{n}/(1 + p\tau_n)$, and

$$\delta n = \frac{1}{1 + p\tau_n} \cdot \frac{\partial n}{\partial V} \cdot \delta V, \text{ etc.}$$

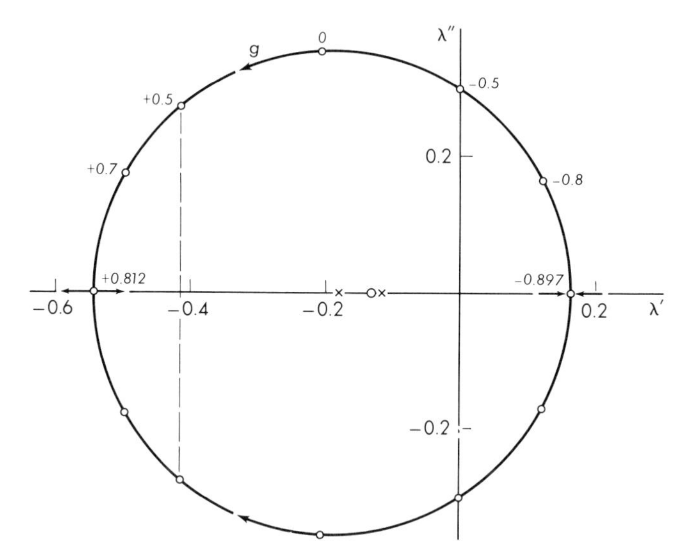

FIG. 3:41. The behavior of the characteristic equation in region of the conjugate complex roots, $\lambda' \pm j\lambda''$, as the conductance g, across the membrane is changed between the approximate values of -0.9 and 0.8. The circle on the real axis is the position of the real root and the crosses are two poles.

Then equation 3:6 becomes

$$g(p) = \delta I/\delta V$$

$$= Cp + \bar{g}_K n^4 + \bar{g}_K n^4 (V - E_K) \frac{4 \dfrac{\partial}{\partial V}(\ln n)}{1 + p\tau_n} + \bar{g}_{Na} m^3 h$$

$$+ \bar{g}_{Na} m^3 h (V - E_{Na}) \left[\frac{3 \dfrac{\partial}{\partial V}(\ln m)}{1 + p\tau_m} + \frac{\dfrac{\partial}{\partial V}(\ln h)}{1 + p\tau_h} \right]$$

or as in equation 3:7

$$g(p) = Cp + g_\infty + \frac{g_n}{1 + p\tau_n} + \frac{g_m}{1 + p\tau_m} + \frac{g_h}{1 + p\tau_h}$$

where g_∞ is the sum of the infinite frequency conductances and g_n, g_m, g_h, are the differences between the infinite and zero frequency conductances. The ionic admittance is thus composed of three reactive admittance components in parallel. Graphical or numerical procedures have been used to estimate the coefficients of each.

Graphical and electromechanical methods, figure 3:42, have been very useful for finding approximate values of the real and imaginary components of each and of the sums. One such result is given in figure 3:43 (Cole, 1955), although the IBM 650 gave better values more rapidly (Chandler et al., 1962).

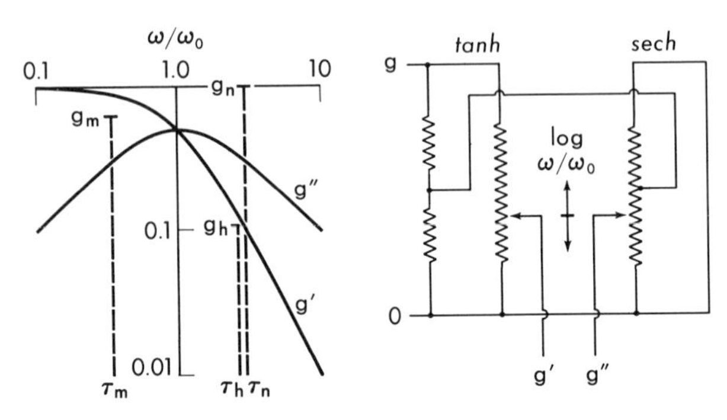

FIG. 3:42. A graphical method for computing membrane ionic admittance, left. The normalized real and imaginary components of a single branch are plotted as g' and g''. The three g coefficients are represented as vertical lines at their characteristic frequencies, $f = 1/2\,\pi\tau$, on a transparent overlay. The components of each conductance are read off at the desired frequencies as the overlay is slid over the master curves. An electro-mechanical calculation of element admittance, $g' + jg''$, right. The coefficient potential is supplied to the function potentiometers and the ganged contacts are driven by a shaft rotation proportional to the logarithm of frequency.

These loci are rather like the pedal curves for a resistive network with two reactances. The reason for this and the basis for an approximation is that τ_n and τ_h are always within a factor of 2 of each other while the coefficients are of the same sign—to give inductive reactances. Conversely, to the extent that a real membrane behaves in the same way, it is not usually possible to differentiate these two time constants experimentally. Furthermore, in most of the membrane potential range the h coefficient is considerably smaller than that for n. By contrast, the values of τ_m are about $\frac{1}{10}$ of τ_n or τ_h, and the reactance is that of a negative inductance or a conventional capacity with a negative real admittance, $g_{\mathrm{Na}}(0) - g_{\mathrm{Na}}(\infty) < 0$.

The ionic conductances computed from the earlier longitudinal impedance measurements were now available for the comparison

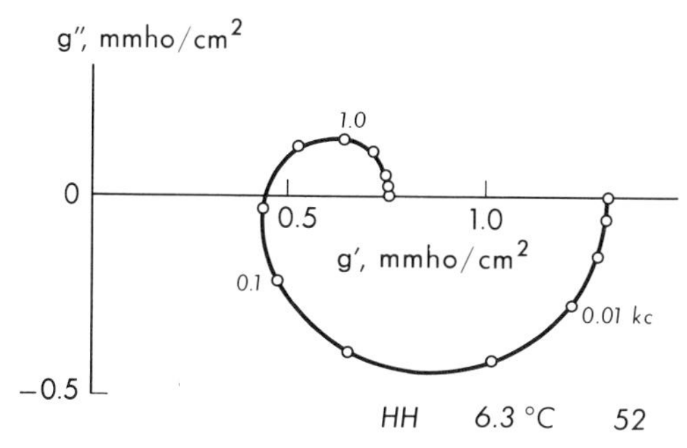

FIG. 3:43. The complex ionic conductance locus of the Hodgkin-Huxley membrane at rest. The low frequency inductive component is mostly for potassium and the high frequency capacitive component comes from the sodium "on."

with these calculations of the HH axon. A marked difference was that the experiments on the more satisfactory axons in sea water showed a capacitive low frequency component rather than the inductive component of the first experiments and of the HH axon. This suggests that the former were relatively hyperpolarized and came into better agreement when depolarized by excess K_e as in figure 1:51. The zero frequency resistance and the inductive reactance decreased as they should if the zero and infinite frequency conductance approach each other.

Voltage clamp data and the HH expressions for them are thus consistent with the earlier linear measurements in containing L or C characteristics in addition to $C = 1\ \mu F/cm^2$. These clamp data furthermore contain the reactances as factors in the time constants of nonlinear processes, and attribute much of the inductive reactance to potassium conductance, as had been surmised long before. They do, however, identify the higher frequency capacitive reactance with the negative sodium conductances and require an inductive reactance component that probably cannot be resolved experimentally. But even in the simplest analytical case of linear approximation the expediency of automatic computation is obvious.

Subthreshold and threshold

As we leave the region of a millivolt or two and approximate linearity, we might creep up into increasing complexity by adding second order terms as has often been done for engineering problems. The next region of physiological significance, the threshold for excitation, may require even another or more, and more complicated, steps in such a process. It is probably expedient, as has been done, to dive into all of the complications of the complete HH equations and let the computer—human such as Huxley or any of the machines—deal with them.

Hodgkin and Huxley showed the existence of a threshold for a sudden depolarization between 6 and 7 mV, and calculated an excitation at the break of a 30-mV hyperpolarization, to agree well with experiments. They inferred accommodation because both the increase of n and the decrease of h during a constant outward current raise the threshold, and also because for a sufficiently slow rise of current the potential will stay close to the steady state path and remain stable.

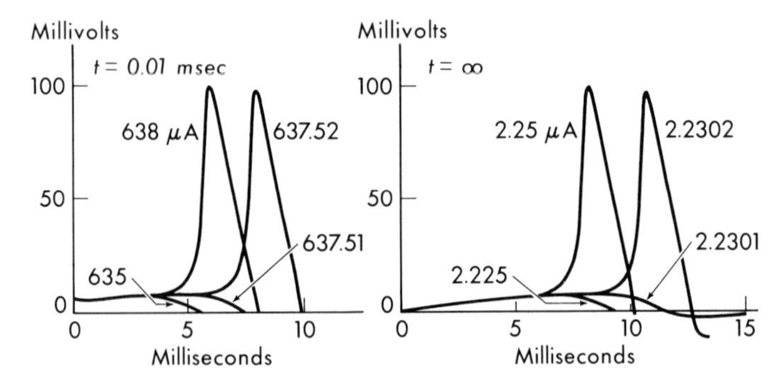

FIG. 3:44. Near threshold responses of Hodgkin-Huxley membrane calculated on SEAC. Left, for a short pulse and right, for a constant current.

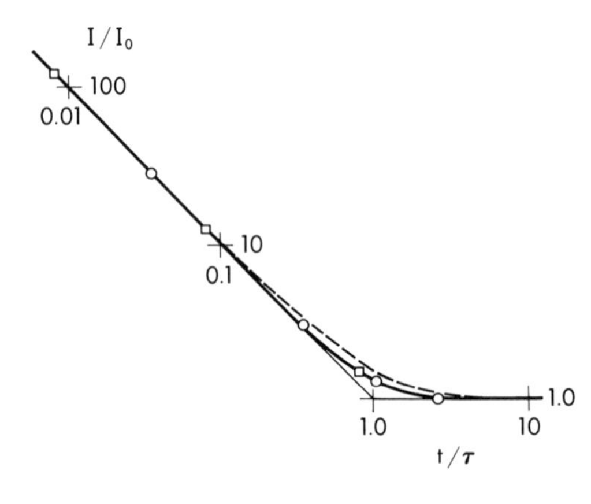

FIG. 3:45. Normalized current I, vs. duration t threshold on logarithmic scales computed for Hodgkin-Huxley membrane. Circles are at 6.3°C, squares at 20°C, I_0 and τ are the respective rheobase and time constant. The dashed line is for a single exponential process.

The threshold relation between a current and its duration was computed on SEAC (Cole et al., 1955) as shown in figures 3:44 and 3:45 and again by FitzHugh (1966). The transition between the short-time constant quantity and the long-time rheobase with $\sigma = 1.317\text{–}1.344$ is definitely below that for a single exponential process. The difference, however, as predicted in the two-factor accommodation formulation is not as much as that found experimentally (figure 3:12).

Considerable efforts to determine the exact nature of the threshold produced trajectories on the phase plane which required a saddle point, or an all-or-none threshold, at a depolarization of about 8.3 mV. This support of my long-time prejudice confounded Bonhoeffer and aroused a firm skepticism in Huxley, while FitzHugh refused to believe it. After Bonhoeffer's death and in quite a different connection it was discovered that a mistake of order had been made in coding, as described by FitzHugh and Antosiewicz (1959). This introduced a spurious saddle point for each of the two indeterminates, 0/0, in the HH axon. Then on recalculation the all-or-none threshold was lost and a graded response took its place—over a stimulus range of about 10^{-8} (figure 3:46)! Except for correcting my mistaken concept and introducing some difficulties in accounting for accommodation phenomena the changes were not significant—at least by comparison with an optimistic 1 percent level of experimental variability. Nonetheless there is certain to be an absolute threshold for the initiation of an eventually uniformly propagating impulse. This must be calculated from the full partial differential equation. But whether or not this absolute threshold

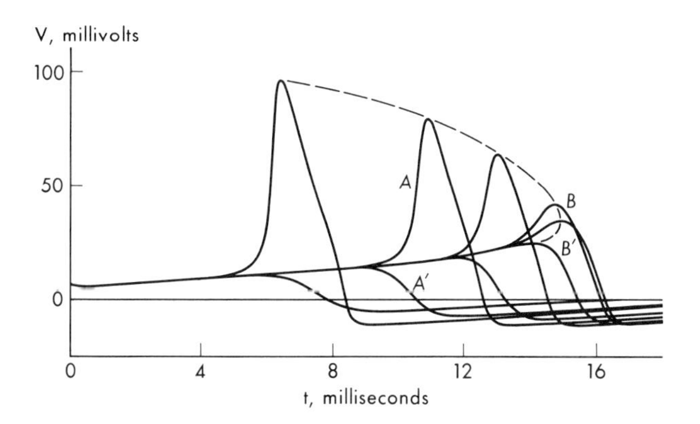

FIG. 3:46. Quasi threshold response computed from *HH* equations after short pulses of current. The difference between the stimuli for curves A and A' was one part in 10^8 and for B and B', one in 10^{14}.

can be simply correlated with some part of the quasi threshold in a space clamp is quite unknown and seems rather improbable.

The form of the computed threshold strength duration curves (figure 3:45) was in good agreement with the excitation data of 1947 and 1948 with Marmont (Bonhoeffer, 1953; Cole, 1955; and unpublished). The constant quantity Q_0 was reached at short times, the transition to the rheobase I_0 was again sharper than possible with the two-factor formulation and had no significant temperature dependence. Furthermore, Q_0 was only slightly temperature-dependent while the calculated $Q_{10} = 1.95$ for the rheobase was in excellent agreement with the mean value $Q_{10} = 2.0$ from experiment. However, the calculated values of $Q_0 = 0.69 \cdot 10^{-8}$ coulomb/cm^2 and $I_0 = 5.6$ μA/cm^2 were decisively less than the corresponding means of $2.1 \cdot 10^{-8}$ coulomb/cm^2 (1947) and 21 μA/cm^2 (1947), 41 μA/cm^2 (1948).

Thus the general nature of the excitation process was well described by the HH axon while the absolute values of the parameters found at Woods Hole, although quite variable, were certainly considerably higher than those given by the equations. Different axons seemed to require different coefficients, but probably not a different formulation.

There have not been computations of variation of Q_0 with external sodium Na_e for comparison with Marmont's (unpublished) expression for our data, $\log Q_0 = A + B/Na_e$. Furthermore, there was neither experiment nor calculation in the long time range of accommodation phenomena, while considerable discrepancies were found in the repetitive responses so fundamental to communication in nervous systems.

Action potentials

Hodgkin and Huxley presented as their first calculation the action potential response to a short stimulus well over threshold (15 mV), which was compared very favorably with a recorded action potential. Although we were not aiming to check either Hodgkin and Huxley or our own computational abilities, it did seem quite obvious that this should be the first chore for SEAC. As repeated on the IBM 704, figure 3:47 shows there was no obvious reason to question either computation; FitzHugh (1960) has given the corresponding analogue records of m, n, and h (figure 3:48). Hodgkin and Huxley (1952d) also noted the delay—the increased latency—without

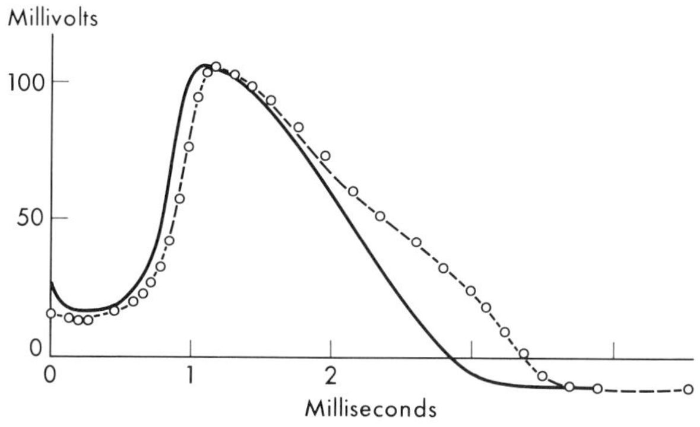

Fig. 3:47. Comparison of squid axon action potentials. Circles, from *HH* calculated curve; dashed line, computed on SEAC; solid line from *HH* curve for real axon.

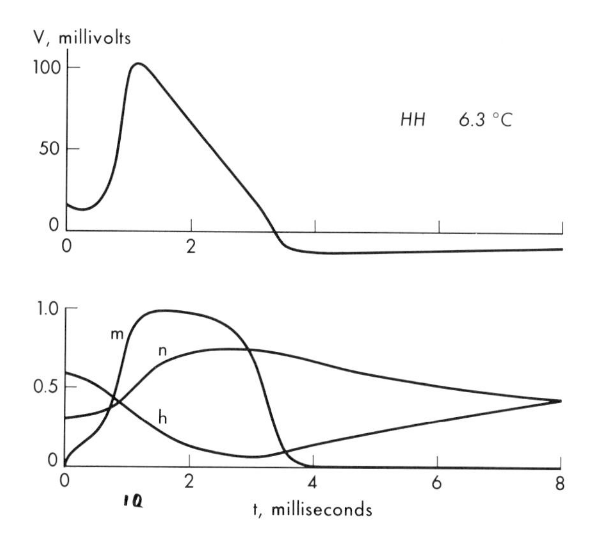

Fig. 3:48. Analogue computations of *HH* membrane. Above, action potential after short pulse. Below, time courses of the parameters m, h, and n.

marked change of form of the potential as the stimulus strength was reduced to near threshold.

Naturally the gratuitous bump on the recovery limb was an interesting curiosity. So far as I know, there has not yet appeared a reason for this rather marked difference from the usual squid action potential. On the other hand, it has been highly prophetic of the tetraethylammonium (Tasaki and Hagiwara, 1957) and heavy metal effects (Spyropoulos and Brady, 1959) which produced a plateau as a gross exaggeration of this characteristic of the model.

Hodgkin and Huxley then traced the effect of a second massive 90 mV stimulus during the response to the first. In the rising phase and well into the return of the potential this produced about the potential change to be expected from the purely passive discharge of the imposed condenser charge corresponding to an absolute, or at least highly refractory period. Then there appeared an increasing short turn toward the sodium potential through a relatively refractory period until recovery was about complete, beyond the valley of the undershoot of the potential. These changes were correlated with the interplay of the increases and decreases of the sodium and potassium currents and the representations of their effects on the phase plane, to give an understanding of the successive phases of the very similar potentials produced experimentally.

Yet another strong physiological demonstration of the power of the HH axon as a representation of a real axon was made in the opposite direction. Hodgkin and Huxley computed the threshold for elimination of the response by an anode—or countershock—in the recovery phase. They had, however, to speed up the m and h processes by a factor of 4 in order to make comparison with frog node data (Tasaki, 1956)—our earlier data on squid being then as now unpublished.

It seems quite impossible that calculations of the effect of changes of external sodium concentration would not agree generally with the observations on the squid axon in a space clamp, figure 3:22b (Cole, 1955), and highly improbable that any usefully significant differences would appear. However, these calculations are so easy and these data so simply and so impressively correlate the spike height with a Nernst equilibrium potential for sodium that they should be done. Perhaps then the rather unimportant differences from the original work of Hodgkin and Katz might be explained as, for example, by the use of sucrose or propagating impulses.

Summary

Upon leaving the linear region, numerical and analogue computations of the space-clamped HH axon provide a wealth of impressive information. These calculations are of particular physiological importance because they reproduce most of the characteristic behavior of real axons in the subthreshold, threshold, and response regions. It may be of considerably more fundamental importance to have this further evidence of the power of voltage clamp data extended both in breadth and depth. The sodium theory interpretation of these clamp data then provides both qualitative and detailed ionic descriptions of the physiological phenomena, but without any particularly impressive tests of the theory. There are considerable numerical differences between the calculations and the experiments, particularly in the threshold region. For the present, at least, these may be ascribed to differences between the axons described by the voltage clamp experiments and those subjected to excitation tests. The theoretical conclusion that the thresholds are not absolutely all-or-none is very interesting but without physiological significance.

Free action potentials

Constant speed

As if they had not already done enough to show that the HH axon and, therefore also, the voltage clamp data contain many of the cherished phenomena of classical electrophysiology, Hodgkin and Huxley then included rather prodigious calculations to show both that the data gave an excellent estimate of the speed of impulse propagation and also, most importantly, independent evidence to support the interpretation of these data in terms of the sodium theory.

The difficulty of analytical solutions for the nonlinear Kelvin cable equation has already been considerably emphasized, and the difficulty of computation will receive yet further attention. In one of their not infrequent masterpieces of understatement, Hodgkin and Huxley explain that it is rather easier to deal with an impulse propagating with uniform speed θ. Some may—and have—quibbled that this is no proof that an impulse ever does or even can achieve this speed. Reserving this objection for further comment later, I only observe now that the speed of a real impulse in a real axon is a quite well-established fact and that the voltage clamp data have so far supported such an assumption. In reverse, I think that the evidence for the reality of a constant speed solution is so strong that I will question it only on the presentation of experimental or analytical evidence purporting to show that such propagation is either impossible or at least questionable.

Then, with uniform propagation as a highly acceptable premise, we rewrite the equations for a free and propagating axon—quite unfettered by voltage, current, or space clamps as in figure 3:49

$$I_g + C\frac{dV}{dt} = \frac{1}{\mathrm{K}} \cdot \frac{d^2V}{dt^2} \tag{2:6}$$

where

$$\mathrm{K} = \theta^2(r_i + r_e).$$

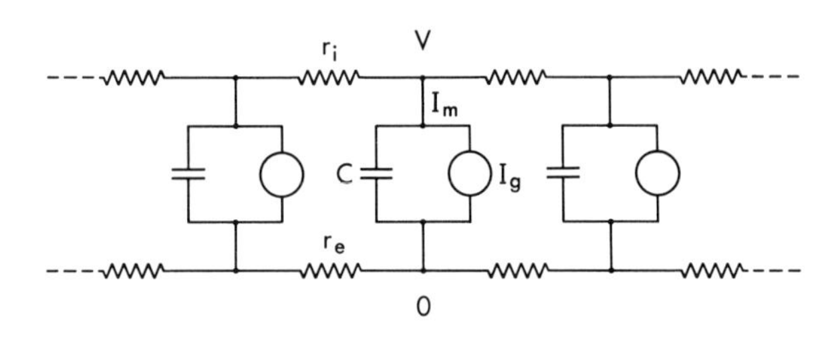

$$I_m = C\frac{dV}{dt} + I_g = \frac{1}{\theta^2(r_i + r_e)} \cdot \frac{d^2V}{dt^2}$$

FIG. 3:49. The cable equivalent of the squid axon and the special, ordinary, differential equation for an impulse velocity θ in which the ionic current I_g may be given by the Hodgkin-Huxley equations.

This is, of course, immeasurably more simple than the partial differential equation of which it is a special case. But there is no basis to estimate θ except from the experiment, and if one insists on theoretical purity, this is indeed heresy. There is, however, the further assumption that the velocity of the propagating potential shall allow it to have a reasonable form and return finally to the resting potential; this value of θ can only be found by trial and error. As high and low values of θ turned the potential rather abruptly toward $+\infty$ and $-\infty$, it was possible to return to a point at which the difference had been negligible and proceed further with an interpolated θ. The process had to be repeated with increasing difficulty, and for this obvious reason Hodgkin and Huxley only carried it to the maximum potential. Later potentials were calculated as in space clamp and corrected by trial and error for the longitudinal current, d^2V/dt^2. Quite understandably, only a few of these free action potentials were computed in the original work. One of these results is shown in comparison, on fast and slow time scales, with recorded potentials in figure 3:50. The forms of the potentials are so nearly the same that one cannot choose which was directly recorded and which was created entirely from voltage clamp data without including other information.

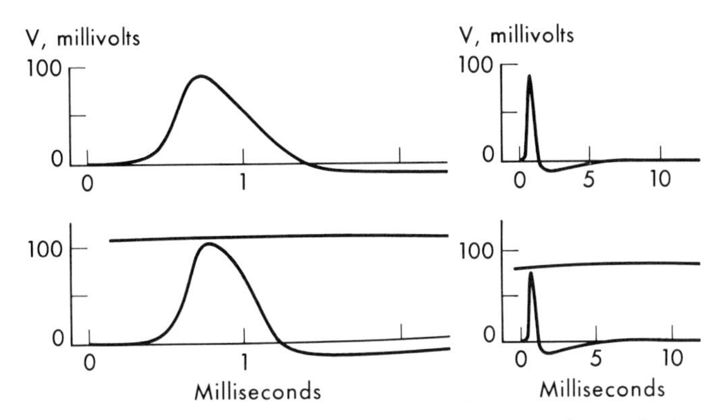

Fig. 3:50. Calculations, above, and experimental records, below, for propagating impulse on fast time scale, left, and slow scale, right.

The tedium of the hand process is well illustrated by the automatic computations of FitzHugh and Antosiewicz (1959) on an IBM 704. Figure 3:51 shows the difference between values of θ which diverge in opposite directions when the computer search had narrowed the margin down to the last unit difference of the machine, $1/10^8$. It was then necessary to invent an interpolation process to carry the solution beyond the machine accuracy. This process was continued until the solution could be connected up with the proper one of the four possible roots of the linear approximation by which it could return to the resting potential.

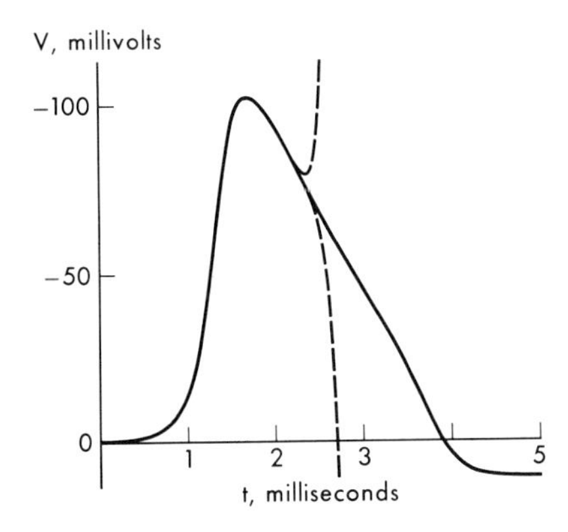

Fig. 3:51. Digital computation of a propagating impulse. The dashed lines are for a difference of speeds at limit of computer, one part in 10^8.

Even though the propagating impulse is the goal and the achievement of peripheral nerve, the value of such calculations may seem somewhat obscure. It is obvious that such fantastic accuracy in the computed impulse speed—about one part in 10^{18}—is of no experimental significance whatsoever. The agreement between 21.2 m/sec and 18.8 m/sec for the experimental and calculated speeds of the impulses at 18.5°C as shown in figure 3:50 is eminently satisfactory in view of the many errors, approximations, and assumptions made along the two paths before they finally converged. But it was only by hand calculations and then by carrying the process through to completion that the last lingering doubts as to the relation between voltage clamp data and normal axon function could be dispelled.

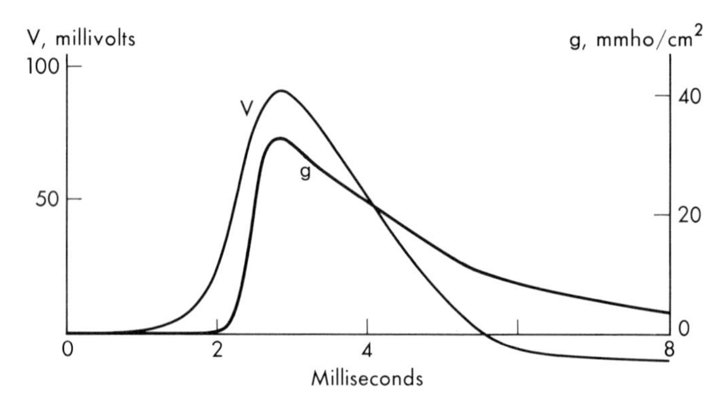

FIG. 3:52. Calculated high frequency impedance change g and propagating action potential V.

With this connection established, much more could be obtained in support of both the machinery of the analysis and the theory of the interpretation of the voltage clamp data. It is not at all necessary that the voltage clamp currents be expressed in terms of conductances. However, the calculations of the propagating impulse include them, as shown in figure 3:52, and the sum compares so well with their direct measurement as to give strong external support to that contained in clamp experiments. The computed maximum conductance increases of 30–50 mmho/cm^2 comfortably cover the reported experimental average of 40 mmho/cm^2 (Cole and Curtis, 1938a, 1939). Strange as it may seem, I found this a considerable relief from the nagging worries that mounted over the decades as our results had not been checked and had indeed been used for at least one calibration. The calculated time course of the conductance increase is also in general agreement with experimental records, although I hope that they may be brought closer in the future. All records as corrected for instrument distortion showed a considerably more rapid rise of the conductance than had the calculation, and I believe that an eventual agreement will be found closer to the experiments than to the calculation. Another obvious difference is the more complicated recovery—perhaps related to the slight shoulder on the calculated action potential. Although squid data are not available, the further calculation (Huxley, 1959a) of $Q_{10} = 1.74$ for the speed of propagation is a reasonable prediction.

It is far from impossible that other conductance components might be used to interpret the clamp data and give as good or even better comparison with direct experiment, perhaps such as Tasaki and Freygang (1955) and Moore and Cole (1963) and others have tried to record. But until proved to the contrary, it seems reasonable to assume that the agreement of the sum with experiment is far from negligible support for the components that make it up.

With the voltage clamp data, their interpretation, and their synthesis to reproduce the propagating impulse, it was then possible to compare the results in other ways with earlier data and to see what we might have been able to do then—if we had been that wise. As before, we can trace the path of a computed propagating impulse as the V vs. I_g locus (figure 3:53) to compare reasonably well with more direct calculations from a real impulse (figure 2:23). Furthermore, we have the sum of the ionic conductances to give us the g_∞ slopes as shown by the dashed lines, and also their intercepts E_∞ to give a complete description of the Thévenin equivalent circuit (Guillemin, 1935),

$$I = g_\infty(V - E_\infty).$$

With more faith in the measurements, and with the insight that was to come later, the 1939 data could have been so expressed. But after the sun of sodium rose in 1947 we had E_K and E_{Na}, and the data could be dissected to give the corresponding component conductances

$$g_K = g_\infty(E_{Na} - E_\infty)/(E_{Na} - E_K)$$
$$g_{Na} = g_\infty(E_\infty - E_K)/(E_{Na} - E_K).$$

The conductance components then permit calculation of the ionic currents, figure 3:54. Here there is not even an independent check

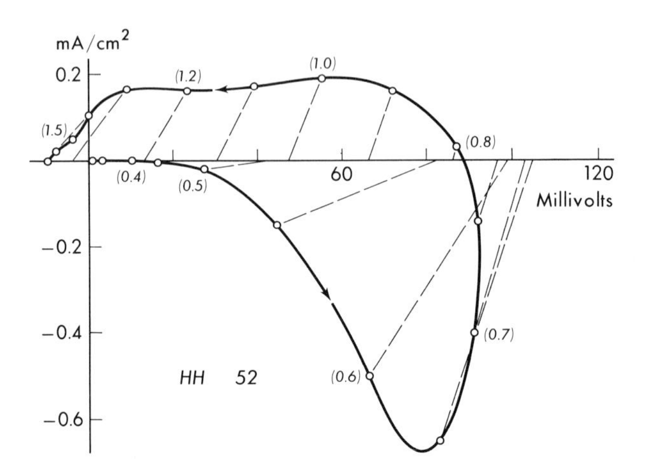

FIG. 3:53. Relation between membrane potential and current during a propagating impulse for an *HH* axon, solid line. Times in msec are given in parentheses. The slopes of the dashed lines are high frequency conductances and the intercepts on the voltage axis are equivalent, Thévenin, emfs.

on their sum. This might be achieved with differential electrodes (Cole and Moore, 1960a) but not from the record of Cole and Curtis (1939) showing $\partial^2 V/\partial x^2$ which relies on the cable equation. But Hodgkin and Huxley go in the other direction. They integrate I_{Na} to give the total gain of sodium and I_K to give the total loss of potassium as 4.33 and 4.26 p mole/cm² impulse, at 18.5°C. These are in gratifyingly good agreement with the values of 3.5 and 3.0 at 22°C obtained by Keynes and Lewis (1951), an agreement of great importance. All of the checks made so far have been essentially electrical and of an internal self-consistency of the system. Here, for the first time, is an independent measurement made by completely separate techniques in no way relying on the internal rules of the complicated electrical system.

Noting that the principal effect of a decrease in temperature was to extend the duration of a propagating action potential, Shanes (1954) promptly undertook a flame photometer measurement of the expected large increase of potassium loss at low temperature as a test of the HH axon. His average value of 11.4 p mole/cm² impulse at 6.1°C is indeed much larger than the 3.7 p mole/cm² impulse at 24°C and may be expressed by the impressive $Q_{10} = 1/1.91$. The

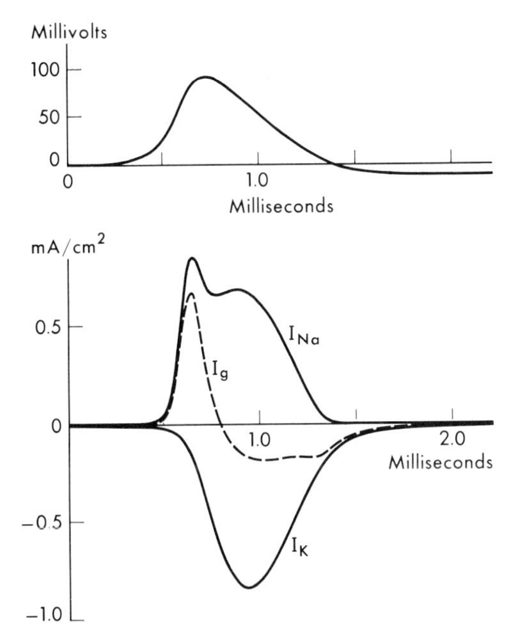

FIG. 3:54. Calculated sodium and potassium currents, solid, and the net current, dashed, as compared with the propagating action potential, above.

loss near room temperature was rather close to the experiments of Keynes and Lewis (1951) and the Hodgkin and Huxley (1952d) calculations of about 4.3 p mole/cm^2 impulse for 18.5°C but the experimental values were significantly high at high temperature and low at low temperature by comparison with the corresponding calculations. The detailed calculations of the entire impulse at 18.5°C (FitzHugh and Antosiewicz, 1959) and at 6.3°C (FitzHugh and Cole, 1964) were only a few percent different from the approximations used by Shanes and from them the Q_{10} for K loss was 1/2.75. The agreement of this additional direct chemical measurement of the K loss with the calculations based on voltage clamp data is further support for the HH axon while the differences in the temperature dependence probably indicate that the HH axon must be expanded and refined beyond its initial 1952 form.

When Huxley (1959a) turned again to automatic computation he investigated the effect of temperature on the form of the action potential, and the velocity in free propagation. The action potentials are in excellent agreement, figure 3:55, with those reported by Hodgkin and Katz (1949b), becoming faster and somewhat smaller at higher temperatures. The velocities increased regularly with temperature and, again in agreement with experiments, no solution could be found above 38°C. Most amazingly, however, the calculations predicted both a second lower velocity and a much lower spike amplitude below the critical temperature. This Huxley (1959b) has discussed in some detail and he concludes there is a real but unstable situation that is probably not to be found experimentally. It is, in fact, the threshold between uniform propagation and a decrementing impulse, as Nagumo et al. (1962) showed. Although the analogy may be quite misleading, I cannot help thinking that the upper velocity is like rolling a pea down the inside of a pipe, and the lower velocity is an even more difficult performance than rolling the pea along the top of the pipe with a straw.

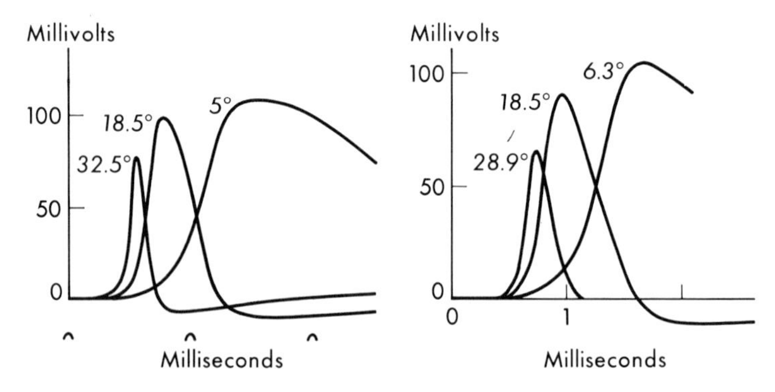

FIG. 3:55. Effect of temperature, Celsius, on the propagating action potential. Left, as measured in a squid axon. Right, as computed from *HH* axon.

Before the importance of the question was fully realized, FitzHugh and I had worried about the adequacy of a space clamp and finally decided to ask the question, "What does it take to stop a propagating impulse?" For this we took an infinite skewered axon and added the shunt conductance g of electrodes, axoplasm, and sea water which would attempt to hold the propagating impulse at the resting potential of the regions ahead and behind the impulse. As g increased, the velocity was regularly decreased, in contrast to a first approximation of Offner, Weinberg, and Young (1942). Then, no solution could be found for $g > 8.6$ mmho/cm^2 and yet the lowest value of θ was 2.2 m/sec. This was indeed puzzling until Huxley's work on temperature suggested reversing the instructions to the IBM 704 in the search for an acceptable value of θ. His discussion and conclusions also seem applicable to the resulting lower limb in figure 3:56.

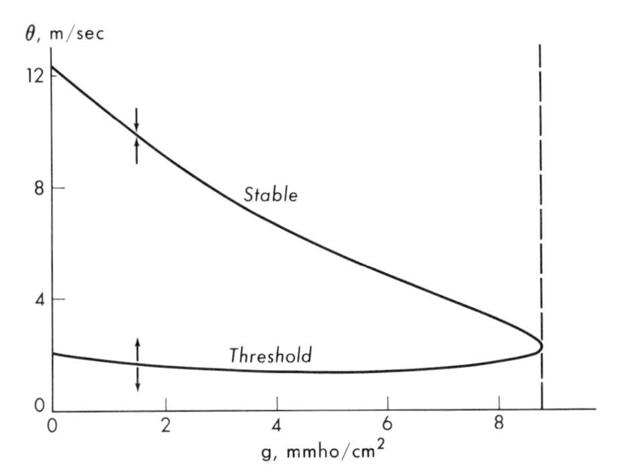

FIG. 3:56. Calculated effect of a leakage conductance on the speed of a propagating impulse. At less than a critical conductance, g_c, vertical dashed line, there was a stable impulse propagation, above, and also an unstable, threshold, impulse, below, that would either grow into stability or vanish. Above g_c there could be only an attenuating disturbance.

It is very interesting to me that the axon can continue to function and produce unexpectedly large action potentials at a useful velocity with perhaps ten times the normal resting conductance. On the other hand, I had expected that the limit might be closer to the minimum membrane resistances in the neighborhood of 10 Ωcm^2; the difference suggests that the limitation is more an increase of threshold than a decrease of stimulus. And, as is to be seen later, this condition is far less stringent than that for an adequate space clamp.

These two examples of the effects of temperature and leakage on the action potentials are nice demonstrations of the safety factor of the axon. At each particular value of T or g, the lower limb, as in figure 3:56 is the threshold between propagation and extinction. The upper limb is the potential available and used in propagation, so the difference between it and the threshold, lower, limb is the excess potential over that just needed for propagation. Then with increasing temperature or leakage the available excess of potential declines to reduce the margin of safety until it vanishes and conduction fails.

This gives us further basis to emphasize that the propagating impulse is strictly an all-or-none phenomenon and again that the continuous response of the space clamped membrane is of far more academic than practical importance.

The detail of the HH axon should make possible the calculation of the heat production during the passage of an impulse. This should give another, independent, and important test of the axon formulation. Although electrical measurements may not be able to distinguish between two equivalent circuits (Guillemin, 1935), heat measurements can in general. It is no surprise to find that the HH membrane produces heat at a rate exceeding that of its Thévenin equivalent as given by the exchange current,

$$\Delta H = (E_{Na} - E_K)^2/(r_{Na} + r_K)$$

where $r = 1/g$ and I_L is ignored.

The measurement of the heat production of excitable tissues has been a very difficult and challenging field. Bernstein and Tschermak (1906) found a cooling during an electric organ discharge equal to the heating in an external circuit. Although this conformed to theory, recent work (Aubert et al., 1959; Aubert and Keynes, 1961) has not only shown discrepancies but also questioned the adequacy of Bernstein's measurements. The careful and complete work of Hill (1965) on skeletal muscle over many years became a classic in spite of a major mistake and, after many futile attempts, he found a heat production accompanying activity in frog nerve (Downing et al., 1926). There was a short initial heat of less than a second and a slow recovery heat persisting up to thirty minutes, and then Beresina and Feng (1932) made the far easier measurements on crab nerves. Hodgkin (1951) estimated that the local circuit current flow inside and outside an axon would give an initial heat about ten times that found. But he countered that according to the ionic theory the heat absorbed by the dilutions in sodium and potassium exchange could balance the external heat production—as Bernstein had proposed.

Hill and his collaborators (Abbott et al., 1958) returned with new and faster instrumentation to the thermal analysis of a crab nerve impulse. They found two phases in the initial heat, a fast production and then a consumption apparently following the electrical recovery. Spyropoulos (1965) used critical thermistors and prolonged action potentials in *Loligo vulgaris* axons to conclude that the heat production covered the time of the action potential with the absorption appearing after the membrane potential had returned to its resting value. Howarth et al. (1965, 1966, 1968) compared the time course of the temperature change with that after a short pulse of high frequency heating and made computer corrections. They concluded that the positive and negative phases both occurred during the action potential and were clearly associated with the electrical depolarization and repolarization of the membrane. They also found these measured heats to be several times larger than predicted by their calculations.

Nachmansohn (1966), and perhaps others, considered this state of affairs so discouraging as to require an approach other than that of Hodgkin and Huxley. It was certainly true that the heat measurements had not given support for the HH axon. But it was also true that more and better experiments would have to be done and that further calculations would probably have to be made before this important line of testing could be used as evidence against the HH formulation of the sodium theory. And indeed an alternative formulation could be no more than an inadequate hypothesis until it could also duplicate all of the many and various accomplishments of the HH axon. Such constructive criticism has yet to appear.

Starting an impulse

Between and beyond the calculations of the decaying spread in a linear axon, of the initiation and blockade of a space clamped impulse, and of uniform velocity in free propagation, there lie the initiation and blockade of propagating impulses in a free axon. It is of vital importance to ask: How does an impulse grow—or die?

There has always been the question of the extent to which formulations of excitation include the behavior of nearby but as yet not excited regions of an axon. Then the liminal length requirement found by Rushton (1937) from the assumption of a critical, sudden threshold change of an emf put the problem in other terms. Our futile efforts to evolve a description of membrane properties from the membrane potentials at the site of successful and unsuccessful stimuli only made answers seem the more desirable. The Hodgkin-Huxley axon provided all the necessary information for calculation of these answers as well as giving them somewhat more importance.

The assumption of a uniform propagation can and has been questioned—"How do you know that such an impulse can be started?" "If it can, how long and how far from the start does it achieve its ultimate shape and velocity?"

The problem is then to solve the complete cable equation, figure 3:34, for the membrane potential in both distance and time as a result of a specified stimulus. The usual procedure is to replace the parabolic partial differential equation 1:31 by a finite difference equation in which the intervals Δt and Δx are small enough to give a useful approximation to a continuous solution. These intervals are not, however, independent, and spurious oscillations may result if they are not properly chosen (Milne, 1953). Furthermore the values of the potential must be available over the rectangle, 0, 0; x_0, 0; x_0, t_0; 0, t_0, if the next stage of the calculation is to be valid to a reasonable fraction of x_0, t_0.

Such solutions have been found for heat flow problems with linear and near linear characteristics by the use of electric simulators (Paschkis and Baker, 1942). However, analogue simulators of the HH axon have not been easily available in sufficient numbers to encourage this approach. Similarly, a digital calculation requires a considerable number of HH computer units to operate in a restricted region while the rest of the rectangle is held at appropriate values. When FitzHugh (1962) considered this problem, he soon found that even a rough approximation would require not only a very large computing capacity but also, for a reasonably useful range x and t, a memory that was entirely beyond anything then available. So he turned to the less demanding problem in medullated axon excitation to be mentioned later.

It was easy to take a sour grape attitude because I could see no real prospect that anything interesting or important might appear even when, only recently, machines and money could be found to work on such problems. This is particularly true when there are so many and much more alluring problems at hand and reasonable progress is being made with simplifications of the ponderous HH structure. Nonetheless, Huxley (1964) with EDSAC III and Cooley et al. (1965) with the resources of IBM, gave tantalizing preliminary reports of strange effects that seemed to have escaped observation in real axons. Cooley and Dodge (1966) and Noble and Stein (1966) reported some of their work, paying particular attention to threshold and repetitive phenomena.

Summary

General analytical solutions of the cable equation with HH membrane conductances have not yet been found practical, and almost all calculations of a free unclamped squid axon have been made with the relatively great simplification of an assumed constant impulse velocity. The excellent agreement between a calculated and an observed velocity and the change of waveform with temperature are still further, and by far the most impressive, demonstration that voltage clamp data contain most of the important facts of nerve function. Such calculations are, however, even more significant in providing tests for the sodium theory interpretations of the clamp data. The sum of the calculated conductance increases for sodium and potassium are about right in shape and amplitude and are somewhat confirmatory. The calculated gain of sodium and loss of potassium during an impulse, and their changes with temperature are in substantial agreement with isotope and other independent measurements.

Considerable and difficult further experimental work will be necessary for a detailed test of the separation of the clamp currents into ionic components. The measurements of the heat production during an impulse do not agree entirely and calculations may not be adequate. But until these chores are completed, the presumption is almost overwhelmingly in favor of the HH analysis.

FURTHER EXPERIMENTS

With all of the obvious near and far goals and with all of the solid achievement to build on in 1952, it would have been all but impossible to avoid starting again to have another "go" at the squid axon. Along with the computer program at the Bureau of Standards on SEAC an experimental program for Woods Hole was soon set up and under way.

Moore and I saw hints that the axial potential electrode of Hodgkin and Huxley might be too far from the membrane. To meet this and also the obvious need for a direct and accurate measure of the membrane potential, we turned to the micropipet electrode of Ling and Gerard (1949). In the development of this approach we found reason to believe that in the living, squirting squid the axon resting potential might be larger than in the usual excised preparation and that the action potential would be without the highly characteristic undershoot seen *in vitro* (Hodgkin, 1958; Moore and Cole, 1960). Then, too, we were able to make a much firmer estimate of the resting membrane potential and the liquid junction correction (Cole and Moore, 1960b) than had been possible so many years before (Curtis and Cole, 1942).

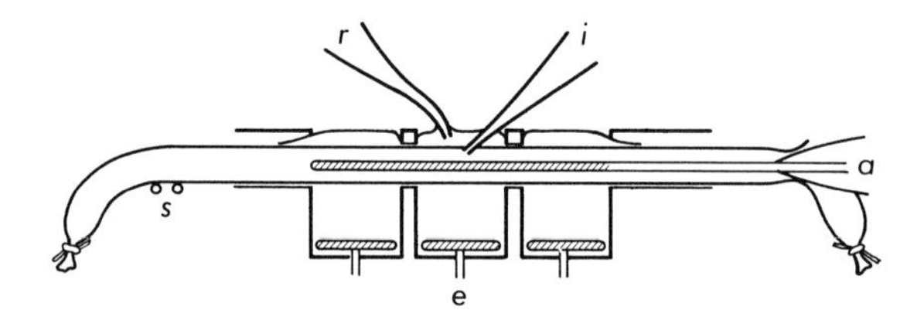

FIG. 3:57. Schematic of axon and electrode arrangement. The potential electrodes are internal *i* and external *r*. The axial current electrode is *a* and the guarded external current electrode is *e*. A stimulus could be applied at *s*.

The micropipet electrode required a return to the early open horizontal measuring cell (figure 3:57) rather than the vertical arrangement favored in Britain or the closed system described by Marmont (1949). The one innovation that has continued to be highly practical and without considerable drawbacks was the use of the wires—in various sizes, metals, and insulations—for axial electrodes. The use of the microelectrode, and subsequently also a larger, similar, outside reference electrode, for measurement and control of the membrane potential, figure 3:58, led to considerable development and invention. Along with other requirements, this resulted in a system that relied heavily on stabilized, operational amplifier units as shown in figures 3:59 and 3:60 and described in some detail by Moore and Gebhart (1962) and Moore and Cole (1963).

POTASSIUM ION CURRENT

One of our first investigations (Cole and Moore, 1960c) was of the potassium ion current I_K. This was given from the HH axon as

$$I_K = \bar{g}_K n^4 (V - E_K) \qquad (3:9)$$

where $$n = n_\infty - (n_\infty - n_0) \exp(-t/\tau_n).$$

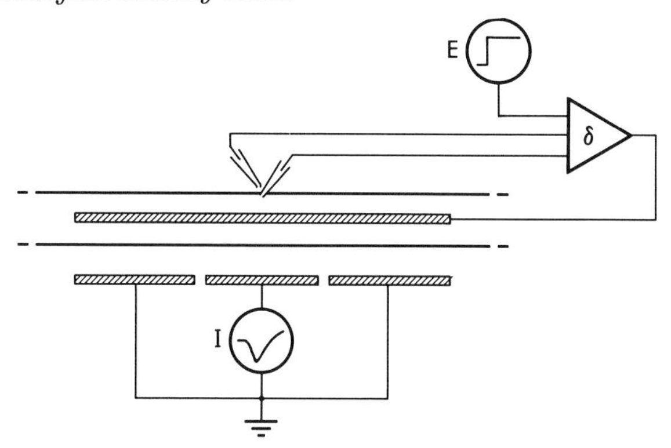

FIG. 3:58. Modified squid axon voltage clamp. The membrane potential, measured between internal, micropipet, and external, reference electrodes was compared with the command E by amplifier δ. The amplifier supplied the necessary current to the axial electrode and the guarded, central electrode, portion was measured, I.

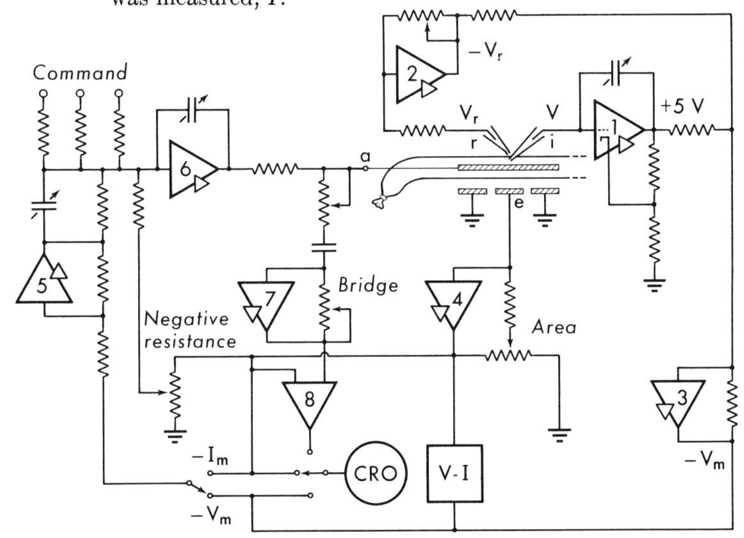

FIG. 3:59. Circuit diagram for a general squid axon voltage clamp equipment. The stabilized operational amplifiers, 1–7, provide for current I_m, or potential V_m control; compensation, negative resistance; change from passive properties, bridge; calibrations, area; and direct recording of V–I characteristics.

For high resting potentials or slight hyperpolarization the starting value $n_0 \to 0$ and a step to $V = E_{Na}$ where the sodium current should be zero at all times

$$I_g = I_K = \bar{g}_K[1 - \exp(-t/\tau_n)]^4(E_{Na} - E_K).$$

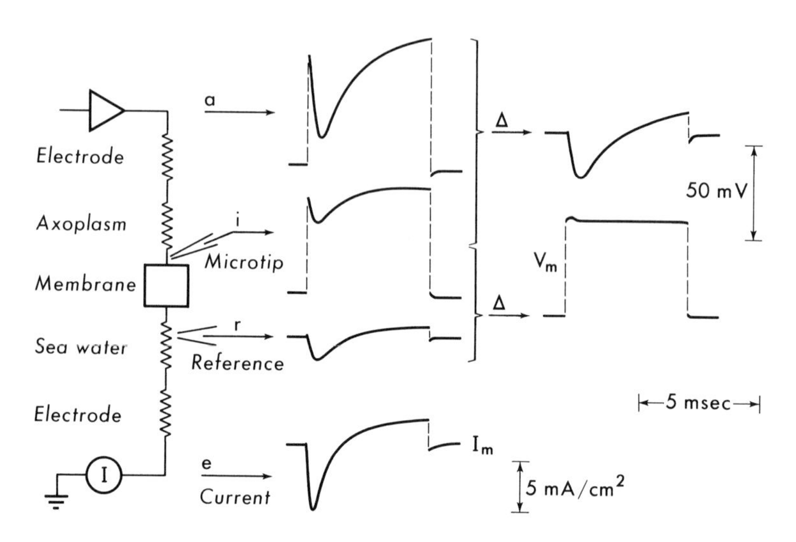

FIG. 3:60. Relations between the current flow through the membrane circuit, left, the measured potentials, center, and the recorded potentials, V_m and I_m.

Data from four axons, plotted as log t vs. log I_g, showed this type of behavior but were far better represented by an exponent of 6 rather than the 4 given by Hodgkin and Huxley. This was no particular surprise as they had noted that their data could be better expressed by a higher power but that the difference did not justify the extra effort—which included the compiling of tables.

FitzHugh pointed out that the HH formulation for I_K was based on the first order time dependence for the single parameter n and that equation 3:9 could be rewritten

$$I_K = \bar{g}_K n_\infty^4 \{1 - \exp[-(t - t_0)/\tau_n]\}^4 (V - E_K)$$

where $t_0 = \tau \ln [n_\infty / n_\infty - n_0]$. Thus for any starting point, $0 \leq n_0 < 1$, the transient currents should be the same except for a translation along the time axis. This was shown to be true for a second step to $V = E_{Na}$ at various times during a first step and at the same time after various first steps as shown in figure 3:61. This same superposition was found, figure 4:48, for steps to E_{Na} from hyperpolarizations far beyond the normal range for which the HH axon had been developed.

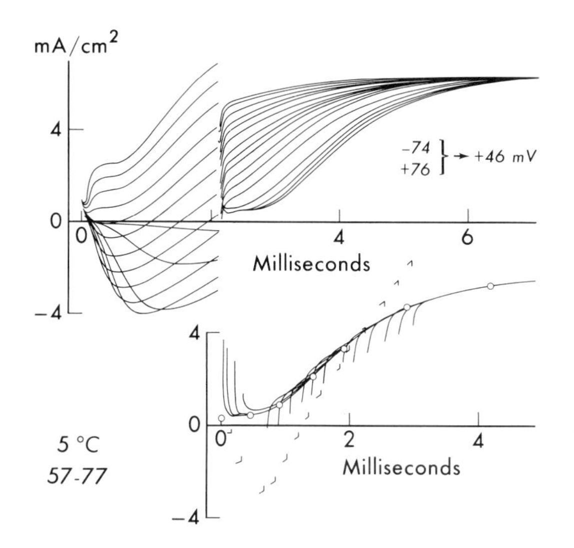

FIG. 3:61. Some tests of statement that the potassium current is a single variable of state.

These new tests of the HH axon formulation thus gave reasonable additional support for the potassium component.

THE NOTCH

By 1954 the essentials of the experimental system as it later evolved were in operation and, as axons with larger resting and action potentials and longer survivals were obtained, the voltage clamp currents—both for the inward early peak and for the steady state—rose above 5 mA/cm² as compared with the corresponding value of 1 to 2 mA/cm² that Hodgkin and Huxley and I had found before. But in these axons with the characteristics of figure 3:62—which we had every reason to believe to be the best, the most powerful, and the most nearly normal—we began to encounter marked variations, as shown in figure 3:63, from the previous patterns. There was neither a simple and obvious explanation for this behavior nor any clue to its origin until Moore suggested that it might be correlated with the surface impedance of the axial internal electrode. And indeed, as I paid more attention to this characteristic and began to improve it by cleaning and platinizing the wires, this new "notch" did not appear—for a while at least. But then it became a continuing struggle between the axons and me for several seasons; when our best axons could produce a notch with the best electrodes I could prepare we knew it was a problem we could no

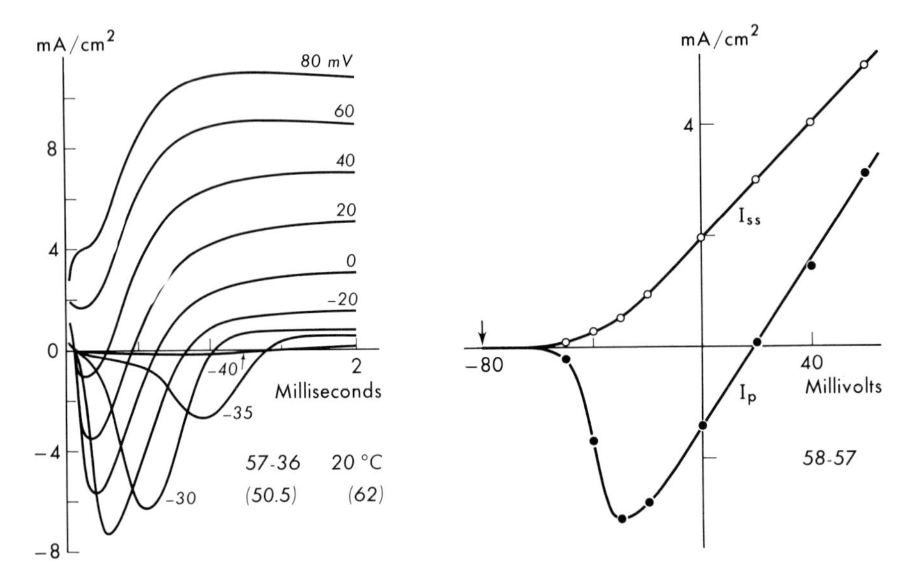

FIG. 3:62. Typical voltage clamp results. Left, membrane current densities for clamp potentials as labelled. Right, peak early current I_p and steady state currents I_{ss} as functions of the clamp potential.

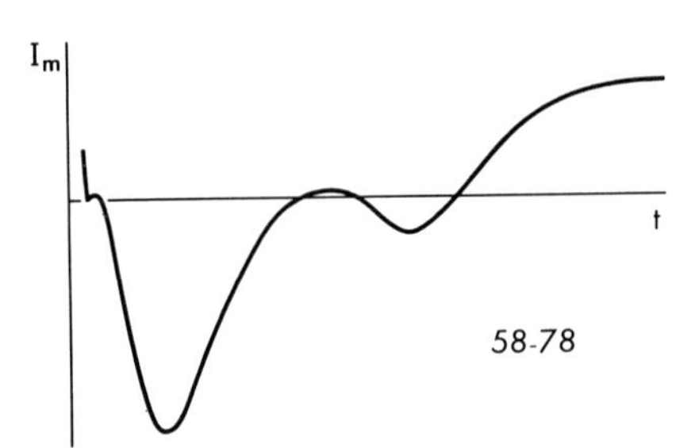

FIG. 3:63. A typical example of the deviations from earlier voltage clamp records. The second minimum was an anomaly referred to as a notch.

longer ignore as only a minor annoyance. By then, Frankenhaeuser and Hodgkin (1957) had also encountered the same problem which they were able to minimize by better isolation between the external measuring and guard chambers and at the ends of the guard chambers. The problem soon appeared to take on a more important and fundamental aspect when Tasaki with Bak (1958) and Spyropoulos (1958) and then also Shanes et al. (1959) not only produced far more bizarre aberrations of the patterns we had accepted as normal, but also suggested that the voltage clamp was of very limited value because they had not been able to apply it to exceptionally strong or even reasonably good, average axons. They thus implied that, for any but obviously decrepit and uninteresting axons, the voltage clamp concept was worthless and the HH formulations and analyses were quite inadequate. I could not deny either of these implications as I came to the very unhappy realization that I, and apparently Hodgkin and Huxley also, had little if any idea as to how close a technique should be to ideal or how far from this ideal we might have been. I did not then know how to write voltage clamp specifications—much less how to check performance against them. But starting with our apparent correlations between electrode impedances, axon power, and the "abominable notches," and with the collaboration of first Taylor and del Castillo and later FitzHugh and Chandler, we began looking for answers. And in spite of rather considerable provocation I think that we were usually able to keep our efforts within the framework of "What is right?" rather than "Who is wrong?"

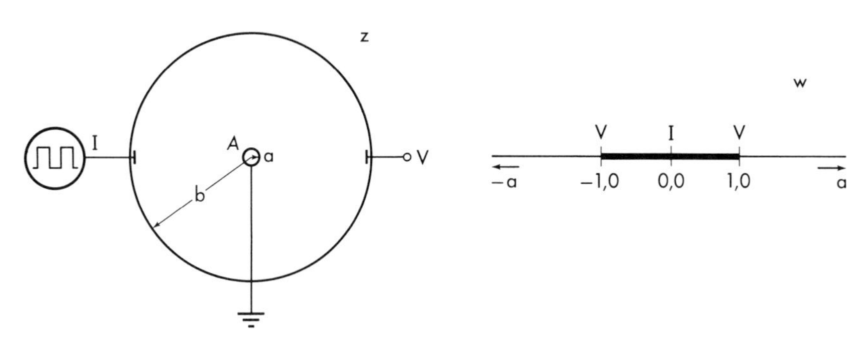

FIG. 3:64. Left, experimental arrangement to obtain axial electrode properties. A square wave of current I applied at the periphery b flowed through the electrolyte to the axial electrode A of radius a. The recorded potential V depended upon the surface characteristics of A and also upon the electrolyte in the beaker. Right, a conformal transformation from the geometric z plane to this w plane gives a direct solution for the contribution of the electrolyte to measured impedance of an axial electrode in the beaker measurement.

One starting point was to find out what happened with high impedance electrodes following a precept Langmuir gave sometime, somewhere, "If you want to find out what's wrong, first see what you can do to make it worse!" This required a measure of electrode impedance which became quite well standardized (Moore and Cole, 1963) as shown in figure 3:64z. The square wave current flow I gave an equivalent resistance between the axial electrode A, of radius a, and the potential electrode V in a beaker radius b, in 0.5M KCl, specific resistance r, over the time range from zero to 10 msec. These resistances from R_0 to R_{10} were most conveniently referred to a cm^2 of axon membrane and in units of Ωcm^2. As better electrodes were made, the electrolyte contribution to this measurement had to be estimated. I did not know how to get an analytical solution to this nice looking two-dimensional potential theory problem. But from the series of images for a square beaker and then by an analogue on Teledeltos paper I was happy to find the approximation for a needle length l,

$$R_e = (r/2\pi l) \ln (b/4a).$$

However, Chandler was sure the problem could be solved by a conformal transformation, and overnight he produced a very neat solution (Kober, 1952).

The original plane z is transformed to the w plane (figure 3:64w) by

$$w = \frac{1}{2}\left(\frac{z}{b} + \frac{b}{z}\right) + 1.$$

Now the current electrode I is at 0, 0, the circular boundary becomes the real axis from -1, 0 to 1, 0 with the potential electrode V at each end, and the center point of A now appears as points at $\pm\infty$. The problem is again symmetrical with no current crossing the boundary wall as required. The potential

$$V = \frac{Ir}{2\pi l} \ln |w| + \text{constant}$$

becomes, at $z = a \ll b$

$$V(a) \approx \frac{Ir}{2\pi l} \ln \frac{b}{2a} + \text{constant}$$

and at the potential electrode $z = b$,

$$V(1, 0) = \frac{Ir}{2\pi l} \ln 2 + \text{constant}.$$

Then the transfer resistance of the electrolyte is

$$R_e = -\frac{V(1, 0) - V(a)}{I} = \frac{r}{2\pi l} \ln \frac{b}{4a}.$$

This was identical with my uncertain and certainly untidy result, and usually gave a correction of 1–2 Ωcm^2.

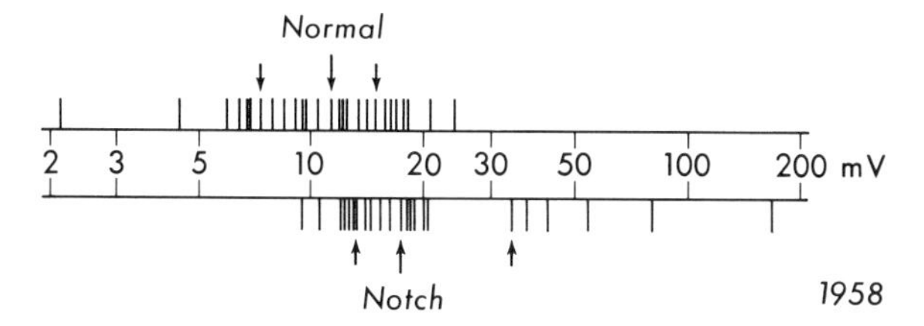

FIG. 3:65. The product of axial electrode resistance and axon maximum inward current as an index of performance. Index on logarithmic scale, center; distribution of experiments with median and quartile arrows shown for normal patterns, above, and notch or other abberant results, below.

Then characterizing the 1958 axons by their maximum peak inward currents I_p and the electrodes by R_0, these two quantities could be rather well correlated with notches as given in Taylor et al. (1960), or by the product I_pR_0, as in figure 3:65 (Cole, 1961b). This index of an averaged maximum potential drop inside and outside the membrane gave quantitative support for the conclusion that a poor electrode in a good axon was much more certain to produce a notch than was a good electrode in a poor axon.

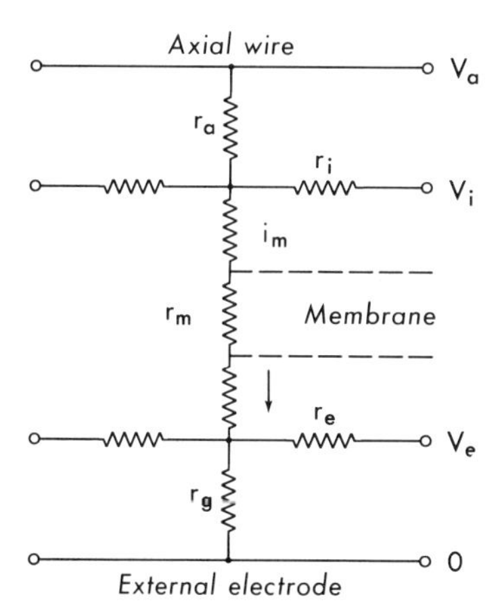

FIG. 3:66. Equivalent circuit for an element of a squid axon with axial wire and external electrodes. Membrane current i_m depends on the electrode contributions to r_a and r_g and on the longitudinal coupling along axoplasm and external sea water r_i and r_e.

A somewhat cautious approach was made to the theoretical problem of the steady state spatial variation for the membrane potential of an axon with an axial electrode, figure 3:66 (Taylor et al., 1960). This led to the equations

$$\frac{d^2V_i}{dx^2} = -r_e[(V_a - V_i)/r_a - i_m]$$

$$\frac{d^2V_e}{dx^2} = -r_e(i_m - V_e/r_g)$$

with the solutions

$$V_i(x) = A \exp(-x/\lambda_1) + B \exp(-x/\lambda_2) + V_i(\infty)$$
$$V_e(x) = C \exp(-x/\lambda_1) + D \exp(-x/\lambda_2) + V_e(\infty),$$

where A, B, C, D depend on boundary conditions and λ_1, λ_2 are the positive roots of

$$\left[r_i\left(\frac{1}{r_a} + \frac{1}{r_m}\right) - \frac{1}{\lambda^2}\right]\left[r_e\left(\frac{1}{r_m} + \frac{1}{r_g}\right) - \frac{1}{\lambda^2}\right] - \frac{r_i r_e}{r_m{}^2} = 0.$$

Some progress was made with this considerably complicated system, but it was also found possible to reasonably approximate it by V_e = constant, and

$$V_i(x) = E \exp(-x/\lambda) + V_i(\infty),\ \lambda = \sqrt{1\bigg/r_i\left(\frac{1}{r_a} + \frac{1}{r_m}\right)}.$$

Here we had a clear expression of the effect of the axial electrode to reduce the usual space constant of the free axon, as was also demonstrated experimentally (Cole and Moore, 1960a) in figure 3:67. Although some more involved situations were worked out analyt-

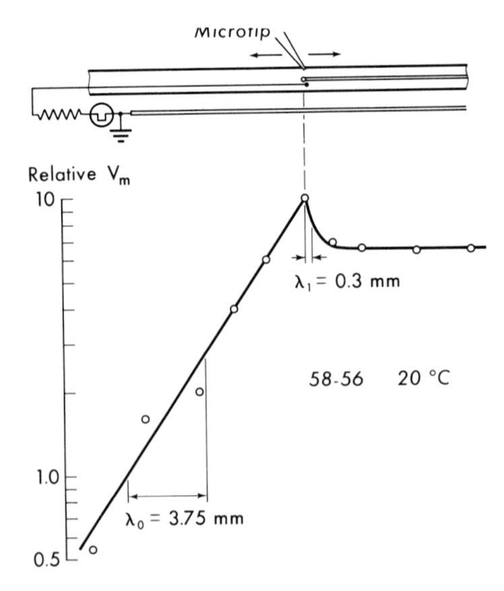

FIG. 3:67. Distribution of membrane potential along regions, right and left, with and without floating axial electrode, for current supplied at center. The solid line was computed from reasonable values of axon and electrode parameters.

ically and by analogue, any comprehensive solutions did not promise to be very useful, particularly not for comparison with the difficult time variations for which real axons had to produce solutions.

Potential and current surveys

Extensive investigations of the internal potentials were made with a potential probe as had been described by Marmont (1949). These patterns varied widely both in distance and time, but usually between the two extremes shown in figure 3:68. The right-hand results, with rather small time variations at one position after a series of potential commands and comparable maximum variations over the measuring region, were characteristic of a low impedance "needle" in a moderately "good" axon and gave the orthodox simple current patterns. The left figure, presenting the time and distance behavior of the internal potential after single, constant potential commands, was found with a rather high-impedance needle in about an average axon and produced a marked notch in the meas-

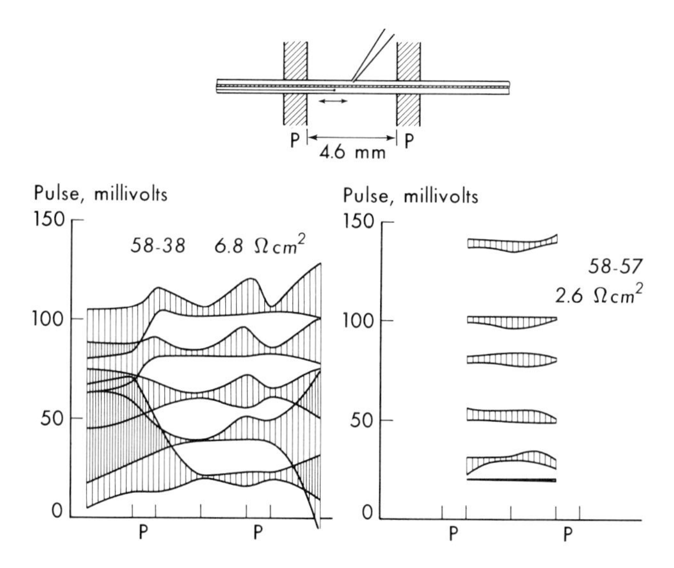

Fig. 3:68. Distribution of potential inside squid axons in voltage clamp measured by explorer from left. The shaded areas show the range of explorer potential during command pulse. The control near the microtip electrode is good in both experiments, but the potential variations over the current measuring region are excessive at left with a poor axial electrode while control is adequate at right with a better electrode.

uring chamber current. But here, and in spite of hills and valleys elsewhere, the potential near the microtip potential control electrode was quite reasonably constant. These and many other similar and intermediate surveys led to the conclusions that: (1) No matter how bad the electrode or how good the axon, the potential near the control point was quite satisfactorily similar to the command; (2) a good electrode and a fair to middling axon usually gave a similarly good potential elsewhere and currents without a notch; and (3) any combination of needle and axon which produced a current notch always showed marked aberrant internal potentials within the measuring region. These results suggested, but certainly did not prove, that we would not have a notch if we could only control the potential over the entire measuring region.

When we were somewhere in this process J. Y. Lettvin paid us one of his all too infrequent visits. He told us that he and Walter Pitts thought that the crux of the uniformity problem we seemed to be facing lay in a demonstration of the size and direction of the external longitudinal current flows near the axon. They further suggested that this aspect of the problem was very easily investigated by measuring the potential between a close pair of electrodes parallel to the axis of the axon and near the membrane. In the ensuing discussion we came to think that this might not be a particularly efficient procedure but I finally realized that such a differential electrode pair oriented perpendicular to the membrane and close to it would give a measure of the local membrane current density. Pleased though we were to have this new tool to survey the distribution of membrane current density over the measuring region, it seemed so simple and so obvious that I wondered why I had not thought of it long before!

We would expect intuitively that the resolution of the differential electrodes would be comparable to their separation. As a simple approximation we consider a current density I flowing from a line source in a plane membrane into a medium of resistivity r and the potential V at a distance ρ from the source

$$V = -\frac{Ir}{\pi}\int_R^\rho \frac{d\rho}{\rho} = -\frac{Ir}{\pi}\ln\frac{\rho}{R} = -\frac{Ir}{2\pi}\ln\frac{x^2+y^2}{R^2}$$

where $V = 0$ at some arbitrary distance $\rho = R$. The field normal to the membrane which would be measured by truly differential electrodes at a distance a from the membrane is

$$Y(x) = -\frac{\partial V}{\partial y}\bigg|_a = \frac{Ir}{\pi}\cdot\frac{y}{x^2+y^2}\bigg|_a = \frac{Ir}{\pi}\cdot\frac{a}{x^2+a^2}.$$

The maximum field at $x = 0$ is

$$Y(0) = \frac{Ir}{\pi}\cdot\frac{1}{a}$$

and at some distance $x = \delta$ the field is half maximum

$$Y(\delta) = Y(0)/2 = \frac{Ir}{\pi}\cdot\frac{a}{\delta^2+a^2}$$

so
$$Y(0)/Y(\delta) = 2 = \frac{1}{a}\cdot\frac{\delta^2+a^2}{a} = 1 + \frac{\delta^2}{a^2}.$$

Then the entire width between half maxima is $2\delta = 2a$ as an estimate of the resolution. The actual situation is considerably more complicated. The axon is cylindrical and the length of the chamber is finite. Although one electrode can be placed quite close to the axon $\rho \approx a$, the other must be far enough away b to give a useful signal—say $\Delta V = 0.1$ mV—for a membrane current density $I_0 = 1$ mA/cm^2. Then proceeding similarly, we find

$$\Delta V = I_0 ra \ln\frac{a+b}{a}$$

and with $r = 20$ Ωcm,

$$\begin{aligned}10^{-4}\text{ volt} &= 10^{-3}\cdot 20\cdot 250\cdot 10^{-4}\ln(a+b)/a\\ &= 5\cdot 10^{-4}\ln(a+b)/a\end{aligned}$$

or $\ln (a + b)/a = 0.2$. This gives $b = 50$ μ, but the electrodes themselves need a diameter of about 25 μ for mechanical strength with reasonable resistance and noise, so the electrode separation cannot go much below 100 μ.

This all looks possible, but it adds up to a combination of variables much too involved for useful analysis, and it is far more practical to work with an analogue. But no analogue—such as a resistor board—can be quite as good as the actual equipment itself, so the most practical solution was to build a reasonable differential electrode pair and determine its characteristics with analogue axons. The sensitivity could easily be calibrated in a trough or with a well-platinized 500 μ wire. The resolution was determined from another 500 μ wire insulated except for a well-platinized stripe 120 μ wide around it (figure 3:69a). The superposition of two of the differential patterns from this arrangement showed the stripe separation which

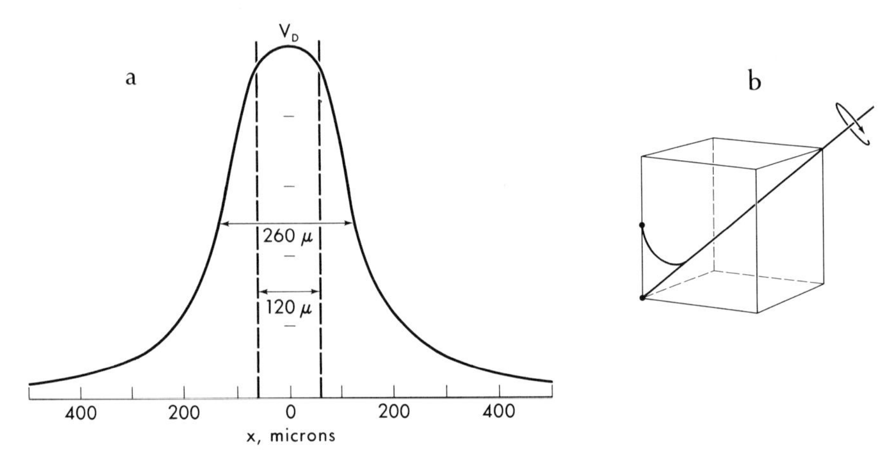

FIG. 3:69. (a.) Resolution of a differential electrode tested near a simulated axon. The electrode output given by solid line as it was moved past the current source shown by the dashed lines. (b.) Arrangement for measuring three orthogonal components of current density by rotation of a dipole on the diagonal axis.

can be resolved by the appearance of a minimum potential between two maxima. For differential electrodes separated by about 250 μ this resolution was found to be about 250 μ, which was more than adequate. A mount was used for these electrodes which gave the three mutually perpendicular components of the current vector by 120° rotations shown in figure 3:69b. Its convenience and neatness more than compensated for the fact that it was mostly used for the component normal to the membrane!

As before, the surveys of potential along the axon ranged from reasonably uniform patterns of potential and current density to extreme variations such as those shown by Taylor et al. (1960), figures 3:70 and 3:71. Again an orthodox pattern would be found in the total measuring chamber current with a uniform distribution, while a notch usually accompanied obvious variations. And again, in the latter case the current pattern close to the control point never had more than a faint suggestion of a notch no matter how extreme the distortions might be elsewhere. Yet another and particularly striking illustration of current density distribution was obtained as a differential electrode was moved at right angles to the axon away from the control point (figure 3:72). Thus as the resolving power of the doublet was reduced, and as it responded to current flow over an increasing length of axon farther from the control point, the notch became much more marked. On the other hand, an

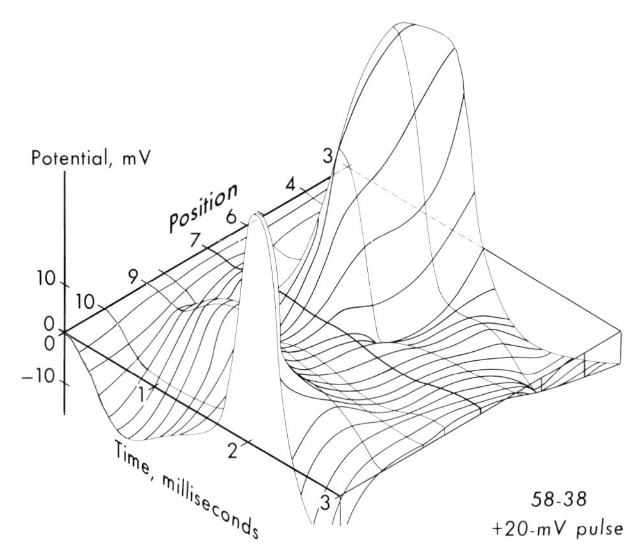

FIG. 3:70. Deviations of potential between internal probe and control electrode at position 7. The hills and valleys extend into the measuring region between positions 6 and 8 although they have little influence at the control point.

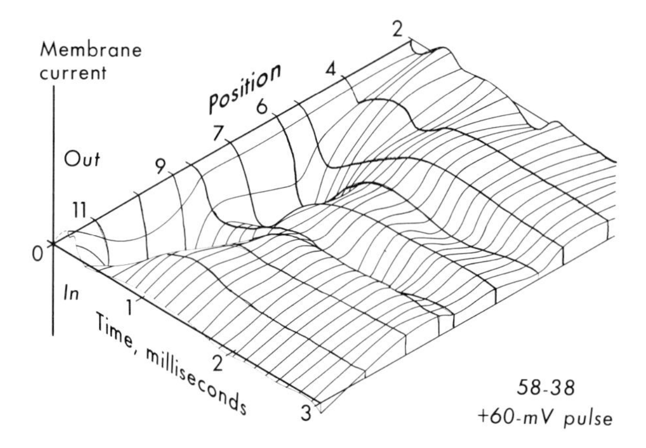

FIG. 3:71. Local membrane current density. The pattern at the control point 7 is normal while the currents at the boundaries of the measuring region, 6 and 8, both show abberations.

extrapolation back to the surface of the axon, at the control point, gave a simple orthodox current pattern.

Although it was quite impractical to make parallel surveys of the potential and current density profiles across the measuring chamber,

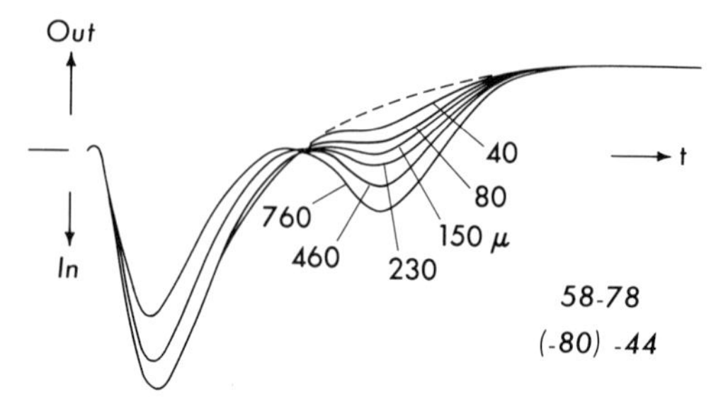

FIG. 3:72. Current densities at indicated distances from outside of an axon along a line from the control point and normal to the axis of the axon. The dashed portion is the extrapolated current density at the membrane.

only the most dedicated dissenter would be able to escape the conclusion that an aberrant current pattern was to be correlated with a poorly controlled membrane potential. Here there was at least a pragmatic answer to the problems of experimental design. When the current density was rather constant in amplitude and form over the measuring chamber, there was reason to expect that the membrane potential was similarly constant and that the total current in the region would not be seriously in error. On the other hand, when the membrane current density varied widely, the membrane potential could be expected to be so far from uniform as to discourage any assumption that the total current was representative of anything useful. Nonetheless, even in this unfortunate situation it was still possible to expect the membrane potential at and near the control point to be close to the command potential and to produce a valid local membrane current density. We could, therefore, have turned to a complete reliance on differential electrode measurements of the membrane current density close to the control point as shown in figure 3:82 except for three factors: (1) The logical chain had a weak link in the correlation between poor potential control and unorthodox current patterns that came dangerously close to the assumption that a notch was an experimental artifact; (2) we were so deeply and satisfactorily committed to the stable measurement of large currents to a sturdy electrode at ground potential that a change to the considerably less satisfactory measurement of a smaller differential potential between fragile and unstable electrodes was not easy to make; and (3) the procedure was not yet based upon any reasonably complete understanding of the behavior of the system or of the actual and allowable deviations from perfection.

Negative resistance

The various data had been accumulating into a bewildering mass and I was completely without any orienting point of view until someone, perhaps Moore, suggested two measures for the negative resistance behavior of a membrane. The first was from the negative slope of the peak inward current vs. clamp potential, usually found between −50 and −30 mV; the other was from the slopes of the increasing inward clamp currents at some fixed time after the application of the increasing potential steps in the same range, such as Hodgkin and Huxley had given. There was no simple, obvious justification for the use of either of these slopes. The first came from the difference of the rather broad maxima in the currents after steps to two different potentials. Even if each inward current maximum were thought of as a quasi steady state, it was rather more difficult to think of the difference between them in the same terms. And not even such a qualitative justification could be found for the use of the isochronal slopes. Hodgkin could only say that their times of from 0.28 to 1.53 msec were chosen because they gave interesting characteristics. Certainly the widespread reproduction of some of these isochronal curves is excellent endorsement of these choices. But both curves gave about the same values of something that had the dimensions of a negative resistance, and these numbers were far more useful than the only alternative—nothing!

An average of the maximum peak slopes corresponded to the negative resistance of about −3.5 Ωcm² with only a few as small as −2 Ωcm², such as shown in figure 3:73. Nominal values for the electrodes, axoplasm, and sea water then gave about 7 Ωcm² for the resistance in series with the membrane between the current electrodes. In a steady or quasi steady state the membrane current I_m and potential V_m would then be provided by an axial electrode potential

$$V_0 = V_m - RI_m$$

as shown by the "load line" of figure 3:73 where the resistance in series with the membrane $R = 7$ Ωcm². At this axial electrode potential the center equilibrium is unstable, but any small depar-

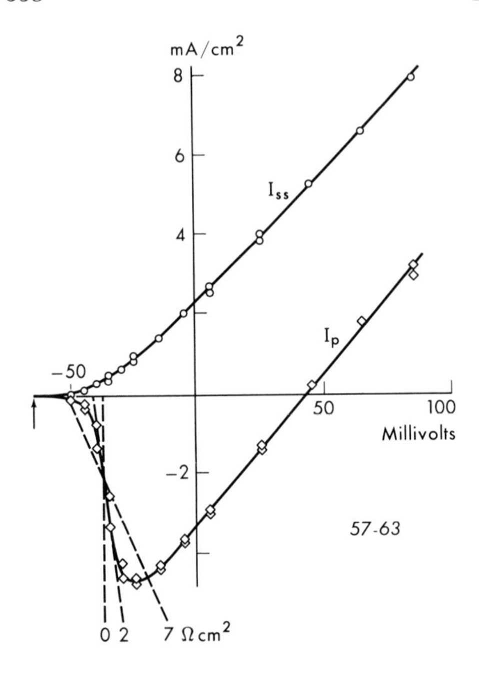

FIG. 3:73. Peak early current I_p and squares, and steady state current I_{ss} and circles, in voltage clamps. In the excitation region, -25 to -50 mV, I_p shows a negative resistance with the dashed line minimum of $-2\ \Omega\text{cm}^2$. The nominal resistance between the axial and outer current electrodes is the indicated $7\ \Omega\text{cm}^2$, but this is reduced to zero by perfect voltage clamping.

ture from it can be corrected by control amplification δ, which provides the effective load line

$$V_0' = V_m - R'I_m$$

where $V_0' = V_m\delta/(1 + \delta)$ and $R' = R/(1 + \delta)$. For large δ $V_0' \approx E \approx V_m$ and $R' \approx 0$; this gives the single stable equilibrium indicated in figure 3:74a. But without control any small departure from the center equilibrium will increase as required by the time constants of the system until it comes to the stable equilibrium on one side or the other, figure 3:74b.

This led Taylor to consider the "two-patch" model of figure 3:75, with one patch under control and the other far enough away to be independent except for the small changes of V_0 necessary to control the fluctuations of the first patch. Since the fluctuations of both patches are spontaneous, they are in no way correlated; the changes of V_0 are in no way related to the fluctuations of the second patch and it goes to one or the other of the stable points. This is well illustrated by the balanced pointer analogy. Although the pointer can be moved or held as desired by appropriate movements of the hand, these movements will in no way be appropriate for a second pointer on the same hand and it will fall to a stable position.

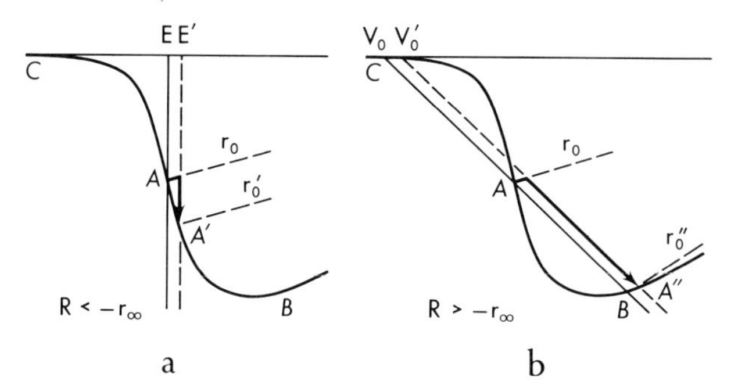

FIG. 3:74. Nature of instability in excitation region considering the negative resistance region CAB as quasi steady state characteristic. An emf E is applied directly in a to produce current A. At a small fluctuation to E' the current first moves along the instantaneous line r_0, and then to point A' as r_0 decreases to r_0'. If in b the potential V_0 is applied through a resistance R given by the solid line there is again an intersection at A. For a variation to V_0' the initial change is along r_0 but as r_0 decreases there is no solution until r_0'' is reached at A''. Thus a is stable and b unstable as shown where $-r_\infty$ is given by the slope at A.

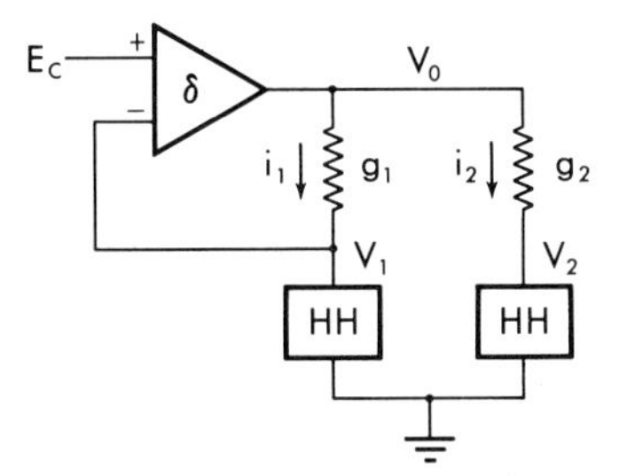

FIG. 3:75. Two patch theory. The potential V_1 applied to a HH membrane is controlled by potential E_C and amplifier δ, but with no coupling the potential V_2 of the second patch may be different, even unstable.

Taylor and FitzHugh (1959) and Taylor et al. (1960), gave this model a much more complete, kinetic test by analogue calculations with the HH membrane equations. They first recorded the $V_0(t)$ which was required to produce and maintain a step to V_1 across HH through a conductance g_1 (figure 3:75). $V_0(t)$ from a curve follower was then applied to HH through the same g_2 and i_2 recorded. As hoped and expected, for each value of V_1 in the critical region there was a minimum value for g which gave the same current pattern in both the controlled and the uncontrolled patches. Below this critical g_c the uncontrolled patch currents showed either notches with an extra inward current component or a "no notch" error of an additional outward current component. Nonuniformity was represented by dissimilar patches. Then squid axon voltage clamp current patterns were to be found which corresponded closely to each of those produced by the uncontrolled patch of this two-patch model.

The two-patch model was then investigated much more thoroughly (Chandler et al., 1962) in order to get an answer to the question, "For what series resistance R_c or corresponding conductance g_c will the second uncontrolled patch be on the boundary between stability and instability at each clamping potential on this first controlled patch?" And there was also the second question, "Do these critical resistances just happen to have any useful relation to the critical resistances arrived at on the assumption that either the peak or the isochronal current vs. potential relation represents the steady state negative resistance property of the axon?"

As mentioned before, the criteria of Hurwitz (1875) or Routh (1892) could be used to demonstrate the presence or absence of a positive root for the linearized HH equation. The locations of the real roots were given at the resting potential as in figure 3:40 and the critical g_c could be determined for them, but the value of g_c for the critical imaginary roots on the root locus of figure 3:41 was not straightforward.

It had been found experimentally that, after holding it at a potential 10 to 30 mV more negative than the resting potential, an axon membrane usually produced larger peak inward currents and had more of a tendency to instability in a routine voltage clamp pulse series (Frankenhaeuser and Hodgkin, 1957; Cole and Moore, 1960a). This was expressed in the HH axon by the initial condition that the inactivation parameter h became approximately unity. It was thus appropriate to investigate the stability problems during clamp pulses starting from -20 mV relative to the resting potential. The three left-hand branches gave only negative roots since the three poles were in the negative half-plane. Calculation of the time course of the right-hand branch gave the location of real roots and indicated the regions and nature of the complex roots for various ranges of the series conductance g. But again this did not lead directly to a value for the critical conductance in the complex root regimen.

Turning then to the complex admittance locus, it was relatively simple to apply the equivalent Nyquist (1932) criterion. For this we apply the Laplace transform

$$\bar{F}(p) = \int_0^\infty \exp(-pt)\, F(t)\, dt$$

where $p = \sigma + j\omega$. Then the generalized admittance of the HH membrane,

$$\bar{A}(p) = \overline{\delta I}/\overline{\delta V}$$

is of the same form as $A(j\omega)$ given by equation 3:8. A perturbation of the axial electrode potential by $\overline{\delta V_a}$ then gives the current

$$\overline{\delta I} = \bar{A}(p) = (\overline{\delta V_a} - \overline{\delta V})g$$

and the overall transfer function from the axial electrode to the membrane is

$$\frac{\overline{\delta V}}{\overline{\delta V_a}} = \frac{g}{g + \bar{A}(p)}.$$

When the denominator

$$g + \bar{A}(p) = 0,$$

the roots now give the required values of p for the transient response. For $\sigma = 0$ and $p = j\omega$ we have the steady state response such as shown in figure 3:76a, and this is extended to $-\infty$ by the addition of the mirror image. The curves of constant ω and constant σ then form a coordinate system and, as shown by Nyquist and Bode, the loci for $\sigma > 0$ lie to the right as frequency increases (figure 3:76b). At $\bar{A} = -g_A$, as shown, we have σ_A and ω_A and the solution of the equation is of the form

$$\exp(\sigma_A t \pm j\omega_A t) \text{ and } \sigma_A > 0$$

so it is unstable. This continues to be the case until $g = g_C$ where $\sigma = 0$ and the solution is

$$\exp(\pm j\omega_C t).$$

Then for $g > g_C$ we have, at g_B for example,

$$\exp(\sigma_B t \pm j\omega_B t) \text{ with } \sigma_B < 0$$

which now decays to zero. Thus g_C is critical, and in this case, has two imaginary roots.

From the computations of the loci at various times after the application of a clamp potential, the critical conductance can be obtained as the left-most intersection of the locus with the axis of reals. These have been plotted for depolarizations of the HH axon from an initial hyperpolarization of −20 mV from the resting potential. The largest, $g_C = 83$ mmho/cm^2, is found during a

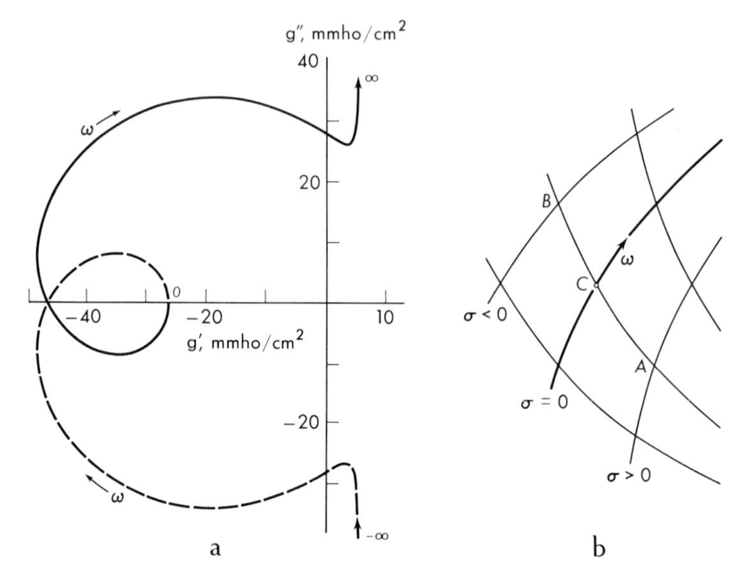

FIG. 3:76. (a.) Complex admittance of *HH* membrane for 0 to ∞ frequency range, solid line, and −∞ to 0, dashed line. A shunt conductance, will move this pattern to the right and if it leaves a negative intersection on the g' axis the system is unstable. (b.) Portion of steady state complex admittance locus, $\sigma = 0$. To the left of increasing ω, the damping $\sigma > 0$ so a point B in this outside region moves to the steady state. Conversely, A on the inside to right is unstable because with $\sigma < 0$ a fluctuation increases indefinitely.

depolarization of +29 mV, so all responses are stable for a series conductance $g > 83$ mmho/cm². Furthermore, by what can so far be considered only a happy coincidence, the critical conductances estimated from the peak and isochronal vs. potential slopes are sufficiently close, as shown in figure 3:77, to be used as considerably better than qualitative guides to stability.

SPATIAL DISTRIBUTION

As circumstantial evidence accumulated, it seemed increasingly certain that membrane regions sufficiently remote from the control point not only could but probably did fail to follow the potential command and so produced anomalous current patterns. Such an explanation might seem to be an adequate reason to reject anomalous current measurements when they included membrane regions more than the millimeter away from the control point that was suggested by the length constants for a skewered, depolarized axon. There was as yet no approach or satisfactory answer to the question,

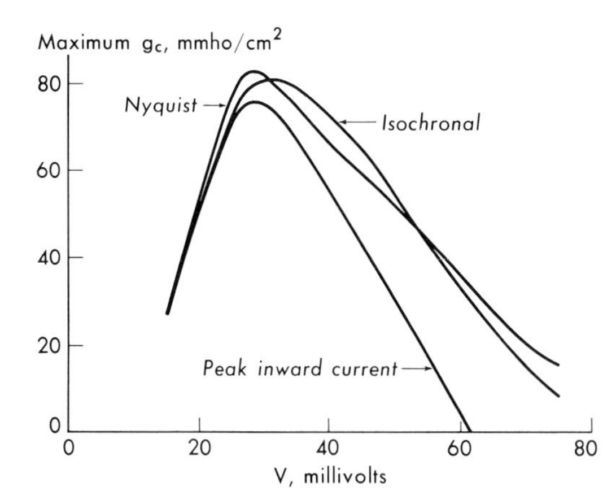

FIG. 3:77. Calculations of the maximum of the critical conductance g_c required for control of *HH* membrane throughout voltage clamps V. The curve for the quasi steady state characteristic at a fixed time, isochronal, is quite close to that given by the Nyquist criterion and the stability requirement for the peak inward current characteristic is not far away.

"How far from the control point will a current measurement be satisfactorily close to that at the control point?" The obvious answer, "Close enough," was not particularly helpful, but I could do no better until after Taylor raised the question as to what the space constant was for a negative membrane resistance. Here the offhand answer, that it ought to be imaginary, was both correct and useful (Cole, 1961a).

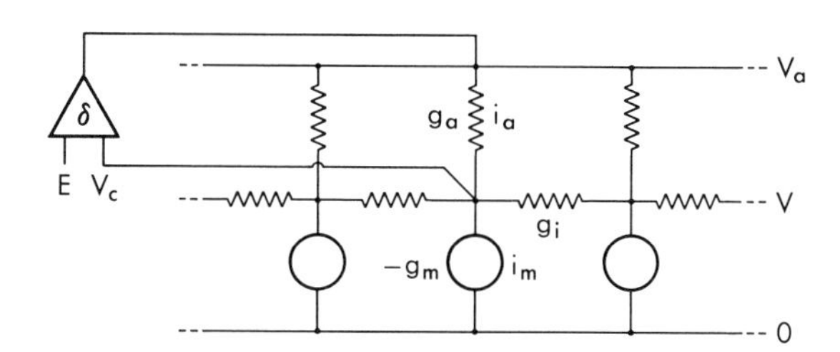

FIG. 3:78. Equivalent circuit for analysis of control of axon membrane potential V in neighborhood of control point at V_c with axial electrode at V_a.

If, as before, we complicate the cable equation by the insertion of an axial electrode, then the membrane potential V is given by:

$$\frac{\partial^2 V}{\partial x^2} = (i_m - i_a)/g_i$$

and figure 3:78 where i_m is the membrane current to the outside at zero potential, i_a is the current from the axial electrode at potential

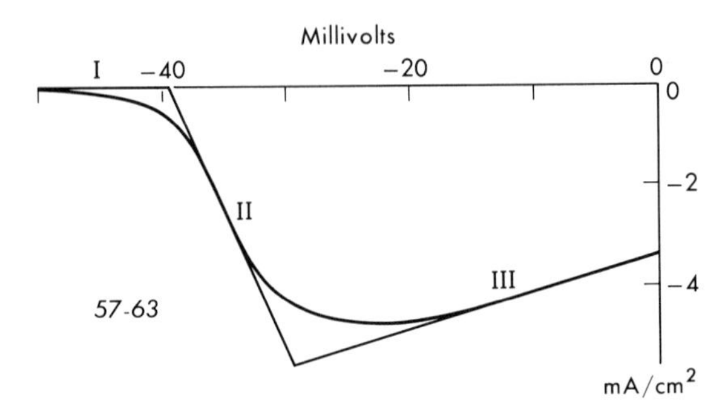

FIG. 3:79. Quasi steady state squid membrane characteristic, heavy curve, as approximated by three linear segments.

V_a, and g_i is the longitudinal conductance of the axoplasm in mho cm. Then we approximate the membrane characteristic by three line segments of figure 3:79 to give about the worst possible case and to permit analytical solutions for each segment. For an electrode and axoplasm conductance g_a,

$$i_a = g_a(V_A - V)$$

and for a membrane conductance g_m,

$$i_m = g_m(V_B - V)$$

where V_B is the potential for which $i_m = 0$. Then in steady state the characteristic equation

$$\frac{d^2V}{dx^2} - \alpha^2 V = 0$$

has a symmetrical solution of the form, equation 1:37,

$$V = A \exp(-\alpha x) + B, \qquad x \geq 0$$

where $\alpha^2 = (g_a + g_m)/g_i$ as before with $g_a + g_m > 0$. If, however, we have $g_a + g_m < 0$ the symmetrical solution is of the form

$$V = C \cos \omega x + D, \qquad x \geq 0 \tag{3:10}$$

and $\omega^2 = (g_m - g_a)/g_i$ for $g_m > g_a > 0$.

The segments I, II, and III are taken to have steady state or "slope" resistances of 1000 Ωcm^2, -2 Ωcm^2, and 6 Ωcm^2. The corresponding conductances g_m for a 480 μ axon are then $1.4 \cdot 10^{-5}$ mho/mm, $-7 \cdot 10^{-3}$ mho/mm, and $2.3 \cdot 10^{-3}$ mho/mm. The longitudinal conductance of the axoplasm is $g_i = 7 \cdot 10^{-4}$ mho mm, while the nominal resistance of 6 Ωcm^2 for the electrodes, axoplasm, and sea water gives the corresponding conductance $g_a = 2.3 \cdot 10^{-3}$ mho/mm.

Without loss of generality we can further simplify the problem by considering only the linear regions I, with a negligible conductance, and II with a conductance $g_m < 0$, figure 3:80. It is also simpler to see the nature of the solutions by representing them on the phase plane with the coordinates V and $y = k\, dV/dx$. Then

$$\frac{d^2V}{dx^2} = \frac{1}{k} \cdot \frac{dy}{dx} = \frac{1}{k^2} y \frac{dy}{dV}$$

and so

$$\frac{dy}{dV} = \frac{(i_m - i_a)/g_i}{k^2 y}.$$

For i_a and i_m which are known functions of V, the solutions that fulfill the boundary conditions may be traced as the lines—or trajectories—on the V, y plane. But for the linear approximations we have the exponentials which are replaced by straight lines, or the hyperbolic sines and cosines by hyperbolae, or the trigonometric sines and cosines by ellipses but, with a proper choice of k, by circles.

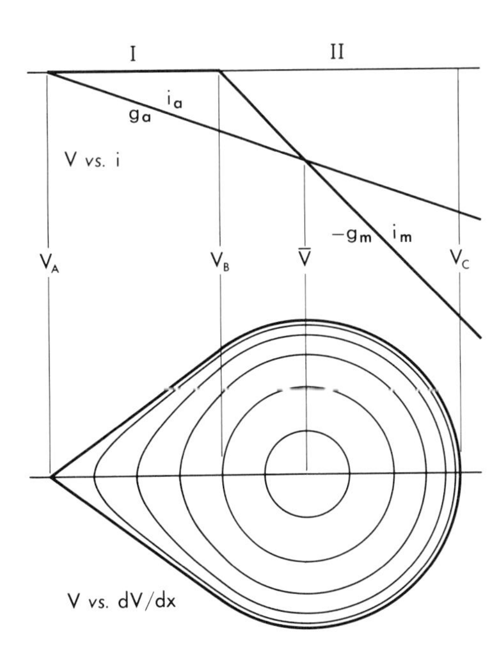

FIG. 3:80. Axon membrane with negative conductance region *II* controlled at potential $\overline{V}$ by axial potential V_A and conductance g_a. Above, membrane characteristic and control load line. Below, phase diagram normalized to show circular loci in *II*, and extended as hyperbolae into *I* for finite axon lengths.

Then for an infinite extent of axon, $V \to V_A$ at $\pm\infty$ on the outer straight lines of figure 3:80 from $V = V_B$. Then in region II as scaled, this solution becomes the outside circle, and is symmetrical about V_C. A long, but finite, axon with dV/dx and y equal to zero at the ends will then give hyperbolae as indicated in region I and join the smaller circles in region II. The solution becomes a complete circle for an axon length

$$2L = 2\pi \sqrt{-g_i/(g_m + g_a)} \tag{3:11}$$

or a semicircle for half this length. But for any length below L there is no solution except $V = \overline{V}$ everywhere, which is then a perfect clamp. These solutions are also shown as functions of the distance x in figure 3:81.

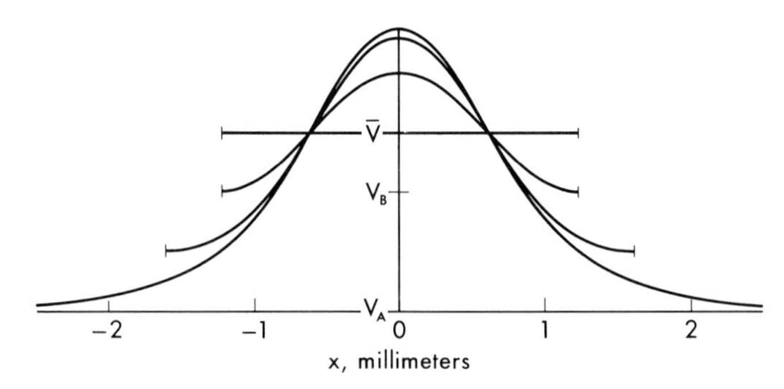

FIG. 3:81. Membrane potential distribution along an axon. The potential is uniform at $\overline{V}$ for the critical length or less, a cosine for $V > V_B$, hyperbolic cosines for longer axons and $V < V_B$, or exponential in the limit.

Returning to the infinite axon, it is seen that the corresponding solution at the control point is $\cos \omega x$, and the errors of potential and of current will only increase slowly for small distances from it. If i_c is the central control point current, the i at x is

$$\frac{i_c - i}{i_c} = \frac{\sqrt{g_m/g_a}}{1 + \sqrt{g_m/g_a}} (1 - \cos \omega x).$$

For nominal characteristics an error of 20 percent appears at 0.3 mm on either side of the center, and the average current density measured over a 0.6 mm stretch will be in error by less than 10 percent.

Estimates of the stability and speed of control require a return to the modified cable equations

$$\frac{\partial^2 V}{\partial x^2} - \alpha^2(V - V_a) = \frac{c}{g_i} \cdot \frac{\partial V}{\partial t}, \qquad V_B \geq V \geq V_A$$

$$\frac{\partial^2 V}{\partial x^2} + \omega^2(V - \overline{V}) = \frac{c}{g_i} \cdot \frac{\partial V}{\partial t}, \qquad V \geq V_B.$$

These have not been satisfactorily solved, although it has been suggested that the solution might be in the form of an error integral of an imaginary argument. Various approximations indicate that with a control amplification δ: (1) The system is stable for $\delta > 1$; (2) the membrane current error at the control point is about $1/\delta$; and (3) the time constant is about $c/\delta g_a$ at the control point and becomes much longer with increasing distance away from it.

The results show a surprisingly low amplification requirement for stability of the spatial pattern and a reasonable requirement for accuracy, but indicate that the speed of response is limited by the external control system rather than by the membrane and electrode system.

Other approximations

There were, however, secondary worries aroused by the first approximations with a steady state negative conductance. The maximum length of axon for which a perfect clamp was assured was 1.0 mm under the assumed conditions. This was much too close to the half circumference of the axon for comfort, and suggested the possibility that the potential might vary around the circumference at the control point as well as along the length of the axon. The worst possibility was that only a small disk near the control point might be reasonably clamped; for this case a circular flat membrane backed by a layer of axoplasm and a plane disk current electrode would be a possible approximation. If I_1 and I_3 are the electrode and membrane currents over an annulus $\Delta\rho$ and I_2 is the radial current at radius ρ,

$$\Delta I_2 = I_1 - I_3$$

and

$$\frac{d^2 V}{d\rho^2} + \frac{1}{\rho}\frac{dV}{d\rho} = \frac{r_2}{\delta}(i_3 - i_1)$$

where V is the approximate membrane potential, r_2 and δ are the equivalent axoplasm resistivity and thickness, and i_1 and i_3 are the electrode and membrane current densities. For

$$i_1 = g_1(V_A - V);\ i_3 = g_3(V_B - V),\ V > V_B$$
$$= 0,\ V < V_B$$

and

$$\frac{d^2V}{d\rho^2} + \frac{1}{\rho}\frac{dV}{d\rho} + \frac{r_2}{\delta}(g_3 - g_1)V = \frac{r_2}{\delta}(g_3V_B - g_1V_A).$$

This is a Bessel equation of zero order, and for $g_3 > g_1$, $J_0(\omega\rho)$ is a solution with $\omega^2 = r_2(g_3 - g_1)/\delta$. For $V > V_B$ and $dV/d\rho = 0$ at $\rho = a$, there is no solution $J_0'(\omega a) = 0$ for $\omega a < 3.83$ except $V(\rho) =$ constant. Thus the limiting radius

$$a = 3.83/\sqrt{r_2(g_3 - g_1)/\delta} \approx 1.7 \text{ mm} \qquad (3:12)$$

for nominal parameters.

Although the plane problem did give a maximum diameter which comfortably overlapped the circumference of the axon, it was still worthwhile to try yet another approximation. This was merely to ignore variations in the x variable and look at the behavior in the plane normal to the axis of the axon. Here the Laplace equation is

$$\frac{\partial^2V}{\partial\rho^2} + \frac{1}{\rho}\frac{\partial V}{\partial\rho} + \frac{1}{\rho^2}\frac{\partial^2V}{\partial\theta^2} = 0$$

with the boundary conditions

$$\partial V/\partial\theta = 0,\ \theta = 0,\ \pi/2;\ i = -(1/r_2)\partial V/\partial\rho = g(V_B - V),\ \rho = a$$

and a solution is

$$V = A \ln \rho/b + B\rho\cos\theta + C.$$

At $\rho = a$,

$$\frac{\partial V}{\partial\rho} + \frac{A}{a/b} + B\cos\theta = -r_2i = -r_2g(V_B - V)$$

so

$$\frac{A}{a/b} + B\cos\theta = -r_2g(V_B - A\ln a/b - C) + r_2gaB\cos\theta \qquad (3:13)$$

and this requires $r_2ga \geq 1$. For $a < 1/r_2g$ there is no solution except $V =$ constant—a perfect clamp. For the nominal values this gave $a = 0.5$ mm for the critical case and once again this seemed reasonably safe.

It would be nice to have obvious, better approximations and even some calculations of three-dimensional cases, although they do not seem essential for the *Loligo* axons and perhaps not for the larger ones of *Dosidicus*.

Clamp errors

It is rather difficult to make blanket specifications for all components of such a system without an equally complete catalogue of the uses to which the results may be put. No matter what the ultimate needs, an ad hoc requirement that the output of the system, the membrane current density, have a reliability of about 5 percent seems reasonable. To demand that all factors be known and reproducible to the extent that the current density has meaning at a 1 percent level might preclude at least some possible and useful experiments because of limitations of equipment or material. At the other extreme, to allow a 25 percent range of variability might excuse a sloppiness that would be an insult to the aims and achievements of both physical and biological sciences.

Even a 5 percent requirement on the membrane current density is rather stringent in the negative resistance region. For a 2.5 mA/cm^2 peak inward current, we would allow only 0.12 mA/cm^2 uncertainty and, for -2 Ωcm^2 in this region, the membrane potential would need to be known and constant to 0.25 mV. Over many years and for many reasons a 1 millivolt limit has seemed to be about all that could be expected. This variability is not yet particularly important except in the negative resistance region; here the scatter of our experimental points often shows that we should do better. Nonetheless a certainty of 5 percent in current and 1 mV in potential seems eminently desirable and perhaps not importantly beyond what we have usually achieved (Cole and Moore, 1960a).

Although we have encountered several problems in systems with uniform axons and uniform electrodes, only a little experience forces one to recognize that both axons and electrodes may show considerable differences between one point and another. If we ignore longitudinal current flows, which have short space constants in our present situation, the potential between the current electrodes

$$V_0 = V_m + R \cdot 2\pi a i$$

where R is the sum of the inside and outside electrode and electrolyte resistances, a is the axon radius, and i is the membrane current density. For small variations we have

$$\delta V_0 = \delta V_m + 2\pi \bar{a}\bar{\imath}\bar{R}\left(\frac{\delta a}{\bar{a}} + \frac{\delta i}{\bar{\imath}} + \frac{\delta R}{\bar{R}}\right)$$

and for V_0 fixed,

$$\delta V_m = -2\pi\bar{a}\bar{\imath}\bar{R}\left(\frac{\delta a}{\bar{a}} + \frac{\delta i}{\bar{\imath}} + \frac{\delta R}{\bar{R}}\right)$$

which becomes

$$\delta V_m = -35\left(\frac{\delta a}{\bar{a}} + \frac{\delta i}{\bar{\imath}} + \frac{\delta R}{\bar{R}}\right) \text{mV}$$

for nominal values of $\bar{a}$, $\bar{\imath}$, and $\bar{R}$. This is a truly appalling result for if we accept the limitation of 1 mV for δV_m it requires

$$\frac{\delta a}{\bar{a}} + \frac{\delta i}{\bar{\imath}} + \frac{\delta R}{\bar{R}} \leq 3 \text{ percent.}$$

Since there is little experience to justify an assumption that any one of these quantities varies less than 3 percent over the length of more than 1 mm, we are faced with the alternatives of measuring a membrane current either over a considerable length—which may or may not correspond to some mean of a rather wide potential variation—or over a very limited region of membrane in which the potential and current density are either demonstrated or expected to be satisfactorily constant.

There remains, however, the unidentified and essentially unmeasured resistance r_S in the gross membrane structure and in series with capacity. This appeared to be reasonably certain at about 4 Ωcm^2 in the original space and current clamp results. It was confirmed by Hodgkin et al. (1952) at about 7 Ωcm^2 and the major portion of it was ascribed to the exterior of the membrane. Our later measurements have been more in the neighborhood of 3 Ωcm^2. At 5 mA/cm^2 this contributes the potential variation of 15 mV that would be intolerable—if anything could be done about it. Even if this resistance is known, the data cannot be corrected for it any better than I did originally (figure 3:19) (Taylor et al., 1960). The compensated feedback can be used in experiments if r_S is known—but no objective measure of it has appeared, except perhaps an infinite frequency resistance such as that determined by Taylor and Chandler (1962) and Taylor (1965).

Further modifications

As the requirements on the voltage clamp technique became better known, there were strong indications that instability was always possible and would lead to intolerable variations of membrane potential beyond lengths of 1 mm or perhaps less. Although it might not be possible to insure that potential control would always be maintained over an indefinite axon length, it appeared to be worth while to do everything practically possible to make the wavelength in the negative resistance region $2L = 2\pi/\omega$ (equation 3:11) as long as possible and to make current measurements only over a small enough region, near the control point, to be constant within 10 or 20 percent. One factor that could be altered was the system of outside current electrodes. If these electrodes could be cylinders with a fixed impedance per unit area, the decrease in sea water impedance as they were brought closer to the axon might offset the increase of impedance corresponding to the smaller electrode area. There was indeed a minimum, but it appeared at an electrode radius less than that of the axon! So then the electrodes should be as close to the axon as possible, but it had become important to have free access to the axon, not only for the micropipet but also for the outside reference electrodes and for the differential electrodes that had come to be the court of last resort, figure 3:82. Also, as the current electrodes came closer it was increasingly important to create a rapid circulation that would reduce the chance of toxicity from the platinized silver that had become the surface of choice (Cole and Kishimoto, 1962).

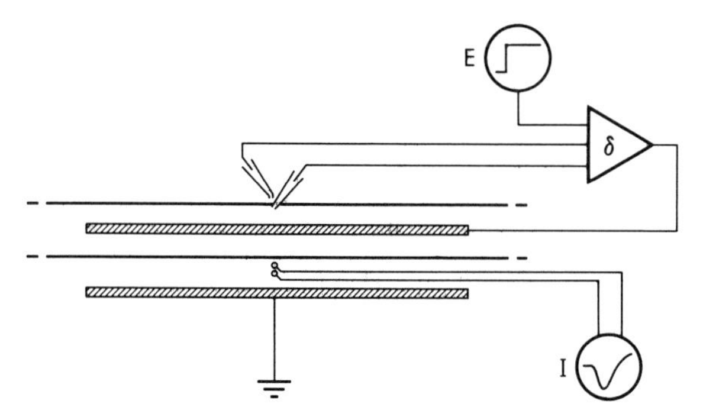

FIG. 3:82. Squid axon voltage clamp arrangement in which current density I is measured near control point by differential electrodes.

All of these factors urged a return to the square trough with vertical electrodes in the sides—such as we had used so many years before. But now it was important to have the trough large enough so that the membrane current density would be sufficiently uniform over the circumference of the axon. This again was a nice potential theory problem; also it was one for which I had no solution. So I set up a Teledeltos paper analogue to find that, for a trough width three times the axon diameter, the maximum and minimum axon current at the sides and at the top and bottom respectively would be less than 10 percent different from the average over the whole periphery. This was a conservative figure because it was based upon equipotentials for the axon and the electrodes, and the impedances of both could be expected to reduce the variation. It was also an eminently practical design in allowing a working space of an axon diameter on each side of the axon, while not excessively critical to the solution level above or the entrance of the circulating electrolyte from below.

Short segments of the vertical electrode walls could then be used for the current measurement, with the equipment grounded on one side. There were, however, no clear criteria as to how short these electrodes must be or how long they could be. The theory was obviously difficult, and no simple analogue attack was to be found. Since the mean distance from the axon to the electrodes was about 1 mm it could be expected that the current from a narrow stripe around the axon would spread over about a millimeter length of electrode. Conversely, although such an electrode would have an "ovido"—a view—of about 2 mm of axon, more than half of the current would come from less than 1 mm of the length. On little more information than this, the first cell with such a measuring electrode length was built, figure 3:83, and tested with a 500 μ wire insulated except for a 100 μ wide stripe around it. About as hoped, with this arrangement the contribution to the measuring electrode was heavily weighted in favor of the central 0.5 mm length of axon and increasingly against the more distant portions of the axon in which potential control might be less effective.

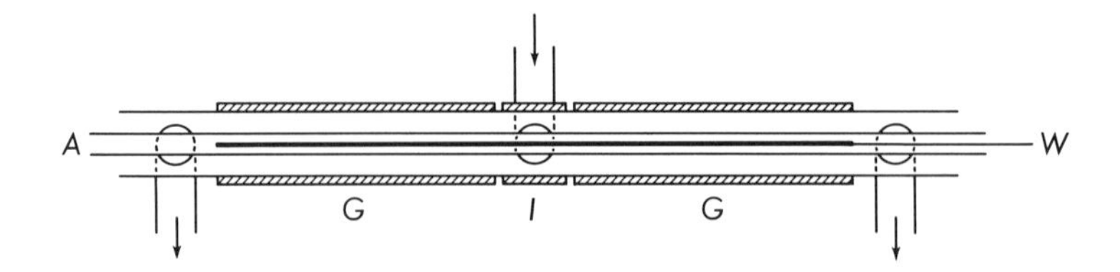

FIG. 3:83. Schematic top view of axon *A* with axial wire *W* and external current electrodes arranged on sides of square section trough. The current is measured through *I*, and *G* are guards. Solution in- and outflows at bottom of trough indicated by arrows.

With the use first of this and similar axon cells and then also with the ruthlessly platinized and hideous axial electrodes promoted by Kishimoto, we did not in years find significant or suspicious variations of membrane current density over the measuring regions or indications of "notch" or "no notch" errors in the total measured currents.

Conclusions

In my first measurements, I was too concerned about correcting for the errors in the ionic membrane potentials and currents (equation 3:1) to pay any attention to the overall characteristics of the experimental arrangement as they appeared at the internal and external electrodes. The importance of the stability of this system was belatedly realized quite a few years later and only then were the data plotted as shown in figure 3:84. The margin of safety indicated by the high negative slope corresponding to less than $-8\ \Omega\text{cm}^2$ showed how near to inadequate and unstable these first experiments had been and how very lucky I was that the electrodes were not a little worse or the axons only a little better!

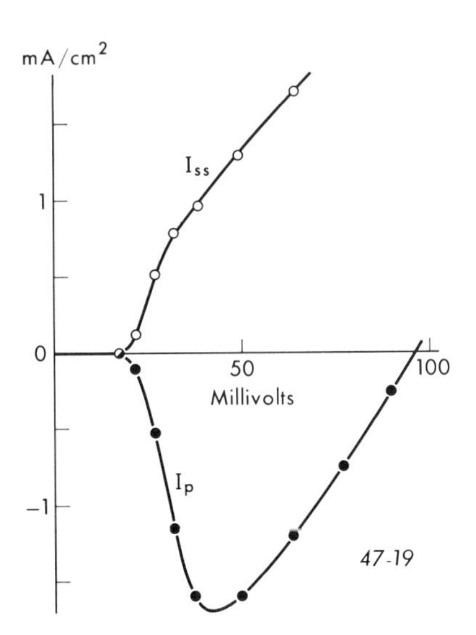

FIG. 3:84. Data of one of the original squid axon voltage clamp experiments without resistance correction. The open circles were approximate steady state currents, and the solid circles for peak inward currents would have become unstable for a better axon.

In having accepted the voltage clamp concept and much of its implementation, Hodgkin, Huxley and Katz had tacitly relied on me for its reliability. They had but a small area of Ag-AgCl for their current electrode, and probably a high resistance from silver to membrane. But, along with the uncertainties of an extended, separate potential electrode, they had at least some insurance against large variations of potential over a considerable part of the membrane area being measured. And they did not have outstanding axons.

In modifying Hodgkin, Huxley, and Katz's modification of my original voltage clamp technique, Moore and I had chosen to rely on the microelectrode for potential measurement and control. As we benefited from its advantages and overcame most of its limitations, we had increasing reason to trust the experimental results—of which those in figure 3:73 were among the best. Although there were decided quantitative differences from the earlier data, the qualitative characteristics were the same.

Thus at least one unanticipated facet of the objectives had been clarified. In common with many other powerful concepts, the voltage clamp had been found to present difficulties and dangers that required significant modifications of technique. On the other hand, these hazards were apparently under control in the earlier measurements; they were not altered qualitatively, and were confirmed except for numerical factors. These factors did not require alteration of the HH expressions or interpretations of the data in any significant way.

As the various stages of the work progressed it became increasingly clear and certain that:

(1) The membrane potential and the membrane current density should be constant over the measuring area to within about 1 mV and about 5 percent respectively.
(2) These specifications could be approximated sometimes over the several millimeter length of axon, and practically always produced the orthodox current pattern in time and potential as found by Cole and by Hodgkin and Huxley—except that they were usually several times larger.

IDENTIFICATION OF MEMBRANE CURRENT COMPONENTS

It was rather well established that, with reasonable care, the so-called voltage clamp did produce measures of ionic current flow across the membrane. These currents had been dissected and described as two independent, early, normally inward sodium and later outward potassium, ion components by Hodgkin and Huxley (1952a,d). This detailed analysis was also dramatically successful in describing most of the important electrical phenomena of nerve.

But even with many and impressive ramifications this analysis was still at best no more than a possible, a sufficient, description of membrane behavior. Given a single experimental fact that it could not explain, it was no longer even sufficient. Another explanation of voltage clamp data might also be a useful and an alternative sufficient description. There is no single necessary condition that can decide between any and all possibilities. Every experimental fact is a necessary condition, although probably the voltage clamp data form the simplest and most comprehensive necessary conditions. Yet no explanation should be expected to span such a mass of information as is now available—even if it is essentially correct, and particularly if it is in a relatively primitive form. Consequently, the success or failure of any particular explanation must be judged by the number and the importance of the phenomena for which it is or is not satisfactory.

The Hodgkin-Huxley description of the sodium and potassium ion currents has been so successful an explanation of a wide range of experimental results that, even if it is wrong, a correct explanation has an enormous handicap to overcome in matching its positive achievements. So it seems far easier and perhaps more effective to test the HH axon by looking for important failures.

Some experiments

The early inflow of sodium and the later outflow of potassium is the simple, fundamental explanation of a propagating impulse. These are calculated in detail by Hodgkin and Huxley, so an independent measure of these net flows during the passage of an impulse is certainly an important, perhaps the most important, ultimate test of the entire HH structure of analysis and synthesis. The average currents are about 0.5 mA/cm^2 over a time of about 0.5 msec and an axon length of 1 cm. A time resolution of 0.1 msec and an equivalent length of 2 mm would then require the analysis of less than $2 \cdot 10^{-14}$ moles, or about 10^{10} ions each of sodium and potassium. This is so far beyond the present sensitivity of any analytical method for either sodium or potassium that an extreme pooling of samples will be necessary, even when the time and space resolution can be achieved. Until such an experiment can be performed we are limited to other, somewhat less complete but in some cases more direct and perhaps as convincing, experiments.

The analysis of ion movements during the passage of a nerve impulse began with a highly improbable experiment by Hodgkin and Huxley (1947). From the increase of membrane conductance of a *Carcinus* fiber in oil they arrived at a potassium loss of 1.7 p mole/cm^2 per impulse and a resting reabsorption of 300 p mole/cm^2 sec. They pointed out that this ion loss n during an impulse would change the potential of a 1.3 μF/cm^2 membrane capacity by an amount

$$\Delta V = ne/C = 150 \text{ mV}$$

which could account for the amplitude of an action potential.

Keynes (1949, 1951) then found the net inflow of sodium with Na^{24} and outflow of potassium with K^{42} for the *Sepia* axon as 3.7 and 4.3 p mole/cm^2 impulse. And finally Keynes and Lewis (1951), with activation analysis, obtained the corresponding net inflow and outflow figures of 3.5 and 3.0 p mole/cm^2 impulse for *Loligo forbesi* axons, and Hodgkin (1951) deduced a sodium entry of 4.5 p mole/cm^2 impulse from the Na^{24} experiments of Rothenberg (1950) and Grundfest and Nachmansohn (1950) on *L. pealii*.

Thus these figures, along with others, for the sodium-potassium exchange in cephalopod axons near room temperature were in the range of 3–4 p mole/cm^2 impulse. The integration of their current vs. time calculations gave Hodgkin and Huxley (1952d) the exchange of 4.3 p mole/cm^2 impulse at 18.5°C. This reasonable agreement with the direct experiments I found an impressive and highly satisfactory support for the analysis.

The charge carried by the postulated sodium and potassium currents had been identified by completely independent, specific, analytical methods as net transport of sodium and potassium across the membrane. Furthermore Shanes (1954) showed by flame photometry that the increase of potassium loss from 9.3 p mole/cm^2 impulse at 24°C to 31.1 p mole/cm^2 impulse at 6°C was in approximate agreement with the calculated increase resulting largely from increased duration of the impulse at lower temperatures.

However, Hodgkin and Huxley (1953) were not satisfied with their identification of the final steady state current in a voltage clamp as a potassium ion current. So they loaded *Sepia* axons with K^{42} and measured the loss accompanying an outward current flow.

This gave Faraday's constant, 96,500 coulombs per mole of potassium, to within 2 percent and adequately identified the steady state current. Hodgkin felt much more confident of the accumulated evidence that the early current was a sodium component; when he told me of the potassium work he objected that it would be far more difficult and not nearly as necessary to make a similar identification of the early component.

It can, of course, be argued that the agreements between the electrical calculations and the chemical data on the total ion transports are no more than presumptive evidence in favor of the validity of the detailed current vs. time relations given by the electrical calculations. And indeed Tasaki with his collaborators insisted that, independent of all other evidence, the agreement of tracer and calculated sodium influx and potassium outflux during an impulse was no proof that the time courses were also as calculated. They struggled to get evidence that all positive tracer ions leave *Nitella* or an axon in similar amounts and rather constantly during the entire course of a propagating impulse (Spyropoulos et al. 1961). And indeed they produced preliminary evidence on the time pattern of the outfluxes from *Nitella* and the tetraethylammonium treated squid axon which indicated that the agreement was fortuitous at best. Some of the results for *Nitella* were questioned by Mullins (1962). Since the HH axon analysis was not necessarily valid for either of these systems and the appropriate analyses as well as influx data were not available, further discussion was best postponed.

Isotope or other chemical analyses of the net flux during a free normal squid impulse are not only difficult—to say the least—but also seem to be an unnecessarily indirect attack on the sodium theory. It is much easier and more cogent to turn back to the original HH analysis of the voltage clamp currents. Some progress has been made by Moore and Adelman (1961), Mullins, Adelman, and Sjodin (1962), and Adelman and Mullins (unpublished), as will be outlined.

Moore and Adelman (1961) somewhat extended the HH experiments to show that the absolute potentials of clamp pulses at which the early transient current reversed followed the Nernst relation as Na_e was varied to give an acceptable value for Na_i. Then they applied short enough pulses to include only sodium current, and at potentials $V > E_{Na}$ to make this net current inward.

The inward transfer of sodium was calculated by $\int I_{Na}\,dt$, and agreed well with that calculated from the increase of E_{Na} observed during the operation of this "electrical pump." But these were again summation experiments and entirely electrical and, although interesting, were not crucial.

It was indeed difficult to trace the time course of both influx and outflux of an ion as for equation 3:15 during a passing impulse. But Mullins, Adelman, and Sjodin (1962) simplified the situation. By using a space clamp they avoided the need of spatial resolution and were able to use a considerable length of axon. They then voltage-clamped the Na^{24} loaded axon and depolarized it in short pulses to above E_{Na} where there should be only an outward sodium current (figure 3:85). The isotope and electric charge leaving the axon agreed quite well, to support the Hodgkin-Huxley interpretation.

Paradoxically, similar experiments with K* for long pulses at E_{Na} where there should be only an outward potassium current were in general agreement, but with some complicating trends that I at least

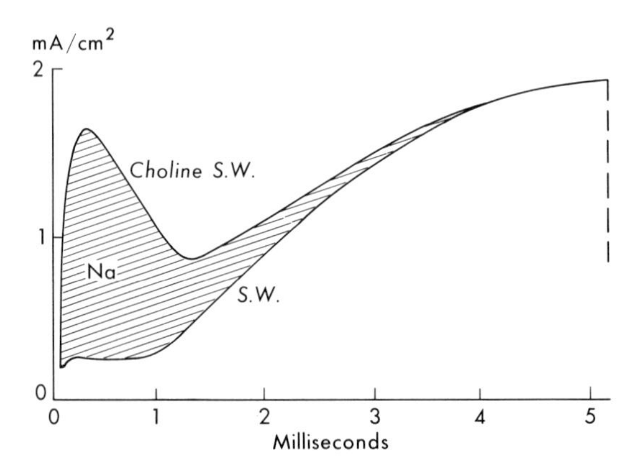

FIG. 3:85. Isotope test for *HH* analysis of voltage clamp data. Clamp at E_{Na} in sea water gave typical pattern *S.W.* and an output of Na^{24} from the loaded axon. The external sodium was replaced by choline and the membrane again clamped at the same potential to give the outward current pattern *Choline S.W.* and an increased release of Na^{24}. The electrical measure of the outflow, given by the shaded area, agreed with the increased loss of tracer to identify the additional current as I_{Na}.

do not yet understand. Mullins, Sjodin, and Brinley (1962) confirmed the original potassium outflow results of Hodgkin and Huxley with the depolarized squid axon, but found that the potassium ion flow was unable to account for hyperpolarizing currents. More complete and careful experiments (Brinley and Mullins, 1965) again gave a potassium current to tracer outflow ratio of 1.0 ± 0.3. But the steady state current is not yet completely explained—at least under some extreme conditions.

I have not examined all of the evidence in detail, but I feel entitled to the conclusion that it is heavily in favor of a separation of the ionic current into sodium and potassium components, much as given by Hodgkin and Huxley. Direct chemical analyses support the other evidence that under voltage clamp the early and late currents are predominantly sodium and potassium respectively. The direct consequence of this separation is the prediction of net ion exchanges during an impulse that are in reasonable agreement with the direct chemical determinations. Direct chemical analyses of the time courses of the net fluxes which could also be predicted during an impulse would indeed be impressive, but the present evidence in this direction is not yet complete or convincing.

Ion flux theory

The many stable and active isotopes that became generally available after World War II allowed tracer investigations to take on a new, and widespread, importance. Among these was the work of Ussing on active and passive ion transport across frog skin. He and Teorell rediscovered Behn's result for the passive, steady state flow of a single ion in electrodiffusion; gave it in more modern terms and under more general conditions; and introduced the idea of separable inward and outward fluxes (Behn, 1897; Ussing, 1949; Teorell, 1949, 1951, 1953).

For a development in simple form, an ion current flow may be expressed as

$$I = une \frac{d}{dx} (kT \ln n + eV) = une\, kT \frac{d}{dx} \ln a$$

where a is an electrochemical activity. By definition $n = a/\xi$ for $\xi = \exp eV/kT$ to give

$$\frac{I}{ekT} \int_0^\delta \frac{\xi \, dx}{u} = \frac{I}{B} = a(\delta) - a(0). \qquad (3:14)$$

The current for one ion will be outward $\overleftarrow{I}_1$ for $n_1(0) = 0$

$$\overleftarrow{I}_1 = B_1 a_1(\delta) = B_1 \xi n_1(\delta)$$

and inward $\overrightarrow{I}_2$ for a second ion with $n_2(\delta) = 0$

$$\overrightarrow{I}_2 = B_2 a_2(0) = B_2 n_2(0).$$

In principle at least the Behn coefficients B_1 and B_2 can be both calculated and measured. If, however, the second ion is an isotope of the first, $u_1 = u_2$, $B_1 = B_2$ and so we have Ussing's relation,

$$\overleftarrow{I}/\overrightarrow{I}^* = \xi n(\delta)/n^*(0)$$

where the asterisks refer to the isotope. The difference of these two fluxes which would be observed by chemical analyses alone is

$$\overrightarrow{I} - \overleftarrow{I}^* = B[\xi n(\delta) - n^*(0)]$$

and so presumably the net flux for normal isotopes on both sides

$$I = \overrightarrow{I} - \overleftarrow{I} = B[\xi n(\delta) - n(0)]$$

which is Behn's result.

In this development we may look upon the product $n \exp eV/kT$ as the electrochemical activity a which replaces the concentration. The resultant net flow is then an extended Fick's law, and the permeability factor is back again to the units of cm/sec.

Hodgkin (1951) used the separate in- and outflux relation of Ussing to compute a relation at equilibrium across a membrane, between these equal fluxes, $\overrightarrow{I} = \overleftarrow{I} = I$ and the conductance G,

$$G = \frac{z^2e^2}{kT} I.$$

This may be a basis for the report by Singer, et al. (1966) that the resistance-flux product was found to be constant for the squid axon membrane.

The propriety of separating a net flux into its inward and outward components has been a matter of some concern. Stephen Yeandle (unpublished) calculated the net fluxes of natural and tracer isotopes directly from the Nernst-Planck electrodiffusion equation. At steady, or quasi-steady, state the original equations

$$I = ue(kT\, dn/dx + neV)$$

were multiplied through by ξ,

$$I\xi = uekT\, d(n\xi)/dx$$

and integrated to give

$$\frac{I}{ekT}\int_0^\delta \frac{\xi\, dx}{u} = \frac{I}{B} = \xi n(\delta) - n(0)$$

where B is the same for normal and tracer isotopes. Consequently

$$\frac{I}{\xi n(\delta) - n(0)} = \frac{I^*}{\xi n^*(\delta) - n^*(0)} = B.$$

However, as before, B can be determined by in- and outflux tracer measurements

$$B = \overleftarrow{I}^*/\xi n^*(\delta) = \overrightarrow{I}^*/n^*(0)$$

and so

$$I = \frac{n(\delta)}{n^*(\delta)}\overrightarrow{I}^* - \frac{n(0)}{n^*(0)}\overleftarrow{I}^*. \tag{3:15}$$

This expression is the same as that obtained by the separation of the net flux of a single ion species into in- and outflux components. So if the separation is to be considered as only an artifice it is at least useful and justified for this example to the extent that a generalized electrodiffusion and the subsequent assumptions are justified.

INDEPENDENCE PRINCIPLE

Instead of relying on the previous results for ion fluxes based on electrodiffusion, Hodgkin and Huxley (1952a) took a more general approach. They assumed only that "the chance that any individual ion will cross the membrane in a specified interval of time is independent of the other ions" and they called this the "independence principle." On this basis, the outflux and the influx

$$\overleftarrow{I} = \overleftarrow{B}n(\delta) \text{ and } \overrightarrow{I} = \overrightarrow{B}n(0)$$

and the net current

$$I = \overleftarrow{I} - \overrightarrow{I} = \overleftarrow{B}n(\delta) - \overrightarrow{B}n(0) \tag{3:16}$$

where the B coefficients are constants depending on the conditions and the potential of the membrane. There would be an equilibrium for the given potential V if $n(0)$ were replaced by $n'(0)$,

$$\exp eV/kT = n'(0)/n(\delta) = \xi$$

at this potential and for $n'(0)$, $\overleftarrow{I} = \overrightarrow{I}$ so $\overleftarrow{B}/\overrightarrow{B} = n'(0)/n(\delta) = \xi$. The Ussing relation then follows,

$$\overleftarrow{I}/\overrightarrow{I} = \xi/\xi_0 = \exp\,(V - E)e/kT$$

for the equilibrium potential E and $\exp\, eE/kT = \xi_0 = n(0)/n(\delta)$.

The net current flow for the single ion is then

$$I = Bn(\delta)[\xi - \xi_0]$$

with $\overrightarrow{B} = B$ and $\overleftarrow{B} = \xi B$. This factor B is given specifically for electrodiffusion and constant mobility by Behn and Teorell and in integral form in equation 3:14 for an arbitrary $u(x)$. The net current flow I' for a changed outside concentration, but at the same potential, is then given by the independence principle

$$I'/I = (\xi - \xi_0')/(\xi - \xi_0) = \frac{\exp\,(V - E)e/kT - n'(0)/n(0)}{\exp\,(V - E)e/kT - 1} \qquad (3:17)$$

and so includes the identical, but more restricted, result given by Behn.

The prediction of ion currents for changes of external concentrations as given by equation 3:17 has been widely used and so often confirmed as to establish this result, and probably also the independence principle upon which it is based, as a necessary condition on any theory of ion flow across membranes.

The sum of the single ion currents may be set equal to zero to give a membrane potential—independent of electrodiffusion or any other specific process. As shown by Patlak (1960), on the basis of the Ussing relation, this results in an equation of the Goldman type. There is thus a possibility that such an equation may be with us for a long time to come.

CONCLUSION

Subsequent calculations and experiments based on the 1952 production by Hodgkin and Huxley did not uncover any major discrepancy; rather they gave more reason to trust it as a description of the behavior of the membrane of the squid giant axon. To a considerable extent some of these investigations are an essential part of the original work and especially so because they uncovered a hazard that was only narrowly avoided.

General Summary

As it stood, there were clear paths from Hermann, Overton, and Bernstein to the considerable clearing in which we found ourselves. Some of the going had been rough; we had been very slow, often stupid and frequently completely lost. But at the end of World War II everything was ready—the squid axon and its passive characteristics, the active conductance increase and the overshoot of the action potential, and the voltage clamp concept.

Hodgkin's idea of an early sodium ion current and the development of the clamp paved the way for a prompt description of the squid membrane in terms of voltage dependent conductances. This description was powerful in that it gave physical reality to nearly all of classical electrophysiology. Although it was certainly limited in scope, and probably not representative in detail, the experimental basis and the interpretations of the data for this description were well confirmed. We could easily be highly satisfied with these achievements, for indeed they mark the end of an era. It would be sensible and logical to close this account in the comfort and security of a goal largely achieved.

I suspect that our present status is analogous to that in physics when the spectroscope was by far the most accurate and powerful instrument available. The Balmer series was the first and a purely empirical generalization for the hydrogen spectrum—perhaps comparable to the normal squid voltage clamp data we have. We may now be in the process of extension and expansion such as led to the general Rydberg equation for a variety of spectra which included the Balmer series as a special case. But Bohr concentrated on the hydrogen atom and, with new ideas from Planck, Einstein, and Rutherford, evaluated the Rydberg constant in fundamental terms.

So the work goes on; some are enlarging the present clearing and making it almost civilized, others are striking out again where there are no paths and where perhaps they can find no guidance. We have no idea of where or what the next major clearing—the nature of the ion permeabilities—may be. So I, too, continue this account with the full expectation of having to stop where there are no stopping points—only the challenges of the unknown.

References

Abbott, B. C., Hill, A. V., and Howarth, J. V., 1958. The positive and negative heat production associated with a nerve impulse. Proc. Roy. Soc. (London) B, *148:* 149–187.

Aubert, X., and Keynes, R. D., 1961. Temperature changes in the electric organ of *Electrophorus electricus* during and after its discharge. J. Physiol. *158:* 17P–18P.

Aubert, X., Fessard, A., and Keynes, R. D., 1959. The thermal events during and after the discharge of the electric organs of *Torpedo* and *Electrophorus.* In Chagas, C. and Paes de Carvalho, A. (eds.): Bioelectrogenesis; proceedings of the Symposium on comparative bioelectrogenesis, New York, Elsevier, p. 136–146.

Bartlett, J. H., 1945. Transient anode phenomena. Trans. Electrochem. Soc. *87:* 521–545.

Behn, U., 1897. Ueber wechselseitige Diffusion von Electrolyten in verdünnten wässerigen Lösungen, insbesondere über Diffusion gegen das Concentrationsgefälle. Ann. Phys. u. Chem., Neue folge, *62:* 54–67.

Bernstein, J., and Tschermak, A., 1906. Untersuchungen zur Thermodynamik der bioelektrischen Ströme. II. Arch. ges. Physiol. *112:* 439–521.

Beresina, M., and Feng, T. P., 1932. The heat production of crustacean nerve. J. Physiol. *77:* 111–138.

Bonhoeffer, K. F., 1953. Modelle der Nervenerregung. Naturwissen. *40:* 301–311.

Brinley, F. J., Jr., and Mullins, L. J., 1965. Ion fluxes and transference numbers in squid axons. J. Neurophysiol. *28:* 526–544.

Chandler, W. K., FitzHugh, R., and Cole, K. S., 1962. Theoretical stability properties of a space-clamped axon. Biophys. J. *2:* 105–127.

Cole, K. S., 1941. Rectification and inductance in the squid giant axon. J. Gen. Physiol. *25:* 29–51.

Cole, K. S., 1949. Dynamic electrical characteristics of the squid axon membrane. Arch. Sci. physiol. *3:* 253–258.

Cole, K. S., 1955. Ions, potentials, and the nerve impulse. In Shedlovsky, T. (ed.): Electrochemistry in biology and medicine. New York, Wiley. p. 121–140.

Cole, K. S., 1961a. An analysis of the membrane potential along a clamped squid axon. Biophys. J. *1:* 401–418.

Cole, K. S., 1961b. Caractéristiques électriques dynamiques de l'axone géant de la seiche. Actual. Neurophysiol. *3:* 251–263.

Cole, K. S., 1965. Electrodiffusion models for the membrane of squid giant axon. Physiol. Rev. *45:* 340–379.

Cole, K. S., and Baker, R. F., 1941. Longitudinal impedance of the squid giant axon. J. Gen. Physiol. *24:* 771–788.

Cole, K. S., and Curtis, H. J., 1938a. Electric impedance of nerve during activity. Nature *142:* 209.

Cole, K. S., and Curtis, H. J., 1938b. Electric impedance of *Nitella* during activity. J. Gen. Physiol. *22:* 37–63.

Cole, K. S., and Curtis, H. J., 1939. Electric impedance of the squid giant axon during activity. J. Gen. Physiol. *22:* 649–670.

Cole, K. S., and Kishimoto, U., 1962. Platinized silver chloride electrode. Science *136:* 381–382.
Cole, K. S., and Moore, J. W., 1960a. Ionic current measurements in the squid giant axon. J. Gen. Physiol. *44:* 123–167.
Cole, K. S., and Moore, J. W., 1960b. Liquid junction and membrane potentials of the squid giant axon. J. Gen. Physiol. *43:* 971–980.
Cole, K. S., and Moore, J. W., 1960c. Potassium ion current in the squid giant axon. Biophys. J. *1:* 1–14.
Cole, K. S., Antosiewicz, H. A., and Rabinowitz, P., 1955. Automatic computation of nerve excitation. J. Soc. Indust. Appl. Math. *3:* 153–172.
Cooley, J. W., and Dodge, F. A., Jr., 1966. Digital computer solutions for excitation and propagation of the nerve impulse. Biophys. J. *6:* 583–599.
Cooley, J., Dodge, F., and Cohen, H., 1965. Digital computer solutions for excitable membrane models. J. Cell. Comp. Physiol. *66* (Suppl. 2): 99–108.
Curtis, H. J., and Cole, K. S., 1940. Membrane action potentials from the squid giant axon. J. Cell. Comp. Physiol. *15:* 147–157.
Curtis, H. J., and Cole, K. S., 1942. Membrane resting and action potentials from the squid giant axon. J. Cell. Comp. Physiol. *19:* 135–144.
Dean, R. B., 1941. Theories of electrolyte equilibrium in muscle. Biological Symposia *3:* 331–348.
del Castillo, J., and Moore, J. W., 1959. On increasing the velocity of a nerve impulse. J. Physiol. *148:* 665–670.
Downing, A. C., Gerard, R. W., and Hill, A. V., 1926. The heat production of nerve. Proc. Roy. Soc. (London) B, *100:* 223–251.
FitzHugh, R., 1960. Thresholds and plateaus in the Hodgkin-Huxley nerve equations. J. Gen. Physiol. *43:* 867–896.
FitzHugh, R., 1962. Computation of impulse initiation and saltatory conduction in a myelinated nerve fiber. Biophys. J. *2:* 11–21.
FitzHugh, R., 1966. Theoretical effect of temperature on threshold in the Hodgkin-Huxley nerve model. J. Gen. Physiol. *49:* 989–1005.
FitzHugh, R., and Antosiewicz, H. A., 1959. Automatic computation of nerve excitation—detailed corrections and additions. J. Soc. Indust. Appl. Math. *7:* 447–458.
FitzHugh, R., and Cole, K. S., 1964. Theoretical potassium loss from squid axons as a function of temperature. Biophys. J. *4:* 257–265.
Frankenhaeuser, B., and Hodgkin, A. L., 1957. The action of calcium on the electrical properties of squid axons. J. Physiol. *137:* 218–244.
Goldman, D. E., 1943. Potential, impedance, and rectification in membranes. J. Gen. Physiol. *27:* 37–60.
Grundfest, H., and Nachmansohn, D., 1950. Increased sodium entry into squid giant axons during activity at high frequencies and during reversible inactivation of cholinesterase. Fed. Proc. *9:* 53.
Guillemin, E. A., 1935. Communication networks. New York, Wiley. Vol. 2, p. 181.
Guttman, R. 1962. Effect of temperature on the potential and current thresholds of axon membrane. J. Gen. Physiol. *46:* 257–266.
Guttman, R., 1966. Temperature characteristics of excitation in space-clamped squid axons. J. Gen. Physiol. *49:* 1007–1018.
Hagiwara, S., and Oomura, Y., 1958. The critical depolarization for the spike in the squid. Japan. J. Physiol. *8:* 234–245.
Hartree, D. R., 1932–33. A practical method for the numerical solution of differential equations. Mem. Manchr. Lit. Phil. Soc. *77:* 91–107.

Hearon, J. Z., 1964. Application of results from linear kinetics to the Hodgkin-Huxley equations. Biophys. J. *4:* 69–75.
Hill, A. V., 1965. Trails and trials in physiology. London, H. Arnold, Part II, Chap. 3.
Hodgkin, A. L., 1938. The subthreshold potentials in a crustacean nerve fibre. Proc. Roy. Soc. (London) B, *126:* 87–121.
Hodgkin, A. L., 1951. The ionic basis of electrical activity in nerve and muscle. Biol. Rev. *26:* 339–409.
Hodgkin, A. L., 1958. Ionic movements and electrical activity in giant nerve fibres. Proc. Roy. Soc. (London) B, *148:* 1–37.
Hodgkin, A. L., 1964. The ionic basis for nerve conduction. Science *145:* 1148–1154.
Hodgkin, A. L., and Huxley, A. F., 1947. Potassium leakage from an active nerve fibre. J. Physiol. *106:* 341–367.
Hodgkin, A. L., and Huxley, A. F., 1952a. Currents carried by sodium and potassium ions through the membranc of the giant axon of *Loligo*. J. Physiol. *116:* 449–472.
Hodgkin, A. L., and Huxley, A. F., 1952b. The components of membrane conductance in the giant axon of *Loligo*. J. Physiol. *116:* 473–496.
Hodgkin, A. L., and Huxley, A. F., 1952c. The dual effect of membrane potential on sodium conductance in the giant axon of *Loligo*. J. Physiol. *116:* 497–506.
Hodgkin, A. L., and Huxley, A. F., 1952d. A quantitative description of membrane current and its application to conduction and excitation in nerve. J. Physiol. *117:* 500–544.
Hodgkin, A. L., and Huxley, A. F., 1952e. Movements of sodium and potassium ions during nervous activity. Cold Spring Harbor Symp. Quant. Biol. *17:* 43–52.
Hodgkin, A. L., and Huxley, A. F., 1953. Movement of radioactive potassium and membrane current in a giant axon. J. Physiol. *121:* 403–414.
Hodgkin, A. L., and Katz, B., 1949a. The effect of sodium ions on the electrical activity of the giant axon of the squid. J. Physiol. *108:* 37–77.
Hodgkin, A. L., and Katz, B., 1949b. The effect of temperature on the electrical activity of the giant axon of the squid. J. Physiol. *109:* 240–249.
Hodgkin, A. L., Huxley, A. F., and Katz, B., 1949. Ionic currents underlying activity in the giant axon of the squid. Arch. Sci. physiol. *3:* 129–150.
Hodgkin, A. L., Huxley, A. F., and Katz, B., 1952. Measurement of current-voltage relations in the membrane of the giant axon of *Loligo*. J. Physiol. *116:* 424–448.
Householder, A. S., 1953. Principles of numerical analyses. New York, McGraw-Hill.
Howarth, J. V., Keynes, R. D., and Ritchie, J. M., 1965. The relation between the initial heat production and the action potential in mammalian non-myelinated nerve fibres. J. Physiol. *181:* 40–42P.
Howarth, J. V., Keynes, R. D., and Ritchie, J. M., 1966. The heat production of mammalian non-myelinated (C) nerve fibers. J. Physiol. *186:* 60–62P.
Howarth, J. V., Keynes, R. D., and Ritchie, J. M., 1968. The origin of the initial heat associated with a single impulse in mammalian non-myelinated nerve fibres. J. Physiol. *194:* 745–793.
Hurwitz, A., 1895. Ueber die Bedingungen, unter welchen eine Gleichung nur Wurzeln mit negativen reelen Theilen besitzt. Math. Ann. *46:* 273–284.
Huxley, A. F., 1959a. Ion movements during nerve activity. Ann. N. Y. Acad. Sci. *81:* 221–246.

Huxley, A. F., 1959b. Can a nerve propagate a subthreshold disturbance? J. Physiol. *148:* 80–81P.
Huxley, A. F., 1964. Excitation and conduction in nerve: quantitative analysis. Science *145:* 1154–1159.
Huxley, A. F., and Stämpfli, R., 1951a. Direct determination of membrane resting potential and action potential in single myelinated nerve fibres. J. Physiol. *112:* 476–495.
Huxley, A. F., and Stämpfli, R., 1951b. Effect of potassium and sodium on resting and action potentials of single myelinated nerve fibres. J. Physiol. *112:* 496–508.
Katz, B., 1939. Electric excitation of nerve. London, Oxford Univ. Press.
Kelvin, Lord (William Thomson), 1855. On the theory of the electric telegraph. Proc. Roy. Soc. (London) *7:* 382–399.
Keynes, R. D., 1949. The movements of radioactive ions in resting and stimulated nerve. Arch. Sci. physiol. *3:* 165–175.
Keynes, R. D., 1951. The ionic movements during nervous activity. J. Physiol. *114:* 119–150.
Keynes, R. D. and Lewis, P. R., 1951. The sodium and potassium content of cephalopod nerve fibres. J. Physiol. *114:* 151–182.
Kober, H., 1952. Dictionary of conformal representations. New York, Dover. Sect. 8.1.
Le Fevre, P. G., 1950. Excitation characteristics of the squid giant axon. J. Gen. Physiol. *34:* 19–36.
Lefschetz, S., 1957. Differential equations: geometric theory. New York, Interscience.
Liapunov, A. M., 1907. Probleme générale de la stabilité du movement. Annales de la Faculté des Sciences de Toulouse, 2nd ser., *9:* 203–474. (Reprinted Annals of Math. Studies, No. 17, 1947, Princeton, Univ. Press.)
Ling, G., and Gerard, R. W., 1949. The normal membrane potential of frog sartorius fibers. J. Cell. Comp. Physiol. *34:* 383–396.
Marmont, G., 1949. Studies on the axon membrane; I. A new method. J. Cell. Comp. Physiol. *34:* 351–382.
Milne, W. E., 1953. Numerical solution of differential equations. New York, Wiley.
Minorsky, N., 1947. Introduction to non-linear mechanics. Ann Arbor, Michigan, Edwards.
Monnier, A. M., and Coppée, G., 1939. Nouvelles recherches sur la résonance des tissus excitables. I. Arch. Int. Physiol. *48:* 129–180.
Moore, J. W., 1958. Temperature and drug effects on squid axon membrane ion conductances. Fed. Proc. *17:* 113.
Moore, J. W., and Adelman, W. J., Jr., 1961. Electronic measurement of the intracellular concentration and net flux of sodium in the squid axon. J. Gen. Physiol. *45:* 77–92.
Moore, J. W., and Cole, K. S., 1960. Resting and action potentials of the squid giant axon *in vivo*. J. Gen. Physiol. *43:* 961–970.
Moore, J. W., and Cole, K. S., 1963. Voltage clamp techniques. In Nastuk, W. L. (ed.): Physical techniques in biological research. New York, Academic Press. Vol. 6, p. 263–321.
Moore, J. W., and Gebhart, J. H., 1962. Stabilized wide-band potentiometric preamplifiers. Proc. Inst. Radio Eng. *50:* 1928–1941.
Mullins, L. J., 1962. Efflux of chloride ions during the action potential of *Nitella*. Nature. *196:* 986–987.

Mullins, L. J., Sjodin, R. A., and Brinley, F. J., Jr., 1962. The correlation between current and potassium ion flux in squid axons. Biophys. Soc. Abstr. FE 10.
Mullins, L. J., Adelman, W. J., Jr., and Sjodin, R. A., 1962. Sodium and potassium ion effluxes from squid axons under voltage clamp conditions. Biophys. J. *2:* 257–274.
Nachmansohn, D., 1966. Chemical control of the permeability cycle during the activity of excitable membranes. Ann. N. Y. Acad. Sci. *137:* 877–900.
Nagumo, J., Arimoto, S., and Yoshizawa, S., 1962. An active pulse transmission line simulating nerve axon. Proc. Inst. Radio Eng. *50:* 2061–2070.
Nastuk, W. L., and Hodgkin, A. L., 1950. The electrical activity of single fibres. J. Cell. Comp. Physiol. *35:* 39–73.
Noble, D., 1966. Application of the Hodgkin-Huxley equations to excitable tissues. Physiol. Rev. *46:* 1–50.
Noble, D., and Stein, R. B., 1966. The threshold conditions for initiation of action potentials by excitable cells. J. Physiol. *187:* 129–142.
Nyquist, H., 1932. Regeneration theory. Bell Syst. Tech. J. *11:* 126–147.
Offner, F., Weinberg, A., and Young, G., 1940. Nerve conduction theory. Bull. Math. Biophys. *2:* 89–103.
Overton, E., 1902. Beiträge zur allgemeinen Muskel- und Nervenphysiologie. Arch. ges. Physiol. *92:* 346–386.
Patlak, C. S., 1960. Derivation of an equation for the diffusion potential. Nature *188:* 944–945.
Paschkis, V., and Baker, H. D., 1942. A method for determining unsteady-state heat transfer by means of an electrical analogy. Trans. Am. Soc. Mech. Eng. *64:* 105–112.
Rothenberg, M. A., 1950. Studies on permeability in relation to nerve function. Biochim. Biophys. Acta *4:* 96–114.
Routh, E. J., 1892. A treatise on the dynamics of a system of rigid bodies. 5th ed. New York, Macmillan, p. 192–202.
Rushton, W. A. H., 1937. Initiation of the propagated disturbance. Proc. Roy. Soc. (London) B, *124:* 210–243.
Shanes, A. M., 1954. Effect of temperature on potassium liberation during nerve activity. Am. J. Physiol. *177:* 377–382.
Shanes, A. M., Freygang, W. H., Grundfest, H., and Amatniek, E., 1959. Anesthetic and calcium action in the voltage clamped squid giant axon. J. Gen. Physiol. *42:* 798–802.
Singer, I., Tasaki, I., Watanabe, A., and Kobatake, Y., 1966. Interdiffusion flux, ionic currents, membrane resistance and impedance in squid giant axons. Fed. Proc. *25:* 569.
Sjodin, R. A., and Mullins, L. J., 1958. Oscillatory behavior of the squid axon membrane potential. J. Gen. Physiol. *42:* 39–47.
Spyropoulos, C. S., 1965. The role of temperature, potassium, and divalent ions in the current-voltage characteristics of nerve membranes. J. Gen. Physiol. *48:* (5, pt. 2) 49–53.
Spyropoulos, C. S., and Brady, R. O., 1959. Prolongation of response of node of Ranvier by heavy metal ions. Science *129:* 1366–1367.
Spyropoulos, C. S., Tasaki, I., and Hayward, G., 1961. Fractionation of tracer effluxes during action potential. Science *133:* 2064–2065.
Tasaki, I., 1956. Initiation and abolition of the action potential of a single node of Ranvier. J. Gen. Physiol. *39:* 377–395.
Tasaki, I., and Bak, A. F., 1958. Discrete threshold and repetitive responses in

the squid axon membrane under "voltage clamp." Am. J. Physiol. *193:* 301–308.

Tasaki, I., and Freygang, W. H., Jr., 1955. The parallelism between the action potential, action current, and membrane resistance at a node of Ranvier. J. Gen. Physiol. *39:* 211–223.

Tasaki, I., and Hagiwara, S., 1957. Demonstration of two stable potential states in the squid giant axon under tetraethylammonium chloride. J. Gen. Physiol. *40:* 859–885.

Tasaki, I., and Spyropoulos, C. S., 1958. Membrane conductance and current-voltage relation in the squid axon under "voltage clamp." Am. J. Physiol. *193:* 318–327.

Taylor, R. E., 1965. Impedance of the squid axon membrane. J. Cell. Comp. Physiol. *66* (Suppl. 2): 21–25.

Taylor, R. E., and Chandler, W. K., 1962. Effect of temperature on squid axon membrane capacity. Biophys. Soc. Abstr. TD1.

Taylor, R. E., and FitzHugh, R., 1959. A source of anomalous current patterns in the "voltage clamped" squid axon. Biophys. Soc. Abstr. G6.

Taylor, R. E., Moore, J. W., and Cole, K. S., 1960. Analysis of certain errors in squid axon voltage clamp measurements. Biophys. J. *1:* 161–202.

Teorell, T., 1949. Membrane electrophoresis in relation to bio-electrical polarization effects. Arch. Sci. physiol. *3:* 205–219.

Teorell, T., 1951. Zur quantitativen Behandlung der Membranpermeabilität. Zeit. Elektrochem. *55:* 460–469.

Teorell, T., 1953. Transport processes and electrical phenomena in ionic membranes. In Butler, J. A. V., and Randall, J. T. (eds.) Prog. in biophys. and biophys. chem., Vol. 3, p. 305–369.

Ussing, H. H., 1949. The distinction by means of tracers between active transport and diffusion. The transfer of iodide across the isolated frog skin. Acta Physiol. Scand. *19:* 43–56.

Wiener, N., 1948. "Cybernetics" or Control and Communication in the Animal and the Machine, 2nd edition, New York, Wiley (1961).

PART IV

Other Membranes, New Techniques, and More Data

Contents to Part IV

Introduction

The long history of electrophysiology is filled with observations on almost every type of excitable tissue. In spite of the many specific characteristics to be found in the overwhelming detail of the literature, the similarities have been very striking—the tissues seemed more alike than different. Some of these common features of excitation and propagation were collected and emphasized over the years between Lapicque's (1926) monograph on the time characteristics and Hodgkin's (1951) review of nerve and muscle in support of the ionic theory. The similar microscopic structures, chemical analyses, and passive electrical characteristics of various muscles and smooth, nonmedullated, nerve fibers also gave reason to suspect a similar basis for the activity. Nonetheless the squid axon, which by 1952 had given a clear, simple, and detailed picture of activity, was most unusual both in its origin as the fusion of many axons and in its consequent size. Its operation might be a unique invention of the squid, but this suspicion of a generality—which for me was an important expectation—obviously should and would be tested as soon as means could be found—unless, of course, the squid results or interpretations fell and the tide of the ionic theory was to be stemmed at its most important source. But an ebb, as of a tide, has not become perceptible—at least in the crests and troughs of a few confusing waves. So voltage clamp measurements and their analysis and consequences have come into increasingly wide usage for a continually widening variety of forms. This exploitation of what seems to be the best current approach can, should, and almost certainly will continue until it reaches a dead end or is superseded. I have been no more than an interested spectator in much of this expansion. It is useful to refer to Noble (1966) for the major developments in the applications of the Hodgkin and Huxley equations to smooth and myelinated axons and, in particular, to cardiac muscle.

The preparations other than the squid giant axon considered here are arranged, somewhat chronologically, in groups according to the various techniques that have been developed for applying the voltage clamp, and interpreting the results. Each of these applications of this rapidly aging concept has usually been based on extensive experience with the preparation, a considerable background of information on the passive electrical characteristics, and an accumulation of questions, particularly about active behavior. It is very interesting that with this combination of circumstances the most complete investigation has been made of medullated axons—which I would have voted the most unlikely to succeed. It is even more important that such extensive and careful work gave us more confidence and faith with which to face the problems that were already appearing in the relatively early results from other cells. We should be able to move faster and more certainly with more experience and as general principles are even more firmly established.

Medullated Axons

The medullated axons are in a separate category of excitable cells. They are in the range of 10 μ in diameter, and covered with a fatty layer, myelin, about 2 μ thick, except for the nodes of Ranvier. These nodes are gaps in the myelin a micron or so wide that appear every few millimeters, as indicated in figure 4:1. Yet with this unique structure such small fibers can conduct impulses in warm-blooded mammals with high velocity, 100 meters per second, and high frequency, 1000 per second, and an efficiency that allows a vast system of rapid communication. And it is this system that made it possible for man to exist and to dominate in his world.

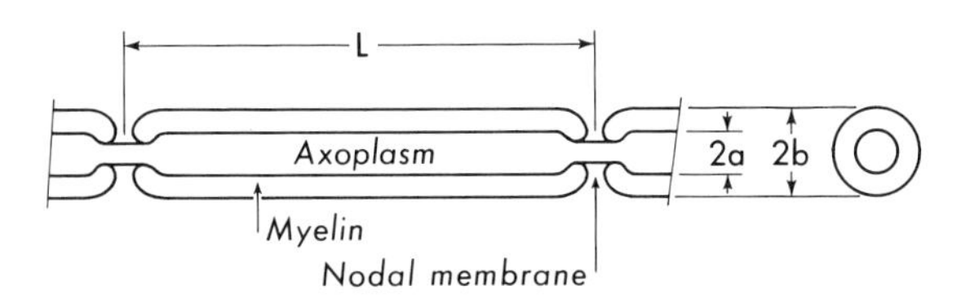

FIG. 4:1. Diagram of a medullated axon showing the passive myelinated internode between two excitable nodes of Ranvier.

But long and intensive efforts were to be expended on this difficult and important problem. Early progress was made mostly by producing and refuting the conclusions from experiments on physiological function. These laid the basis for increasingly careful and complete analyses from, first, the passive, physical characteristics to, finally, a satisfactorily complete description of the parameters responsible for excitation and propagation.

SALTATORY THEORY

The local circuit idea for nerve conduction had appeared about 1900, along with those of the membrane and of potassium and sodium; but explicit quantitative confirmation did not come until the present era of biophysics. So the extension and application of the idea to the medullated axon by Lillie in 1925 may be looked upon as almost contemporary with the earlier ideas. He was working with the passive iron wire analogue suggested by Ostwald (1900) and found that the activity moved at a higher speed when segments of the wire were covered by glass tubing. So he proposed that the myelin was an entirely passive insulation which forced the current from an active node to flow some distance ahead to stimulate the next node.

This important idea was first put into analytical form by Rashevsky (1933a,b) as an extension of his formulation of impulse propagation in smooth axons. Even at this stage, node by node conduction was quite an abstract vision only remotely connected—if at all—to the facts of an axon. Later, with the mounting evidence for Lillie's model and with the new fact of a conductance increase in *Nitella*, Offner et al. (1940), applied this phenomenon of excitation to the theory of saltatory conauction.

In the United States such a saltatory process received strong support when Erlanger and Blair (1934, 1938) showed that the potentials from a single axon were blocked in a step-by-step fashion which corresponded to a node-by-node blockade. However, it was from Japan and the Japanese that the evidence came which made saltatory conduction seem highly probable. Kato (1934) had pioneered in the use of single medullated axons and Tasaki (1953) carried on with it, essentially in parallel with the exploitation of the squid axon. He and his collaborators developed the powerful single and double air gaps for the isolation of nodes and internodes, and they did many ingenious experiments with a skill that has not been matched. They produced evidence that excitation occurs at the nodes (Tasaki, 1939c; Tasaki and Mizuguchi, 1948) and that blockade of propagating impulses is also localized at the nodes (Tasaki and Takeuchi, 1941, 1942). Tasaki (1939a,b) demonstrated a capacity and a conductance for the internode by means of currents flowing through it from adjacent nodes and made preliminary estimates of these parameters and of the resistance of the axis cylinder. The current at a node was first outward, as in stimulation, and then it reversed as an excitation which could serve to excite the next node (Tasaki and Takeuchi, 1941). Then, finally, Tasaki and Mizuguchi (1949) connected up with the squid axon directly by demonstrating an impedance decrease accompanying the impulse at a node.

However, there were still objections to the saltatory theory. Perhaps conduction took place entirely in the axon and within the myelin. If so, then all of the observations at the nodes would be quite unnecessary and entirely coincidental accidents in the noded structures, and the same unknown process would allow for rapid conduction by the myelinated fibers of the central nervous system which were reported to be without nodes. Hodgkin (1951) reviewed the histological evidence to conclude that the nodes had not been noticed, so these central axons were not an exception. If indeed the impulse conduction process were entirely confined to within the myelin sheath, it should not require the external conducting part of the local circuit. On the other hand, if the process were blocked at one node, an external local circuit should not be able to carry the impulse past the blockade.

Huxley and Stämpfli (1949b) produced a block by an oil barrier around an internode and then, following the classic experiment of Osterhout and Hill (1930) on *Nitella*, they restored conduction by a salt bridge across the barrier. Tasaki, and others, had difficulty in establishing such a block but, after demonstrating the great power of an axon to conduct and the assistance of slight conductances and capacities, Tasaki (1959) concluded that "conduction does depend on the electric pathway outside the myelin sheath." He had already shown that an impulse could travel across one and sometimes two insulated, narcotized nodes with the addition of an external path—which should not be effective if it had no part in the conduction process (Tasaki, 1953).

It should be noted that Lorente de Nó has challenged this entire structure of saltatory conduction. With Condouris (1959) he contended that the nodal membrane of isolated axons did not behave in the all-or-none fashion and, at length with Honrubia (1965–1966), he revived the discarded concept that the internodes played an active part in impulse conduction.

Passive characteristics

It was clearly established that what happened at the nodes of Ranvier was the crucial and unanswered question in the propagation of an impulse in a medullated axon. But to answer this question it was again necessary to build, with analysis and experiment, step upon step, beginning with the passive, linear characteristics of this more complicated structure.

The addition of the passive myelin internode is a major and important modification of the uniform cable model for the smooth axons. For an elementary approximation we may consider the steady state distribution without sheath leakage—the same problem that was considered by Jeans (1911) for a telegraph line with periodic leaks, as Taylor (1963) pointed out. It has since been vastly extended to include lump-loaded telegraph and telephone lines and finally the amplifier-loaded submarine cable. Yet for all of their range, speed, and reliability these means of communication seem rather less important—except perhaps in dollar valuation—than the medullated nerve fibers.

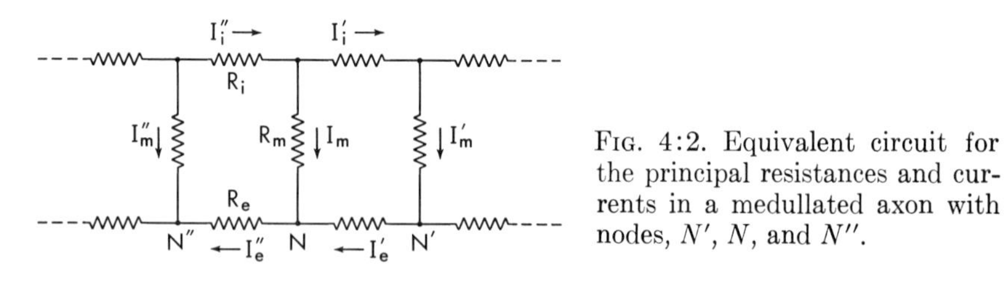

FIG. 4:2. Equivalent circuit for the principal resistances and currents in a medullated axon with nodes, N', N, and N''.

The potentials and currents may be set up as for a uniform cable but they are left in the form of finite differences. With a constant net current flow $I_i' - I_e' = I_i'' - I_e''$, the circuit of figure 4:2 leads to the equation for the membrane currents (Taylor, 1963),

$$I_m' - 2\phi I_m + I_m'' = 0$$

where $\phi = 1 + (R_i + R_e)/2R_m$. This is the standard finite difference equation analogous to the uniform cable and the solution (Norlund, 1924) is

$$I_m(N) = I_m(0) \exp(-wN)$$

where N is the node number and the attenuation

$$w = \cosh^{-1}\Phi.$$

Tasaki and Takeuchi (1941) estimated the axoplasm and node resistances. They used as experimental values a resistance of 40 MΩ between the axoplasm and the outside medium for a long axon and an attenuation factor of 0.3–0.5 for the potential between adjacent nodes. The ladder theory then gave the two resistances as nearly equal at 20 MΩ and the axoplasm and node parameters as 100 Ωcm and 10 Ωcm^2 respectively.

From this simple beginning, the other passive parameters may be added as they seem important and as we are willing to face the rapidly increasing complexities of solution and calculation. Much of such work has been carried out in communication engineering and some of these results have been found useful, but we can look forward to an almost complete reliance upon machine calculation in the future.

Experimental evidence was produced showing a current flow across the myelin—contrary to the simple idea of saltatory conduction that flow was restricted to the nodes. But this current was shown to be through the entirely passive conductance and capacity of the myelin internode.

At about this point Berne joined with Cambridge to start the investigations which led to as complete descriptions for frog and toad nodes as for the squid axon. Hodgkin provided the simple concept and Huxley and Stämpfli (1949a,b) proceeded rapidly with both power and delicacy to a general coherent support for the saltatory theory.

The experiment was one of Spartan simplicity. A single medullated axon was threaded through a glass capillary with a 40 μ hole and a 0.7 mm effective length. The potential difference V_e between the two ends of the capillary was recorded for a conducted impulse. These records were taken at intervals between the nodes as the axon was pulled through the capillary hole.

From the dimensions of the axon and the hole and the resistivity of the external solution r_1, the average current I_e was

$$V_e = \frac{lr_1}{\sigma} I_e = R_e I_e$$

where σ and l are the cross section and length of the external solution and R_e its resistance. Then, since there was no net current flow between the ends of the capillary, the average internal current

$$I_i = -I_e = -V_e/R_e.$$

If indeed the primitive saltatory theory applied, I_i should be constant along each internode during each impulse. This was found to be at least approximately true as shown in figure 4:3 for four internodes. Now with the time course of an average I_i known over three regions of each internode, the increment of potential ΔV over each region Δx at a given time,

$$\Delta V = R_i I_i \, \Delta x$$

and

$$V(x) = V_0 + R_i \sum I_i \, \Delta x.$$

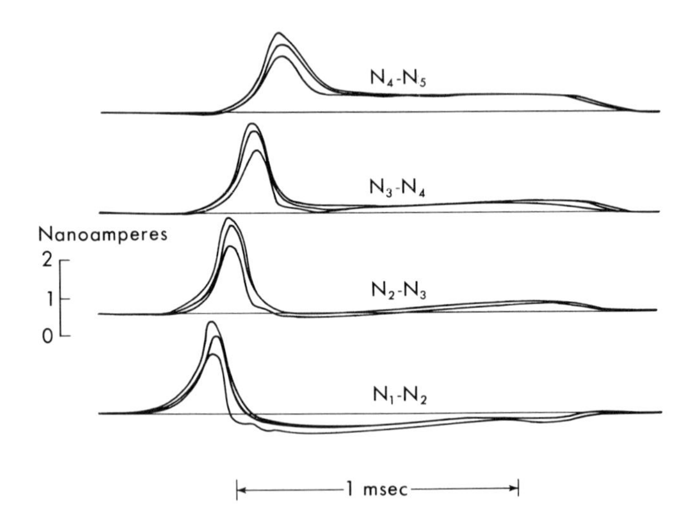

Fig. 4:3. Longitudinal currents in internodes as impulse propagates along axon with nodes N_1 to N_5. The peak currents appear at about the same time in each internode and with definite intervals between the maxima in successive internodes.

The approximate action potential was then obtained from the series of $I_i(t)$ records at t = constant to give $V(x)$ along the axon, and as it varied with time, figure 4:4, from the speed of 23 m/sec.

There were, however, systematic differences between the $I_i(t)$ records for different regions within the same internode which might be attributed to mostly outward capacitive and conductive currents across the myelin sheath. The equivalent circuit of figure 4:5 may be used as the simplest approximation, so

$$I_S = I_i - I_i'' = I_e' - I_e''.$$

The average of these differences within a single internode and over several internodes gives the myelin current of figure 4:6. The three phases of the time course correspond to outward capacitive and conductive currents during the rising action potential, a pre-

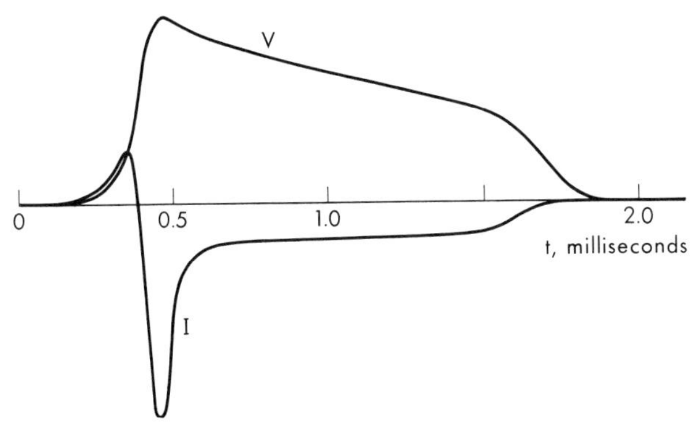

FIG. 4:4. The potential and current across a nodal membrane during a propagated impulse. The large inward current is characteristic of activity and stimulates the next node.

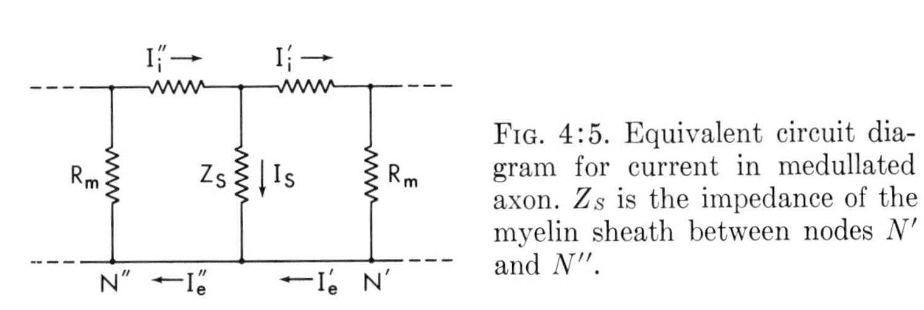

FIG. 4:5. Equivalent circuit diagram for current in medullated axon. Z_S is the impedance of the myelin sheath between nodes N' and N''.

dominant outward conductive current during the slow recovery which is replaced by the increased capacitive current during the final rapid stage of recovery.

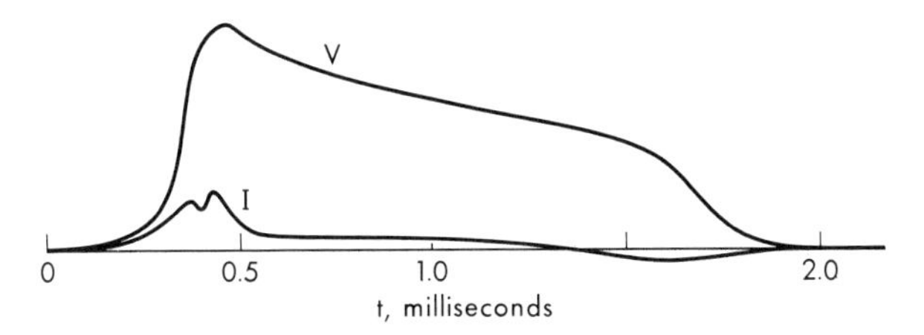

FIG. 4:6. The potential and current across the myelin sheath of an internode during a propagated impulse correspond to its passive characteristics.

In the same way the current at a node I_m was found from the difference of the potential records made on each side of the node to give the average current pattern of figure 4:4. The difference between the node and the internode currents and the similarity between the node current and the membrane current during an impulse in a smooth axon are both striking. Taken together they strongly support the saltatory theory that the internode is purely passive and that the activity confined to the node is at least of the same kind as for other axon membranes.

The quantitative analysis used by Huxley and Stämpfli to obtain the capacity and the conductance of the internode was novel. It is particularly interesting as a sound, simple method which averages out unexplained and perhaps incidental details. We have the myelin current density,

$$i_m = g_m(V - V_0) + C_m\, dV/dt$$

where g_m and C_m are for a unit area and V_0 is the rest potential. Then at a time t_0 before the impulse and at t_∞ after it is past, $V = V_0$ and the peak potential V_p is at the time t_1,

$$\int_{t_0}^{t} i_m\, dt = g_m \int_{t_0}^{t} (V - V_0)\, dt + C_m[V(t) - V(t_0)]$$

and

$$g_m = \int_{t_0}^{t_\infty} i_m\, dt \Big/ \int_{t_0}^{t_\infty} (V - V_0)\, dt \tag{4:1}$$

and further,

$$C_m = \left[\int_{t_0}^{t_1} i_m\, dt - g_m \int_{t_0}^{t_1} (V - V_0)\, dt\right] \Big/ [V_p - V_0]. \tag{4:2}$$

The integrals were evaluated numerically in terms of α, the unknown ratio of the axoplasm resistivity to that of the Ringer's solution (90 Ωcm). This factor was first estimated to be 1.4–1.8 but later measurements of the action potential gave basis for the choice of $\alpha = 1.2$. Then the capacity of the myelin was $2.5 \cdot 10^{-9}$ F/cm² and the resistance was $1.6 \cdot 10^{5}$ Ωcm², with the corresponding dielectric constant of 5.4 and resistivity of $8 \cdot 10^{8}$ Ωcm.

Tasaki (1955) used his double airgap to measure directly the transverse currents in a millimeter length of axon. Colliding impulses from both ends of the axon essentially eliminated spatial gradients of current and potential. Adaptations of the integral analysis of Huxley and Stämpfli, equations 4:1 and 4:2, were used to give the capacity and resistance of internode myelin again and of stretches with an inactivated node. The capacity of the myelin at $5 \cdot 10^{-9}$ F/cm² was about as before, that estimated for the node was about 5 μF/cm²; neither of these values was altered by several changes in the outside solution. The resistances of the myelin and the normal resting node were about 10^{5} Ωcm² and 10 Ωcm² respectively, to again confirm earlier estimates. Furthermore each of these was found to depend on some of the external agents applied.

The capacity and resistance of the myelin as determined by Huxley and Stämpfli (1949b), along with the resting parameters of the node redetermined by Tasaki (1955), were all assembled by Hodgkin (1964) in table 4:1. These led to a dielectric constant of

TABLE 4:1

ELECTRICAL CHARACTERISTICS OF A FROG MYELINATED FIBER

Fiber diameter	14 μ
Thickness of myelin	2 μ
Distance between nodes (L)	2 mm
Area of nodal membrane (assumed)	22 μ^2
Resistance per unit length of axis cylinder (r_i)	140 MΩ/cm
Specific resistance of axoplasm (r_2)	110 Ωcm
Capacity per unit length of myelin sheath (C_S)	10–16 pF/cm
Dielectric constant of myelin sheath (ϵ)	5–10
Resistance x unit length of myelin sheath (r_S)	25–40 MΩcm
Specific resistance of myelin sheath (r_4)	500–800 MΩcm
Capacity of node (C_m)	0.6–1.5 pF
Capacity per unit area of nodal membrane (c_3)	3–7 μF/cm^2
Resistance of resting node (r_m)	40–80 MΩ
Resistance x unit area of nodal membrane (r_3)	10–20 Ωcm^2
Action potential	116 mV
Rest potential	71 mV
Peak inward current density	20 mA/cm^2
Conduction velocity	23 m/sec

5–10 and a specific resistance of $0.5–0.8 \cdot 10^9$ Ωcm for the myelin sheath, and with the node capacity and resistance of 3–7 μF/cm^2 and 10–20 Ωcm^2 give a complete description of the axon as a passive segmented cable.

From the x-ray diffraction of myelin by Schmitt et al. (1941) and the electron microscope evidence of Fernandez-Moran (1950) it seemed clear that this sheath was made of layers each about 80 Å thick. Hodgkin (1951) then pointed out that each of the 250 layers in an average sheath would have a capacity of about 0.6 μF/cm^2 and a resistance of 600 Ωcm^2. These values were so close to the resting parameters of many cell membranes as to suggest that they were not accidental.

EXCITATION AND PROPAGATION

Our ignorance of the performance of a node of Ranvier in propagating an impulse in a medullated axon had begun to dissipate when

Tasaki and Takeuchi (1941) had shown, by blockade of a node, that the electrical resistance of the plasma membrane at the beginning of activity was negligibly small as compared with the resistance at rest. Then from the similarity of the action current and decrease of high-frequency impedance at a node, Tasaki and Mizuguchi (1949) concluded that these two measurements were but different expressions of the membrane process.

Huxley and Stämpfli (1949b) point out that at the node the initial outward current of figure 4:4 may represent a passive charging of the node capacity. This is followed, beginning at about the inflection of the potential, by the inward current characteristic of the excitation process. Then there is the long slow recovery phase in which the small outward ionic current is largely masked by the capacitive component. The final fast recovery is brought about by the last burst of outward current to completely recharge the capacity of the node.

This was much the same as had been found for the squid axon, and the correspondence between the two membranes became much more compelling.

Dimensional analysis

There are, however, some interesting and useful things that can and have been done without formal analysis or computation—as indeed Nature seems to have done them. Even without specific information on the behavior of the nodes, Rushton (1951) was able to generalize it by a dimensional analysis considerably beyond the start made by Huxley and Stämpfli (1949b). From the geometry and performance of various medullated axons, he established by physical reasoning, similar to that of equation 2:4 given by Rosenblueth et al. (1948) and Hodgkin (1954) for a smooth axon, a high probability that all known nodes were very similar in their excitation and response characteristics.

Rushton first gave the internodal distance L as a scaling factor between different axons. Then if the similar membrane potentials at a node produce similar current densities, each nodal current is proportional to its area σ. A corresponding membrane potential between adjacent nodes is then proportional to the axoplasm conductance and so to a^2/L, where $2a$ is the internal diameter of the internode. The change of axial current across a node is equal to the nodal current and so to a^2/L^2. The leakage across the myelin is proportional to effective path length, so the potential varies as $1/\ln(b/a)$ for an external myelin of diameter $2b$ and the capacitive current during a change of potential has the same factor. Then with a proportionality constant k_1^2

$$L^2/b^2 = k_1^2(a^2/b^2)\ln(b/a)$$

and L/b is a maximum for $a/b = 1/\sqrt{e} = 0.6$. Data were available showing that a/b was between 0.47 and 0.74 for two thirds of the fibers of a nerve so the expected variation of L/b was 5 percent. Hodgkin (1964) has modified this to include the effect of the node capacity to give an optimum $a/b = 0.7$, in good agreement with the data he quotes.

For a given external diameter $2b$ there is an optimum core diameter $2a$ which gives a maximum for the electrotonic spread or the space constant λ, where λ^2 = core conductance/sheath admittance,

$$\lambda = k_2^2\, a^2 \ln b/a,$$

and k_2^2 is a constant containing only the specific constants of the axoplasm and the myelin. Thus

$$\frac{\lambda}{k_2 b} = \frac{a}{b}\sqrt{\ln\frac{b}{a}} = \frac{L}{k_1 b}$$

so an optimum for L is also an optimum for λ; this is reasonably supported by microscopic observation. The scaling factor, proportional to L, is thus a basis for the comparison of different axons; to the extent that it is adequate it requires that the conduction velocity in these axons be proportional to the internodal distance or similarly to the axon diameter.

From this analysis Rushton gave a basis for the observations that:

(1) The space constant, the velocity, and length of the impulse varied as the axon diameter;
(2) the spike duration, chronaxie, summation time, and refractory period were nearly independent of axon diameter; and
(3) the threshold predictions were less satisfactory but not unreasonable.

Finally he pointed out that, as in figure 4:7, Nature may have chosen axon conduction speed and diameter as a design criteria. Where a speed less than 2 m/sec is adequate, a smooth axon takes less space and is used exclusively. For all faster communication the

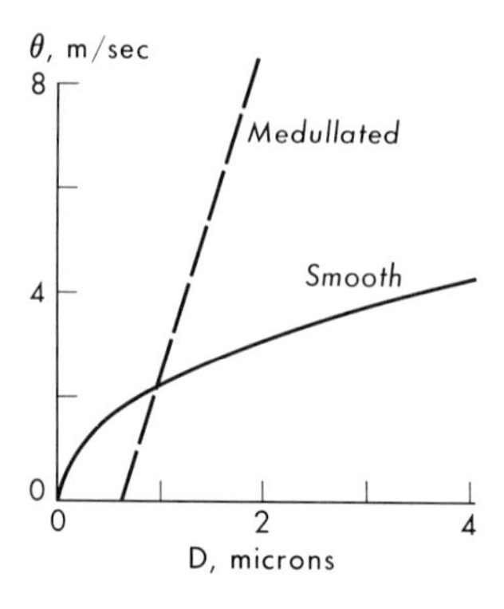

FIG. 4:7. Impulse speeds θ in axons, with and without active nodes and medullated internodes, as they depend on axon diameters D.

medullated axon is the better choice, and at top speeds it is the only practical possibility.

However, FitzHugh (1968) reviewed the problem of Nature's design of medullated axons more powerfully and in more detail. He points out that such obvious desiderata of design as high conduction speed, economy of energy, and high margin of safety cannot all be maximized. Until there is more and more satisfactory information from at least one animal it is not possible to estimate the relative importance of these—and perhaps other—factors. So FitzHugh raises unanswered questions about both Rushton's assumptions and his conclusions.

Some years ago at Princeton, Marston Morse asked what a nerve minimized—perhaps an integral? May it now be possible to invoke entropy changes and communication theory?

NATURE OF EXCITATION

In the early 1950's the passive parameters of a medullated axon were reasonably well known, there was practically conclusive support for the saltatory explanation of impulse conduction, and there was a high probability that the excitation processes at the nodes of all such axons were very similar to each other.

The next and—from the point of view of both general and mammalian physiology—highly significant development was the answer to a crucial question. Is there any reason to expect that the results from the squid membrane are in any way related to impulse transmission in the medullated axons?

There was however, in 1951, no evidence for the nature of the excitation at the nodes of medullated axons. Both the resting potential and the action potential were quite uncertain in both relative and absolute values. There were some indications of an effect of external potassium on the rest potential, but no hint of a relation between the action potential and external sodium for medullated axons. With the evidence for the sodium theory from the squid axon (Hodgkin and Katz, 1949) and the sartorius muscle fiber (Nastuk and Hodgkin, 1950) the suspicions, as well as the hopes, of a similar explanation for the node were high. Yet there was no likely experimental method available to provide the data that were then so obviously of immediate importance.

So Huxley and Stämpfli (1951a,b) took another new step. Instead of measuring the axial current flow by the potential drop across a known resistance, they introduced a known potential and used the criterion of no axial current flow to balance the unknown potential at a node. This was no more than an elementary potentiometer in principle, but its use is an excellent demonstration of both the difficulty and the power of adequate application of simple physical concepts to biological systems. In figure 4:8, N is the normal node under investigation; it is connected by the axial resistance R_i to the node N' which is inactivated in KCl. Huxley and Stämpfli had turned to the glass capillary when they had several difficulties with an oil barrier in their saltatory experiments. But here the oil barrier between C and D was very effective in creating the high external resistance which would produce a useful potential for current flow between these points. All that remained was to bring this current to zero by applying the potential V to balance the effective emf E of the node.

First the rest potential V_0 was balanced to give an average of -71 mV. This was applied continuously to help maintain the node in useful condition for several hours. Then for changes of external

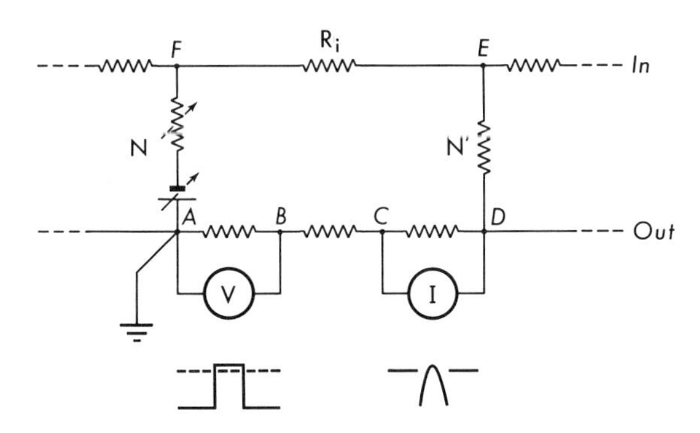

Fig. 4:8. Experiment to measure rest potential and maximum of action potential at node N by applying potentials V to balance current I to zero.

potassium, shown in figure 4:10a, the rest potential varied much the same as it had in the squid axon (Curtis and Cole, 1942), and could be expressed by the Goldman equation as had the squid data (Hodgkin and Katz, 1949). Thus potassium was largely responsible for the rest potential.

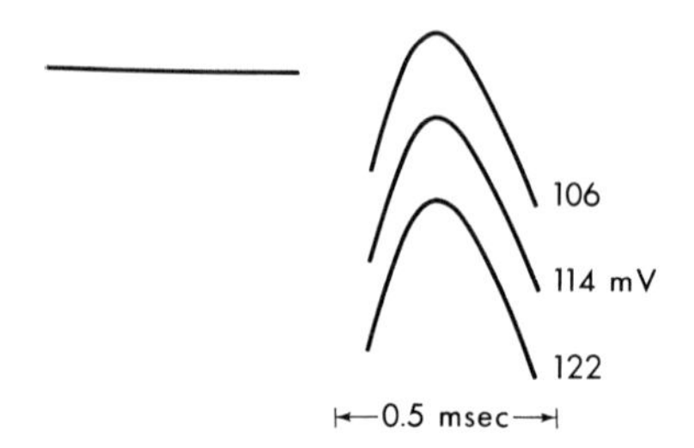

FIG. 4:9. Action potential of a node is balanced by a potential of 111 mV.

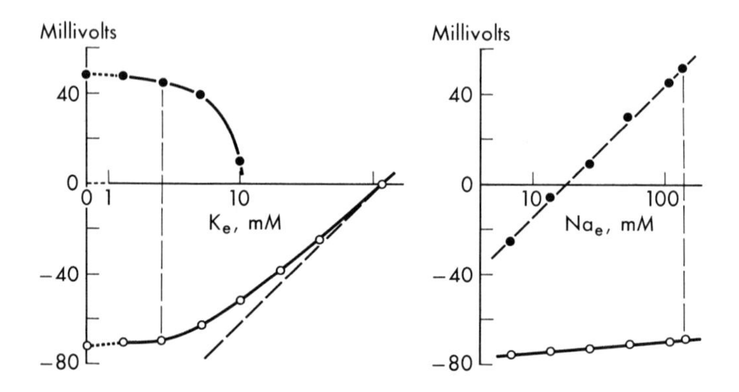

FIG. 4:10. Effect of external ions, potassium, left, sodium, right, on behavior of frog node. Resting potential, open circles, and action potential, solid circles. Vertical dashed lines are at normal ion concentrations, slant dashed lines are for changes of equilibrium potentials vs. ion concentrations on logarithmic scales.

The node was excited by a short superposed square wave which was then adjusted in amplitude until the node current was zero at the peak of the action to give V_p as in figure 4:9. For this Huxley and Stämpfli gave an average value of 116 mV so this was yet another example of an action potential overshoot. The maximum of the action potential followed the external sodium ion concentration, as in figure 4:10b, according to the Nernst equilibrium potential, with but small changes of the rest potential. Thus there was basis to believe that the sodium permeability at the spike was relatively high as compared to other ions.

So the membrane of the node was added to those of the squid axon and the sartorius fiber to form the firm core for Hodgkin's (1951) first sodium hypothesis. The conclusions of this seventy-page review, "The Ionic Basis of Electrical Activity in Nerve and Muscle," are quite cautious and mostly qualitative. The seventeen items of his Summary cover a vast amount of material; I quote only two of them:

"2. In a wide range of excitable tissues the resting membrane potential is of the order of 50–100 mV and the action potential of the order of 80–130 mV. . . .

15. The permeability changes during the action potential probably consist of a rapid but transient increase in the permeability to sodium and a delayed increase in the permeability to potassium."

Although there was no obvious reason to question the Huxley and Stämpfli results, the method—for all of its elegance—was ponderous and limited to manual point-by-point measurements. A direct recording would be much more useful—and perhaps more convincing to some. This was to be achieved by other methods for again increasing the effective external resistance of an internode R_e of figure 4:11. Stämpfli (1954) replaced the external electrolyte by low-conductance iso-osmotic sucrose solution. Tasaki and Frank (1955) used an air gap and an electrostatic shield, driven at the output potential, to reduce the capacitive load on the internode resistance R_i. The next step, by Frankenhaeuser (1957a), was a return to the original Huxley and Stämpfli idea, with electronic control as shown in figure 4:12. In this, any current flow around the circuit was balanced by the amplifier output V to give the potential difference across the node N to within 1 percent. In spite of some differences, these three experiments essentially agreed in confirming the conclusions of Huxley and Stämpfli.

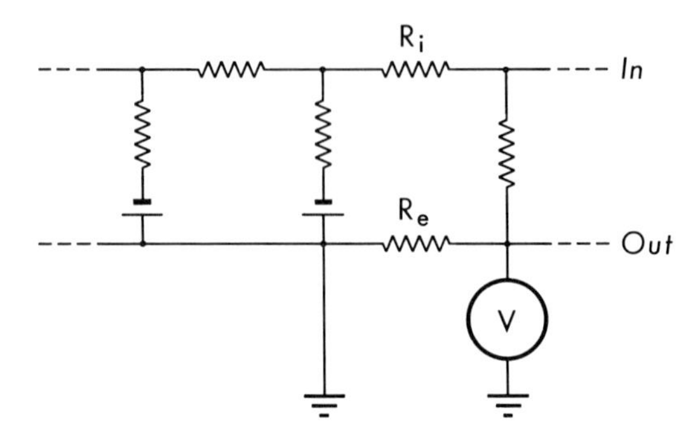

FIG. 4:11. As the external resistance R_e, between an active node, center, and an inactivated node, right, is increased, the potential V approaches the internal potential of the active node.

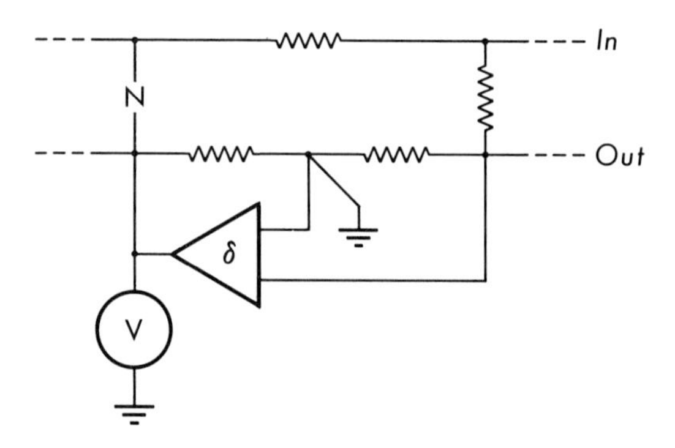

FIG. 4:12. Current flow from active node N through inactive node, right, is balanced by amplifier δ to give V, the potential of N.

CLAMPING A NODE

Hodgkin (1951) had set the stage for a widespread application of the sodium hypothesis, so the Hodgkin and Huxley quantitative description of the sodium and potassium permeabilities for the squid axon membrane could not but raise the question as to the possible similarity of other excitable membranes. One immediate, and intermediate, answer to this question was given by a participant at Cold Spring Harbor in 1952. "This (the HH analysis and synthesis of voltage clamp data) is certainly interesting—it is obviously a possible and even a probable description of the nerve-like behavior of the squid axon. It is, however, entirely certain that this behavior and this analysis are in no way related to the operation of the different (and obviously more important) frog nerves." Such a firm opinion, the strong expression of it, and the later conclusion that it was probably wrong are in no way unique in electrophysiology, particularly so in the development of an understanding of conduction in the medullated axons. I prefer to emphasize the successful steps and the grandeur of the achievement rather than the controversy—perhaps because I have been but an enthusiastic spectator. To others of us there seemed little reason to expect that the characteristics of the node under voltage clamp would be qualitatively different from those of the squid axon (Cole, 1949), or that the interpretations of such data would not be similar to those for the squid axon (Hodgkin and Huxley, 1952b).

It was not until probably 1955 that John W. Moore and I decided, independently, simultaneously, and to our great surprise, that a node had been designed for voltage clamping. We, as others, had found the limited and uncertain squid season at Woods Hole a considerable handicap in the investigation of membrane phenomena and mechanisms. And to realize that the node, with its natural insulation on both sides, presented a limited area of membrane over which current density and potential must be rather uniform was quite a shock after the many difficulties of geometry, electrodes, and electronics encountered in achieving this same objective with the squid axon. It was simple to show by the conventional theory of artificial cables that an internode was too long to be represented by a single T section but that a three section network was a reasonable approximation. In our worry about how an impulse starts in an HH axon, FitzHugh found that our resources in machine and money were not adequate for computation on the problem. However, he did find that the initiation of propagation could be investigated for a medullated axon with the known internodes and somewhat modified HH characteristics at the nodes. In this FitzHugh (1962) used the three T section approximation for the internode and showed that a four section was not detectibly different. Although other aspects of these calculations will appear later, there was then basis to represent a medullated frog axon by the equivalent network of figure 4:13.

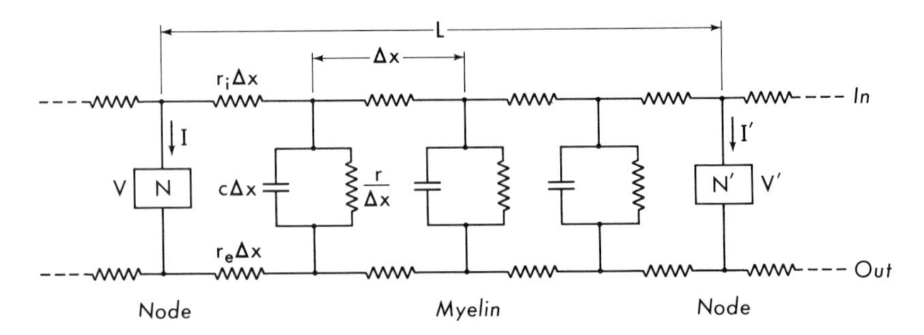

FIG. 4:13. Equivalent circuit elements for simulation and calculation of medullated axon. Three filter sections represent the passive myelinated internode between two node membranes.

With such an artificial medullated axon, it was found that the problems of voltage clamping were not simple. The control gain was extremely high because the internode internal resistance was very high, and with such gain only a slight stray coupling was disastrous. Nonetheless, major design problems were brought under control. Then, following Tasaki's extensive and successful use of air gaps to isolate a node by decreasing the leakage along the outside of the myelin on either side, this very promising approach to voltage clamp investigations was used by del Castillo et al. (1957), Tasaki and Bak (1958a), and Ooyama and Wright (1962). Finally we heard that, after his work on squid at Plymouth with Hodgkin, Frankenhaeuser had returned to the node and was in the process of voltage clamping it with a superior technique at Stockholm. Since we respected Frankenhaeuser highly and he was far ahead of us, and because we had so much work to do with squid axon, we regretfully put our brilliant idea on the shelf.

MEASUREMENTS AND ANALYSIS

After the preliminary announcement (Frankenhaeuser and Persson, 1957) Dodge and Frankenhaeuser (1958) described the experimental arrangement shown schematically in figure 4:14. Amplifier 1 measures the potential V across node N through the inactivated node N' as before. Current I through N is introduced through the inactive node N''. Then either V or I are controlled by amplifier 2 to follow the command C.

They emphasized that oscillations and all-or-none responses (Tasaki and Bak, 1958b) were regularly traced to insufficient

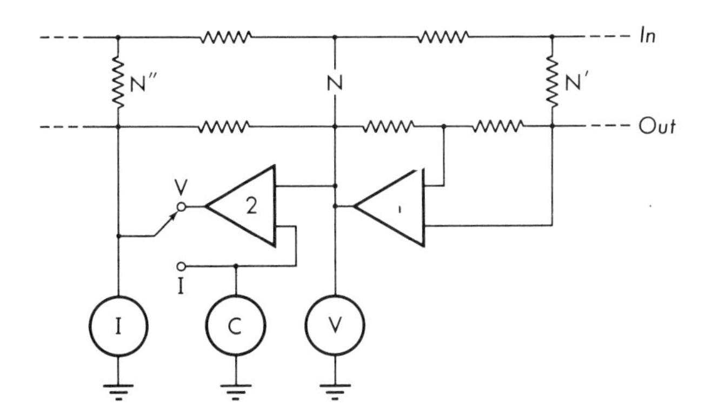

FIG. 4:14. Potential V and current I properties of node N are measured through inactivated nodes N', N''. The command C controls either V or I by amplifier 2.

stabilization of the membrane potential. In preliminary experiments, Moore and del Castillo (1959) used the Frankenhaeuser approach and obtained evidence in agreement with Dodge and Frankenhaeuser (1958). They encountered and, for the most part, overcame the antagonistic demands for high control amplification with the high degree of output-input coupling elimination, as emphasized by Dodge (1963). I am not at all sure that potential control was established, even to the extent that had been needed to justify the HH analysis, but it was made highly probable.

First with Dodge (1958, 1959) and later alone, Frankenhaeuser (1959, 1960, 1962a,b,c, 1963a,b) investigated the frog node and the large node of the African toad, *Xenopus*, in detail. It was shown, first, that the nodal membrane currents under voltage clamp are essentially the same as for the squid axon, as seen in figures 4:15 and 4:19 and, second, that the analysis of the sodium and potassium components of the ionic current following Hodgkin and Huxley were qualitatively the same as for the squid membrane although with significant differences.

They again identified the early current as sodium by choline substitution and with only minor discrepancies. However, the conductance $g_{\mathrm{Na}} = I_{\mathrm{Na}}/(V - E_{\mathrm{Na}})$ at inward peak was no longer independent of V, so they resorted to the Goldman permeability P_{Na}, equation 2:26, as a more adequate representation of the data,

$$I_{\mathrm{Na}} = P_{\mathrm{Na}}\mathrm{Na}_e \frac{e^2V}{kT} \frac{\exp\left[(V - E_{\mathrm{Na}})e/kT\right] - 1}{\exp\left[Ve/kT\right] - 1} \qquad (4{:}3)$$

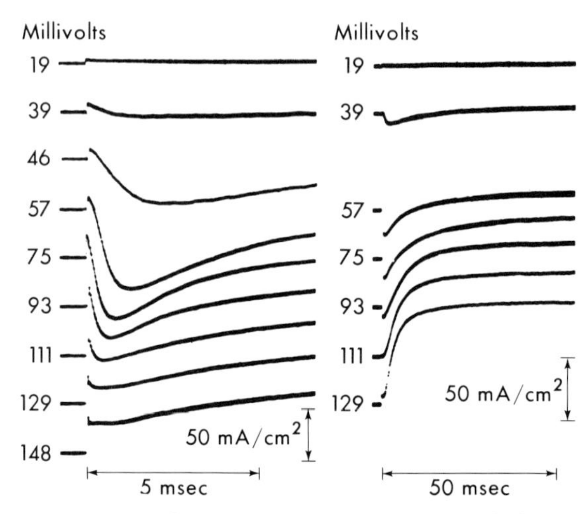

FIG. 4:15. Voltage clamp currents for two different frog nodes on fast and slow time scales. The indicated potential steps are from −15 mV (left) and +5 mV (right) from the resting potentials.

which is a steady state expression. The late and the steady state currents were identified as potassium (Frankenhaeuser, 1962c) by changes of K_e and similarly expressed as

$$I_K = P_K K_e \frac{e^2 V}{KT} \frac{\exp (V - E_K)e/kT - 1}{\exp Ve/kT - 1}. \qquad (4:4)$$

Furthermore, the I_K and I_{Na} separation procedure of Hodgkin and Huxley was not directly applicable for the transients.

Frankenhaeuser and Huxley (1964) summarized the voltage clamp data for the *Xenopus* node and mentioned some uncertainties under fifteen items. They then produced "standard data" in the form of the HH axon, but added a small nonspecific delayed current I_d, and used the permeabilities P_{Na}, P_K, P_d instead of conductances. The action potential computed from these data corresponded well with the experimental ones, while the effects of arbitrary variation of the standard parameters showed little basis to modify the approximations made to the clamp data. Frankenhaeuser (1965) continued the computations to investigate the behavior near threshold, the strength-duration relation, and the effect of variations of some parameters. The effects of current on the after potential, and of changes of temperature and external sodium were also computed. The overall conclusions from these London and Stockholm computations were that there was no essential difference between the toad node and the squid membrane in either data or interpretation and that the node voltage clamp data satisfactorily represented classical experiments.

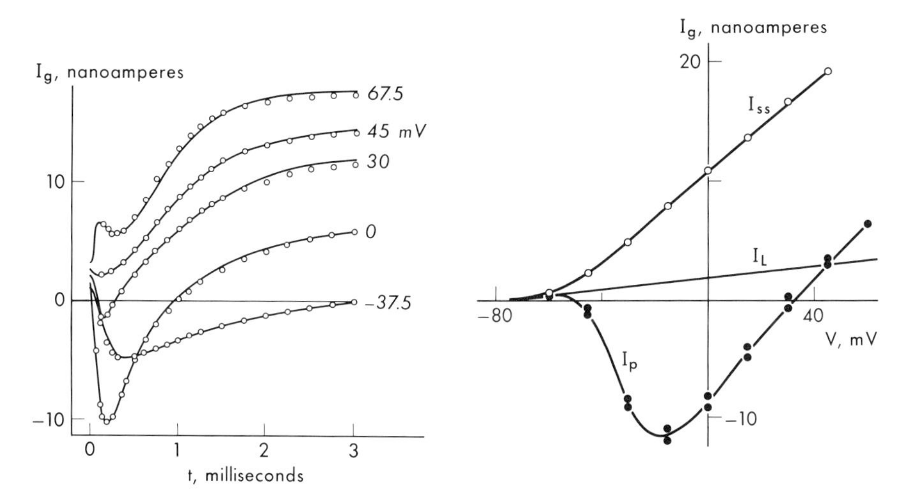

FIG. 4:16. Frog node clamp currents. Left, circles, after clamps to the indicated potentials; the curves are given by modified *HH* empirical equations. Right, early peak currents I_p solid circles; steady state currents I_{ss} open circles; initial leakage, I_L.

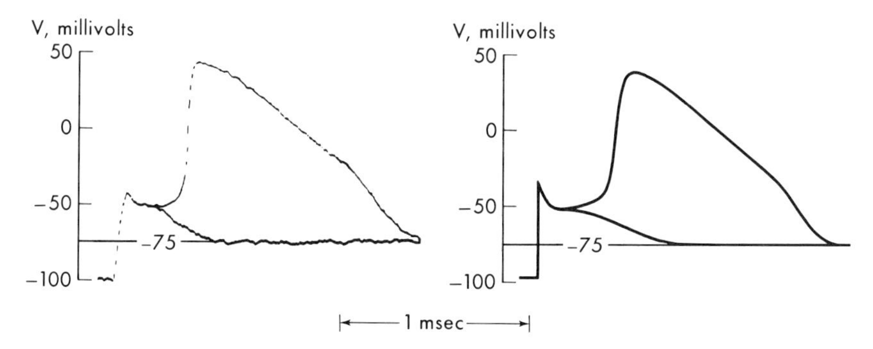

FIG. 4:17. Frog node potentials after near to threshold current pulses. Left, as observed; right as computed from empirical equations.

In extending the work on the frog node such as shown in figure 4:16, Dodge (1963) also modified the HH axon by introducing the permeability P_{Na} but approximating P_K by g_K and retaining the m, h, and n variables, as seen in figure 4:58. The nodal subthreshold and

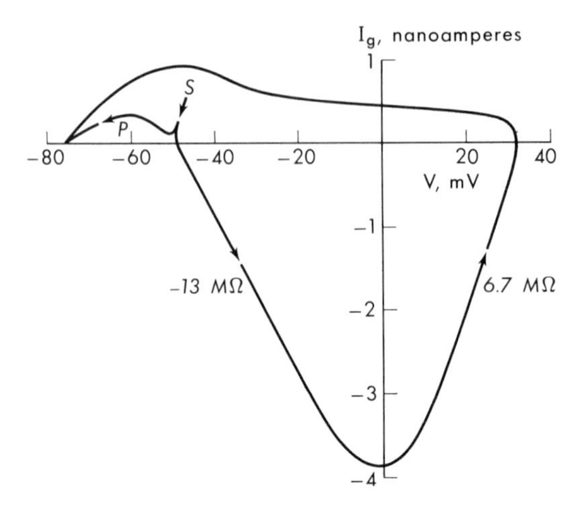

FIG. 4:18. Frog node membrane potential and ionic current after near threshold stimuli S. The passive subthreshold response P returns directly to the rest potential but in activity the path first invades the other three quadrants.

action potentials calculated on this basis are shown in figure 4:17 to be very similar to those from experiment, and the impedance changes during an action and the potential responses during a second pulse were well predicted. Dodge (1961) further showed his calculations, from data, of the ionic currents I_g after subthreshold and threshold short stimuli. These are given as trajectories on the V, I_g plane of figure 4:18. A comparison with the similar representations for squid membranes again emphasizes the similarities. However, for a nominal node area of 22 μ^2 the resistances of about $-3\ \Omega\text{cm}^2$ during excitation and 1.5 Ωcm^2 during the height of activity are rather higher than expected from some other measurements.

Dodge conceived and built a prototype on-line voltage clamp-to-computer system for Hille, his successor at Rockefeller University. In final form (Hille, 1966, 1967) 2000 digital points at 50 μsec intervals for frog nodal membrane potential and current from each of two successive runs were reduced by various averages to 100 numbers and transferred to magnetic tape, which was read out in figure 4:19.

It is of particular interest that the electrical estimates of sodium gain are in good agreement with chemical analyses (Hurlbut, 1958) at $2.6 \cdot 10^{-17}$ mole/node impulse. Such a medullated axon has a

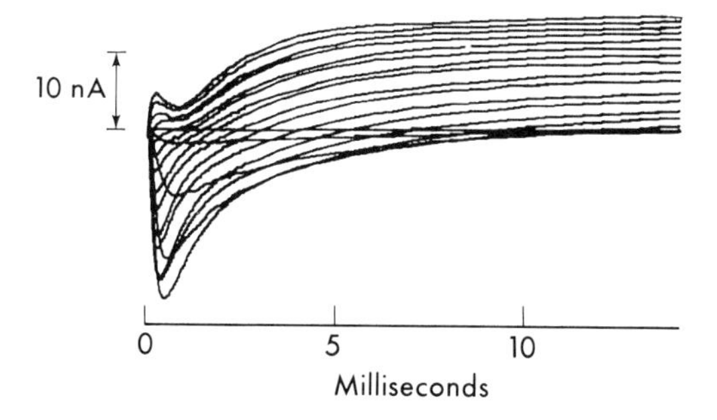

FIG. 4:19. Frog node voltage clamp currents, recorded, averaged, and presented to plotter in digital form, in Ringer's solution from −67.5 to 67.5 mV in 7.5 mV steps.

conduction velocity about the same as a squid axon at room temperature. This costs the squid axon a sodium for potassium exchange of about $6 \cdot 10^{-13}$ mole/cm impulse while the frog axon does it for only $1.3 \cdot 10^{-16}$ mole/cm impulse. So the new, improved model is far more economical in the transmission of an impulse—by a factor of 5000—and in its space requirement—by 1000 times. And this has been achieved by 200 to 300 layers that are suspiciously close in their electrical properties to the squid axon membrane!

Since the close similarity of the frog node to the squid axon has been established, we are in a position to look critically at the calculations of FitzHugh (1962) on the initiation of propagation in a medullated axon as realistic rather than hopeful and prophetic as they were when they were undertaken. As a result he showed that a uniformly propagating impulse could be initiated in such a theoretical axon, and he was able to conclude that the computed action current curves agree fairly accurately with the published experimental data from frog and toad fibers.

SUMMARY

The problem of understanding the nature of impulse conduction in a medullated axon has been rather apart, different, and considerably more difficult than that in the comparatively exotic—and perhaps eccentric—giant axon of the squid. But devotion and a considerable measure of genius has brought the problem to the level reached for the squid membrane. It has been demonstrated that the Lillie idea—along with those of Hermann, Bernstein, and

Overton—was essentially correct. Excitation is propagated node to node by the passive cable conduction of the internode. The passive, linear parameters of the internode as well as those of the node have been measured. Once again, the voltage clamp concept has been applied and, once again not without difficulties that required critical analysis and powerful solution, it has formed the basis of a description of the ionic permeability of the nodal membrane.

The conclusion that the node is qualitatively the same as the squid membrane and only an order of magnitude or so different quantitatively is impressive. The internode that is but a roll of a few hundred layers of membrane with the characteristics of a squid membrane calls for respect for the research and development which produced it. The conclusive demonstration that the amazing untilization of these few raw materials has been the basis of such an improved communication system—with a thousand fold less plant and even more reduction in operating expense—makes it even more imperative that we understand the fundamental principles that have been used in such an effective design and construction.

More Clamping Techniques and More Membranes

CHANGES OF SQUID AXON TECHNIQUE

Modifications and improvements must be expected as a continuing process. Adelman and Taylor (1964) found improved survival with the end ligatures and the axial electrode puncture of the axon covered with iso-osmotic sucrose. This was done in the chamber of figure 3:83 by an inflow of sucrose at both ends which shared the exit channels with the experimental solutions entering at the center. The micropipet potential electrode served very well for a decade, but it was a limiting component in the response time of the potential control feedback system. And there were considerable suspicions of damage to the axon at the point of puncture. These difficulties led Armstrong and Binstock (1964) to incorporate both the internal potential and current electrodes into the original 100 μ glass pipet, and the arrangement of figure 4:20 evolved. Although this introduces an undesirable axoplasm resistance in the control circuit, the total is not much worse than that of the troublesome, considerable resistance closely associated with the membrane capacity. In an apparently reluctant return to voltage clamping, the Cambridge–Plymouth axis continued with the vertical arrangement (Chandler and Meves, 1965). They merely cemented a 10 mm long platinized platinum current wire I to the 100 μ capillary internal potential electrode V (figure 4:21). They further measured the current

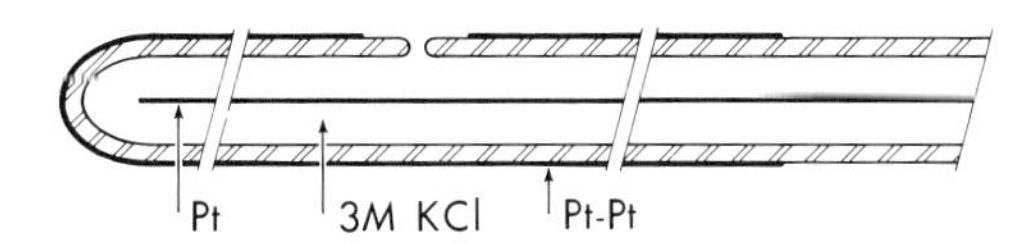

FIG. 4:20. Modified squid axial electrode. The outside coating on the 100 μ glass tube is the current electrode, the axoplasm potential is measured through the pore with a 3 M KCl calomel cell, and the internal wire lessens transient artifacts.

density by the potential ΔV between a pair of wire electrodes in the form of semicircles about 1 and 5 mm from the center of the axon and in the plane perpendicular to its axis. Although this arrangement appears to give a 1.7 mm separation of half maxima for a narrow stripe of axon there are no obvious indications of inadequate control. A further step may now be to use the reference electrode also as a local electrode for the measurement of current density with probably somewhat better resolution.

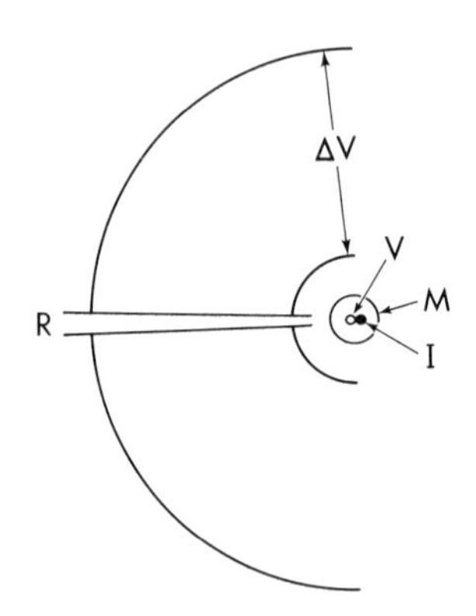

FIG. 4:21. Modified squid voltage clamp arrangement in horizontal section. Potential of membrane M is measured and controlled with an internal axial pipet V and an external reference pipet R. The clamp current is supplied by I and the density near the control point is measured by the potential between arc electrodes in the external solution.

A voltage clamped squid axon had been a very attractive preparation for measurements of radioactive tracer ion influx and outflux experiments (Mullins, Adelman, and Sjodin, 1962), but it became necessary to measure the axon activity during an experiment. This required a reasonable stretch of clamped axon but also precluded the use of the usual control microelectrode. For this purpose Adelman (1963) used an axial current wire from one end and a sucrose-guarded potential electrode at the other end. Calculations of this sugar-wire arrangement indicated that the accuracy of the potential measurement was quite limited, and experimental tests showed that the stability requirement was even more severe.

PSEUDO-NODES

Many of us had long been concerned with the unavoidable effects of an external current leakage path on outside measurements of resting and action potentials as, for example, in Cole and Curtis (1938). Here we had predicted a membrane potential

$$V_m(x) = (1 + r_i/r_e)V_e(x)$$

where $V_e(x)$ was measured outside and r_e and r_i were the uniform outside and inside longitudinal resistances. This uniformity was not easy either to achieve or to be certain of and the situation was impossibly complicated (Taylor, 1963) otherwise.

We had also been somewhat anxiously looking for a way to break our addiction to the squid axon. *Sepia* was not available in the Americas while *Carcinus* seemed too much of a come down in size. Although we were not at all sure how they might be useful, the 100 μ *Homarus* axons introduced by Tobias and Bryant (1955) were very interesting. Quite apart from the fact that we all ate lobster on every possible occasion, Dalton started work on this preparation. He found it very useful (1958, 1959), but we did no more about clamping than to despair of getting a 20 μ axial electrode into it. There were drooping troubles with a 100 μ wire and this would get worse as the fourth power of the diameter; also the buckling under an axial load was a new hazard.

Errors from leakage current flow over the surface of the myelin internodes of a single medullated axon had been controlled and minimized by Tasaki with his single and double air gaps, and then by the counter emf arrangement that had been used most effectively by Huxley and Stämpfli in clinching the saltatory explanation of conduction in these axons. Stämpfli (1954) turned very successfully to the use of iso-osmotic sucrose to replace the air gaps—which were reputed to reach their best performance only when Tasaki himself fanned them. Sucrose was used under Vaseline in our laboratory without apparent anomalies (Whitcomb and Friess, 1960). The double gap insulation was reminiscent of Hodgkin's old (1938) favorite three electrode technique for the *Carcinus* fiber, also used recently by Wright et al. (1955). Here the center of the axon dipped into saline solution and regions on both sides were raised into the oil above by the forceps holding the ends. But, as many have found, the longitudinal conductance of the saline solution trapped in between the oil and the axon membrane was comparable to that of the axoplasm.

Wright with Tomita (1962, 1965) isolated 100 μ gaps of various blue crab and lobster axons between segments which were washed with sucrose before they were put under oil. Long survival times without evidence of abnormal polarization were obtained.

LOBSTER AXON

Against this background, it should have been completely obvious that almost any "smooth" axon could be "medullated" with sucrose or electronically. And our friends across the street, Julian and Goldman, reported on the sucrose procedure for the lobster axon in 1960 and on its use in their investigation of mechanical stimulation in 1962. With our collaboration, they first made a full report on the action potentials, and then on their voltage clamp results with the arrangement of figure 4:22 (Julian, Moore, and Goldman, 1962a,b).

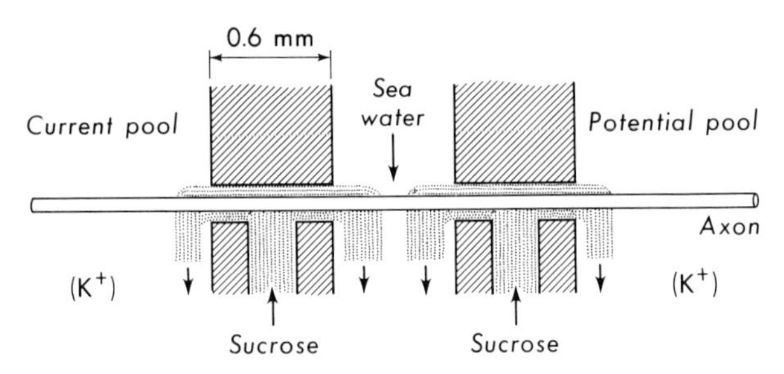

FIG. 4:22. Sucrose gap technique for clamp of lobster axon without internal electrodes. The flowing sucrose limits the active membrane area, center, and allows current application left, and potential measurement, right, through inactivated regions.

I had been somewhat apprehensive of the effect of external sucrose on an internode, in spite of Beutner's old evidence that ions did not cross apple skin into outside distilled water, and Frankenhaeuser (1957) rather obviously preferred to use his electronic "insulation." But I was not prepared for the resting and action potentials of the lobster axon that were 20–60 mV higher with sucrose gaps than for a microtip or for the evidence that the sucrose was responsible.

Taylor and Adelman (unpublished) explored the potentials across and near the boundaries of the three-phase system, sea water–squid axon–sucrose. A large potential across the sea water–sucrose interface and a high gradient of potential along the surface of the axon in the sucrose near the sea water interface strongly suggest an axon hyperpolarization by large liquid junction potentials as shown by Blaustein and Goldman (1966). Although this sucrose effect needs to be better understood the results otherwise seem highly satisfactory. The resting membrane resistances were quite variable about 1 $k\Omega cm^2$. The resting potential followed the Nernst relation for K_e changes from 10–500 mM and was zero at the latter value corresponding to the expected K_i. The impedance during excitation, however, returned to the resting value earlier than for the squid membrane, figure 4:23.

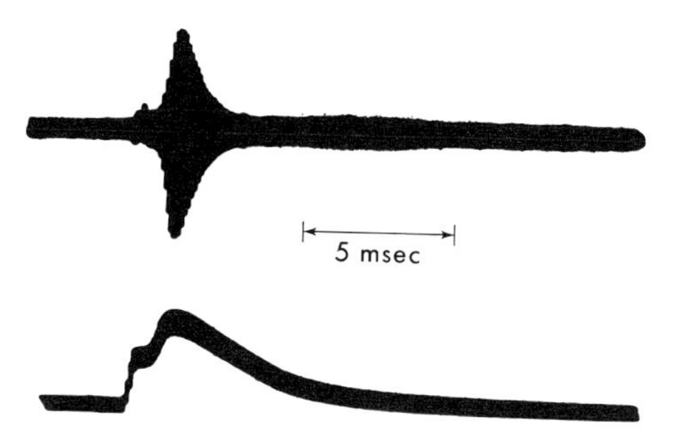

FIG. 4:23. Lobster axon impedance change at 8 kc, above, and action potential, below, in sucrose gap.

It was nicely shown (figure 4:24a) in a voltage clamp that the bizarre current patterns with a low control amplification became conventional—dare I say orthodox?—as the amplification was increased, as had been shown for the frog node and squid axon (Moore and del Castillo, 1959; Cole and Moore, 1960a). Also, the pattern remained constant for narrow saline gaps but deteriorated rapidly as they were made more than 75 μ wide (figure 4:24b). This certainly seems good presumptive evidence of adequate clamping but I felt rather uncomfortable. Some additional experimental or theoretical work to show the accuracy of the measured membrane potential, the variations of potential over the measuring region, and the conditions for the onset of instability should be reassuring. Nonetheless, these voltage clamp results, including those of figure 4:25, are very similar to those for the squid and frog and toad node membranes. It can only be expected that detailed analysis would not show the lobster membrane to differ in any important way.

SQUID AXON

The effect of external sucrose on the squid axon was as marked and as unknown as it was for the lobster axon but the simplicity of experiment and the absence of obvious or mechanical insult to

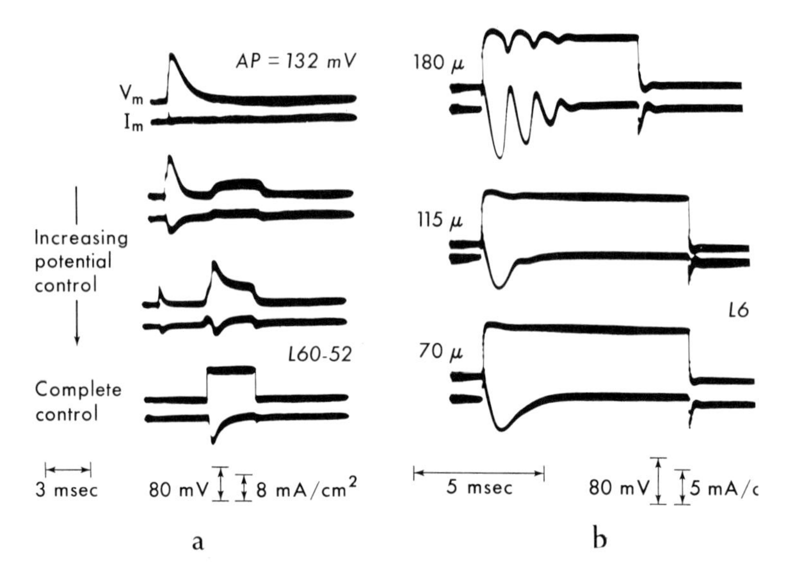

FIG. 4:24. (a.) Effect of potential control on lobster axon potential and current. In all records an adequate, short, current stimulus was first applied. Without potential control, top, this gave an action potential. Next lower record, with a weak control, the potential and current interacted to seriously modify both this response and a later potential command. Under stronger control, next to bottom, the stimulus response was reduced but the potential command was not strictly enforced. When the potential followed the command, bottom, the characteristic clamp current response appeared. (b.) Voltage clamp behavior for the different lengths, at left, of lobster axon in gap between sucrose interfaces. The axon diameter was 125 μ. For a length about half the circumference, the potential control was not adequate and the current oscillated vigorously. Some of this behavior still appeared at a length about equal to the diameter, center. The potential control was nearly complete with an even shorter exposed axon, bottom, and the current became typical.

the axon made it a very attractive approach. However, it seems not to have been tried until Guttman (1962). The threshold behavior of a space-clamped axon as a function of temperature was measured for long stimuli, rheobase, and extended to short times (Guttman, 1966). The results were found to be in general agreement with early data and predictions of the HH axon (Cole, 1955) and extensions by FitzHugh (1966b). The procedure was then used for voltage clamping of the squid membrane in an investigation of the action of tetrodotoxin by Narahashi et al. (1964). In these investigations there was no test of the accuracy of the potential measurement, the uniformity of the membrane current density, or of stability, and for these reliance had to be placed on calculations. On the other hand, Guttman (1966) found that use of the apparent membrane area gave unreasonably high values for the membrane capacity, and she estimated the effective area from the measured capacity!

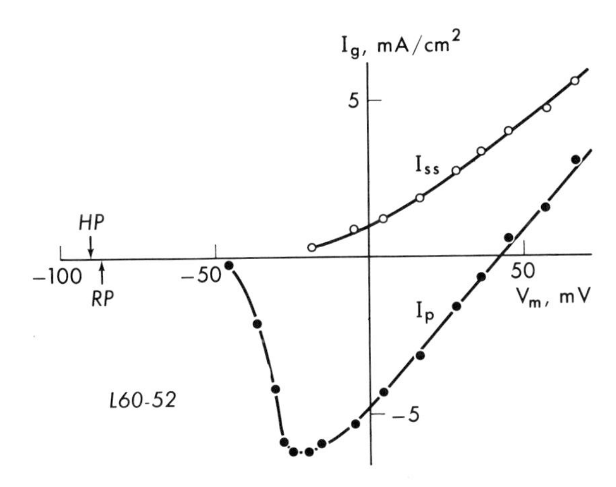

FIG. 4:25. Lobster axon membrane current densities in voltage clamps. The solid circles are for the peak early currents and the open circles are in steady state.

MUSCLE FIBERS

A sodium dependence had been found for the action potential (Nastuk and Hodgkin, 1950) but there were differences from nerve membranes. Single skeletal muscle fibers came into increasingly wide use following the work of Ramsey and Street (1940). Many attempts were made to voltage clamp these fibers but they were mostly unsuccessful and unpublished, partly because the fibers do not like sucrose. A careful and thoroughgoing investigation was undertaken by Frankenhaeuser based on an Israeli frog, because of easy dissection, and his electronic insulation which allowed normal external media for the fibers. He spoke privately of some success and in a first installment (Frankenhaeuser et al., 1966) the action potentials were measured and it was estimated that voltage clamping was feasible.

TWO-NEEDLE CLAMPS

Spherical cells and approximations

Central cells

As the use of the intracellular microelectrode spread over a wide range of preparations it was often almost as easy to insert two electrodes as one, and control of current or potential became possible. In this way Hagiwara and Tasaki (1958) voltage clamped the squid stellate ganglion. Hagiwara and Saito (1959a,b) found that the voltage clamp currents in central cells of the puffer fish and a mollusk were similar to those of the squid membrane and were to be interpreted similarly in terms of sodium and potassium components. Then Tauc (1955), and with Frank (1964) took further advantage of the potentialities of *Aplysia.*

However, things were not so easy in higher animals. Eccles with his collaborators (Brock et al., 1952) achieved a major advance when they simplified the problems of the central nervous system by intracellular recording with a microcapillary electrode. From this, increasingly powerful attacks were made on motor horn cells with various electrophoretic injections near the soma from needles with up to seven barrels (Salmoiraghi and Bloom, 1964; Curtis, 1964). A single microtip in a bridge circuit worked amazingly well as a current clamp (Araki and Otani, 1955; Frank and Fuortes, 1956). The simple double-barreled electrodes were obviously all that was necessary for direct current or voltage clamping. Terzuolo and Araki (1959; Araki and Terzuolo, 1962) started voltage clamping with side by side electrodes as Frank et al. (1959) placed the potential capillary inside the one for current. There was capacitive coupling—cross talk—between these channels, and it was nicely neutralized electronically (Frank et al., 1959).

Purkinje fibers

Along with peripheral nerve and skeletal muscle, cardiac muscle has been a major field of electrophysiology. But following Einthoven, cardiology came to depend upon the electrocardiogram as a principal means of diagnosis; this was for long the only important use of bioelectricity in medicine. I was introduced to the problems of understanding the electrocardiogram by H. B. Williams, my chief at Columbia. Then through Williams, I began a long and inspiring tutorial under Frank N. Wilson of Michigan. As head of the Heart Station, Wilson had taught himself calculus and potential theory. He set up and solved many problems (Wilson et al., 1933) of current flow and potential distribution in the heart and the rest of the body of the kind that are still being worked on. He always asked me many and more searching questions than I could possibly answer. Many of these questions he answered himself and, eventually realizing my limitations, he sent me Gardner and Barnes (1942) soon after publication—because I should learn to use the Laplace transform—and then von Neumann and Morgenstern (1944)—because it was such exciting new mathematics! Wilson made many contributions to the practice of medicine, and many investigators and clinicians all over the world became his devoted disciples, as seen in the symposium arranged by one of them, H. H. Hecht (1957). But as far as I know he did not hear of the microelectrode intracellular potentials from heart muscle nor of the flows of sodium and potassium in and out of the squid axon. I wish I could have told Wilson of these new things and tried to answer all the questions he would have asked.

When I challenged the cardiologists (Cole, 1957) to give an ionic explanation of their membrane potentials, Walter Woodbury promised to use a voltage clamp when I would tell him how! Then apparently Deck et al. (1964) as well as Hecht et al. (1963) succeeded with the Purkinje fibers, and Deck and Trautwein (1964) gave some analysis of the ion flows.

Here a two-needle clamp was used on a short 2 mm region between two ligatures. The reported currents are apparently free from effects of nonuniformity and are in reasonable accord with the many observations on unclamped preparations. The membrane characteristics returned to normal in the vicinities of the ties in 20 percent of the preparations, and tests of the uniformity of membrane potential were satisfactory over most of the potential range. So their assumption of an equivalent spherical cell is probably valid. Hecht et al. (1964) have also been successful with a short segment between healed-over cut ends, and Trautwein et al. (1965) merely pinched off a section with fine forceps.

Such treatments of a tissue seem thoroughly insulting, to say the least. However, Hodgkin and I (1939) found indications that the squid axon membrane resistance at and near a ligature was high and that the membrane was not severely injured in this region.

Also, Guttman (1941) obtained electrical evidence for the formation of a membrane of sorts after a transverse cut of frog sartorius muscles. Deck et al. judged 20 percent of their ligated preparations to have "healed over" well enough to give significant results, although less than 10 percent of all the preparations gave normal action potentials. They found average resistances of 70 kΩ before ligation and 210 kΩ after ligatures were placed an average of 1.7 mm apart. For the first measurement we have, from cable theory,

$$R_\infty = \frac{1}{2\pi a}\sqrt{\frac{r_3 r_2}{2a}} \text{ and } \lambda = \sqrt{\frac{a r_3}{2 r_2}}$$

and for the second, assuming uniform current density

$$R_1 = r_3/2\pi a l$$

where r_3 and r_2 are membrane and myoplasm resistivities in Ωcm^2 and Ωcm respectively, a is the effective fiber radius, and l the separation of the ligatures. Then $R_1/R_\infty = 2\lambda/l$, and for the nominal $\lambda = 2.0$ mm (Weidmann, 1952), this ratio of 2.35 is comparable to the experimental average of 3.7.

The performance of the control equipment was somewhat limited. It could charge the membrane capacity in 2–3 msec through about a 10 MΩ microelectrode but then it was unable to clamp the fiber membrane for the first 10 msec, presumably because of its low resistance in this period.

As hoped and expected, the currents under voltage clamp were

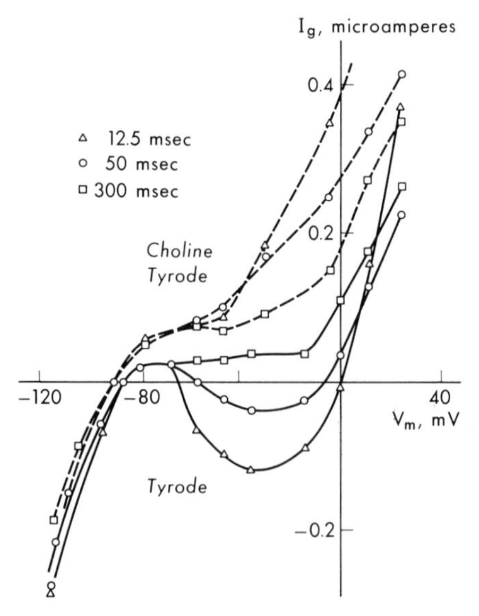

FIG. 4:26. Purkinje fiber preparation in voltage clamp. The results are presented as currents I_g for a membrane area of 0.87 mm² at fixed times, shown by different symbols, after change of membrane potential to V_m. These isochronal points are connected by solid lines for Tyrode's solution and by dashed lines after choline substitution for sodium.

both similar to those of nerve membranes and also significantly different from them, figure 4:26. At 12.5 msec the inward sodium current shows a quasi steady negative resistance of about $-15\ \Omega\text{cm}^2$. With internal and external resistivities of 100 Ωcm and an equivalent sphere radius of 0.3 mm, this resistance is safely above the critical value given by equation 4:7 as $-4.5\ \Omega\text{cm}^2$, and stability is to be expected.

Cylindrical cells

The solutions of potential and current distributions in cylindrical cells with positive membrane resistances had always seemed possible. The wide variety of actual combinations of internal and external electrode arrangements as well as variations of the membrane resistances often made the solutions so complicated as to require considerable simplifying assumptions or analogue solutions. With a negative membrane resistance, the boundary conditions at and near the ends of internal and external electrodes were not satisfactorily resolved, but were avoided by the axon and electrode terminations already discussed.

NITELLA

There had been a personal report (Nastuk, unpublished) of internal resting and action potential measurements made from one end of a *Nitella* cell. But it had not been possible to maintain chronic electrodes in the cell sap for either potential measurement or current supply so that the squid axon voltage clamp arrangement was not a promising possibility. However, the micropipet electrode had been nicely used to show—and confirm—that the internal protoplasmic interface supported a passive potential with activity entirely located at the external surface of the protoplasm (Walker, 1955; Findlay, 1959; Kishimoto, 1959). Building on this, Hope (1961) and Findlay (1961) reported two-needle voltage clamp experiments on *Nitella* and *Chara* in which they assumed that the clamp was adequate and uniform over the cell surface because of the high conductivity of the cell sap. This assumption was quite reasonable for a resting membrane and the consequent long space constant. But, for considerable depolarizations, the reduced membrane resistance made the assumption seem questionable. The still unresolved and potentially more dangerous situation with a negative membrane resistance, as well as the availability of the squid measuring chambers with a short, external current measuring electrode, led Kishimoto (1964, 1965) to use a two-needle arrangement with a local current density measurement on *Nitella*, figure 4:27. There was no obvious evidence of an overt instability in the current records, but the inward current increased so rapidly with increasingly depolarizing pulses as to approximate a threshold and produce at most a nearly negligible negative resistance in this region, figure 4:28. If, indeed, this arrangement gave results with any uncertainty of interpretation, it seemed quite difficult to place as much confidence in current measurements which included membrane regions even more remote from the control electrode.

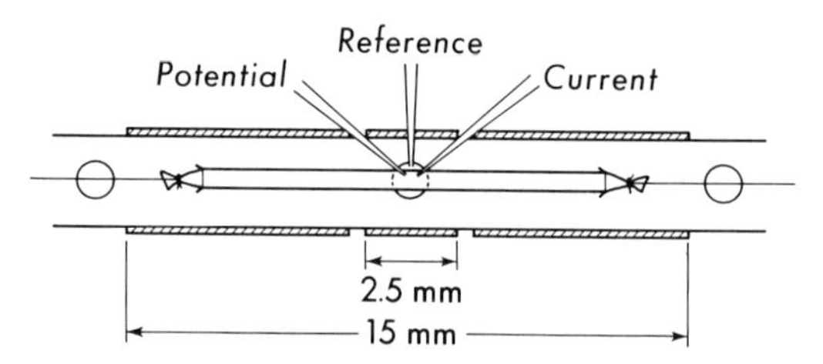

FIG. 4:27. Trough voltage clamp arrangement for *Nitella*. The potential between an internal micropipet and external reference electrodes was controlled by current through yet another micropipet in the central region. Current in the central region was measured through the short external electrodes.

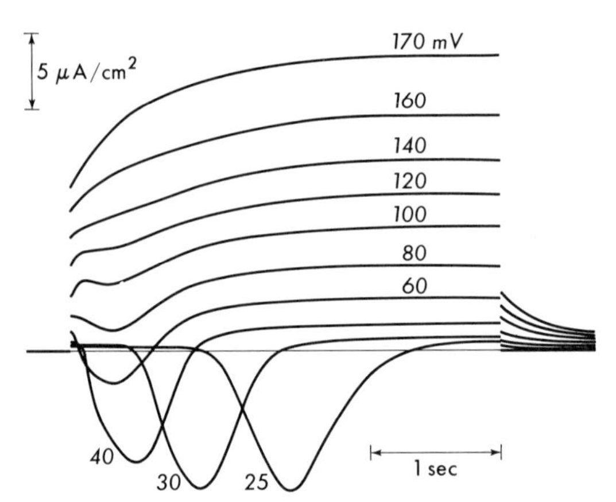

FIG. 4:28. Current densities for *Nitella* after clamps to potentials above rest as designated.

Since the restricted current measurement did not show any evidence of contributions from propagating action potentials in one or the other or both directions, there was at least a chance that this simple arrangement might be used with axons that were considerably more powerful and difficult to control. A single experiment with a squid axon was so completely discouraging that it was not repeated or tried on any other axon. But it would have been nice if it had seemed to work—although proof of the adequacy of control might have been a considerable chore that then could not have been ignored!

Then Adrian et al. (1966) clamped a sartorius muscle fiber near one end by applying feedback to the three electrode arrangements used by Adrian and Freygang (1962). As can be shown near this natural termination there can be good spatial uniformity with either a positive or a negative conductance.

OTHERS

Barnacle muscle

Nature has been very versatile in providing extreme variations of structure to meet special demands for function, and at the same time very generous to investigators in helping them use their very limited facilities as effectively as possible. Another spectacular example of this appeared in the giant muscle fibers of a barnacle, 1 to 2 mm in diameter and 4 to 5 cm long. With an internal capillary electrode inserted into a cut end and externally insulated with oil, Hagiwara et al. (1964) found that the resting potential behaves mostly as a potassium electrode. Elimination of internal calcium by perfusion changed the characteristic graded response of crustacean muscles to the all-or-none response typical of propagating axons. The action potential was, however, independent of Na_e, and changed by 29 mV per decade change of Ca_e (Hagiwara, 1965). This and an increased Ca^{45} influx of 50–100 p mole/cm^2 per impulse along with activity in external barium and strontium strongly supported the conclusion of Fatt and Ginsborg (1958) and of Parnas and Abbott (1964) that the response in several crustacean muscles is an increase of permeability to these divalent ions.

Hagiwara et al. (1964) also reported briefly on potassium currents in a voltage clamp and, presumably with adequate control, found a quasi steady state negative conductance for calcium.

The alternative experimental arrangement used by Hagiwara et al. (1964) is the spectacular and delightfully direct attack that many of us have thought of—but only as pure fantasy—for the squid axon. They simply cut out a centimeter length of the fiber, slit it lengthwise, and spread it as in a sandwich between two thin glass plates with opposing one mm holes. The membrane was still functional and comparable to the product of the less brutal procedure but, most unfortunately, adhering internal material and the usual slow deterioration prevented it from becoming the method of choice.

Frog skin and toad bladder

The time-honored frog skin is becoming a much more interesting and useful preparation as more recent concepts and techniques are being applied to it. Steady state electrical and isotopic measurements have been used extensively to investigate the sodium pump (Ussing, 1952). Along with the somewhat simpler toad bladder, it is found that the epithelial, "exterior," surfaces are more permeable to sodium than to other ions (Koefoed-Johnsen and Ussing, 1958). The bladder emf and resistance are about equally divided between the two faces of the cells in this layer (Frazier, 1962) whereas about half of the emf is reported to be at the inner surface of the frog skin (Lindemann, 1965). At least the cells of the outer layer of the frog skin are so tightly fused together (Choi, 1963) that they may be virtually equivalent electrically to a single membrane covering the surface. And indeed Finkelstein (1964) reported skin capacities of 2–5 $\mu F/cm^2$ and resistances of 0.2–1.0 $k\Omega cm^2$ which I confirmed from earlier data (Cole, 1932)! The capacities of the bladder were in the same range, with resistances of 0.75–2.5 $k\Omega cm^2$ where Loewenstein et al. (1965) showed strings of electrically connected cells. Finkelstein first found an action potential response for frog skin in external lithium (1961) and then for frog skin and toad bladder in external sodium (1964). During these responses the skin was approximately balanced, in a bridge at 1 kc, by an increase of resistance without change of the parallel capacity. It seems possible from earlier data (Cole, 1932) that these measurements represent an infinite frequency approximation to a linear resistance, but the demonstrations of thresholds with constant current strongly suggest at least a quasi steady state negative resistance. Lindemann (1965) has shown an essentially steady state negative resistance under voltage clamp, although the time resolution of the clamp and the significance of inside vs. outside surfaces make direct comparisons with other excitable membranes difficult. There is no information yet about the space clamp stability, but excitation in the frog skin seems certainly to involve the rapid increase of something called a resistance—resulting in a near steady state negative resistance region—probably in a single membrane.

Electroplax

The electric organ of the electric eel provides a single cell preparation which Grundfest and his collaborators have been investigating. The single electroplax is a disk about 60 μ thick, δ, with an internal specific resistance r_2 of 100 Ωcm^2. One face is completely passive with a low effective membrane resistance $1/g = 0.1\ \Omega\text{cm}^2$. The opposite active face has the usual resting potential of -85 mV and resistance of 4–5 Ωcm^2. Normally the synaptic endings on this face depolarize it to contribute to the spectacular organ discharge, but after these endings are inactivated with curare the plax displays the conventional threshold and action potential of nerve membranes with a stimulating current. A microelectrode was placed just inside the active membrane. Its potential and the voltage clamp currents were again quite similar to those of axon membranes except that the steady state rectification was anomalous (Nakamura et al., 1965). The peak of the transient inward current gave a quasi steady state negative resistance, $-1/g_3$, averaging $-1\ \Omega\text{cm}^2$.

Hille et al. (1965) gave a preliminary report on skate electroplax. With reduced chloride, this preparation showed a threshold and spike potential. This persisted in an attempted voltage clamp presumably because of spatial instability.

Quasi and sundry clamps

It is axiomatic that the smaller the effective resistance of a potential source applied to a cell membrane the faster will be the initial charging of the membrane capacity and the more nearly will the subsequent potential approach that of a perfect clamp. The membranes of the cells in a suspension or a tissue may be rather effectively clamped by an externally applied potential because of the relatively low resistivities of the usual external and internal electrolytes, say r_1. For example, the membrane charging time τ_2 in a suspension of cells, radius a, is about

$$\tau_2/\tau_1 = ar_1/r_3$$

where $\tau_1 = r_3c_3$ is the membrane time constant. This is, however, of little use since the potential changes will be in opposite directions on the two sides of each cell.

The same principle underlies the arrangement developed by Strickholm (1961), who placed a pipet with a smooth tip about 100 μ in diameter against a muscle fiber membrane. A current flow through a relatively small leakage resistance R_p between the pipet and the membrane can produce a nearly constant potential V across the membrane, figure 4:29. With a constant current I_0 applied through the pipet resistance R_S we have from the potential V_0 at this point,

$$V = V_0 - R_S I_0$$

and the membrane current

$$I = (1 + R_S/R_p)I_0 + V_0/R_p.$$

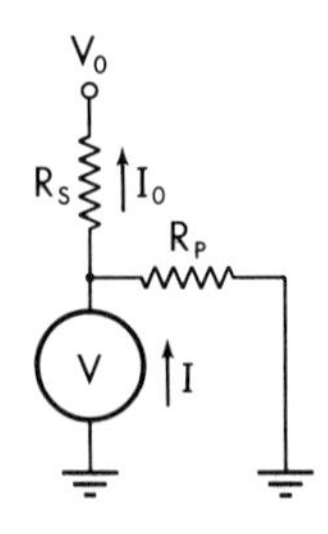

FIG. 4:29. Equivalent circuit for investigation of membrane characteristics with an external capillary electrode to give a quasi voltage clamp.

Such clamp currents showed no outward steady state nor a reversal of the early phases in the frog sartorius fibers and are not yet explained. A crab muscle fiber, however, conformed in general to the orthodoxy of the squid (Strickholm, 1963). The stability of this system may be difficult to predict and it has not yet been possible to use it on a lobster axon.

There is no place within the framework of adequate voltage clamps, as we have defined them, for arrangements and procedures which are obviously and even powerfully unstable. The two-needle clamp which Weidmann (1955) applied to the Purkinje preparation probably controlled the membrane potential reasonably well in the immediate vicinity of the potential electrode, but the membrane current density could only be expected to vary widely between this position and the more distant points. I have no idea whether or not such data can be analyzed practically as we did for the squid axon or if short external electrodes would be a solution. Nonetheless, this preliminary clamp was an interesting and helpful experiment. Patience with imagination and skill may yield useful additions to the understanding of some such situations.

Although I have strongly preferred to apply power to control a system, I cannot but admire a subtle approach such as was used by Bennett and Grundfest (1966) on the electric organ. Here they applied two steps of current to the membrane, the first to bring the potential into the unstable range and the second to reach the vicinity of an unstable point and find the direction of departure from this point. As shown in figure 4:30, enough of such data well locates both the critical points and the negative resistance characteristic.

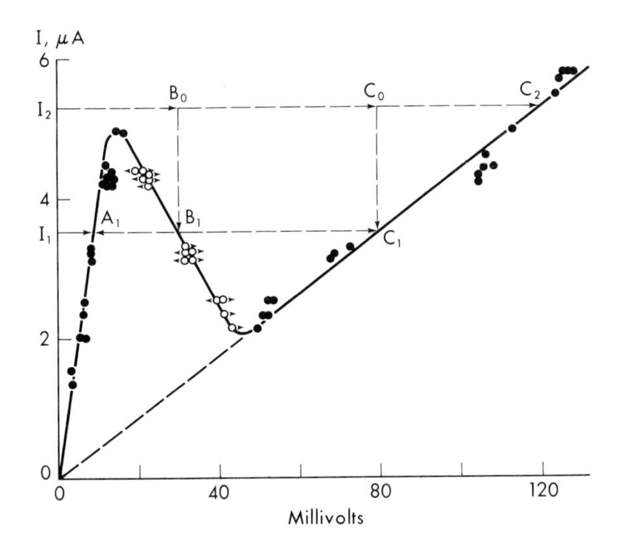

FIG. 4:30. Characteristic of a depolarized electroplax membrane in current clamp. Solid circles are steady state points. Application of I_1 gave points A_1 up to maximum of limb A and points C_2 on limb C for larger current I_2. The initial application of I_2 until C_0 was reached followed by I_1 gave C_1 and so the rest of the C limb. A shorter application of I_2, to point B_0, followed the change to I_1, leads to a point such as B_1 on limb B. According as the point is on one side or the other of this unstable limb it goes to stable points A_1 or C_1 as indicated by the open circles and the directions of take off.

I have often been asked about the advantages and disadvantages of control waveforms other than steps of potential. Probably the question can only be effectively answered in terms of the results desired and obtained with the various possibilities. My original aim was only to look for the stability to be predicted at constant potential for an N type negative resistance characteristic and to prevent the participation of a capacitive current. A potential step was the obvious choice, and it has been so successful that little consideration has been given to other possibilities.

A ramp of potential changing linearly with time has been found useful, particularly at low speeds. Here the capacitive current is constant and may be negligible and near to steady state current—potential characteristics are most readily obtained. An example was shown for the squid membrane (Cole and Moore, 1960a). Although more difficult to interpret, higher speed ramps have also been used for an electroplax (Bennett and Grundfest, 1966) and Purkinje fiber (Trautwein et al., 1965). In the latter case, the slow ramp showed an unexplained hysteresis.

Our long experience had impressed us with the power of sinusoidal analysis. This was tried in 1947 under current control. It was apparent, however, that such precision was not immediately needed and that the cost in axon survival time was prohibitive. With further experience and information, a limited application of alternating current in combination with a step potential shows considerable promise (Cole and Moore, 1960a).

INSTRUMENTATION

Although our review of voltage clamp techniques (Moore and Cole, 1963) was somewhat out of date by the time it appeared, there have been no radical changes in the objectives or design of the electronics which made the voltage clamp possible. There has, however, been a gradual increase in the use of solid state components as they have become more available and more reliable. Combinations of computer techniques and computer use were inevitable, with Hille (1966, 1967) showing the way, figure 4:19.

SUMMARY

Following the work on medullated axons, lobster and squid axons were similarly voltage clamped with sucrose for pseudo-internodes. Intracellular microelectrodes for current and potential gave useful clamps for near spherical central cells and tied-off Purkinje fibers. *Nitella*, electric organs, frog skin, and barnacle muscle were also clamped with various techniques. These all showed an early transient inward current which could be expressed as a quasi steady state negative resistance. This was usually sodium, but in some cases chloride or calcium. The delayed outward current was again usually potassium. There were, however, unique features in almost every experiment, some certainly characteristic of the membrane, others perhaps reflections of technique.

Some Theory and Calculation on Clamps

The several new and different techniques for the application of the voltage clamp concept have each had their advantages and their limitations. As each new arrangement has come along the first—and often major—problem has been to get it working. It would have been helpful to have had some idea of the requirements for uniformity of potential and current, accuracy of their measurement, and stability of the system. This information is all the more important when data are in hand, particularly so if the data have a novelty that may be a "discovery." Such precautions are not easy to take. Practical limitations with the help of experience and intuition may completely dictate initial design. Experimental tests are usually difficult and often impossible. Theory and calculations are always incomplete, uncertain and tedious. And besides, there are always so many more experiments to be done. However, there are some primitive approaches that have been helpful, and there are also some obvious omissions of entirely unknown importance.

NODES AND PSEUDO-NODES

An axon is here considered, as in figure 4:31, completely insulated so that there is no current flow along the outside or across the surface membrane—except for a central band of width δ. A current flow I_0 is introduced at one end and a potential V_0 is measured at the other end.

The simplest approach is that of the uniform cable with an axoplasm resistance r_i Ω/cm, and a membrane resistance r_m Ωcm. Then, equation 1:34,

$$\frac{d^2V}{dx^2} - \frac{V}{\lambda^2} = 0; \qquad \lambda^2 = r_m/r_i$$

and from equation 1:37

$$V = A \cosh x/\lambda + B \sinh x/\lambda.$$

In order that no current flow beyond the node, let

$$dV/dx = 0 \text{ at } x = \delta$$

and

$$V = V_0 \cosh (\delta - x)/\lambda.$$

The maximum potential error, at $x = 0$, is

$$\frac{V(0) - V_0}{V_0} = \cosh \delta/\lambda - 1 \approx \delta^2/2!\lambda^2.$$

The total current is

$$I_0 = \int_0^\delta i_m \, dx = \frac{V_m}{r_m} \sinh \delta/\lambda \approx \frac{V_0}{r_m} \delta(1 + \delta^2/3!\lambda^2).$$

For a potential variation of 5 percent,

$$\left(\frac{\delta}{\lambda}\right)^2 = \frac{r_2}{r_3} \cdot \frac{2\delta^2}{a} < 0.1$$

or

$$\delta < 0.25 \sqrt{ar_3/r_2}$$

and, as a particular example, for nominal $r_2 = 50$ Ωcm and $r_3 =$ 1000 Ωcm² at rest,

$$\delta < \sqrt{a}.$$

Then for squid $\delta < 0.16$ cm, lobster $\delta < 0.07$ cm, *Xenopus* $\delta <$ 0.03 cm, which are all reasonable and practical. In the active region we take a nominal $r_3 = 2.5$ Ωcm² and here $\delta < 0.05\sqrt{a}$. For squid $\delta < 80\ \mu$, lobster $\delta < 35\ \mu$, *Xenopus* $\delta < 15\ \mu$, which indicates a need for caution for the larger axons and better analysis where $\delta < a$ and the radial current flow becomes important.

The excited membrane region with the negative resistance $-r_m$ is described by equation 3:10

$$\frac{d^2V}{dx^2} + \omega^2 V = 0, \qquad \omega^2 = r_i/r_m$$

and

$$V = C \cos \omega x + D = V_0 \cos \omega(\delta - x).$$

The potential error at $x = 0$ is

$$\frac{V(0) - V_0}{V_0} = \cos \omega\delta - 1 \approx -\omega^2\delta^2/2!.$$

The current is

$$I_0 = \frac{V_0}{\omega r_m} \sin \omega\delta \approx \frac{V_0}{r_m} \delta(1 - \omega^2\delta^2/3!).$$

Choosing a nominal $r_3 = -2.5\ \Omega\text{cm}^2$ and 5 percent potential variation, we have the same limitations on δ as for the active region and the same need for more complete analysis. At least with such restricted regions the feedback requirements for stability seem conventional but neither these nor more extensive lengths of exposed membrane have been considered with any care.

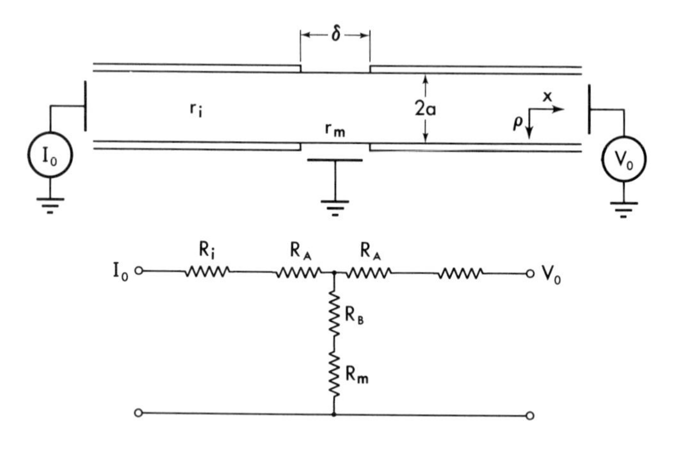

FIG. 4:31. Above, geometry of nodes and pseudo nodes, center, with current applied at left end and potential measured at right. Below, equivalent circuit for a short node with the distribution of current and potential in the axoplasm given by elements R_A and R_B.

It has been found before that the cable equation becomes an uncertain approximation for $\lambda < 2a$. The same difficulty was encountered long before in the design of electrodes for the buret cell used in the sea urchin egg impedance measurements but only resolved intuitively. Now we turn again to the Laplace equation for the cylinder,

$$\frac{\partial^2 V}{\partial \rho^2} + \frac{1}{\rho} \cdot \frac{\partial V}{\partial \rho} + \frac{\partial^2 V}{\partial x^2} = 0.$$

The appropriate solutions are of the form

$$V = AI_0(\gamma\rho) \sin \gamma x$$

where I_0 is a Bessel function and A and γ are to be determined.

Our first question was that of the effect of radial current flow giving a difference between the potential at the inner surface of the nodal membrane and the measured potential V_0 for *Xenopus*. In the problem worked out by Chandler (unpublished), these solutions were combined by integration to meet the condition that the current density $\bar{I}$ be uniform over a small node width δ,

$$\bar{I} = \frac{1}{r_2}\left(\frac{\partial V}{\partial \rho}\right)_{\rho=a}.$$

The first approximation for the Bessel functions could be interpreted to give $R_A = r_i\delta/2$ (figure 4:31), which might have been guessed. Then a second approximation gave the potential V_0 at $\rho = a/\sqrt{2}$ and the difference from V_a at $\rho = a$ gave R_B which was expressed as

$$R_B = ar_2/4 \ \Omega\text{cm}^2.$$

With the same nominal $r_2 = 50$ Ωcm this radial contribution to the apparent membrane resistance is, for squid, $R_B \approx 0.3$ Ωcm²; lobster, $R_B \approx 0.06$ Ωcm²; and *Xenopus*, $R_B \approx 0.01$ Ωcm². It is completely negligible for the node and lobster and no more than 10 percent for squid.

As we extend the length of the pseudo-node to obtain more, and more easily measured, current the problem and its solution are considerably more complicated. Perhaps they can be calculated by superposition, although this has not been tried. Further approximations to reality using resting, active, and negative membrane resistances promised so much difficulty of analysis and calculation that I turned to the resistor board analogue.

RESISTOR ANALOGUES

The resistor analogue is quite versatile and convenient and has been widely used in the solution of potential theory problems (Liebmann, 1953; Soroka, 1954). In a steady state a linear array approximates the cable equation, and a square, cylindrical, or cubic array may be used as the approximations to the corresponding two- and three-dimensional Laplace equations. Various boundaries and the Poisson equation have been used and, with condensers, transients may be obtained in real time.

For the current at 0 in the circuit of figure 4:32 to be zero,

$$\frac{V_P - V_0}{R_1} + \frac{V_Q - V_0}{R_2} + \frac{V_R - V_0}{R_3} + \frac{V_S - V_0}{R_4} = 0 \qquad (4:5)$$

and for all R equal this is the approximation to the Laplace equation

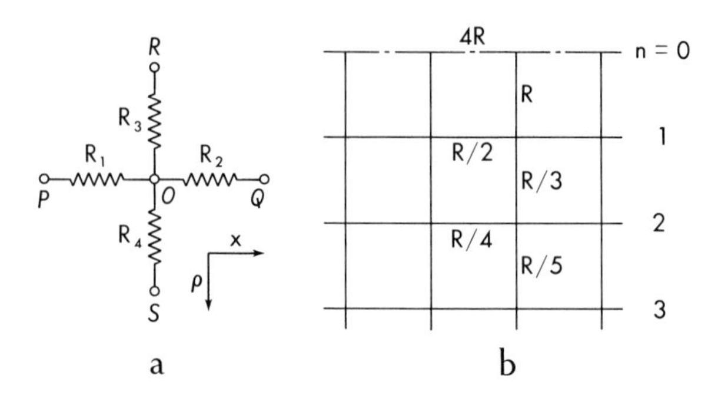

FIG. 4:32. (a.) Resistors at a junction 0 to give finite difference approximation to Laplace equation. (b.) Resistor analogue network for cylindrical potential distributions symmetrical about the dashed axis.

in a plane. In cylindrical coordinates with symmetry about the x axis we need (de Packh, 1947)

$$\frac{V_P - V_0}{a} + \frac{V_Q - V_0}{a} + \left(\frac{1}{a} - \frac{1}{2r}\right)(V_R - V_0)$$
$$+ \left(\frac{1}{a} + \frac{1}{2r}\right)(V_S - V_0) = 0 \qquad (4:6)$$

where a and r are the finite difference units in the x and ρ directions. We will have a solution when we equate the coefficients of equation 4:6 to the resistors in equation 4:5 except at $\rho = 0$. Then at $\rho = na$ for $r = a$,

$$\frac{1}{R_1} : \frac{1}{R_2} : \frac{1}{R_3} : \frac{1}{R_4} = 2n:2n:(2n - 1):(2n + 1).$$

If $R_{0,1} = R$, the radial resistors decrease by the factors 1, $\frac{1}{3}$, $\frac{1}{5}$ . . . and the longitudinals by $\frac{1}{2}$, $\frac{1}{4}$, $\frac{1}{6}$ . . . at distances $n = 1, 2, 3. \ldots$ At $n = 0$ we use symmetry to find the $R_0 = 4R$ for a factor of 4.

The first check—against the approximate theory—gave the equivalent membrane resistance for a 100 μ pseudo-node of a squid axon to be 0.14 Ωcm^2 (figure 4:33). This difference from the value of 0.19 Ωcm^2 for a very narrow node is in the expected direction.

A calculation for a 2 Ωcm^2 node 400 μ wide along a squid axon gave almost a two to one variation of current density, with the potential lead corresponding to the membrane potential at the center of the node. By scaling these analogue calculations to the

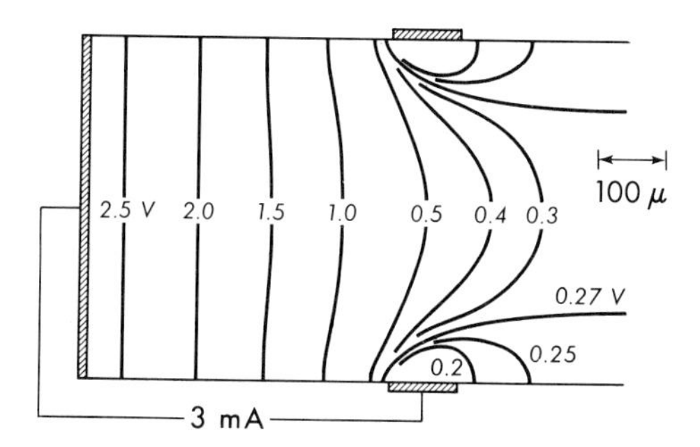

FIG. 4:33. Potential distribution given by resistor analogue for current entering squid axon at left end and leaving through 100 μ pseudo node. The potential measured at right end is about 0.27 *V* above the internal potential at the pseudo node.

lobster axon we find that for an 80 μ node the current density variation is the same, and the potential lead error is 0.34 Ωcm².

Only scattered returns are available for a negative membrane resistance or for the more realistic membrane analogue with two segments—one high resistance and the other negative. A negative resistance element is easily produced by a conventional two-transistor circuit, and the two segment characteristic requires an additional diode. As predicted, the current density variation along the "node" is reversed from that for a normal membrane resistance, but the potential error has not been obtained.

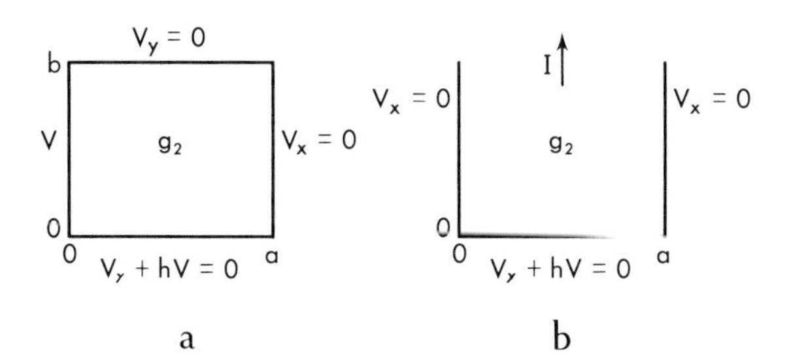

FIG. 4:34. Two plane potential theory problems as simple illustrations for the result of a negative conductance boundary, bottom, on potential and current distributions.

In the midst of all this confusion I found it necessary to turn to the simpler problem in a plane—corresponding to a diametral slice through the axon. This, too, proved difficult, and I retreated further to the simple rectangle of figure 4:34a, with a constant potential applied to one side and "membrane" at the base. This is a classical problem in potential theory treated in general by Carslaw and Jaeger (1947, Sect. 63, II) but I did not follow it through for the negative radiation case, $h < 0$. Instead, I only found a solution to the primitive problem of current flow up a long strip with a negative resistance at the lower boundary (figure 4:34b). Here the Laplace equation has a solution

$$V = A \exp(-\alpha y) \cos \alpha x + By + C$$

for a surface conductance g_2 and a negative edge conductance $h = -g_3$ if $\alpha = \pi g_2/g_3$.

This solution, corresponding to that of the squid axon with an axial electrode, was confirmed on a resistor analogue with series resistors at $y = 0$ to conform to the boundary condition. However, the stability of this and higher order solutions was not tested, nor have I climbed back up the ladder toward a solution of the original problem.

SUGAR-WIRE CLAMP

It was obvious that the current electrode and electrolyte resistances must be low enough to stabilize the most powerful membranes if we are to clamp the axon adequately over the major portion of its length. Near each end—of which the potential end is considerably more important—some four different space constants may be involved as the wire end is moved from sea water across the interface into the sucrose regions (Adelman, 1963). An analytical solution is not simple. The 50 $\mu \times 100\ \mu$ calculations have been carried through for three membrane resistances: zero, $+3.5\ \Omega\text{cm}^2$, and $-2\ \Omega\text{cm}^2$. The zero resistance case was deceptively simple and the $+3.4\ \Omega\text{cm}^2$ case could be approximated by reasonable values for the space constants with a second order cable. The $-2\ \Omega\text{cm}^2$ calculations were considerably more frustrating and less satisfactory, and no detailed conclusions have yet been evolved. About all that can be said is that the potential error changes so rapidly as the wire tip moves past the interface from sea water to sucrose that the only practical experimental procedure is probably to place it well into the sucrose region.

Also—as might be expected—all phenomena so far noted for the $+3.5\ \Omega\text{cm}^2$ case are inverted with the $-2\ \Omega\text{cm}^2$ membrane.

TWO-NEEDLE CLAMPS

I had rather narrowly escaped the disaster of nonuniformity in the squid axon when I could certainly not have recognized it, and a survey of either the potential or the current density over the membrane was often so difficult as to be virtually impossible—particularly in functional central cells. So it seemed quite important to make some estimate of the conditions under which spatial instability of other cells could be expected.

One of the simplest arrangements was a flat membrane, such as frog skin, toad bladder or an electroplax, between two current electrodes and voltage clamped at one point. The disk model developed as an approximation to the squid axon was directly applicable to these preparations. A uniform potential and current density were to be expected over the membrane surface for a radius a, given by equation 3:12,

$$a < \bar{a} = 3.83/\sqrt{r_2(g_3 - g_1)/\delta}.$$

For the electroplax (Nakamura et al., 1965) we find the critical radius $\bar{a} = 0.3$ mm. Since the effective radius of the preparation was 0.6 mm, it is not surprising that evidence for propagation over the surface was found in an unclamped case.

Spherical cells

The first approximation was obviously the assumption of a current source at the center of a sphere, with potential measuring and control at one point of the surface membrane. This had, just as obviously, the same simple solution for axial symmetry in three dimensions as had been found for the symmetrical two-dimensional problem, given by equation 3:13, except that it could easily be extended to include the external medium. It appeared that perfect control was the only theoretical possibility for

$$a < r_3/(r_1/2 + r_2) \qquad (4:7)$$

where a is the radius of the sphere, r_1 and r_2 the external and internal specific resistances, and $-r_3$ the membrane negative resistance for a unit area. With this, it seemed quite reasonable to expect that an approximately spherical and uniform cell less than 2 mm in diameter would be adequately controlled.

The analysis was, however, only a little less primitive than the assumption of perfect control under any conditions, so I persuaded Wilfred Rall to turn his considerably more extensive and more experienced talents onto the problem as follows. The solution of the Laplace equation for axial symmetry is given in terms of zonal harmonics $P_n(\cos\theta)$,

$$V_i = \frac{Ir_2}{4\pi\rho} + \sum_0^\infty A_n\rho^n P_n(\cos\theta) \qquad \rho < a$$

$$V_e = \sum_0^\infty B_n\rho^{-(n+1)} P_n(\cos\theta) \qquad \rho > a$$

for a current I from the central electrode. At the membrane, $\rho = a$,

$$i_3 = -\frac{1}{r_1}\frac{\partial V_1}{\partial\rho} = -\frac{1}{r_2}\frac{\partial V_2}{\partial\rho} = \frac{V_m}{-r_3} + C\frac{\partial V_m}{\partial t}.$$

For steady state $\partial V_m/\partial t = 0$ and

$$A_0 = -\frac{Ir_3}{4\pi a^2} + \frac{I(r_1 - r_2)}{4\pi a};\ A_n = 0,\ n > 0$$

$$B_0 = \frac{Ir_2}{4\pi};\ B_n = 0,\ n > 0$$

so the only steady state solution is the uniform clamp,

$$V_m = -Ir_3/4\pi a^2.$$

This solution is not necessarily stable since perturbations with A_n, $B_n \neq 0$ may not decay in time. For each n the boundary conditions become

$$-\frac{nA_n}{r_1}a^{n-1} = \frac{(n+1)B_n}{r_2}a^{-(n+2)}$$

$$\frac{(n+1)B_n}{r_2}a^{-(n+2)} = \frac{A_n a^n - B_n a^{-(n+1)}}{-r_3} + C\left[a^n\frac{dA_n}{dt} - a^{-(n+1)}\frac{dB_n}{dt}\right].$$

The elimination of A_n and dA_n/dt and their replacement give

$$\frac{1}{A_n}\frac{dA_n}{dt} = \frac{1}{B_n}\frac{dB_n}{dt}$$

$$= -\frac{1}{Cr_3}\cdot\frac{(n+1)r_3 - ar_1 - \frac{n+1}{n}ar_2}{ar_1 + \frac{n+1}{n}ar_2} = -\frac{1}{\tau_n}.$$

The coefficients A_n, B_n expand or decay exponentially according as the right-hand term $\tau_n < 0$ or $\tau_n > 0$. The condition for stability is then

$$a < r_3/[r_1/(n + 1) + r_2/n].$$

Since this condition is most stringent for $n = 1$, we require again, for stability of the uniform clamp, that

$$a < r_3 \Big/ \left(\frac{r_1}{2} + r_2\right).$$

This is the same as the criterion for uniformity, equation 4:7. These are even more restrictive, as might have been expected, than the first approximation, equation 3:13, in which the external field was ignored.

Rall further pointed out that if the current electrode is off center the steady state solution contains higher order terms and the space clamp is not uniform over the sphere. The case of a double-barreled intracellular electrode will be even more complicated. Also, the effect of feedback on the stability of nonuniform potential distributions has not been investigated. Consequently the condition of equation 4:7 is to be considered only as a guide rather than a guarantee of adequate clamping in space and time.

Cylindrical cells

The voltage clamp techniques used on *Nitella* and *Chara* by Findlay (1961) and Kishimoto (1961, 1964) were not obviously adequate. Some analysis was needed as a critique and a guide for the design and interpretation of these experiments, and there was the hope that the two-needle arrangement might work for other long cells—particularly squid and other axons. The problem was essentially the same as that of what happens beyond the end of an axial wire. For the squid axon this had been found to have a complexity that was considerably more costly than the importance of its contribution justified within the range of the axial electrode. But now the situation was reversed—what happened beyond the end of the hypothetical axial electrode was of paramount importance.

We start again with the cable equation for the membrane potential V and current i_m and longitudinal resistance r_i,

$$\frac{d^2V}{dx^2} = r_i\, i_m$$

and again use the solutions: For a passive membrane resistance $g_m > 0$

$$V = A \exp(-\alpha x) + B, \qquad \alpha^2 = r_i g_m; \qquad (1:35)$$

for an excited membrane resistance $-g_m > 0$

$$V = C \cos \omega x + D, \qquad \omega^2 = r_i g_m. \qquad (3:10)$$

First, terminating the cell at $x = 0$ so $dV/dx = 0$ let $V = \overline{V}$. Then

$$\overline{V} = C + D$$

and for the potential V_0 at the control point x_0,

$$V_0 = C \cos \omega x_0 + D.$$

The membrane characteristic $i_m = g_m(V_B - V)$ for $V > V_B$ and the original equation gives

$$-C\omega^2 \cos \omega x = r_i g_m(V_B - V).$$

At $x = 0$

$$C = \overline{V} - V_B,\ D = V_B$$

and $\overline{V} = (V_0 - V_B)/\cos \omega x_0 + V_B$ to give

$$V = \frac{V_0 - V_B}{\cos \omega x_0} \cos \omega x + V_B. \qquad (4:8)$$

These solutions are nicely presented on the phase plane (figure 4:35).

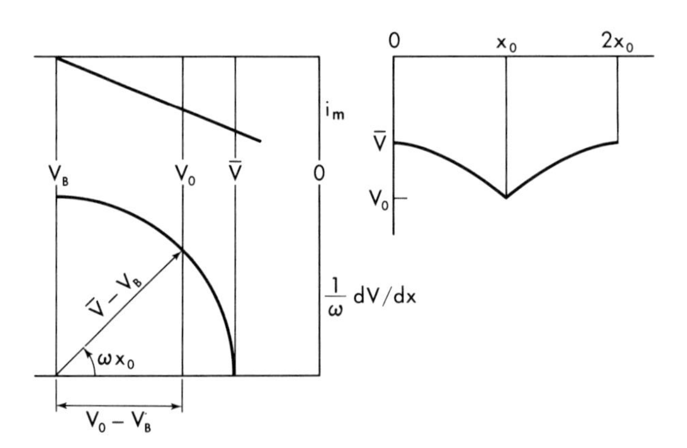

FIG. 4:35. Potential distribution along an axon with the negative conductance characteristic at top left. The potential at the center, x_0, is V_0 and the potentials on each side are given on the phase plane, lower left, and along the axon, right.

The total current I_0 at x_0 from an axon length $2x_0$,

$$I_0 = -\frac{2}{r_i}\frac{dV}{dx} = 2\sqrt{g_m/r_i}\,(V_0 - V_B)\tan \omega x_0.$$

If indeed the half length of the cell x_0 is so small that $\omega x_0 < 0.5$, the total current I_0 will give a passable approximation to g_m. Nominal values for active *Nitella* of r_i = 56 kΩ/cm, r_m = −16 kΩcm then require $2x_0 < 2.5$ mm. Here a single microtip with a bridge or separate current and potential microtips are at least possibilities. The nominal values for an active squid axon require $2x_0 < 0.3$ mm, which is less than a diameter and not promising.

When we measure the average $\bar{\imath}_3$ externally over a restricted length 2δ at x_0, the error Δ gets steadily worse as x_0 is increased:

$$\Delta = \frac{\omega\delta}{2}\cdot \tan \omega x_0.$$

With $\omega x_0 = \pi/4$ as an example for *Nitella*, $2x_0$ = 8 mm and 2δ = 2 mm for $\Delta = 0.1$, which is quite reasonable. But for squid, $2x_0$ = 480 μ, 2δ = 120 μ so that the experiment is impractical and the approximations are not valid.

Enlarging the region $2x_0$ and/or increasing the central potential V_0 will eventually take the boundary potential $\overline{V}$ beyond the negative resistance excited region to enter the positive resistance active region. The algebra and the phase plane of the combined trigonometric and exponential solutions then become more complicated than was found before, figure 3:80 (Cole, 1961).

A similar analysis can be applied to the arrangement used by Adrian et al. (1966) to clamp the end of a sartorius fiber. Let the electrode at x_0 from the end of the fiber have a potential V_0 and another electrode at $x_0/2 = l$ be at V_1. Then by equation 4:8

$$V_0 - V_1 = (V_0 - V_B)\cos(\omega x_0/2)/\cos \omega x_0.$$

The membrane current at $x_0/2$ corresponding to V_1,

$$i_m = -\frac{\omega^2}{r_i}\cdot\frac{V_0 - V_B}{\cos \omega x_0}\cos\frac{\omega x_0}{2}$$

$$\simeq \frac{2(V_0 - V_1)}{3r_i l^2}\left(1 - \frac{\omega^2 l^2}{2}\right)$$

for ωl small and the correction term is less than 5 percent if

$$\omega l = \sqrt{r_i g_m}\, l < 0.3.$$

With an 80 μ fiber, $r_2 = 250\ \Omega$cm, $r_i = 5\ \mathrm{M}\Omega$/cm and for $l = 125\ \mu$, this accuracy calls for $g_m < 0.13$ mmho/cm or $r_3 > 200\ \Omega\mathrm{cm}^2$. Since the electrode separation is already comparable to the fiber diameter, more careful analysis may be necessary if lower negative resistances are encountered.

RÉSUMÉ ON CLAMPING

There may be a dozen definitely different techniques for application of the voltage clamp concept to more than as many different cell membranes. Some cell-technique combinations are obviously impossible while no single one is without its difficulties. In general, the current patterns are similar enough that one is inclined to believe that they indicate adequate clamping, while variations are to be suspected. Certainly one of the most insidious problems is that of knowing what the potential is across the membrane conductance structure and that the membrane current is measured over an area where the potential is reasonably uniform. But without independent confirmation, any conclusions are certainly rather less than logical and may be quite wrong.

Independent experimental test of the clamp adequacy is by far the most convincing evidence. This is often virtually impossible to get and usually so difficult as to preclude doing any other experiments. Then it becomes all the more attractive to investigate the effects of various parameters in a theoretical model as a guide to the requirements for an adequate experiment. And besides, an experimental physicist always feels better if he has some theory to lean on—no matter how inadequate it may be. As we see the various experimental problems now, the theory is rather obvious and straightforward. It is, however, utterly impossible to get any analytical solution while computer solutions are rather less useful even if one had the million dollars or so it might take to get a few of them. So we have to guess at the most important factors, idealize them, and ignore everything else.

Quite a bit of this piecemeal work has been done. It is usually rather tedious and anything but spectacular, but the various pieces are beginning to make an interesting and useful structure. For instance, our simple Kelvin cable model of an axon well deserves the respect and familiarity that it has achieved. But if we put an HH membrane on it we have only begun to find out how an impulse starts or is blocked in a squid axon, although a good start has been made for the far simpler medullated axon.

It is quite useful to simplify the membrane characteristics by ignoring all but the peak of the early current and the steady current as functions of the clamping potential. We have found some basis to consider the early peak as a steady state characteristic, although it is obviously no more than a quasi-steady state at best. This is still an unpleasant problem so we fence it in some more and approximate it by piecewise linear segments. Now—whereas we have usually put the resting membrane resistance of 1000 Ωcm^2 into the cable equation—for enough depolarization we shift to a resistance of 10 Ωcm^2 and match up exponential or hyperbolic functions at the transition. This would be only messy and tedious except that the Kelvin equation may not be appropriate because there will be radial potential differences for so short a length constant; we should look at a more complete Laplace field equation. Here a couple of standard analytical solutions and some calculation show that what may be called the "internal space constant" is about equal to the axon radius. Then the analysis gets so cluttered up with constants that we cannot tell much about it without some kind of numerical or analogue calculation. But we can still guide our intuition in some problems with a little more complicated internal network. The problems are obviously much the same in the quasi-steady state, active part of the sodium characteristic.

As we turn to the excited, negative resistance region we face something new. Now trigonometric sines and cosines replace exponentials and hyperbolic sines and cosines. For one thing, this means that boundary conditions need more careful attention—if only because things do not oblingingly become negligible at large distances. An impressive example of this is found in the point control of an axon with an axial current electrode.

Before this could be considered, the equivalent internal circuit structure had to be modified again to permit longitudinal axoplasm coupling current. Then for a reasonably probable combination of electrode and axon properties the potential at the control point could be held as close as desired, but exceeded useful limits at more than a few tenths of a millimeter away on each side. However, most unexpectedly, it was found that perfect clamping—and perfect clamping only—was possible as the axon length was reduced to about a millimeter on each side of the control point. Since this length was comparable to the axon perimeter, it became important to consider a disk that was a thin transverse section of an axon, insulated on each plane face with the short cylindrical face of membrane clamped at one point. Here again it was found that only a perfect clamp was possible for an axon diameter of about 2 mm and less. Another possible approach was to look at a circular patch on the axon surface and centered on the control point. This solution in Bessel functions was again a perfect clamp for a patch diameter up to 2 mm—more than enough to wrap around a 0.5 mm diameter axon. Thus these three solutions, although separately quite artificial, combine to give some basis for confidence that a short segment of axon may be perfectly clamped.

The same kind of question arises in the clamping of spherical cells such as may be used to approximate the central cells of Eccles, Hagiwara, Frank, and Bennett. The problem is formally similar to that of the disk section, with the conclusion that nothing but perfect clamping is to be expected for a cell diameter less than about 2 mm. Another interesting and perhaps useful problem is that of a flat cell with a membrane on each flat face and the potential and current electrodes near each other as has been used to represent an electroplax and frog skin experiments. Here again, perfect clamping is to be expected for a sufficiently small disk of tissue insulated on the circumference, but only some of the numerical values are at hand.

The node clamp with either sheath and electronic or sucrose insulation on each side is an obviously important problem, and it was very much hoped that once again perfect clamping might appear below some predictable node length. The Kelvin cable with a negative membrane can give only a cosine variation of potential and current with distance from the potential end of the node. The node length over which the potential and current variation may be acceptable is again of the order of an axon diameter. But it varies as the square root of the diameter, being almost a diameter for lobster, rather less for squid, and considerably more for toad and frog nodes. Once again, however, the internal space constant is comparable, and here it cannot be ignored because it creates an error in the membrane potential measurement. The second order cable model used before shows the error to be small for the lobster but significant for squid and utterly negligible for natural nodes. Since the approximate cable had been developed for other purposes, more detailed calculations were made at 50 μ intervals along radii and 100 μ intervals of length. These were reasonable cosinusoidal if—in the squid—an effective length 50 μ longer than actual was used and the potential error was then found to be no more than 1 mV.

Analysis of the recently developed sugar-wire clamp for the squid axon has so far been considerably less satisfactory, and in this respect comparable to experiments with it.

One very important drawback to almost all of these calculations—for the point-wire clamp, the node and pseudo-node clamp, and the sugar-wire clamp—is that as yet they give no certain indication of the instabilities that are thought to be the bases of the anomalous current patterns found as one departs from near to ideal conditions. Crude analyses and some tests of accuracy and stability made it probable that these data were reasonably reliable.

But, in spite of all the uncertainties, it has become probable that almost any membrane can be voltage clamped and will yield information not yet otherwise available.

Separation of Ionic Current Components

Early measurements of current flow after a sudden depolarization, or change to a less negative potential difference across the membrane of the squid giant axon had three principal components. The first nearly exponential current with a time constant of the order of 10 μsec could well be attributed to the charging of a passive membrane capacity of about 1 $\mu F/cm^2$. In the usual physiological range, this is followed first by an early inward current which reaches a peak value in tenths of a millisecond. After the peak, the current returns more slowly to zero and reverses to become a steady state outward current after a few milliseconds.

It seemed rather certain that this current complex following the charging transient was to be ascribed to ion flow, but the identification of the ions was neither simple nor obvious. It was not to be obtained by these electrical measurements alone, since they gave only the net movement of charges without any identification of the ions carrying the charges. Following the first solution of this problem by Hodgkin and Huxley (1952a) and their impressive demonstrations of its fundamental importance, the results of many and different experiments have been increasingly interpreted in their terms. But only Frankenhaeuser and Dodge, working on frog and toad nodes together and separately, have shown the patience and skill necessary to repeat and improve on the original analyses.

Squid membrane currents

By diluting the sodium ion concentration outside the axon with choline, and with the very reasonable assumption that the flow of a particular ion species would vanish at its equilibrium potential, Hodgkin and Huxley established that early currents were sodium and late currents were potassium. Thus by clamping the membrane at the potential E_{Na} at which there was no sodium current, the potassium current I_K appeared uncontaminated (figure 3:28). And so I_K could be obtained at each E_{Na} as Na_e was changed. Then I_{Na} for normal external sea water was found at each of these potentials by subtraction, $I_{Na} = I_g - I_K$. In the course of an experiment on a single axon only a few changes of external solution could be made and considerable survival time of the axon was lost in finding $V = E_{Na}$. Then Hodgkin and Huxley further assumed that the time course of each of these two components was the same at a given membrane potential while the amplitudes depended on the ion concentrations

$$I_{Na}(t, V, E_{Na}) = F_1(t, V) \cdot F_2(V, E_{Na})$$
$$I_K(t, V, E_K) = G_1(t, V) \cdot G_2(V, E_K).$$

At each potential V they computed an interpolation between I_g at two different values of E_{Na} to give I_K for both and I_{Na} for each.

They then had the time courses of both the sodium and potassium currents at each potential for which there were records at two values of E_{Na}, corresponding to two different external sodium ion concentrations.

This was indeed an arduous and sometimes an unsatisfactory procedure, as Hodgkin and Huxley made clear and we abundantly confirmed. But they approximated their inconsistencies to give average behaviors for I_K and I_{Na}.

These currents obviously depend on the driving forces, which can be expected to be $V - E_K$ and $V - E_{Na}$, and the permeabilities, which were defined as

$$g_K = I_K/(V - E_K) \text{ and } g_{Na} = I_{Na}/(V - E_{Na}).$$

These permeabilities, expressed as electrical conductances, were shown to be continuous through a sudden change of clamp potential and constant for such changes. Thus the conductances, g_K and g_{Na}, found by the choline separation were the instantaneous infinite frequency, or chord, conductances, and indeed the direct electrical measures, such as shown in figure 3:30, confirmed these identifications. The experimental results could then be represented by the circuit of figure 3:24 in which C, E_K, and E_{Na} were known constants and g_K and g_{Na} were known functions of time at fixed membrane potential as shown in figure 3:25. There was also the third, small but important, linear conductance given by E_L and g_L which was not identified and is not represented on figure 3:24.

The direct electrical method for separation of current components has been extended and widely used for squid, for nodes, and for other preparations (cf. Dodge and Frankenhaeuser, 1959; Armstrong and Binstock, 1964, 1965). The control potential is changed in three steps. After a usual hyperpolarization, or conditioning pulse, the experimental level is established. The initial conduction current, either extrapolated back through the capacity charging transient or estimated after the transient, is taken as the leakage I_L. Then at appropriate times the potential is stepped to either E_K or E_{Na}. To the extent that these parameters are known and constant, the instantaneous current is I_{Na} at E_K or I_K at E_{Na}, and the time course of each of these components is determined.

Yet another experiment gave basis to separate g_{Na} into two factors—one the early rise, or "on," and the other a slower "off" process. In a steady state this latter, inactivation, completely blocked a sodium current at depolarizations to potentials beyond -20 mV but allowed a maximum current after the membrane was hyperpolarized to -85 mV or more. This and the kinetics were measured by the maximum of the peak inward current.

At this point the experiment and the manipulation of the experimental data comes to an end. These are fundamentally voltage clamp current data separated, by further experiment, into potassium and sodium components. Any further representation and approximation, such as by empirical or theoretical mathematical expressions, or further interpretations, in terms of membrane permeability processes, in no way affect the position of the circuit and the components of figure 3:24. These remain as experimental facts subject to revision only by the weight of more and better or other experimental evidence.

NODAL MEMBRANE CURRENTS

Dodge and Frankenhaeuser (1958) emphasized the striking similarity between their records of current flow at a single node of Ranvier of the frog axon and those in the squid membrane. They listed eight points of similarity in the data, but there were also differences. The most obvious were that the ionic current densities appeared to be much larger, the peak of the early current vs. membrane clamp potential relation was decidedly curved on both sides of the zero current point, and the steady state currents were indeed steady state. They then established (Dodge and Frankenhaeuser,

1959) by *e*-fold dilutions of external sodium with choline or sucrose that this point of zero early current corresponded to E_{Na}. Although the conductance, $g_{Na} = (V - E_{Na})/I_{Na}$, for the *Xenopus* node was not conveniently constant, they were able to express the data by a peak sodium permeability P_{Na} which was calculated from the Goldman equation 4:3. The time course of these sodium permeabilities was followed by measures of the instantaneous currents produced on return of the membrane to near E_K, where $I_K = 0$. Following Hodgkin and Huxley, the characteristics of sodium inactivation were measured by the peak inward current at about the potential for its maximum.

Frankenhaeuser then varied K_e to show that the steady state current was mostly potassium (1962a), and the time course of the current decay on repolarization in high K_e was the same as the fall of the instantancous currents in low K_e (1962b): again these could be expressed as Goldman permeabilities (1962c) from equation 4:4.

Dodge returned to the frog medullated axon and presented (1961, 1963) similar and detailed work on the node of this thoroughly familiar and eminently reputable, not to say standard, preparation. The early current was once more identified as sodium—by its absence at an E_{Na}, given approximately by Na_e at normal and diluted values, and by confirmation of the independence principle for the peak values, figure 4:16. The late current was similar to that for *Xenopus* and was similarly attributed to potassium.

But there were difficulties. The original choline procedure for separating I_{Na} from I_K in I_g was modified because the steady state current again was affected by the external sodium concentration. Accordingly an additional coefficient of proportionality between the currents at long times in high- and low-sodium media was found. This, with the coefficient at short times where I_K was assumed negligible, gave first I_K and then I_{Na}. Then a different approach was used. For the squid membrane,

$$I_K = A \cdot T(t/\tau)$$

where the amplitude A and the time constant τ depend on the membrane potential alone while the time function T is the same for all potentials. To the extent that this is a valid expression, T may be found at E_{Na} and, with appropriate changes of A and τ to express I_K at long times for other values of V, it will give the entire time course of I_K at each V. These changes of scale were made with log t vs. log I plots, figure 4:36, to give potassium currents in excellent agreement with this component as calculated by choline separation.

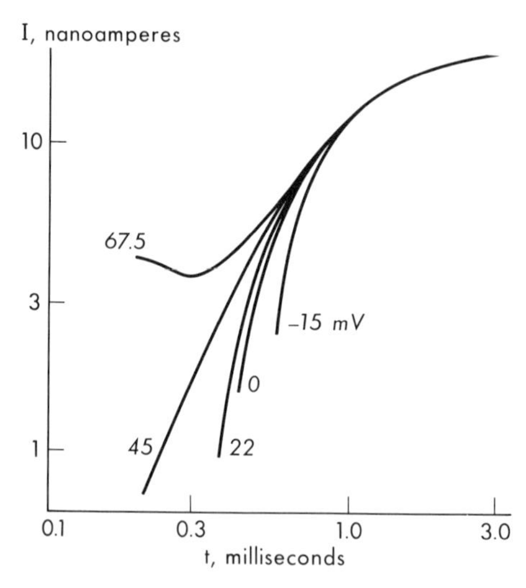

FIG. 4:36. Frog node currents vs. time in voltage clamps as indicated. The variables are on logarithmic scales for I_K in a clamp to 45 mV. The curves at other potentials have been translated to coincide for long times thus giving τ_K and I_{Na}.

The sodium components had then to show as good agreements; they were well confirmed—even if not entirely reliably—by extrapolated instantaneous currents upon return of the membrane potential to near E_K as Dodge and Frankenhaeuser (1959) had done. Next it was shown that the so-called leakage, linear instantaneous, currents I_L had an E_L which coincided with E_K and changed with it as K_e was changed from normal to iso-osmotic, figure 4:48. Dodge thus proposed that I_L was an instantaneous potassium current.

Except that Dodge was unwilling to estimate a membrane area for the frog node, the similarities between this node and the squid membranes were far more remarkable than the differences.

Later, however, Hille (1966) reverted completely to the HH formulation for his computer assisted analysis of frog node voltage clamp currents. His tape recording served as the input to the computer which removed the linear leakage current, fitted the late outward current to an HH fourth power function to give I_K, and subtracted I_K to leave I_{Na}. The five HH parameters, plus a relaxing leak, were then determined from these derived data. This was not a complete answer to my prayer for a presentation and a record of these parameters within a second after the clamp pulse on the axon. But it was a very substantial first installment.

PURKINJE FIBER MEMBRANE CURRENTS

Turning now to the other published current component separations, that of Deck and Trautwein (1964) on the Purkinje fibers was in a preliminary but promising phase. The early inward current was already established and diminishing at 12.5 msec for depolarizations from -88 mV and to between -70 and $+10$ mV to give the basis for excitability. However, this inward current only declined slowly with a time constant of 100 msec and did not reverse at -40 mV for about 150 msec (figure 4:26), giving some basis for the slow recovery in this potential region. This is possible because, as was known for skeletal muscle (Katz, 1948, 1949), the steady state rectifier characteristic shows little outward current in this range and the high currents appear for hyperpolarization beyond the resting potential as seen in figure 4:26.

Deck and Trautwein relied entirely on the differences of the ionic currents in a voltage clamp between a normal sodium Tyrode and a choline Tyrode medium. They first find a chloride current which reasonably conforms to their assumption that it is represented by a conductance independent of voltage and time. On the assumption that the steady state current in both solutions is $I_g = I_K + I_{Na}$, they arrive at the sodium current by difference, although Noble (1966) doubts that this procedure gives a good measure of I_{Na}. Then, defining these conductances as did Hodgkin and Huxley, they obtain g_{Na} and g_K as functions of membrane potential and time, figure 4:37. Although there is no independent evidence that such conductances are indeed measurable and linear chord conductances, these are perfectly acceptable as definitions and the means of expression of the experimental data.

At rest these measures produce a membrane resistance of 2200 Ωcm^2, primarily an expression of potassium movement, in excellent agreement with Weidmann's (1952) value of 1900 Ωcm^2. For increasing depolarizations the value of g_{Na} at 12.5 msec steadily increases but is, as the authors note, only about 3 percent of the value found previously at the inaccessible shorter times. As these conductances decrease steadily but slowly to near zero at all potentials,

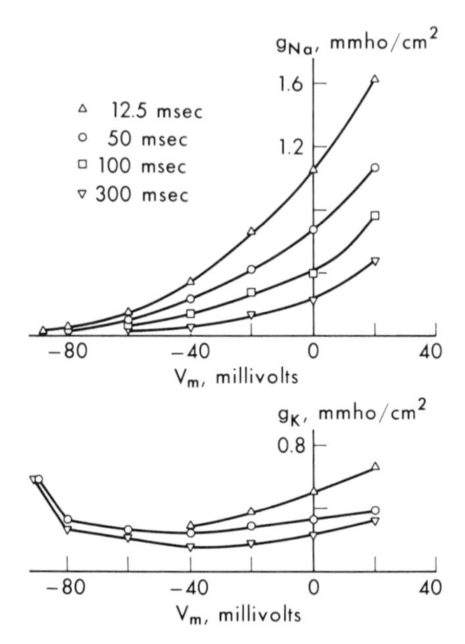

FIG. 4:37. Calculated values of g_{Na} and g_K for Purkinje fiber membrane at times, shown by symbols, after changes of membrane potential to V_m.

the potassium conductance remains relatively constant—independent of both membrane potential and time. These data have not been given empirical analytical expression, but it is nonetheless apparent that other experimental data are reasonably in accord with them while some hypotheses are not.

NITELLA AND OTHERS

Excitation and propagation in *Chara*, a fresh-water plant related to and somewhat similar to *Nitella*, was found by Gaffey and Mullins (1958) to be accompanied by a twenty fold increase of chloride outflux and they explained the electrical activity as analogous to squid with an early outflow of chloride followed by a potassium outflow. These data were confirmed in *Nitella* by Kishimoto (1959) and Mullins (1962). Although Hope (1961) and Findlay (1961) attributed the early inward currents in their voltage clamp and excitation to calcium, they came to agree with Kishimoto (1964) that these currents were indeed chloride as had come to be expected. Nonetheless, it has been found (Kishimoto, 1965) that internal calcium is essential for activity.

The choline separation procedure for eliminating at least the inward I_{Na} has proved useful. It has been considerably supported in work on the eel electroplax (Nakamura et al., 1965) by identical results with tetrodotoxin, a clam muscle toxin, and procaine.

Somewhat similarly, Lindemann (1965) has relied on the conclusion that the antidiuretic hormone increased the sodium permeability of the frog skin, to interpret his ionic currents in voltage clamp as being primarily those of sodium ion flow.

MORE ANALYSIS TECHNIQUES

The original Hodgkin and Huxley choline analysis of the components of ionic current in a voltage clamp was not only inspired and useful but it also produced the independently measurable physical conductances, g_{Na} and g_K. It has also been helpful in separation of the sodium and potassium contributions to the ionic current in the nodes of the medullated axons of toad and frog. However, we had difficulties enough with this, as an experimental routine, to easily dissuade us from any extensive use of it. And for investigation of another independent variable, this procedure required a long sequence of solution and potential changes that most axons were quite unable to survive. Nonetheless, a decisive separation of these components becomes a crucial step in the description of ionic behavior of a membrane as everything else becomes relatively simple— or at least fast and straightforward. We have tried other possibilities; although they have neither been reduced to routine, nor used extensively, nor compared with choline or other procedures, they are not without promise and I do not choose to ignore them.

We first tried superposing small short alternating square pulses on the command step potentials of the voltage clamp. If the capacity transient were fast enough, the following current pulse would measure the quasi steady state instantaneous conductance before the change of potential could cause an appreciable change of current pattern, and the alternation of the pulses would minimize any cumulative effects. This technique was applied by others, particularly Tasaki and his collaborators (Tasaki and Freygang, 1955). In practice, the length of the initial transient was marginal and variable because of other circuit effects, and there was no obvious test of the approach to quasi steady state measurement. The next step was the introduction of an additional circuit which created a Wheatstone bridge (figure 3:59) (Cole and Moore, 1960a) to balance and measure linear components of the membrane. A balance by a resistance and capacity first gave the parameters for the resting membrane, then it canceled the capacity transient after a new clamp potential; later, to the extent that the membrane capacity remained constant, the changes of membrane resistance could be

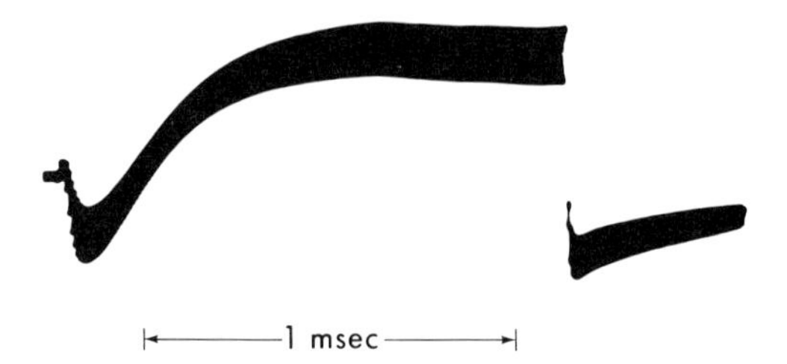

FIG. 4:38. A high frequency potential of a few mV was added to the voltage command and the membrane current at the holding potential balanced out. The change of the membrane high frequency conductance during a clamp then produced an envelope on the usual clamp current record.

followed by the bridge unbalance. All of these various reasons made it quite advantageous to replace the "rick-rack" perturbation by a small, high-frequency sinusoid to give records such as shown by Tasaki and Spyropoulos (1958) and Moore and Cole (1963) (figure 4:38) for squid and by Julian, Moore, and Goldman (unpublished) for lobster. Then we have

$$I_g = g_{\mathrm{Na}}(V - E_{\mathrm{Na}}) + g_{\mathrm{K}}(V - E_{\mathrm{K}})$$
$$g = g_{\mathrm{Na}} + g_{\mathrm{K}}.$$

Insofar as the conductance measurement g is an adequate approach to infinite frequency and the emfs E_{Na} and E_{K} are known and constant, I_{Na}, I_{K}, g_{Na}, and g_{K} are determined. Also the data are simply presented on the g, I plane (figure 4:39) and the individual ion conductances and currents appear as components on the oblique coordinate system between the limiting tangents $V - E_{\mathrm{K}}$ and $V - E_{\mathrm{Na}}$.

An interesting and ingenious suggestion for separation of the sodium and potassium components of the clamp currents was made by FitzHugh. It had been shown (Cole and Moore, 1960b), that I_{K} at E_{Na} was delayed progressively without change of form with an increasing amount of previous hyperpolarization V_H, figure 4:43.

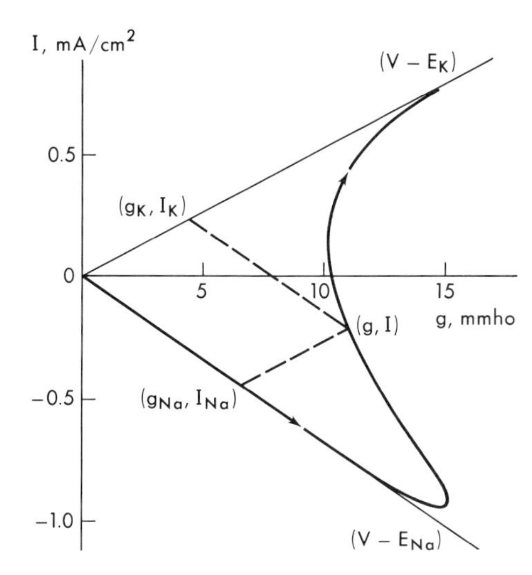

FIG. 4:39. The changes of membrane current and conductance follow the path indicated by arrows during a voltage clamp. The asymptotes as shown at short and long times then become non-orthogonal axes for the resolution of the current and conductance, by the dashed lines, into their sodium and potassium components.

Similarly, as shown in figure 4:40, an increase of V_H delays the current after the early inward peak, but it has no detectable effect on the current before the peak. Then making the obvious assumption that when the peak current is at its maximum value for both values of V_H, $I_{Na}(t)$ is unaffected by the change of V_H while $I_K(t)$ is changed to $I_K(t \pm \tau)$, FitzHugh pointed out that both I_{Na} and I_K were to be found from these data as shown in figure 4:41. At an initial V_H,

$$I_g = I_{Na}(t) + I_K(t),$$

and with an increased V_H',

$$I_g' = I_{Na}(t) + I_K(t - \tau).$$

For small τ,

$$I_K(t - \tau) \approx I_K(t) - \frac{dI_K}{dt} \cdot \tau,$$

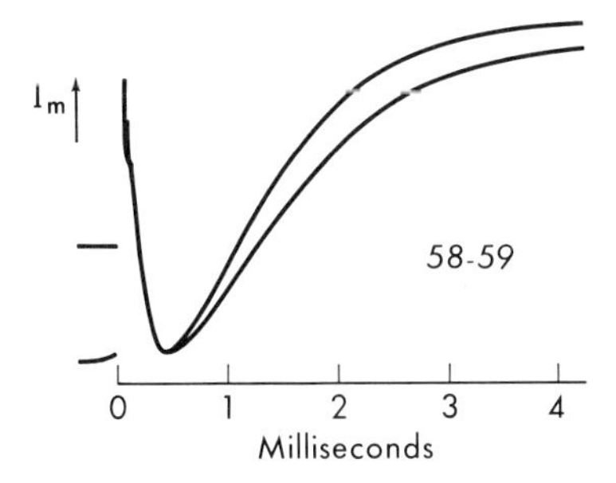

FIG. 4:40. Voltage clamp current, right, delay after a more negative prepotential.

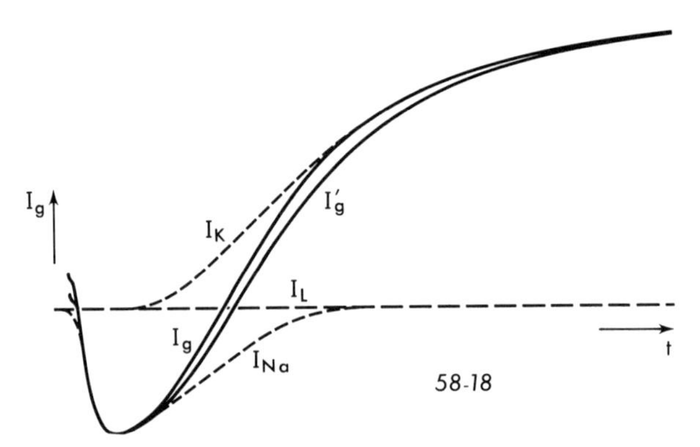

FIG. 4:41. A voltage clamp from a holding potential which saturates the inward peak current, gives the conduction current I_g, with the leakage I_L as a base line. A more negative holding potential produces delay, attributed to potassium, after the inward peak I_g'. An analysis of these curves gives the I_K and I_{Na} components of I_g.

and

$$\Delta I_g(t) = I_g(t) - I_g'(t) \approx \frac{dI_K}{dt} \cdot \tau.$$

The area between the two traces,

$$\int_0^\infty \Delta I_g \, dt = \tau \, I_K(\infty),$$

to give τ, and

$$I_K(t) = \frac{1}{\tau} \int_0^t \Delta I_g \, dt.$$

This method has only been tried casually a few times and has not been compared with other procedures.

Another approach was a brute force fitting of the voltage clamp currents by arbitrary numerical functions without relation to any possible analytical expressions. Following Hodgkin and Huxley,

$$I_g = \bar{I}_{Na}(V)M(t/\tau_m)H(t/\tau_h) + \bar{I}_K(V)N(t/\tau_n)$$

where $\bar{I}_{Na}$ and $\bar{I}_K$ are amplitudes depending upon V alone and M, H, and N are independent of V except insofar as it determines the respective time constants τ_m, τ_h, and τ_n. A primitive electromechanical curve fitter was built in which M, N, and H were generated by tapped and padded potentiometers. For resolution and convenience of varying the respective time constants these potentiometers were driven according to $\theta = R(\log t - \log \tau)$, where $\log t$ is the common shaft rotation, R is a gear ratio and $\log \tau$ is a rotation offset. This then suggested the translation of the usual oscilloscope records or the direct recording of membrane current patterns on a log time x-axis; both have been tried. Then at $V = E_{Na}$, where $\bar{I}_{Na} = 0$, $\bar{I}_K$, N, and τ_n could be determined and for other V, $\bar{I}_K$ was known and τ_n could be approximated at intermediate and long times. This is essentially a restatement of FitzHugh's proposition that I_K is a single parameter function which was tested by Cole and Moore (1960b). It was very gratifying and encouraging to find that Dodge (1963) and Hille (1967) were able to use the same procedure, as shown in figure 4:36, for the frog node with excellent results. However, since the synthesizer had an independent $\bar{I}_K$ control, the log scale was used only for the time variable. One example of a direct log t recording is shown in figure 4:42a. Here, as is rather usual for the squid axon (Cole and Moore, 1960a), the early peak and the steady state currents are not only linear in clamp potentials above -20 mV but have the same slopes in this region, as seen in the

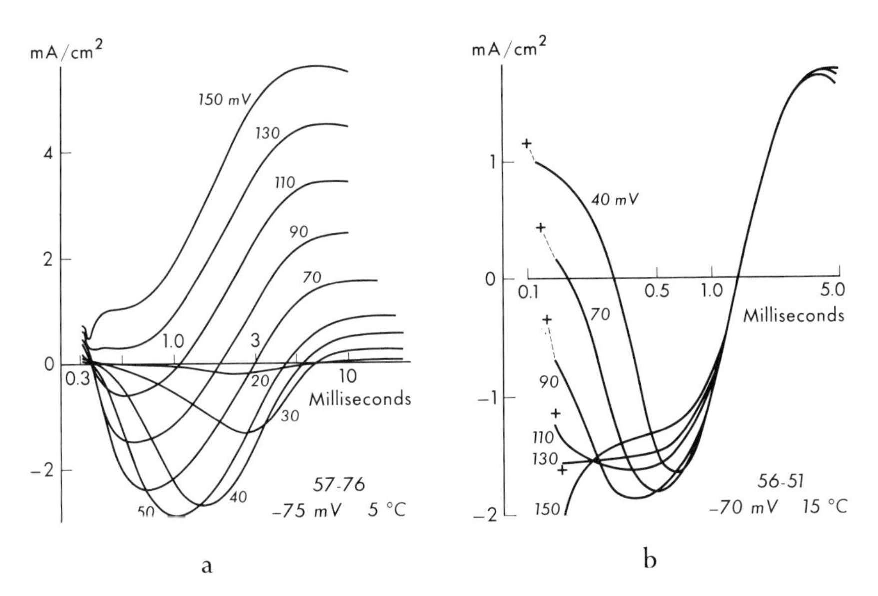

FIG. 4:42. (a.) Voltage clamp currents after indicated changes of potential as recorded on a logarithm of time scale. (b.) Voltage clamp currents, recorded on a logarithm of time scale. These are superposed at the longer times on the I_K at 130 mV. The initial points are shown by crosses. The time parameter of I_K is given the horizontal translation and I_{Na} by the vertical difference from I_K.

superposition of figure 4:42b. It seems apparent not only that such a presentation allows a relatively simple separation of the current components but also that it may be the method of choice for recording all such data.

COMMENT

Separation of the components of ionic currents is a crucial step in the analysis of voltage clamp data. Even at best it is laborious and often uncertain. There is a great temptation to draw only qualitative conclusions and so escape the tedium that went into the contributions of Hodgkin and Huxley and of Frankenhaeuser and Dodge.

Experimental techniques for separation of ionic components of the membrane currents in voltage clamps remain much the same as those originally used on the squid membrane. Reliance on choline substitution for sodium was, however, being supplemented more and more by the use of second steps to potentials such as E_{Na} or E_K. Then the suppression of I_{Na} by the spectacular poisons, tetrodotoxin and saxitoxin, became an accepted and widely used procedure. Other techniques so far show much more promise than performance. It is most regrettable that only the nodes have received careful attention, comparable to the original squid work—which by now has been superseded in experimental procedure and should be in mathematical expression.

It seems important that both the experiments and their analysis be somewhat standardized and simplified. It should be no surprise that computer techniques and programs have been under development to give easier, faster and, we hope, better results.

Some of the New Data

The introduction of the HH axon was followed by a period of comparative quiet, as was probably quite appropriate for various reasons—surprise, bewilderment, consternation, and not a little disbelief. This work was far from being all things to all men. Questions were asked and had to be answered, and the world of ion permeability has become far noisier than ever before. Perhaps this is only because there are more living in it, but perhaps also because there has never before been so much in common. The result is a vast amount of new information with a discouraging lack of progress in the understanding of the problems as posed by Hodgkin and Huxley in 1952, not to mention the lack of as complete and authoritative a description of the squid axons with properties that have since come to be more typical.

I will not attempt any comprehensive or critical review of the newer results—which are only new at the moment or until they are superseded or swamped by the oncoming wide variety of efforts. This is properly material for the subsequent volume that should be done at the end of the next era when order has replaced chaos and leads clearly to the understanding of ion permeability. But I do feel that it is tremendously exciting to try to guess the nature of this goal and to try to ask good questions about it; that the aftermath of the past era is a good preface and introduction to the next.

In our present phase we are deluged by titles "The Effect of . . .", and all I can do is to order some of these topics entirely artificially until a better pattern appears. So I chose first purely physical variables—time, potential, temperature; then some alterations of the external environment—normal ionic constituents and pharmacological agents; finally, at least a mention of changes of the internal environment, from injections to the replacement of almost all of the axoplasm.

PHYSICAL PARAMETERS

Potassium space

With Frankenhaeuser, Hodgkin returned to unfinished business (1956) and gave an explanation for the slow drooping of the steady state potassium currents in a voltage clamp. I had encountered this in 1947 but found that it was explained by electrode polarization. However, with the electrode effect mostly, if not entirely, eliminated, the larger outward currents declined at higher temperatures and longer times. This we also had been finding—in spite of our preoccupation with the validity of our clamp technique. In earlier work Hodgkin and Huxley (1947) had estimated the potassium accumulation between a propagating axon and the oil around it. Perhaps this led Frankenhaeuser and Hodgkin to interpret their data as an accumulation in a space some 300 Å thick outside the axon. This space was bounded by an indifferent barrier to give a time constant of about 30 msec but there was also a possibility that this represented the interstices in the Schwann cell layer. Much of our squid data showed an effective E_K of about -20 mV, which is consistent with the outflow of potassium, and this interpretation, as has been supported by Villegas et al., 1962. Dodge and Frankenhaeuser (1958) note the absence of droop for I_K at the frog node and correlate it with the comparatively loose Schwann structure. However, I_K remains constant for at least 100 msec at the lobster axon (Julian et al., 1962b) where the Schwann cells are said to be as dense as for the squid axons.

Hyperpolarization

The study of the effect of a steady base for the membrane potential—the holding potential V_H—for the squid axon began in an obvious way. With our microelectrode measurement of the membrane potential we could choose any holding potential and were not restricted to the resting potential. This gave far more reproducible clamp currents and for a longer time than before. We naturally chose V_H hyperpolarizing enough to achieve a maximum peak inward current, the HH $h \approx 1.0$, partly on the premise that the hardest driven membrane gives the most information and partly because the results were the most uniform and consistent. As the axon deteriorated, the holding current—probably also the leakage current—increased and increasingly polarized the internal, axial, current electrode until it finally bubbled and terminated the experiment. So we looked for a prepulse that would be as effective and less drastic (Cole, 1958). The inward peak current rose rapidly, in 5 to 10 msec, to reach 90 percent of the maximum, for a prepulse hyperpolarization of about -30 mV, but it took seconds to achieve the remaining 10 percent of the I_g for continuous hyperpolarization. As we also explored the effect of the amplitude of V_H, we found that the potassium current at near E_{Na} was increasingly delayed as V_H was increased to as much as -150 mV below the resting potential. This was shown (Cole and Moore, 1960b) in figure 4:43, and after a pulse of $+60$ mV (figure 4:40), the current after the inward peak was also delayed. The half maximum delay of as much as 0.7 msec at E_{Na} was far beyond anything that could be expressed by the HH n^4 approximation. It should not have been surprising, and certainly was in no way discouraging, to find that the HH axon was inadequate under conditions so far removed from those near the normal for which, after all, it had been produced.

Temperature

The temperature coefficients of the time constants, τ_n, τ_m, and τ_h, for the potassium current and the two components of the sodium

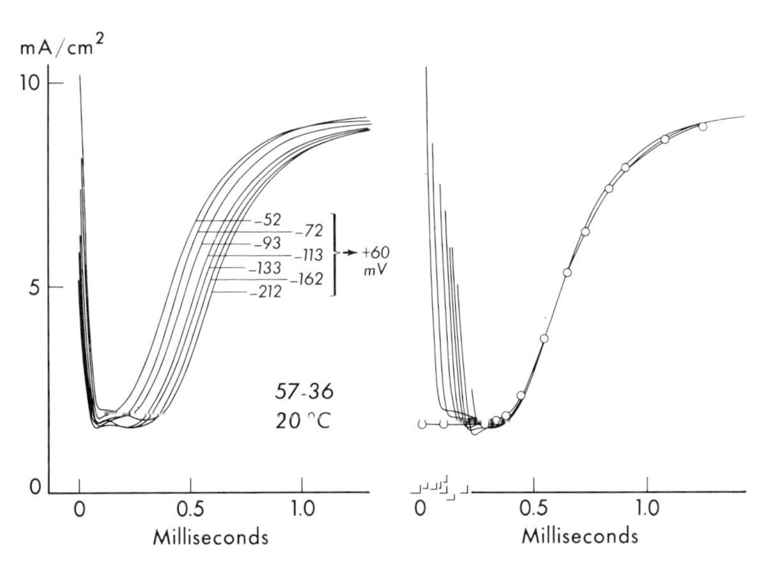

Fig. 4:43. Squid giant axon membrane current density after clamp to near E_{Na} from the different holding potentials as shown. Left as recorded, right superposed to show delay effect of hyperpolarization. Circles are given by an empirical equation.

current were found in the range $Q_{10} = 2.7 - 3.5$, and $Q_{10} = 3$ was used by Hodgkin et al. (1952). Although I do not know of any careful and extensive work to improve upon this value, neither have I seen a cogent reason to question it. Also, the mean coefficient for the conductances g given as $Q_{10} = 1.0 - 1.5$, has been reasonably confirmed (Moore, 1958). These latter data can be equally well interpreted as a temperature coefficient of 3.1 percent per °C of the value at 15°C or as $Q_{10} = 1.4$. Such a variation of conductance was rather reasonably ignored by Hodgkin and Huxley in their first—and far from insignificant—approximation. Their presentation of the data and the calculations from them were considerably simplified. But the fact that this approximation cannot always be ignored is shown for one axon in figure 4:44 where the early peak and steady state currents are rather temperature-dependent.

However, Frankenhaeuser and Moore (1963) have examined the separate rate constants, the α's and β's, of the toad node. Here they find the values for Q_{10}:

	m	h	n
α	1.8	2.8	3.2
β	1.7	2.9	2.8

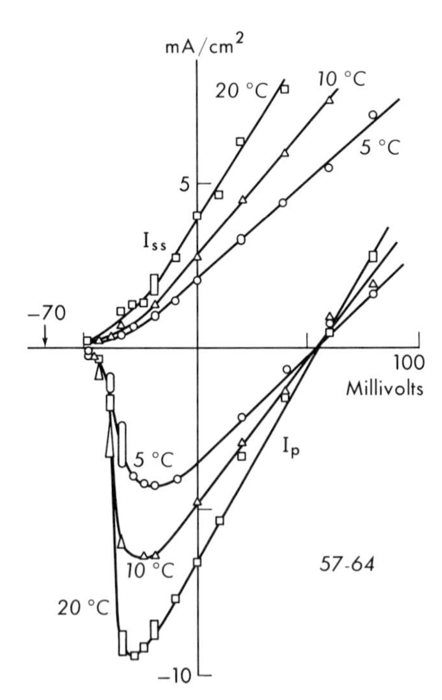

FIG. 4:44. Squid axon membrane current densities at three temperatures under voltage clamps. I_p curves are for early peak currents and I_{ss} for steady state.

The coefficients for h and n, the sodium "off" and the potassium "on," are close to the accepted values and within the limits given for the squid membrane, whereas m, the sodium "on," for the node is much less affected by temperature and is outside the range for the squid. This suggestion of a link between the potassium "on" and sodium "off" and that these slow processes are different from the fast sodium "on" is certainly interesting, but not yet of any particular significance. It is also interesting but rather less provocative that the permeability constants of the node, corresponding to the conductances of the squid membrane, have similar values of 1.3 and 1.2 for the Q_{10} for sodium and potassium respectively. These coefficients are well within the range for aqueous electrolytes.

The effects of temperature become much larger near 0°C and are particularly striking in the range of a few degrees below zero. Spyropoulos (1961, 1965) has investigated many of the interrelations with external potassium and divalent ions in the frog node and *Nitella* as well as the squid axons. However, in normal sea water at −1.2°C, he shows the steady state N characteristic of figure 4:45 for a squid axon.

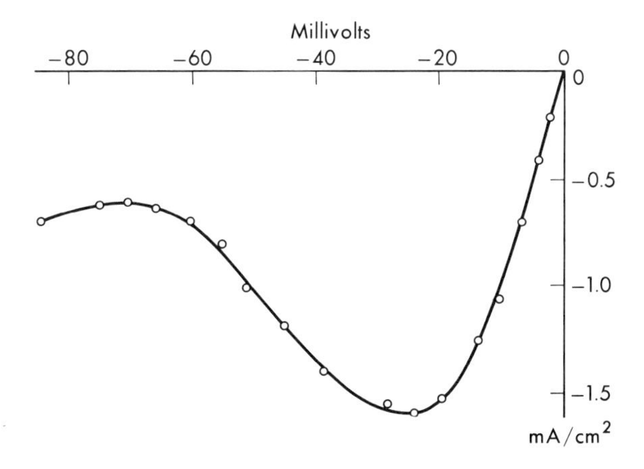

FIG. 4:45. A steady state squid axon membrane current vs. potential characteristic in sea water at −1.2°C.

Early measurements of the threshold strength-duration relation for the squid axon in the space clamp furnished considerable support for the HH axon when it was first computed for this problem on SEAC (Cole et al., 1955). The forms of the transitions from constant quantity Q_0 at short durations to the rheobase I_0 for long pulses were reasonably consistent. So were the temperature coefficients which were quite small for Q_0, and with a Q_{10} for rheobase of 2.0, experimental, and 1.83, computed. The absolute values of these parameters showed much larger differences. At 20°C the Q_0 thresholds corresponded to an average potential change of 21 mV as against the calculated 5.6 mV. At 6.3°C the SEAC value of 6.4 mV confirmed the calculation of 6–7 mV by Hodgkin and Huxley (1952b) where they show an actual threshold between 9.8 and 12 mV. Then Sjodin and Mullins (1958) found a decrease of threshold with increasing temperature at the intermediate duration of 1 msec; Hagiwara and Oomura (1958) showed a threshold change of membrane potential V_T which was independent of stimulus duration; and Guttman (1962) gave a $Q_{10} = 2.3$ for the coefficient of increase of rheobase with temperature in a sucrose clamp.

Guttman (1966) continued with a more extensive investigation of the current, time, and temperature excitation threshold behavior of the squid axon in the sucrose space clamp. The transition from constant quantity to rheobase was again sharper with $\sigma = I(\tau)/I_0 = 1.38$ at $\tau = Q_0/I_0$ than for the Rashevsky-Monnier-Hill two-factor formulation, where $\sigma \geq 1.445$. Q_0 was approximately independent of temperature, and Q_{10} for I_0 averaged 2.35. However, she found $Q_0 = 1.5 \cdot 10^{-8}$ coulomb/cm², a mean threshold potential change V_T about 17 mV, decreasing slightly at longer times and lower temperatures and the mean rheobase was 12 μA/cm² at 20°C. Once again detailed experimental data indicated that the original HH axon was useful in general, but needed some modification in order to be a more complete and adequate representation of at least these real axons.

In the region of long-time threshold effects, Frankenhaeuser and Vallbo (1965) showed a hoped for correlation between the classical accommodation and the inhibiting parameter h. This was for the toad node.

EXTERNAL MEDIUM

Inorganic ions

Turning to intentional alterations of outside ions, we confirmed (Moore, 1958) that lithium could be substituted for sodium with only a small loss of effectiveness (figure 4:46). The fact that what had been E_{Na} was unchanged strongly suggested that this potential did not arise from a reaction because the chemistry for lithium is so different from sodium. With the early currents also very much the same, it seemed much easier to rely on a diffusion explanation in which the mobilities of these two ions were somehow and somewhere effectively the same. Then Dodge (1961, 1963) not only confirmed the observations with lithium but showed that ammonium also may replace sodium.

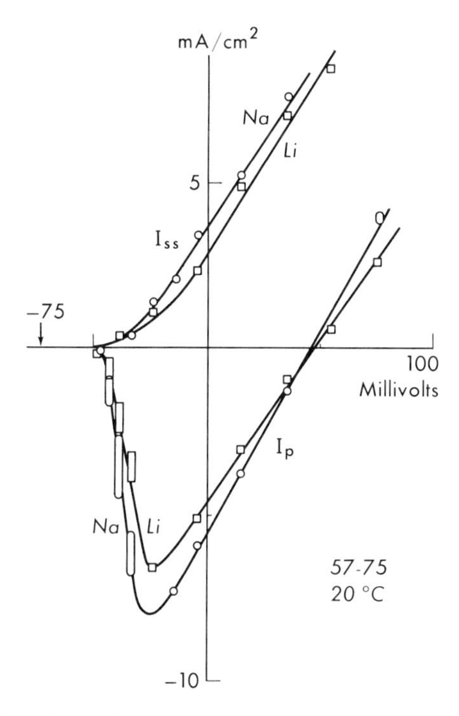

FIG. 4:46. Squid axon membrane current densities in voltage clamp for normal artificial sea water, Na, and for lithium replacement of sodium, Li. I_p are early peak currents, I_{ss} are steady state.

The effect of external potassium ion concentration was appallingly unexpected to me. I had been pleased by our early squid result (Curtis and Cole, 1942) that $V_R \approx E_K$ in high K_e. I was rather disgusted that in this region of simplicity the axon did not conduct an impulse, but I came to accept iso-osmotic K_e as a neutralizer and an inactivator giving $V_R \approx 0$. But I was entirely unprepared for the excitation and propagation in iso-osmotic potassium that Stämpfli (1958) and Mueller (1958) uncovered in a frog axon, and then Segal's (1958) production of the same thing in the squid axon. Moore (1959) quickly applied the power of the voltage clamp to show that the squid membrane potassium permeability resulted in the near steady state negative resistance region for -50 mV $> V > -100$ mV shown in figure 4:47. This experimental result explained the excitability for the squid membrane in high K_e quite satisfactorily, and should have served to shift the emphasis from the two stable states to the unstable state lying between them. But further, this behavior was entirely predictable from the HH axon.

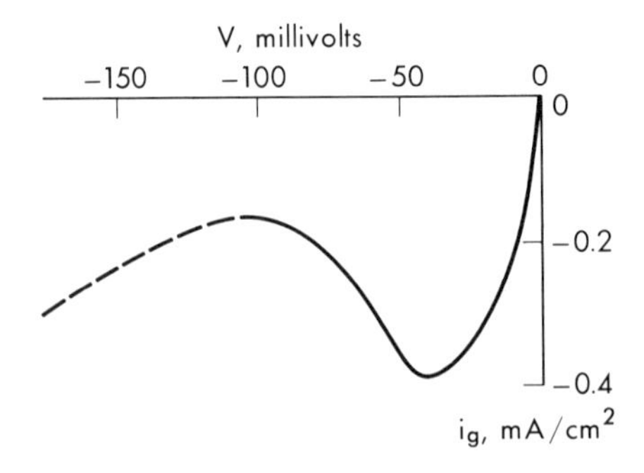

FIG. 4:47. Near steady state characteristic of squid axon membrane in iso-osmotic KCl and normal $CaCl_2$, solid curve. At the larger potentials the current increased from the dashed line towards a breakdown.

In his work on delayed currents in the toad node, Frankenhaeuser found a similar steady state negative characteristic in iso-osmotic K_e. This also was well expressed by the node equations and the independence principle (Frankenhaeuser, 1962c; Dodge, 1963) as is shown in figure 4:48.

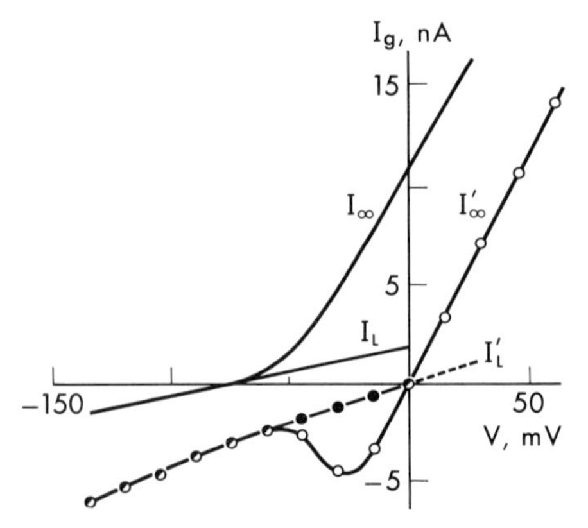

FIG. 4:48. Frog node membrane currents, Ringer's solution, above, and with potassium replacement of sodium, below. I_L and I_L' are the respective instantaneous leakage currents, I_∞ and I_∞' the steady state currents. Currents estimated by independence principle are solid circles for leakage and open circles for steady state.

The lobster axon in the sucrose voltage clamp was similarly found (Julian et al., 1962b) to have a negative resistance region in iso-osmotic KCl as shown in figure 4:49. Furthermore in this region from about −75 mV to −130 mV the currents were again potassium, and were considerably more constant than for squid.

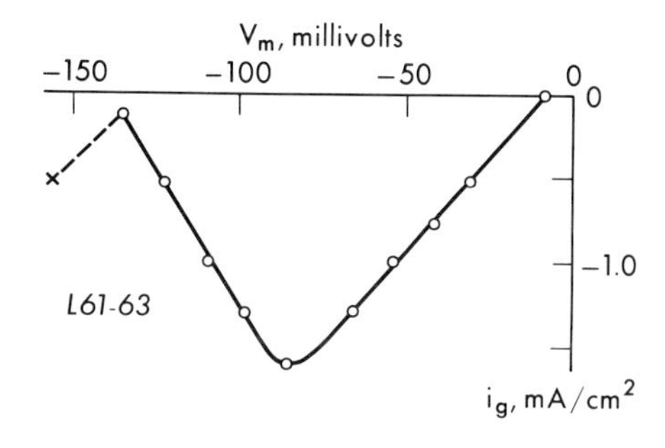

FIG. 4:49. Steady state lobster axon membrane characteristic in iso-osmotic KCl with normal $CaCl_2$, solid circles. The cross shows an initial current which increased towards breakdown.

Hagiwara and Saito (1959a,b) gave similar negative resistances in high K_e for several central cells, but varying from steady state to rapid attenuation. Later Ehrenstein and Gilbert (1964, 1966) worked on a modifying variable with a time constant of minutes for squid. There has been a variety of such effects in electric organs, serving as examples of the diversity of ionic behavior. For the eel electroplax, Nakamura et al. (1965) show the two steady state components of figure 4:50. Both are zero at the resting potential V_R which varies about as E_K for increasing K_e. Both reach steady state values in less than 0.3 msec. The A component is linear in $V - V_R$ and otherwise independent of K_e. The B component conductance is linear for $V - V_R < 20$ mV and increases with K_e but decreases rapidly to become negligible for $V - V_R > 100$ mV. However, the overall transition between high and low conductance is in the right direction and steep enough to give a mild negative conductance region, 40 mV $< V <$ 90 mV.

The steady state potassium conductances were as gratifying to me as they were surprising. Soon after I had met the nerve impulse face to face I became very discouraged by its increasingly obvious complexity and by my inability to ask simplifying questions. The concept of excitation was consoling, and I became obsessed with the aim of making it stand still in space and time so that I could look at it and work on it. With Curtis (1940) and with Baker (1941) I had tried to sneak up on excitation or to hold it, but it always left

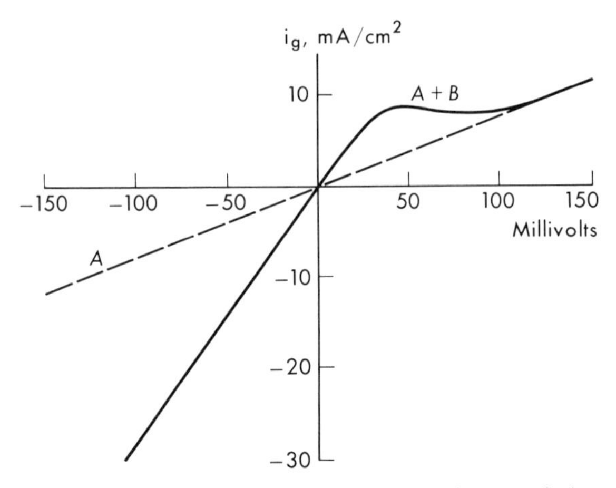

FIG. 4:50. Steady state eel electroplax characteristic. The solid line, $A + B$, for normal saline. The dashed line A was approached by caesium replacement of potassium.

us behind. Unpublished experiments with Curtis showed that it always happened at about the same membrane potential, no matter what the time course of the stimulating current; this was at least some encouragement. No further progress was made until Marmont (1949) and I made the impulse stand still. There were differences from a propagating impulse but they were not obvious; the axon was still and most disappointingly in control. The potential control did indeed prevent excitation and showed that its cause was a negative resistance. But this negative resistance was only fleeting, and I finally resigned myself to a view that for some reason the membrane could not afford to maintain such a costly state. Even though it took nearly twenty years to reach this goal I am now convinced—where I had but hoped before—that this steady state potassium negative resistance as an origin of excitation is the best key to the mechanism of the ion permeability underlying it.

Hodgkin and Huxley used the leakage current as a slightly adjustable parameter and then apparently hoped that it might be a chloride current. It has often been estimated as a linear extrapolation of initial currents from hyperpolarizing pulses, as appears to be true for the nodes. Dodge (1963) concludes that this leakage is practically all an initial potassium current for both frog and toad nodes, figure 4:48. Adelman and Taylor (1961, 1962, 1964) find the leakage to be more than the linear prediction at about the normal

E_{Na} in the squid membrane, although they agree (1964) that it is very probably potassium. To the extent that this initial current is potassium, it may have to be considered as an integral part of the slower potassium phenomena. To the extent that it is nonlinear it represents a process with a time constant considerably less than 100 μsec as was found for the sodium permeabilities of the nodes.

These problems may not be simple to solve and the answer may only become obvious from work on some form other than the squid axon that has served so nobly. From what is so far known, I for one feel required to assume that the negative resistance region, in the several membranes so far showing it, is the result of a process with a millisecond time constant which blocks the permeability process. Yet the several faster nonlinearities now known, with time constants of perhaps much less than 10^{-4} sec, also need to be explained.

The action of external calcium ions on the excitability of irritable membranes has long been recognized as one of the most spectacular and probably one of the most important components of this process (Heilbrunn, 1952; Brink, 1954). I was most deeply impressed by our own measurements (Cole and Marmont, 1942; Cole, 1949) which showed a dramatic effect of external calcium on the ion conductance of squid membranes, and I looked forward to an incorporation of these results into the HH axon—if not a prompt explanation. It should not have been surprising for us to find that, as we were embarked on this objective, Shanes et al. (1959) were making a similar effort. Although it was rather discouraging that the work of Frankenhaeuser and Hodgkin (1957) appeared in the midst of ours, we were in the position of having confirmed the essentials of their results. We had not, however, arrived at their rather amazing generalization that an e-fold change of Ca_e resulted in a 5 mV shift of all of the HH axon parameters on the membrane potential axis. Even more provocative were their indications that external calcium, if only in very low concentration, was essential for excitation. Then, with extreme precautions against calcium, Frankenhaeuser (1957b) found the frog node failed to function in less than a minute and was restored by a trace, $Ca_e < 0.01$ mM.

The Gilbert and Ehrenstein experiments (1965) on the effect of Ca_e on I_K in iso-osmotic K_e are at least approximately explained by such a calcium shift, figure 4:51. But they also give a considerable increase of conductance for reduced Ca_e in the region of $V = 0$ and more than an indication of a symmetrical effect for $V > 0$.

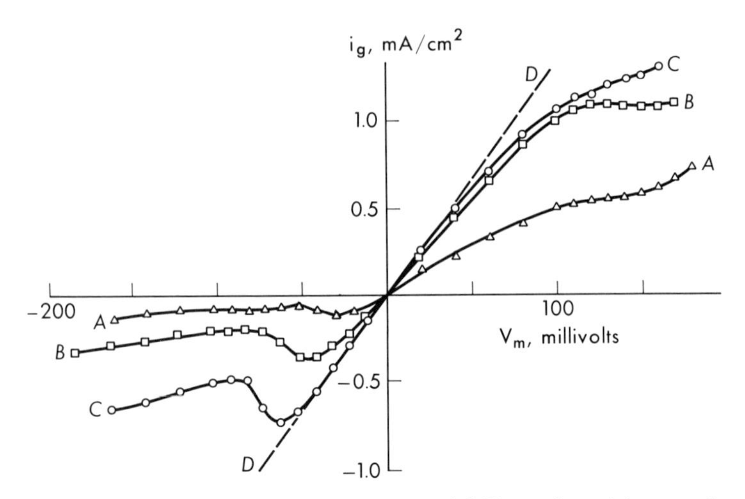

FIG. 4:51. Changes of membrane potential V_m, and resulting steady state currents i_g for *Loligo* axon with intact axoplasm in following external solutions:

Curve	K(mM)	Ca(mM)	Mg(mM)	V_R(mV)
A	440	160	50	1
B	440	10	50	6
C	550	0	0	0

Dashed line *D* shows a possible limiting conductance, about 30 percent larger than found in similar axons perfused with KF.

In the squid axon the calcium ion seems likely to be an external agent. The membrane permeability is slight but clearly increased in activity over the resting value (Hodgkin and Keynes, 1957). Such calcium as does enter appears to be firmly bound and this helps to interpret the finding that the net resting influx of sodium varies somewhat inversely as Ca_e (Adelman and Moore, 1961). Yet in crustacean muscles the divalent cations are actively involved, producing near steady state negative resistance excitation phenomena and action potentials corresponding to their concentrations (Fatt and Katz, 1953; Fatt and Ginsborg, 1958; Hagiwara et al., 1964). Detergents had long been used in red cell investigations and Kishimoto and Adelman (1964) applied them to the squid axon. A nonionic detergent gave a reversible and nearly specific decrease of sodium conductance, while charged detergents generally reduce both g_{Na} and g_K irreversibly.

The skate electroplax (Hille et al., 1965) was not successfully voltage-clamped, but the excitation behavior in reduced Cl_e was quite satisfactorily interpreted in terms of a negative conductance characteristic for the chloride ion!

Partial replacement of external sodium chloride by sucrose was found to decrease both the steady state outward current and the early peak current (Adelman and Taylor, 1964). The steady state current was almost pure potassium, and the decreases were explained as an increase of the series resistance r_S of the membrane. This increase of resistance was proportional to the specific resistance of the outside solution and led to a value in sea water of 3 Ωcm^2. This change of r_S, along with the independence principle, also gave an explanation for the early peak currents. These currents were identified as entirely sodium, and the authors show that for it, equation 3:16,

$$I = \overleftarrow{B}\,\mathrm{Na}_i - \overrightarrow{B}\,\mathrm{Na}_e$$

where $\overleftarrow{B}$ and $\overrightarrow{B}$ are independent of concentrations. From this they conclude that the permeability mechanism—such as across a distribution of barriers—is independent of the concentrations at the boundaries of the membrane. They further gave reason to believe that the potassium mechanism is a separate entity.

Drugs and poisons

It seemed obvious to me that much of pharmacology would have to be rewritten in terms of the pertinent ion permeabilities—sodium and potassium for peripheral nerve. Hopefully also, this should provide clues to the mechanisms. Even in the absence of an understanding of such permeabilities a further breakdown of the components of drug action should help in the design of new agents. In advance of all but the most fragmentary of information, Shanes (1958) undertook to predict the relative roles of potassium and sodium for excitable cells and it will be most interesting to see how successful this courageous work was.

The role of the greatest importance, as far as investigation goes, comes not from the action but from the lack of action of the choline ion. As introduced by Lorente de Nó, it has been an outstandingly successful, practically inert, substitute that has been almost invaluable in investigations of other monovalent cations.

The homologous series of alcohols has long been an attractive approach to problems of narcosis, and ethyl alcohol was an almost necessary, as well as convenient, starting point. In a preliminary report (Moore, 1958) we showed that this alcohol acted at 3–5 percent concentration to reduce the sodium conductance $\bar{g}_{Na}$ to about 5 percent of normal without significant effect on the potassium conductance $\bar{g}_K$, and Armstrong and Binstock (1964) confirmed this conclusion. They went on to show that the sodium effect with the octyl (figure 4:52), *n*-propyl, and methyl alcohols was about the same, at the same molar activities as Ferguson (1951) had recognized. However, at the meeting of the Biophysical Society when Binstock and Armstrong (1963) gave their results, Moore and Ulbricht (1963) reported an increase of $\bar{g}_K$ and no significant change of $\bar{g}_{Na}$ for ethyl alcohol. This aroused some consternation in those immediately involved and a considerable interest of others at the crowded session. The important difference in the experiments was that our laboratory had used the axial wire and microelectrode voltage clamp technique, and this was the first work reported from Duke using the sucrose gaps on the squid axon. If this is indeed the reason for the divergent results, it still is not explained (Moore et al., 1964).

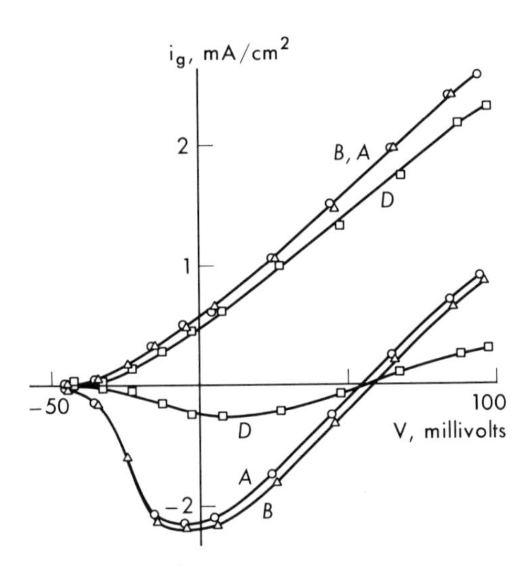

Fig. 4:52. Squid axon membrane current densities under voltage clamp *B* before, *D* during, and *A* after exposure to 0.5 mM octyl alcohol. Lower curves are peak early currents, and upper are for steady state.

Among the tools of membrane research the tetraethylammonium ion (TEA^+), another agent that Lorente de Nó had made important in neurological investigations, became one of the most interesting in an interplay between heart muscle and nerve. The external potentials from heart have been intensively used since Einthoven and the anomalous electrocardiogram pattern was widely investigated. The internal microelectrode finally showed (Coraboeuf and Weidmann, 1949; Draper and Weidmann, 1951) the usual initial sodium spike and proved that it was followed by a long plateau of potential. It was interesting that Fatt and Katz (1953) could produce this in crustacean muscle with TEA, Tasaki (1953) did it with several agents on the frog node, as did Spyropoulos (1956) by repetitive excitation. Then Spyropoulos and Brady (1959) found the same behavior with Ringer's solution which was traced to nickel and chromium from a new stainless steel water supply system. As the result of a complete voltage clamp investigation, Hille (1967) concluded that TEA acted on the maximum potassium conductance $\bar{g}_K$ without affecting the time constant τ_K or the sodium parameters $\bar{g}_{Na}$, τ_n, τ_h or the leakage g_L.

Another interesting and certainly highly significant agent is the puffer fish poison, tetrodotoxin, and the identical trichatoxin. This has long been known and apparently is found in many forms over much of the world. When purified it is the most powerful known poison that is not a protein; specifically, it blocks frog sciatic nerve conduction at a microgram per liter or $3 \cdot 10^{-9}$ M concentration. The composition and most of the crystal structure has been determined (Buchwald et al., 1964). Narahashi et al. (1964) investigated the effect of this poison on the squid axon with the voltage clamp. They found that at 10^{-10} M concentration the inward sodium current was eliminated without markedly changing the potassium current. I was very hopeful that the tetrodotoxin might be analogous to tetraethylammonium and have no effect on outward sodium current. However, results from Columbia (Nakamura et al., 1964) showed that there was no early outward current at potential pulses to $V > E_{Na}$, besides confirming the previous information that the inward I_{Na} was abolished at $V < E_{Na}$. Then titrating tetrodotoxin absorption with lobster nerves, Moore et al. (1967) found an average spacing of almost 3000 Å adequate for blockade. So this compound, too, should be a very helpful tool.

These investigations have given eight minutes or less as the time for complete blockade even at the lowest concentrations. But in addition to this, Guttman (unpublished) found the reversible recovery of the rheobase threshold from tetrodotoxin to have a half time of about six minutes.

Hille (1966) investigated the effect of tetrodotoxin (figure 4:53), prochlorperazine, xylocaine, and other local anesthetics on the frog node with his automated procedure. He found $\bar{g}_{Na}$ to be selectively, nearly reversibly, reduced in seconds. The conclusion was "that the sodium, potassium, and leakage permeabilities were chemically independent, probably because their mechanisms occupy different sites on the nodal membrane."

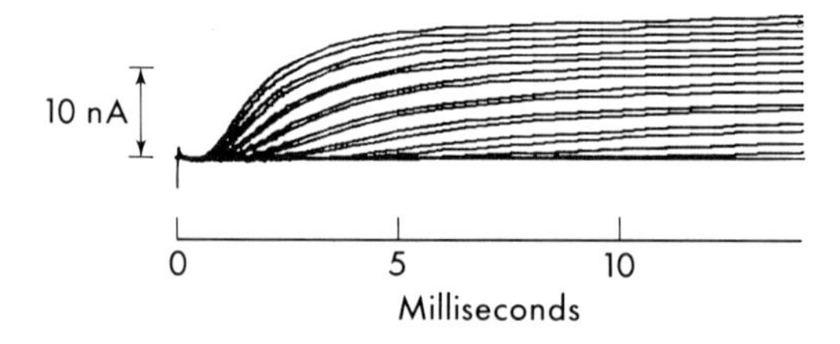

Fig. 4:53. Frog node voltage clamp currents. Same node as figure 4:19 but in $3 \cdot 10^{-7}$ M tetrodotoxin.

Another new and similar compound is saxitoxin which was isolated from clams and mussels (Schantz, 1960) and is of comparable potency.

I have been very impressed by the ability of axons to conduct impulses under extremely unfavorable conditions, and I had come to believe that the peripheral axon was the strongest link in the nervous system. Now we have a challenge not only to find how a nerve works but also to save victims of puffer fish poison!

Procaine is one of the most widely used and investigated of narcotics, and an understanding of its action in terms of ion permeabilities was of great interest and potential importance. Again Shanes et al. (1959) and our laboratory (Taylor, 1959) undertook this work with the voltage clamp on the squid membrane, independently and nearly simultaneously, and arrived at essentially the same conclusions. Although it was not practical to separate the sodium and potassium conductances throughout their time courses,

Taylor was able to show that the primary effect of procaine was to reduce the sodium conductance, and so reduce the excitability of the axon. At the same time, however, there was a decrease of potassium conductance, tending to make the membrane more excitable. Although this antagonism was not enough to prevent narcosis, it did reduce the effectiveness. But probably any agent acting only on the sodium process would have such undesirable effects elsewhere in a mammal as to make it worthless in clinical medicine. If this is generally true, we may not learn much about nerve membrane function from the pharmacopoeia nor add much to the pharmacopoeia as we learn more about nerve membranes. However, procaine seems to block sodium specifically and completely in eel electroplax (Grundfest; personal communication), and an antagonistic action between procaine and calcium has been described for the lobster axon (Goldman and Blaustein, 1965). A report by Ulbricht (1965) indicates that veratridine allows some sodium flow with a long duration. This suggests a specific effect on the steady state sodium inactivation parameter h.

It seems well established that the sodium hypothesis of Hodgkin is quite adequate to explain squid axon behavior under normal and some near normal physiological conditions. The sodium inflow, as described by the Hodgkin-Huxley equations, is then both necessary and sufficient. For other membranes and under other conditions such a sodium flow may not be necessary; an impressive list was assembled by Tasaki et al. (1965). In addition to a dozen items from the literature, they rank in order three dozen organic cations, from seven categories, for their effectiveness externally in producing squid action potentials without sodium. Then, under extreme conditions, Tasaki et al. (1966) produced activity in squid axon with calcium as the only external positive ion.

Tobias (1960) showed that external proteases were entirely without effect on the functioning of lobster axons. But phospholipases almost immediately and irreversibly stopped the electrical activity and soon the normal passive membrane parameters disappeared.

Extensive and impressive investigations of ion and drug actions at the outside of the motor horn cell have been made with microprobes. These had as many as six channels besides the potential tip which was sometimes intracellular. I am not able to review the results and their interpretations, such as Salmoiraghi and Steiner (1963) and Curtis (1966), except to comment that the ionic theory for the squid membrane has been generously applied and modified as given by Eccles (1965).

INTERNAL MEDIUM

Injection

The intentional and direct modification of the axoplasm of the squid axon by injection was first started as a test of the HH axon. The results were however rather controversial because of the long space constant of the resting membrane and the long stretch of an axon involved in an impulse (Grundfest et al., 1954; Brady et al., 1955). Hodgkin and Keynes (1956) put together a set of rules for adequate injection experiments and implemented them with an automatic arrangement which introduced solution to replace the volume of the capillary as it was withdrawn. They were thus able to confirm the prediction of the decrease of the action potential with an increase of the internal sodium ion concentration.

The pronounced differences between the action potentials of smooth axons and skeletal muscle fibers on the one hand, then of the nodes of medullated axons, and finally of cardiac cell membranes, must reflect considerable differences in membrane characteristics. The rising phases of the action potentials were all quite similar and the maximum potential changes, the spike heights, were all near to the sodium potentials. Therefore presumably sodium was the principal component of the early, excitation, phase of the activities, but there was no obvious explanation for the various recoveries.

The first hopeful indication came in the production of a marked plateau in the recovery phase for the node. But these agents were not effective on the squid axon. It was, however, very gratifying when Tasaki and Hagiwara (1957) produced a similar pattern by application of TEA to the inside of the squid axon and it was suggested that this acted to block an outflow of potassium. Armstrong and Binstock (1965) investigated the effect of internal TEA on the squid membrane in some detail. They found after injection to about 40 mM the inward and outward sodium currents and the inward potassium current to be relatively unchanged, whereas the outward potassium current was reduced practically to zero. This was an adequate basis for the prolonged recovery phase. They found a half time for this action of TEA to be about five minutes and they also showed a fast rectification in high external potassium.

Armstrong (1966) continued with TEA injections of the *Dosidicus* axon down to 0.3 mM. These gave the time course of the blockade to outgoing potassium and allowed a more complete description of the interactions between what had first been a TEA diode and the normal potassium gate.

The barnacle giant muscle fiber was massively injected by Hagiwara (1965) by pumping in electrolytes until the fiber was two to four times its normal volume. Changes of internal and external potassium and chloride agreed in making the gradients of both of them responsible for the resting potential and in particular V_R = 58 log ($K_e + Cl_e$) + constant. After the Ca_i was reduced and the membrane became excitable, the spike was completely independent of Na_e. However, both calcium and potassium were implicated in this process by another nice relation—that the overshoot V_0 = 29 log (Ca_e/K_i) + constant.

PERFUSION

After I had learned the delicacy of marine egg cells from embryologists I was often amazed by the indignities that these and other cells would and would not tolerate. Blinks (1930, 1932) impaled the giant plant cells, *Valonia* and *Halicystis*, on capillary tubes so they could survive and permit replacement of the cell sap. Our attempts to put a microcapillary electrode into heart muscle culture cells had promptly killed them (Hogg et al., 1934). I was not particularly hopeful that our first internal electrode would do anything more than kill the squid axon. I was very uncertain about the Ling-Gerard microelectrodes for muscle as I watched them being developed, and then for other cells—the squid axon in particular. I have not been at all happy about tying or cutting Purkinje fibers. I had watched the axoplasm being squeezed out of the squid axon for chemical analysis, but I never had any passing thought about the membrane as it was thrown away—except that it must be ruined. Perhaps Hodgkin had other thoughts. As he was resting his eyes during the cleaning of an axon at Woods Hole in 1938 he had prophesied that he would someday perfuse the inside of the axon. Then in 1948, after he and Katz had been unsuccessful, he promptly joined Baker and Shaw at Plymouth when they had found that the membrane still functioned after being run over by a miniaturized lawn roller. They (Baker et al., 1961) showed first a small delayed action potential corresponding to the high internal resistance in the rolled-over region. This returned to near normal when axoplasm was rolled into it from an adjacent region of the axon, figure 4:54. They then reinflated the axons with potassium sulfate to find that they were usually and in most respects nearly normal.

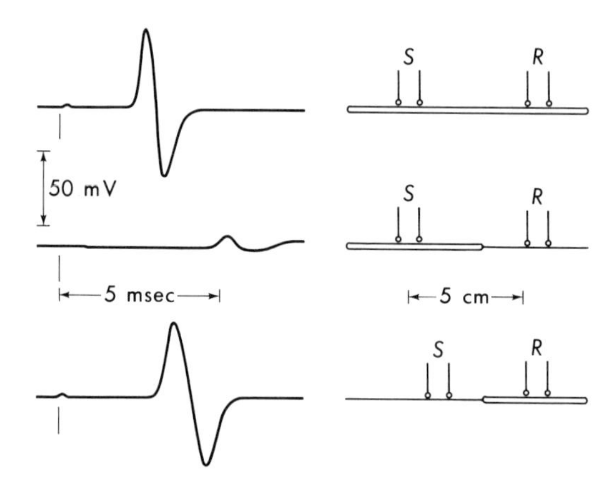

FIG. 4:54. Original experiments on squid axoplasm extrusion. *S*, stimulating, and *R*, recording electrodes gave normal record for normal axon, top. After axoplasm was rolled out from *R* region, center, the impulse was delayed and the external potential was reduced by the high internal resistance. After axoplasm was rolled from *S* region to *R* region, the conduction time was still long because of the flattened length, but the potential returned to full amplitude in restored region.

In general, the internal perfusion was found to be more difficult with the smaller axons available at Woods Hole. Nonetheless, and at about the same time, the group with Tasaki (Oikawa et al., 1961) succeeded in washing the bulk of axoplasm out of these axons and similarly replacing it. A gentler hydraulic erosion procedure (figure 4:55) was evolved by Adelman and Gilbert (1964) and found reasonably satisfactory for voltage clamp work. These groups established that axons with 95 percent of the axoplasm replaced by an inorganic potassium solution can function for hours, a few hundred thousand impulses, and with a solution flow up to 150 times the axon volume. Then Tasaki et al. (1965) perfused *Loligo* with several proteases and Huneeus-Cox et al. (1966) used cysteine in both *Loligo* and *Dosidicus* axons. These apparently removed the last remnants of internal material from the membrane, and without adverse effects.

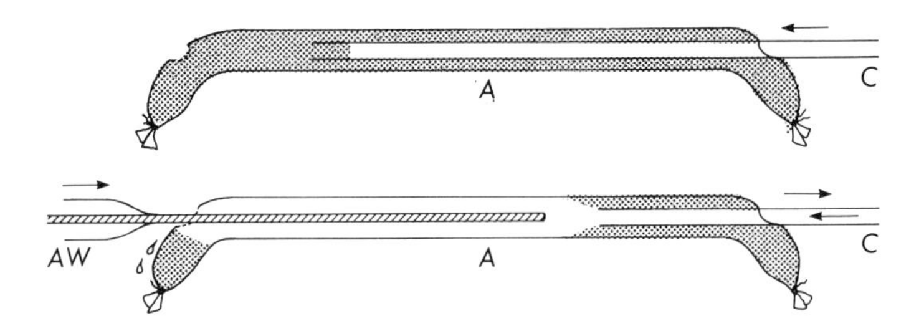

FIG. 4:55. Squid axon perfusion. A capillary *C* is inserted into the axon *A* from right, above; flushed out, withdrawn slowly as perfusate erodes axoplasm, below, and is followed by axial wire *AW*.

This dramatic perfusion development opened a whole new and probably crucial field of investigation. So long as the interior of the axon was inaccessible for more than minor changes of composition and concentration almost anything could be explained away by some disgusting postulate that could not be either confirmed or contradicted. Now, however, except for less than 5 percent of the axoplasm, the last retreat into confusion was cut off and new avenues of information opened.

This perfusion field became understandably increasingly active as more research groups entered it. Variations of the potassium-sodium ratios in iso-osmotic perfusion electrolytes gave essentially normal or predicted resting and action potentials for squid axons. Thus it seems highly improbable that there is any effective source of energy, other than the electrochemical, involved in the electrical behavior of this axon. The field was soon extended to include the giant barnacle muscle fiber (Hagiwara et al., 1964) and *Nitella* (Kishimoto, 1965).

In general, the initial conclusions have been confirmed, but with strange situations such as low ionic strength, and strange substances such as glutamate, results strange to the Hodgkin-Huxley axon and the concepts underlying it have appeared. No doubt we should expect more confusion before we come into a bigger and better state of order.

Ionic strength

Narahashi (1963) perfused an axon in sea water with sucrose and 6 mM potassium to find the expected resting potential near zero and an action potential with the plateau encountered in cardiac fibers. But the peak of the action potential of 60 to 100 mV was about the same as for 540 mM potassium inside—an overshoot with a vengeance and a threat to the sodium theory. An answer to the question whether this was an effect of potassium, or sodium lack, or of low ionic strength has been given by Chandler et al. (1965). They found that the h_∞ vs. V curve was shifted by about 60 mV towards positive potentials by 6 mM potassium in sucrose, but was returned to the original position by an electrolyte, such as choline chloride, instead of the sucrose. They therefore concluded that ionic strength was the principal factor and proposed the zeta potential as an explanation. This has been so well known and for so long—Helmholtz gave the qualitatively correct interpretation—that I have often had to explain that it was negligible in usual biological electrolytes (Cole and Moore, 1960b)!

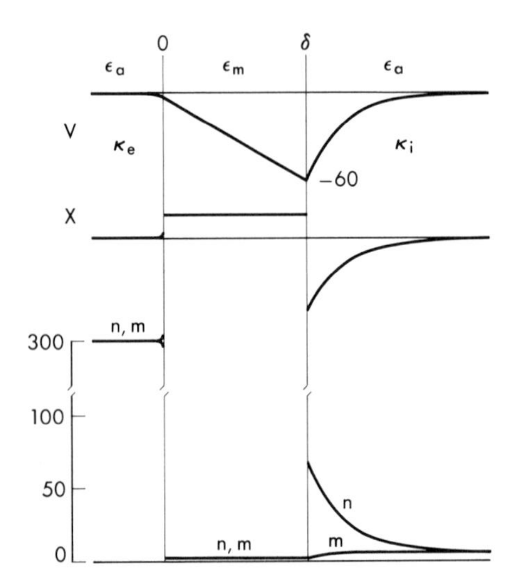

FIG. 4:56. Surface charge at inside of membrane and ion cloud in inner dilute electrolyte. The potential V, field X, and ion distributions m, n, across membrane, 0 to δ, and adjacent electrolytes.

The nature of the zeta potential effect is given in figure 4:56. There is a charge $\bar{\sigma}$ at $x = \delta$, the inside of the membrane and the net charge in the nearby electrolyte gives rise to a field and potential difference. It is found that $V(0)$ is negligible and that the Debye linear approximation is within a factor of two for $V(\delta) = 100$ mV.

The ion concentration within the membrane is so small as to make the Debye κ virtually zero, $\kappa\delta \ll 1$. So we take inside the axon

$$\frac{d^2V_i}{dx^2} = \kappa_i^2 V_i$$

where $\kappa_i^2 = 8\pi n_i e^2/\epsilon_a kT$ and n_i and ϵ_a are the internal concentration and the aqueous dielectric constant. Then the field at the outside of the membrane $X(\delta+) = \kappa_i V(\delta)$, and in the membrane $X(\delta-) = X_m = -V(\delta)/\delta$. Since the Gauss law gives $4\pi\bar{\sigma} = -\epsilon_m X(\delta-) + \epsilon_a X(\delta+)$, where ϵ_m is the membrane dielectric constant, then

$$\frac{\bar{\sigma}}{V(\delta)} = C_m\left(1 + \frac{\epsilon_a}{\epsilon_m} \cdot \kappa_i\right)\delta$$

where C_m is the membrane capacity of 1 μF/cm^2. Then for the corresponding $\delta/\epsilon_m = 10$ Å and $\epsilon_a = 80$,

$$\frac{\bar{\sigma}}{V(\delta)} = C_m(1 + 800\ \kappa_i)$$

where κ_i is in A^{-1}. For a concentration of 6 mM, $\kappa_i = \frac{1}{30}$ A^{-1} and

$$\frac{\bar{\sigma}}{V(\delta)} = 25 \cdot 10^{-6} \text{ coulomb/volt cm}^2.$$

Then at $V(\delta) = 60$ mV,

$$\bar{\sigma} = 1.5 \cdot 10^{-8} \text{ coulomb/cm}^2 = 10^{12} \text{ electrons/cm}^2$$

which is an entirely reasonable value for adsorbed surface charges as found long ago (Abramson and Müller, 1933) and gives a charge separation of 100 Å. On the other hand, a similar charge in normal ionic strength would give only 6 mV.

This is a very heartening proposal. I like it and I hope it can be proved. It would help to strip the best measurements now available down to the actual surface of the membrane—as so many of us have been trying to do for so long. May it not also help to explain indiscriminate shifts of other membrane characteristics along the potential axis—such as suggested by Huxley (Frankenhaeuser and Hodgkin, 1957) for external calcium?

Voltage clamp measurements with internal perfusion of 300 mM KCl plus sucrose gave evidence that potassium can flow outward through sodium channels. Chandler and Meves (1965) found the usual early inward sodium current below 60–70 mV, where it disappeared. But for higher potentials a similar current came out from the sodium free interior. Internal rubidium and caesium gave the same effect but smaller. Calculating permeabilities from the equilibrium potentials they concluded that the sodium system discriminates between the alkali metal ions in the ratios

$$Li:Na:K:Rb:Cs::1.1:1:1/12:1/40:1/61.$$

Continuing with this idea that, within limits, any channel can carry any ion, Binstock and Lecar (1967) presented a rather amazing example. There had been scattered evidence that ammonium was one of the many somewhat satisfactory substituents for sodium in normal nerve activity. When it was substituted for external sodium, the fast transient voltage clamp conductance for the squid membrane was about a third of normal. But when internal potassium was replaced with ammonium by perfusion the slow, steady-state conductance was nearly half of normal. Thus ammonium could be rather well carried by both the fast, normally sodium, and the slow, normally potassium, processes. To complete the picture, Binstock and Lecar showed internal TEA to block the outflow of ammonium as it did for potassium and external tetrodotoxin to block the inflow of ammonium as it did for sodium. Thus the fast and slow processes were not only able to carry the same ion but even with the same ion each process still reacted highly specifically to the two agents.

Tasaki with collaborators reported extensive surveys of the behavior of squid axons perfused internally with a variety of electrolytes, and summarized them with Singer (1965). They compared different perfusates by the times of exposure during which the axons maintained excitability and reasonable action potentials in a space clamp. In this way they found that NaCl, KBr, and KCl were much more favorable for survival in concentrations of 50–100 mM than of 400–600 mM. The negative ions were then ranked in an order of favorability which was the same whether the potassium, rubidium or caesium salts were used:

$$\begin{aligned} F > \text{phosphate} &> \text{glutamate, aspartate} > \\ &\text{citrate} > \text{tartrate} > \text{propionate} > \\ &\text{butyrate} > \text{sulfate} > \text{acetate} > Cl > \\ &NO_3 > Br > I > SCN. \end{aligned}$$

Similarly the positive ions were arranged in the sequence:

$$Cs > Rb > K > NH_4 > Na > Li \gg \text{divalent cations.}$$

Quite apart from Tasaki's macromolecular interpretation of these data and its ultimate significance, this work made a very important contribution. The greatly increased axon survival times with CsF or even KF perfusates made them the standard reference solutions in almost all later experimental work.

Rojas and Ehrenstein (1965) perfused the giant axon of the Chilean squid, *Dosidicus gigas*, with 600 mM KF. With external sea water the normal resting and action potentials of −52 and 100 mV respectively and the voltage clamp currents were reasonably conventional. The sea water was then replaced with 600 mM KCl to give a zero resting potential. In voltage clamp the potential-current relation became nearly linear after 5 msec and usually showed a slight inflection at −70 mV after 30 msec. The authors interpret this result as an indication that some component in the normal axoplasm of the *Loligo pealii* caused the negative resistance segment found by Gilbert and Ehrenstein (1965), as was later confirmed (Lecar et al., 1967).

Adelman et al. (1965a,b) investigated the effects of potassium-free internal solutions at low ionic strength on axons with K free artificial sea water outside. The action potential plateau was increased to many seconds, and such spikes might appear about once a minute over long periods. This behavior they ascribe to an increase of the sodium inactivation time τ_h without appreciable change of τ_n. Conventional experimental techniques were used to measure τ_h which behaved in the usual manner for $V < 0$, but turned and rose rapidly to values as high as 10 sec for $V > 0$.

Moore (1965) gave a preliminary report on experiments with Takata on the effect of internal perfusion with tetrodotoxin. In the complete report (Narahashi et al., 1966) a concentration of 10^{-6} M was without significant effect on either sodium or potassium currents when applied to the inside of the membrane but the sodium current was blocked after external application at 10^{-8} M. These results suggest that the sodium blockade occurs at only the outer surface of the membrane and that perhaps the membrane is relatively impermeable to the toxin. Rojas and Luxoro (1963) reported that internal proteases were highly effective in destroying function in *Dosidicus* but Tasaki et al. (1965) maintained function in *Loligo* after axoplasm removal by low exposures to several proteases. Huneeus-Cox et al. (1966) found internal oxidizing and reducing reagents to be relatively innocuous for *Loligo* while the adverse action of the mercaptids showed that the integrity of the sulfhydryl bonds was important to axon activity.

Spike overshoot

But Tasaki with Takenaka (1963, 1964) and Luxoro (1964) produced yet another problem for Hodgkin's ionic theory. With internal perfusates of fluoride, glutamate and aspartate they again obtained action potentials which exceeded the sodium potentials. There are many reasons for an action potential to be smaller, but for it to be too large is a dramatic performance, to say the least. This Hodgkin himself well realized, and used effectively, when he and Katz (1949) increased the external sodium concentrations to increase the action potential—in accordance with the theory and in spite of the injurious effect of the hyper-osmotic medium.

However, Chandler and Hodgkin (1965) were able to produce a similar effect, and once again it was to be explained in purely physical terms. With the usual British vertical cannulated axon of figure 4:57a, an action potential at the tip was transmitted along the capillary and measured as V by an electrode at the top. As in our first internal electrode measurements, the effect of the capillary transmission line with its internal resistance and wall capacity make it unpleasant to compute the outside potential when it is everywhere the same. But it has not yet been possible to compute a traveling potential from a recorded $V(t)$, and we could only make approximate empirical—and in at least one case, probably wrong—corrections. This, however, seems to have been a minor difficulty in the present case.

In the simple equivalent circuit, figure 4:57b, the membrane potential at V_1 gives the current

$$I = (V - V_1)/R$$

along the capillary. This current flows across the capillary wall in the cannula as

$$I = C\, d(V - V_2)/dt$$

for the cannula potential V_2. Then

$$\tau\, dV/dt + V = \tau\, dV_2/dt + V_1 \qquad (4:9)$$

where $\tau = RC$. As an example Chandler and Hodgkin used square waves, $V_1 = \overline{V}(t)$, and a delayed $V_2(t) = \overline{V}(t - t_0)$. Here the Fourier

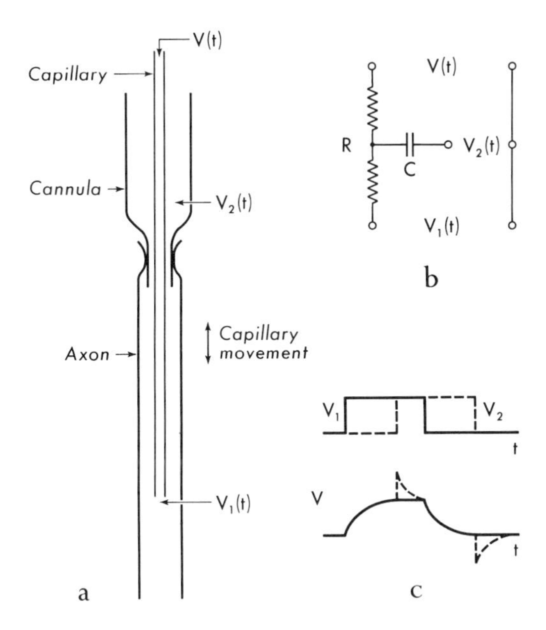

FIG. 4:57. Origin of cannula artifact. A potential such as V_1 at the tip of the glass capillary is conducted through the electrolyte resistances R, to produce an output V lower right, solid line. As this same potential propagates upward, it produces the later potential V_2 in the cannula. The capacitive current through C for the wall of the capillary in the cannula gives an additional output component to V, lower right, dashed lines.

transforms 448, 601, 603.1 (Campbell and Foster, 1931) of equation 4:9 give

$$V/\overline{V} = 1 - \exp(-t/\tau),\ t < t_0$$

to give the solid line of figure 4:57c and then

$$V/\overline{V} = 1 - \exp(-t/\tau) + \exp[-(t - t_0)/\tau],\ t > t_0$$

gives the additional dashed component. This component had been recognized in the past as a "cannula artifact" and is an example of long known voltage doubling devices. Returning to equation 4:9 it is apparent that both the initial distortion and the subsequent artifact are reduced as R is made small so $\tau\, dV/dt \ll V$.

Chandler and Hodgkin showed that the cannula component might be placed anywhere on the $V(t)$ curve for an axon by moving the capillary tip, and that it is virtually eliminated by a platinum wire in the capillary—essentially a "space clamp" at high frequencies. They further give reason to suspect a similar phenomenon in Tasaki's experimental arrangement. It is indeed tragic if, once again, the simple physical behavior of an experimental system has been interpreted—and to any degree accepted—as undermining an altogether admirable and very useful correlation of a considerable range of physiological fact.

COMMENT

Effort seems to me to have been mostly devoted to preliminary and superficial experiment and analysis in new fields. Much new information was available on the external and internal effects of inorganic and organic ions and considerable evidence was accumulated to support some hopeful generalizations:

The internal and external membrane boundaries are quite different in their responses to many agents and various aspects of possible lipoprotein structures are probably involved.

Almost all membranes show at least some slight permeability to almost all ions. These permeabilities may be divided on a temporal basis into two categories according as they are changed rapidly or slowly after a change of membrane potential.

The fast and slow processes are so largely independent of each other as to suggest that they may be spatially separated but the possibility of coupling remains.

In the absence of divalent ions at least the slow and probably the fast, permeabilities are expressed as linear conductances. External calcium produces nonlinear and negative steady state conductances of the slow process of the squid membrane.

This blockade by calcium and, to a lesser extent, other divalent ions, gives at least a possible phenomenological basis for excitability by the slow and, probably also, the fast processes.

However, none of these developments has yet given a clear indication of the nature of membrane ion permeability and its control. But they seem only the necessary excursions in many directions that may not have to go on much longer or much farther before a distinct goal appears and effort can reasonably be concentrated in its direction.

Changing the Empirical Equations

It is almost impossible for me to weight the relative importance of the various components of the Hodgkin and Huxley work of 1952. The identification and separation of the sodium and potassium components of the ionic current under voltage clamp were the initial steps, without which none of the subsequent procedures would have been possible. These steps remain as essentially the end of experiment and the beginning of test and interpretation, but much of the impact and the significance of the work would not have appeared without the next step—the reduction of these data to analytical form.

The expression of the clamp data in the system of equations conveniently called the HH axon, is admittedly purely empirical, only approximate, and of limited scope. It has no necessary connection with the mechanisms of ion permeability. It may represent the mean of widespread data and may depart from this noticeably. The data which it represents are extensive but far from exhaustive, and new data do not necessarily conform. Nonetheless, the HH axon is a remarkable feat both as curve fitting and as an instrument of prediction.

As the HH axon was found to be less than perfect for old facts of other excitable membranes besides the squid axon and for numerous new experimental results on a wide variety of membranes, the urge to tinker was irresistible. How could anyone be expected not to try extending the scope of the equations—particularly if he had a computer or wanted one? So the development of the original equations and some of the modifications have been interesting explorations—if only to illustrate the bruises that come from fumbling in the dark without even the misleading light of an impossible theory.

ORIGINAL SQUID AXON FORMULATION

As to curve fitting, the procedure and the results of Hodgkin and Huxley (1952b) are entirely unorthodox and are looked at with both amazement and admiration by trained mathematicians. But the process was guided by the intuition and the hopes of very imaginative and experienced investigators and many factors entered into the choices of the analytical forms as they were finally presented. For some of these choices I know the basis, but for others I can only guess.

One impressive fact was that conductances changed continuously at a change of potential but currents did not. Incorporating the principle that the currents were proportional to the electrochemical potential $V - E$ as the driving force on an ion, the conductance g appeared as a permeability to give equation 1:27,

$$I_g = g_{\mathrm{K}}(V - E_{\mathrm{K}}) + g_{\mathrm{Na}}(V - E_{\mathrm{Na}}).$$

Then the time decay of currents toward zero after removal of the potential were nearly exponential, as were the rise and fall of the peak inward currents. These behaviors must have urged the use of first order differential equations. The expression for the sigmoid rise of currents and conductances as a power function was probably guided strongly by the rapid fall of the maximum conductances as the membrane was polarized toward and beyond the resting potential. These decreases at a rate approaching $1/e$ for 6 mV suggested either a four valent particle or a probability of four independent events.

Much though they liked a polyvalent carrier, Hodgkin and Huxley settled for the probability parameters n, m, and h of single events. These parameters could then follow first order chemical kinetics, equations 3:3,

$$dn/dt = \alpha_n(1 - n) - \beta_n n$$

with forward and reverse rates α_n and β_n to give

$$n = n_\infty - (n_\infty - n_0) \exp(-t/\tau_n)$$

where n_0 and n_∞ are initial and final values and the time constant $\tau_n = 1/(\alpha_n + \beta_n)$. Now a conductance proportional to n^4 was quite satisfactory. It could be thought of as the probability for an open potassium path requiring four available sites. For a return from n_0 to a negligible n_∞,

$$n^4 = n_0 \exp(-4t/\tau_n)$$

to give the observed exponential tail at a return to the resting potential. A depolarizing step from n_0 negligible then showed

$$n^4 = n_\infty{}^4[1 - \exp(-t/\tau_n)]^4$$

a sigmoid rise approximating the data. A higher power than four would sometimes have given a better representation (Cole and Moore, 1960b) but it was inconvenient; perhaps, too, the authors were reluctant to abandon their fourfold prejudice.

Turning to the sodium current, the slow facilitating factor h, most inappropriately called inactivation, followed an exponential time course so it could be used directly, with h_0 and τ_h known. The fast activation m behaved much the same as n, but m^3 could be used and the complete expression m^3h was then of the fourth degree and analogous to n^4. The separation of m and h was a log-log graphical routine to find τ_m/τ_n and $m_\infty{}^3h_0$ and so obtain m_∞ and τ_m.

There had been, somewhere in this procedure, either the assumption or the discovery that the rate constants α and β, the corresponding steady states $\alpha/(\alpha + \beta)$, and time constants $1/(\alpha + \beta)$ were functions of the instantaneous potential V, and of V alone. Thus the kinetics of n, m, and h and of the consequent g_K, g_{Na} are also functions of V, and of V alone. This is a truly remarkable situation which must have, it seems to me, a high significance for the mechanisms of the permeabilities. True, as ionic concentrations are changed, it has been found necessary to modify coefficients as functions of E_K and E_{Na}. For this an independence principle equation 3:17 analogous to that given by Goldman has been found useful. But the numerical values for the six functions, the α's and β's, have only to be approximated as functions of V. There are doubtless many possible analytical forms with as yet little basis for choice between them except convenience. I suspect that the form chosen by Hodgkin and Huxley was again a shocking innovation, but it was again a result of experience and hope. The use of exponentials reflected the recognition of electrochemical factors, whereas the use of quotients could be based on familiarity with the Goldman formulation and the possibility that it may in some way be a valid approximation to processes of a living membrane.

It is certainly a delightful expression of confidence that after fifteen years these same equations are regarded as the standard axon. Yet I feel entitled to complain again that these equations have not been brought up-to-date except by minor and superficial adjustments.

MODIFICATIONS FOR NODES

It is abundantly clear that the first and the most complete new data and the most expert tinkering were given by Frankenhaeuser and his colleagues for the nodes. Dodge and Frankenhaeuser had found that the sodium currents could not be expressed by the simple linear conductance as for the squid membrane. The instantaneous current was not proportional to the change of membrane potential, but the relation could be expressed by a Goldman permeability P_{Na}. This was disconcerting, to say the least, because I was still hopeful that electrodiffusion was important for potassium and that in this respect, at least, sodium was somewhat similar to potassium. But entirely aside from mechanism, it still may be true that nonlinear systems require a finite time to become nonlinear. Whatever the nonlinear process, it was certainly complete in 10–100 μsec and expressible by a steady state permeability relation. This situation seemed quite acceptable to Frankenhaeuser (1960), and he was concerned as to why the squid membrane was not the same as the *Zenopus* node. Then analysis of the later currents showed them to be mostly potassium, but with some indications of a sodium component. However, the instantaneous relation was nonlinear and had again to be expressed by a permeability P_p. Although the potassium permeability depended on the n^4 parameter as for the squid, the sodium was best given by m^2h instead of m^3h and these factors were separated as had been done before.

Frankenhaeuser and Huxley (1964) assembled all of the toad node voltage clamp results in the form of standard data and showed that these computed to a good approximation of the nodal action potential. After giving the components during the standard action—m, h, n, and d; P_{Na}, P_{K}, and P_d; I_{Na}, I_{K}, I_d, and I_L—they proceeded to show the results of arbitrary changes of C_m, P_{K}, P_d, P_h, and P_{Na}. This led to the very reasonable—and gratifying—conclusion that the voltage clamp measurements, and their approximation by empirical equations, produced better action potentials than were to be had by such simple tinkering.

When Dodge (1963) devoted himself to the orthodox and almost completely noncontroversial frog node, he had still to express the sodium currents as Goldman permeability functions of the membrane potential. Yet he found that the kinetics of this process for this membrane were better expressed in the form m^3h as for the squid membrane. The same experiments as on toad and squid gave the same inactivation phenomena for the frog. Then he could reasonably approximate the potassium behavior by a simple conductance and a probability parameter, which also appeared as n^4 as for the *Zenopus* node and the squid membrane. Later he approximated some of these nodal clamp current data by only the HH conductances g_{Na} and g_K. The particular result for one frog node is shown in figure 4:58 in terms of steady state and time constant parameters; it may be compared with the similar presentation for the squid in figure 3:32.

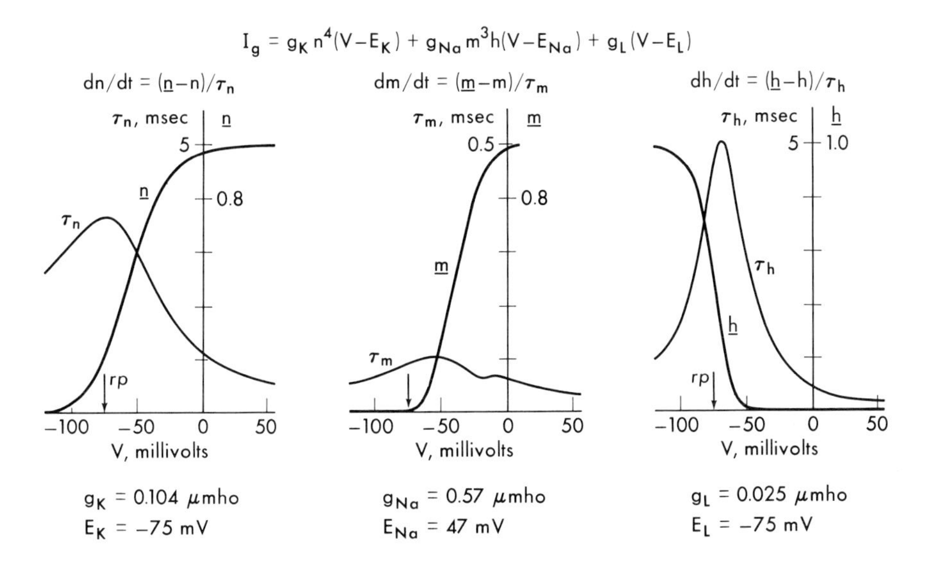

FIG. 4:58. Equations representing the electrical and ionic behavior of a frog node.

Thus we have an amazing similarity for the gross phenomena of the squid membrane and the node membranes of nerves of a toad and a frog. The differences are significant but they in no way vitiate the primarily important conclusion that flows of sodium and potassium ions dominate excitation, its propagation as a nerve impulse, and recovery. Yet these very differences may lead us to an understanding of the processes by which the major ion components do or do not participate in the physiological performance of these membranes.

IMPROVEMENTS

It is entirely proper to improve the HH axon to include known facts, and it is particularly appropriate to do so when these become important in the normal operations of the axon. The two improvements that have received the most attention are the temperature coefficient of the membrane conductances and the larger conductances usually found in presumably better axons.

In general the effects of temperature have been dealt with as originally outlined by Hodgkin and Huxley. Rather than to change all of the α's and β's or all of the τ's, it is far simpler to change only C_m and the time scale by the corresponding factor ϕ. Thus at a temperature T, C_m becomes ϕC_m, the unit of time becomes $1/\phi$ msec where, $Q_{10} = 3$, $\phi = 3^{(T-6.3)/10}$. Huxley (1959) extended this dimensional approach to include the effect of three factors on the speed of propagation θ:

(1) acceleration of permeability changes by factor ϕ
(2) increase of membrane capacity by factor γ, and
(3) increase of conductances by factor η.

He then computed a function of $\eta/\gamma\phi$ for a direct calculation of the change in θ. Including the conductance change with a $Q_{10} = 1.5$, he arrives at the reasonable temperature effect on θ with $Q_{10} = 1.74$, although he notes the absence of squid axon data.

We improved the axon both for conductance and temperature change (FitzHugh and Cole, 1964) in order to get a better comparison with Shane's (1954) potassium loss experiments. Here we let

$$\bar{g}_K = 36\eta \text{ and } \eta = 1.14[1 + .0585(T - 6.3)]$$

on the basis of a typical axon (Moore, 1958). As a result the calculated Q_{10} was increased from 1/2.75 to 1/2.24 as compared to the experimental 1/1.91, and the estimated losses became about 50 percent higher than were measured. We can explain away such differences by observing that different axons under different conditions are quite apt to have different properties. But it is rather more objective merely to claim that the improved axon is in reasonable agreement with this experiment.

The differences between the experimental and calculated space clamp excitation threshold characteristics led FitzHugh (1966b) to test the effect of improving the axon. Expressing the conductance factor as

$$\eta = A[1 + B(T - 6.3)]$$

where $A = 1$ and $B = 0$ for the HH axon and values up to $A = 4$ and $B = 0.061$ have been measured. Various combinations of these parameters were used for computation.

The constant quantity Q_0 was found to have a minimum at a temperature depending on values used for A and B. This effect had not been noticed in the earlier calculations or experiments, but had been found for some axons by Guttman (1966).

FitzHugh explains this by first the domination of τ_m at low temperatures, which is next replaced by $\tau_V = RC$ as the limiting factor as the temperature is raised until finally τ_h and τ_n are small enough to be effective in increasing the threshold at higher temperatures. However, Guttman's average experimental value for Q_0 was about twice the mean value of $7 \cdot 10^{-9}$ coulomb/cm^2 given by calculations.

The minima still persisted for the 1 msec stimulus calculations. The threshold currents for a propagating impulse reported by Sjodin and Mullins (1958) were about 50 percent higher, and tended toward a minimum at a temperature considerably above the values computed for $A = 4$, $B = 0$.

The rheobase increases with temperature for all combinations of A and B, and the data of Guttman are well represented for $A = 4$ and $B = 0.061$. The bluntness of transition, $\sigma = I/I_0$ at $t = Q_0/I_0$, varies from 1.31 to 1.34 for $A = 1$, $B = 0$ over the temperature range, and is not much different from the mean experimental value of 1.38. These correspond to oscillations of the membrane potential and FitzHugh (1966b) extended the Rashevsky–Monnier–Hill two-factor theory of excitation to include such cases.

The various improvements of the HH axon were not completely successful in explaining the threshold characteristics of some real axons. They probably will not be until all data are taken on the same axon under the same conditions, as Vallbo (1964) did for the toad node. The present items of difference are not of a kind or an extent that should encourage such an effort!

OTHER MODIFICATIONS

It seems to me that the most powerful modification of the HH axon was for the effects of calcium on the squid membrane. The early behavior of the real axons with changes of Ca_e, were quite prophetic of the later results of Frankenhaeuser and Hodgkin (1957), which they summarized by an equivalence of a 3–5x change of Ca_e and a 10 mV change of membrane potential.

The impedance loci for decreased Ca_e expanded most dramatically to approach an undamped oscillation as the zero frequency resistance decreased, roughly corresponding to a decrease of 20 mV of the membrane potential. Conversely, the zero frequency resistance increased and the low frequency behavior became that of an increasingly large capacitive reactance at high Ca_e as was to be expected if the potential were made more negative to go beyond E_K.

In unpublished computations, Chandler has shown that this expression of the action of Ca_e is indeed consistent with the early membrane conductance measurements (figures 1:49 and 1:50). Huxley (1959) made other computations on the limits of stability and on repetitive responses as the result of calcium removal. These also require an acceptance of the Frankenhaeuser and Hodgkin results and the challenge they offer for explanation.

In order to study the nature of the behaviors of unstable systems, FitzHugh (1955) first showed the topological similarity of several mathematical models that had been proposed to describe the axon membrane and emphasized the quasi threshold phenomena. Turning to the HH axon, he worked on the four dimensional phase space of V, m, n, and h but, finding it quite unwieldy, he restricted himself to reduced systems, particularly for the pair of variables V and m, which could be represented on a plane (FitzHugh, 1960). An analogue computer then showed three singular points for the squid membrane. After investigating the effects of various constant values of h, the three-dimensional reduced system, V, m, h, was considered. Finally, as a particular example, FitzHugh attempted to modify the complete HH system to reproduce the experimental results of Tasaki and Hagiwara (1957) after injection of tetraethylammonium into the squid axon. This situation, in which a long plateau of potential and decrease of conductance appeared during recovery, was particularly interesting because of its similarity to that of cardiac muscle membranes. With an increase of the potassium "on" time constant by 100x and a decrease of the sodium "off" time to one third, the HH axon did reproduce most of the membrane potential behavior with TEA. However, in the plateau region the calculated membrane conductance remained at a high level rather than falling below the resting value as had been observed before.

Rather than to analyze the probably unnumbered analogues, models, and nerve data, FitzHugh (1961) studied their common salient features as a topological problem of stability and kinetics. This led him to develop the very general, simple, and useful analytical BVP models which include many and important physiological characteristics of these systems. Certainly a specific model for ion permeabilities should do as well but I would not expect it to have a similar analytical form.

There have been other rather arbitrary alterations of the original equations for various reasons such as simplicity of form, ease of computation and better representation of observed phenomena (Goodall and Martin, 1958; Agin and Rojas, 1963) and there will certainly be many more.

The most sweeping generalization and diversification of the HH axon has been the extension to include a fourth parameter—a potassium inactivation corresponding to that for sodium. This was proposed by Grundfest (1960) and has been illustrated often since for various membranes. The characteristics of these membranes have, however, not been reduced to analytical form comparable to the squid and node membrane expressions, so this additional factor has remained a comparatively qualitative expression of the observations.

In our hyperpolarization experiments (Cole and Moore, 1960b) we had established the adequacy of a single parameter for the representation of I_K, but had found the n^4 of HH quite inadequate to represent the delay caused by this considerable departure from the normal range of axon operations. The urge to tinker was irresistible. In the absence of a permeability theory, any one analytical expression is as good or as bad as any other only to the extent that it approximates the data. The HH power function of n had been very useful, so I merely extended it from the exponent of 4 to an exponent of 25, although this seemed a rather oblique approach to a simple time delay.

Somehow this n^{25} seemed rather absurd to many where n^4 had not, so it provoked alternatives to express the rather impressive data. Hodgkin and probably many others have tried without published results. Rosalie Hoyt (1963) entered the field with another approach in which she uses a numerical $g_K(n)$ rather than an analytical relation n^4 between the conductance and the first order variable such as n. A second order equation for sodium and a similar fitting procedure are followed. The results are good agreements with published squid data and give a basis for comparison with a statistical treatment of the membrane processes. By contrast, on the basis of their cardiac experience Johnson and Tille (personal communication) assumed $g_K = \bar{g}_K \cdot n(t)$ and modified the time course of $n(t)$ to follow

$$dn/dt = n[(\omega - n)^2 - \beta]$$

where ω and β are given as empirical, numerical functions of V. At the extreme, FitzHugh (1965) gave a statistical treatment equivalent to n^∞.

It seemed abundantly clear that the analytical formulations of the voltage clamp current for the squid membrane—and following it, for the toad and frog node membrane—had been based on definite, and perhaps astute as well as shrewd, ideas of the underlying mechanisms of ion permeability. Although there seemed to be no definite evidence against these basic assumptions of mechanism, there was yet no more support in favor of their existence. As a consequence I looked for a less dedicated and a noncommittal mode of expression of the experimental results which would be a challenge to any theory of ion permeability and neutral against all of them.

This approach became highly specific in the design of the sordid details of an electromechanical curve fitter—the Bethesda gear box. As John Hervey pointed out to me, potentiometers producing fast components of the ionic current, such as m^3, the sodium "on," could contribute as soon as they had a significance of 0.01 and could be cut out after they reached 0.99 of their ultimate. Similarly the slow components, h the sodium "off" and n^4 the potassium "on," would enter later, progress more slowly, and would become ineffective at a much later time. The speeds were to be taken care of by the ratios of the gears driving the padded potentiometers, so the potentiometers were only to cover the range from 0.01 to 0.99 of their respective chores. These chores were to approximate everything from $\exp(-t/\tau)$ to perhaps $[1 - \exp(-t/\tau)]^{25}$. Exponential powers of 3 and 25, $1 + \tanh x$, a diffusion transient, and FitzHugh's (1965) function have all been plotted on an $x = \beta \log(t + \tau)$ abscissa in figure 4:59 where β is an empirical gear ratio to produce a constant slope dy/dx at $y = 0.5$. It is quite astounding that this range of functions—and undoubtedly a myriad of others—scarcely departs from the mean by more than a few percent of the maximum.

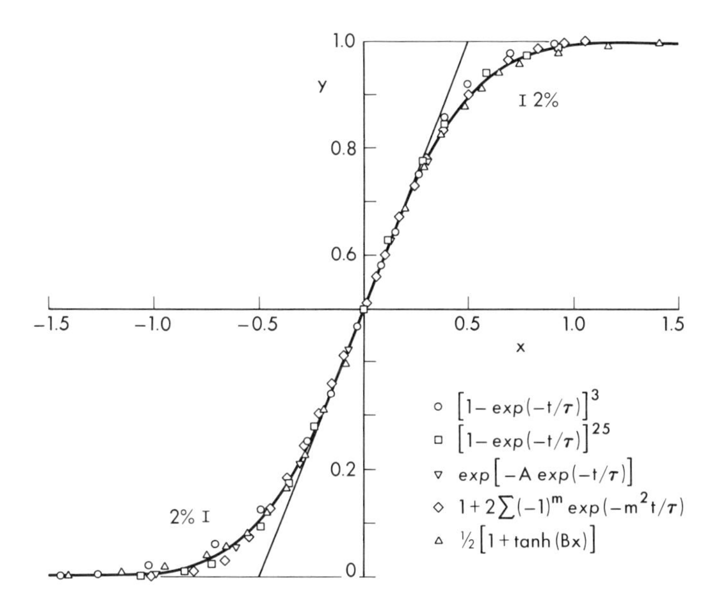

FIG. 4:59. Some functions which can be made to approximate each other by modification of the logarithmic time scale x, to give about the same maximum slopes. The solid curve is empirical, but symmetrical.

This is no particularly important or fundamental observation. It is operationally quite significant because it indicates that the padding of the tapped gear box potentiometers for m, n, and h may be identical, or at least nearly so. On the other hand it does present a norm and an indication of the deviation from it to which any theory for the kinetics of membrane ion permeability can most reasonably be expected to conform. Between these two extremes of utility, this similarity of the group of functions may suggest yet another and more convenient empirical formulation of the voltage clamp current data.

I am quite certain that immeasurable ingenuity and effort have been expended in various attempts to find modifications of the squid membrane characteristics which might express the behavior of muscle, particularly cardiac muscle (Noble, 1966). This is indeed a heroic, not to say a sacrificial, task in that the result can be no more significant than the phenomena for which it is sufficient. If experiment can show that a poor guess was made the effort was wasted, but if experiment confirms the guess it is the experiment which is relied upon—with a bare chance of a passing obeisance to the guess.

It has become obvious that skeletal and cardiac muscle membranes must be represented quite differently from axon membranes. Katz (1948, 1949) found that the sartorius steady state current-potential relation favored inward current, and Hutter and Noble (1960) obtained the same result for Purkinje membranes. This has been widely termed an anomalous rectification—not necessarily because of its difference from the squid membrane but because it is contrary to the simple electrodiffusion model for potassium. Brady and Woodbury (1960) incorporated such a potassium rectification in their model for the frog ventricle along with added very slow components for both sodium "on" and "off." George and Johnson (1961) adapted FitzHugh's (1960) work to the rabbit ventricular fibers. Noble (1962) assumed a fast and a very slow component for g_K and modified the sodium inactivation from the HH form to give a good correspondence with the observed behavior of Purkinje fibers. Then, with a beginning of voltage clamping on the Purkinje fiber (Deck and Trautwein, 1964), we moved out of the region of conjecture into modifications of the HH axon based on experiment (Noble, 1965) and on to the next problems of the explanation of the facts of ion conductances as a function of membrane potential.

The similarity of the h and n parameters has been quite obvious and intriguing, but Hodgkin and Huxley did not feel justified in making a simplification by combining them. The accuracy of expression of the voltage clamp data was significantly reduced and also no basis was available to couple the sodium and the potassium systems. These variables are, however, sufficiently alike that, for mathematical and physiological purposes, FitzHugh (1961) could combine them into a slow, refractory variable $w = (n - h)/2$ by projection. A similar projection combined the fast excitatory variables V and m as $u = V - 36\ m$. These two variables then well described the physiological behavior for short and long stimuli, anode break, and repetitive response of the HH axon on the u, w plane. However, these behaviors were entirely analogous and amazingly similar to what happened on the x, y plane for

$$dx/dt = c(x - x^3/3 + y + z)$$
$$dy/dt = -(x - a + by)/c \tag{2:33}$$

where x corresponds to u and y to w; a, b, and c are constants; and z corresponds to the membrane current I. This system is both an extension of the cubic potential vs. current relation used by van der Pol in his investigations of relaxation oscillations and a simplification of the qualitative description given by Bonhoeffer for the iron wire and, by analogy, for nerve. So FitzHugh has very nicely called this the BVP (Bonhoeffer-van der Pol) model and he discussed its properties in some detail (1961).

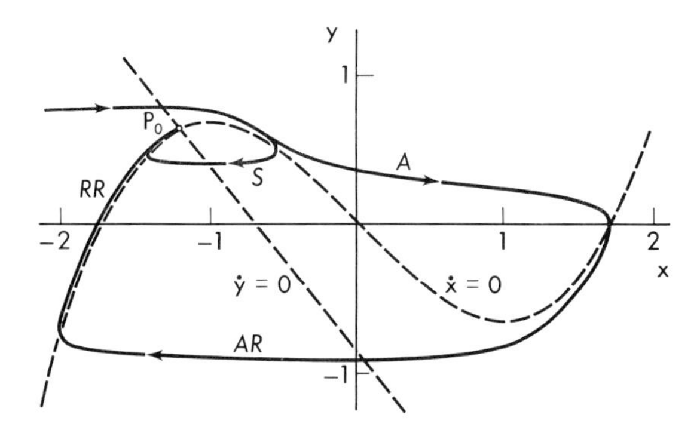

FIG. 4:60. Physiological state diagram for *BVP* model after a short stimulus. x is the fast excitatory parameter and y is the slow refractory variable which were derived as approximations to the *HH* equations. A is the active state, AR and RR are absolutely and relatively refractory, and S is subthreshold.

The orthocline, $dx/dt = 0$, is the cubic characteristic from van der Pol while the straight line for the orthocline, $dy/dt = 0$ is the simplification of the saturating characteristic assumed by Bonhoeffer (figure 2:70). A short stimulus z gives a momentary displacement of the cubic, figure 4:60, and a threshold that is not strictly all-or-none. The response to a constant z was shown in figure 2:71. For large enough z the singular point turns to become an unstable focal point with a stable limit cycle representing repetitive discharge. Recently Huxley (unpublished) has arrived at an analytical solution for the speed of a propagating impulse for the BVP model without recovery as a function of the saddle point potential and for the form of the rising phase (FitzHugh, 1968).

It is very interesting that so simple a model can approximate the HH axon so successfully and that it shows so good a correspondence to the hard-won concepts of classical electrophysiology. That the simplification is not without its drawbacks is best seen in the analogy to the voltage clamp in figure 4:61. This x clamp gives an immediate inward current which proceeds exponentially to the steady state outward current. A blunting of the onset of the early current would improve this representation, figure 4:64.

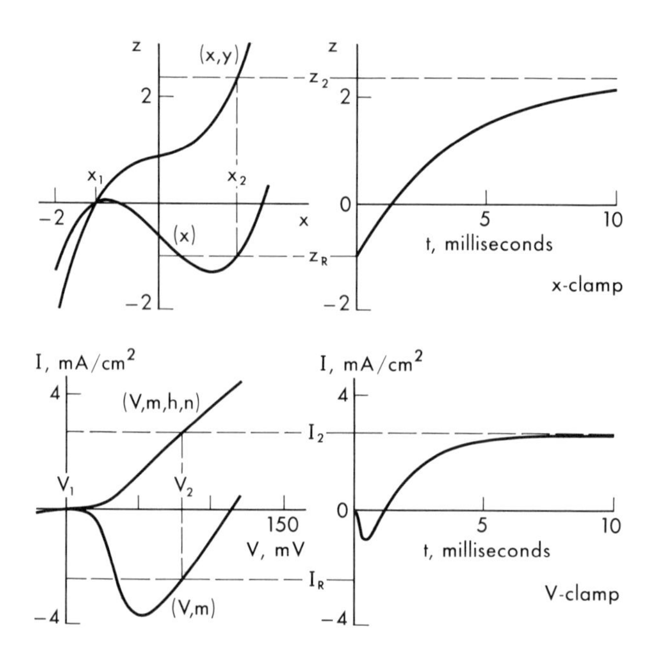

FIG. 4:61. A comparison of the *BVP* model, above, and the *HH* equations, below. At left are peak and steady state currents, z and I, as functions of the parameters x and V. At right are the responses of the two systems to an x-clamp and the corresponding V-clamp.

The next steps were considerable shocks—particularly to those like me who are fond of circuits and really prefer them to differential equations. Nagumo et al. (1962) first pointed out that, except for constants, equations 2:33 also represented the circuit of figure 2:72

where TD had a cubic, negative resistance characteristic.

This analogue was thus a complement to my early potassium model. It gave the excitation and propagation corresponding to the rapid sodium negative resistance characteristic but left the potassium components R and L constant and linear as in figure 4:62. Conversely, below threshold the cubic could be approximated by a constant shunt resistance in figure 2:46 where the now non-linear R, with L, approximated the dominating potassium conductance. Furthermore we had now come around the complete cycle to arrive back at the three-segment model of so long ago, figure 2:65. This model had then been sheer hypothesis and its success was no better than the hypothesis. But with the many intervening steps and years it does indeed now relate directly to the reality of sodium ion flow!

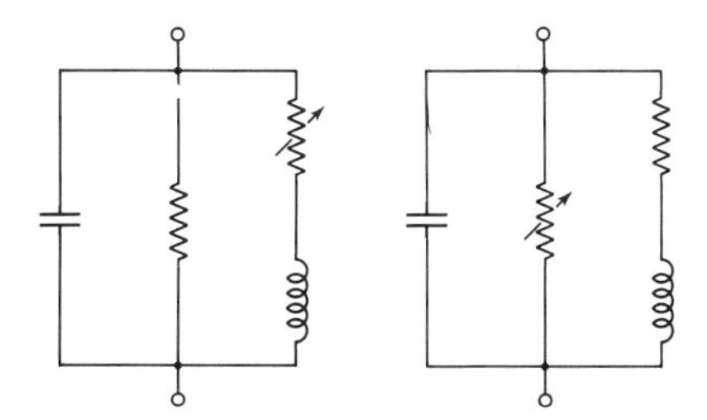

Fig. 4:62. An early potassium model for subthreshold behavior of squid axon membrane, left, a tunnel diode equivalent of the BVP approximation of the HH equations, right.

Nagumo then included this cubic membrane in the cable equation and, eliminating y from equations 1:31 and 2:33, set up the single partial differential equation,

$$\frac{\partial^3 V}{\partial t \partial x^2} = \frac{\partial^2 V}{\partial t^2} + \mu(1 - V + \epsilon V^2)\frac{\partial V}{\partial t} + V$$

where V is now a normalized linear function of the previous variable x; x is now a distance; and μ and ϵ are positive constants. With a digital computer they showed that above threshold a stimulus would approach a limiting form and speed. There were two solutions at constant speed, one the normal stable signal and the other a slower threshold response which was unstable, figure 4:63. A further shock was that Nagumo invoked the Esaki, tunnel diode (1958) that many of us had hoped we could use as an analogue for sodium. But he used it as a circuit element to approximate the cubic component of FitzHugh's BVP model. He then built a nine segment axon

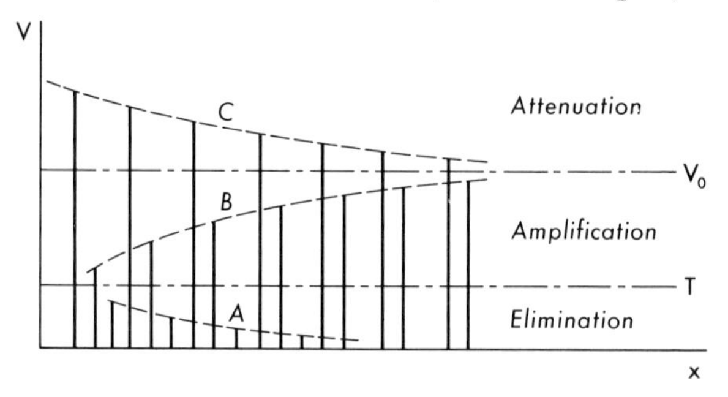

FIG. 4:63. Calculations for the initiation of a propagating *BVP* impulse, maximum potential *V*, vs. distance from stimulus *x*. Below threshold *T* the impulse fades away, *A*. Above threshold, it grows, *B* to achieve the same steady state form, speed, and peak potential *V* as after a very large stimulus *C*.

analogue, demonstrated the emergence of the final impulse from various stimuli, the annihilation of colliding impulses, and the continuing circulation of an impulse around a storage ring, corresponding to fibrillation. This must have been an instructive and exciting adventure for an engineer. But it is interesting to note that the engineer did not use the technique of the physiologist to start his circulating impulse, whereas the physiologist could not have used the procedure of the engineer. It should also be possible, if not relatively easy, to build an analogue of a myelinated axon with these elements for the nodes and an artificial cable for the myelin internodes. This analogue is not simple to put into practice, and it is most effective on a very fast time scale. However, FitzHugh (1966a) constructed a three segment transistor analogue for instruction purposes which operates on a physiological time scale.

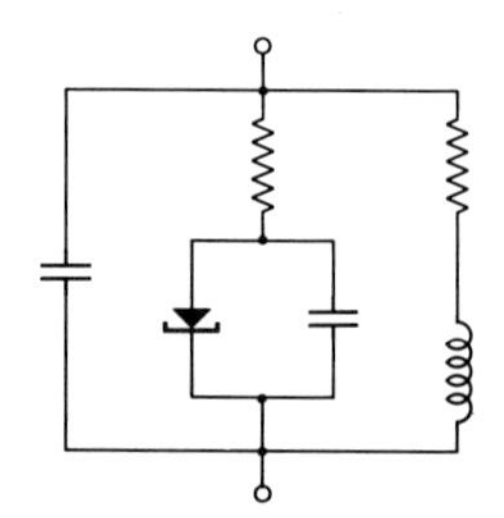

FIG. 4:64. A suggested improvement of the tunnel diode and *BVP* analogies.

It is obvious that both of these approaches would be better if the analogue of the onset of the sodium current were delayed as in the circuit of figure 4:64. Perhaps such a complication in favor of realism may be possible and useful for both, but I have made no attempt to find if the stability requirements for the tunnel diode allow this improvement. However, we can see the correspondence between this analogue and the HH axon as shown in figure 3:24. The n and h characteristics are to be chosen as the mean values independent of potential, while the variation of the m and τ_m can be better approximated by the analogue.

SUMMARY

A few new formulations for some of the limitations of the HH work appeared in a decade and a half, some generalizations and simplifications of it were attempted, and some modifications suggested for the description of excitable membrane other than those of axons.

Conclusion

It is very impressive that the fundamental ideas of Hodgkin and Huxley have been extended without much difficulty to the nodes of medullated axons and—less completely—to other axons.

It is very impressive that the voltage clamp concept—with various and not always certain techniques—has been applied to an uncounted number of excitable cells and membranes and with various results.

It is very impressive that the increasingly rapid and wide application of the voltage clamp concept and the ionic interpretations of the unique data have met with only a few serious alterations and objections. There is no indication that limits to the voltage clamp have been reached either in range or depth and it may remain for some time as an important approach to the fundamental problems of ion permeability.

As a consequence, it seems reasonably probable that the next major advance will have to be a spectacularly successful model of the mechanisms by which ions do—or do not—pass through living membranes. Perhaps such a model could best be based on processes which will produce either fast or slow conductance changes and also negative conductances with the aid of calcium.

References

Abramson, H. A., and Müller, H., 1933. The influence of salts on the electric charge of surfaces in liquids. Cold Spring Harbor Symp. Quant. Biol. *1:* 29–33.

Adelman, W. J., Jr., 1963. A sucrose gap-axial wire voltage clamp. Fed. Proc. *22:* 174.

Adelman, W. J., Jr., and Gilbert, D. L., 1964. Internally perfused squid axons studied under voltage clamp conditions. I. Method. J. Cell. Comp. Physiol. *64:* 423–428.

Adelman, W. J., Jr., and Moore, J. W., 1961. Action of external divalent ion reduction on sodium movement in the squid giant axon. J. Gen. Physiol. *45:* 93–103.

Adelman, W. J., Jr., and Taylor, R. E., 1961. Leakage current rectification in the squid giant axon. Nature *190:* 883–885.

Adelman, W. J., Jr., and Taylor, R. E., 1962. Further evidence for leakage current rectification in the squid giant axon. Fed. Proc. *21:* 355.

Adelman, W. J., Jr., and Taylor, R. E., 1964. Effects of replacement of external sodium chloride with sucrose on membrane currents of the squid giant axon. Biophys. J. *4:* 451–463.

Adelman, W. J., Jr., Dyro, F. M., and Senft, J., 1965a. Long duration responses obtained from internally perfused axons. J. Gen. Physiol. *48:* (5, 2) 1–9.

Adelman, W. J., Jr., Dyro, F. M., and Senft, J., 1965b. Prolonged sodium currents from voltage clamped internally perfused squid axons. J. Cell. Comp. Physiol. *66:* (Suppl. 2) 55–63.

Adrian, R. H., and Freygang, W. H., 1962. Potassium and chloride permeability of frog muscle membrane. J. Physiol. *163:* 61–103.

Adrian, R. H., Chandler, W. K., and Hodgkin, A. L., 1966. Voltage clamp experiments in skeletal muscle fibers. J. Physiol. *186:* 51–52 P.

Agin, D., and Rojas, E., 1963. A third order system for the squid axon membrane. In 16th Conference on engineering in medicine and biology. Proceedings, p. 4–5.

Araki, T., and Otani, T., 1955. Response of single motoneurons to direct stimulation in toad's spinal cord. J. Neurophysiol. *18:* 472–485.

Araki, T., and Terzuolo, C. A., 1962. Membrane currents in spinal motoneurones associated with the action potential and synaptic activity. J. Neurophysiol. *25:* 772–789.

Armstrong, C. M., 1966. Time course of TEA$^+$-induced anomalous rectification in squid giant axons. J. Gen. Physiol. *50:* 491–508.

Armstrong, C. M., and Binstock, L., 1964. The effects of several alcohols on the properties of the squid giant axon. J. Gen. Physiol. *48:* 265–277.

Armstrong, C. M., and Binstock, L., 1965. Anomalous rectification in the squid giant axon injected with tetraethylammonium chloride. J. Gen. Physiol. *48:* 859–872.

Baker, P. F., Hodgkin, A. L., and Shaw, T. I., 1961. Replacement of the protoplasm of a giant nerve fibre with artificial solutions. Nature *190:* 885–887.

Bennett, M. V. L., and Grundfest, H., 1966. Analysis of depolarizing and hyperpolarizing inactivation responses in gymnotid electroplaques. J. Gen. Physiol. *50:* 141–169.

Binstock, L., and Armstrong, C. M., 1963. Effect of a series of alcohols on voltage clamped squid axons. Biophys. Soc. Abst. WC3.

Binstock, L., and Lecar, H., 1967. Ammonium ion substitutions in the voltage clamped squid giant axon. Biophys. Soc. Abstr., WC4.

Blaustein, M. P., and Goldman, D. E., 1966. Origin of axon membrane hyperpolarization under sucrose-gap. Biophys. J. *6:* 453–470.

Blinks, L. R., 1930. The variation of electrical resistance with applied potential. III. Impaled *Valonia ventricosa*. J. Gen. Physiol. *14:* 139–162.

Blinks, L. R., 1932. Protoplasmic potentials in *Halicystis*, II. The effects of potassium on two species with different saps. J. Gen. Physiol. *16:* 147–156.

Brady, A. J., and Woodbury, J. W., 1960. The sodium-potassium hypothesis as the basis of electrical activity in frog ventricle. J. Physiol. *154:* 385–407.

Brady, R. O., Spyropoulos, C. S., and Tasaki, I., 1958. Intra-axonal injection of biologically active materials. Am. J. Physiol. *194:* 207–213.

Brink, F., 1954. The role of calcium ions in neural processes. Pharm. Rev. *6:* 243–298.

Brock, L. G., Coombes, J. S., and Eccles, J. C., 1952. The recording of potentials from motoneurones with an intracellular electrode. J. Physiol. *117:* 431–460.

Buchwald, H. D., Durham, L., Fischer, H. G., Harada, R., Mosher, H. S., Kao, C. Y., and Fuhrman, F. A., 1964. Identity of tarichatoxin and tetrodotoxin. Science *143:* 474–475.

Campbell, G. A., and Foster, R. M., 1931. Fourier integrals for practical applications. Bell Tel. Sys. Monograph B-584. (Reprint 1948, New York, Van Nostrand).

Carslaw, H. S., and Jaeger, J. C., 1947. Conduction of heat in solids. (2nd edition 1959) Oxford, Clarendon Press.

Chandler, W. K., and Hodgkin, A. L., 1965. The effect of internal sodium on the action potential in the presence of different internal anions. J. Physiol. *181:* 594–611.

Chandler, W. K., and Meves, H., 1965. Voltage clamp experiments on internally perfused giant axons. J. Physiol. *180:* 788–820.

Chandler, W. K., Hodgkin, A. L., and Meves, H., 1965. The effect of changing the internal solution on sodium inactivation and related phenomena in giant axons. J. Physiol. *180:* 821–836.

Choi, J. K., 1963. The fine structure of the urinary bladder of the toad, *Bufo marinus*. J. Cell Biol. *16:* 53–72.

Cole, K. S., 1932. Electric phase angle of cell membranes. J. Gen. Physiol. *15:* 641–649.

Cole, K. S., 1949. Dynamic electrical characteristics of the squid axon membrane. Arch. Sci. physiol. *3:* 253–258.

Cole, K. S., 1955. Ions, potentials, and the nerve impulse. In Shedlovsky, T. (ed.) Electrochemistry in biology and medicine. New York, Wiley, p. 121–140.

Cole, K. S., 1957. Beyond membrane potentials. Ann. N. Y. Acad. Sci. *65:* 658–662.

Cole, K. S., 1958. Effect of polarization on membrane ion currents. Biophys. Soc. Abstr., p. 22, I2.

Cole, K. S., 1961. An analysis of the membrane potential along a clamped squid axon. Biophys. J. *1:* 401–418.

Cole, K. S., and Baker, R. F., 1941. Transverse impedance of the squid giant axon during current flow. J. Gen. Physiol. *24:* 535–549.

Cole, K. S., and Curtis, H. J., 1938. Electric impedance of *Nitella* during activity. J. Gen. Physiol. *22:* 37–64.

Cole, K. S., and Hodgkin, A. L., 1939. Membrane and protoplasm resistance in the squid giant axon. J. Gen. Physiol. *22:* 671–687.

Cole, K. S., and Marmont, G., 1942. The effect of ionic environment upon the longitudinal impedance of the squid axon. Fed. Proc. *1:* 15–16.

Cole, K. S., and Moore, J. W., 1960a. Ionic current measurements in the squid giant axon membrane. J. Gen. Physiol. *44:* 123–167.

Cole, K. S., and Moore, J. W., 1960b. Potassium ion current in the squid giant axon; dynamic characteristic. Biophys. J. *1:* 1–14.

Cole, K. S., Antosiewicz, H. A., and Rabinowitz, P., 1955. Automatic computation of nerve excitation. J. Soc. Indust. Appl. Math. *3:* 153–172.

Coraboeuf, E., and Weidmann, S., 1949. Potentials d'action du muscle cardiaque obtenue à l'aide de microélectrodes intracellulaires. C. R. Soc. Biol. Paris *143:* 1360–1361.

Curtis, D. R., 1964. Microelectrophoresis. In Nastuk, W. L. (ed.) Physical techniques in biological research, New York, Academic Press, Chap. 5, p. 144–190.

Curtis, D. R., 1966. Synaptic transmission in the central nervous system and its pharmacology. In Rodahl, K., and Issekutz, B., (eds.) Nerve as a tissue, New York, Harper and Row (Hoeber), p. 321–338.

Curtis, H. J., and Cole, K. S., 1940. Membrane action potentials from the squid giant axon. J. Cell. Comp. Physiol. *15:* 147–157.
Curtis, H. J., and Cole, K. S., 1942. Membrane resting and action potentials from the squid giant axon. J. Cell. Comp. Physiol. *19:* 135–144.
Dalton, J. C., 1958. Effects of external ions on membrane potentials of a lobster giant axon. J. Gen. Physiol. *41:* 529–542.
Dalton, J. C., 1959. Effects of external ions on membrane potentials of a crayfish giant axon. J. Gen. Physiol. *42:* 971–982.
Deck, K. A., Kern, R., and Trautwein, W., 1964. Voltage clamp technique in mammalian cardiac fibres. Arch. ges. Physiol. *280:* 50–62.
Deck, K. A., and Trautwein, W., 1964. Ionic currents in cardiac excitation. Arch. ges. Physiol. *280:* 63–80.
del Castillo, J., Lettvin, J. Y., McCulloch, W. S., and Pitts, W., 1957. Membrane currents in clamped vertebrate nerve. Nature *180:* 1290–1291.
de Packh, D. C., 1947. A resistor network for the approximate solution of the Laplace equation. Rev. Sci. Inst. *18:* 798–799.
Dodge, F. A., 1961. Ionic permeability changes underlying nerve excitation. In Shanes, A. M. (ed.) Biophysics of physiological and pharmacological actions, Washington, D. C., © Am. Assoc. Adv. Sci., 1961, Pub. No. 69, p. 119–143.
Dodge, F. A., 1963. A study of ionic permeability changes underlying excitation in myelinated nerve fibers of the frog. (Thesis) New York, The Rockefeller Institute.
Dodge, F. A., and Frankenhaeuser, B., 1958. Membrane currents in isolated frog nerve fibre under voltage clamp conditions. J. Physiol. *143:* 76–90.
Dodge, F. A., and Frankenhaeuser, B., 1959. Sodium currents in the myelinated nerve fibre of *Xenopus laevis* investigated with the voltage clamp technique. J. Physiol. *148:* 671–676.
Draper, M. H., and Weidmann, S., 1951. Cardiac resting and action potentials recorded with an intracellular electrode. J. Physiol. *115:* 74–94.
Eccles, J. C., 1964. The physiology of synapses. Berlin, Springer, New York, Academic Press.
Ehrenstein, G., and Gilbert, D. L., 1964. Effect of high potassium solutions on the potassium current of squid axon. Biophys. Soc. Abstr. FF6.
Ehrenstein, G., and Gilbert, D. L., 1966. Slow changes in potassium permeability in squid giant axon. Biophys. J. *6:* 553–566.
Erlanger, J., and Blair, E. A., 1934. Manifestations of segmentation in myelinated axons. Am. J. Physiol. *110:* 287–311.
Erlanger, J., and Blair, E. A., 1938. The action of isotonic, salt-free solutions on conduction in medullated nerve fibers. Am. J. Physiol. *124:* 341–359.
Esaki, L., 1958. New phenomenon in narrow germanium *p-n* junctions. Physic. Rev. *109:* 603–604.
Fatt, P., and Ginsborg, B. L., 1958. The ionic requirements for the production of action potentials in crustacean muscle fibres. J. Physiol. *142:* 516–543.
Fatt, P., and Katz, B., 1953. The electrical properties of crustacean muscle fibres. J. Physiol. *120:* 171–204.
Ferguson, J., 1951. Relations between thermodynamic indices of narcotic potency and the molecular structure of the narcotic. In Cent. Natl. Recher. Scient., Mechanisme de la Narcose, Colloq. Intern. No. 26, Paris, p. 25–39.
Fernández-Morán, H., 1950. Sheath and axon structures in the internode portion of vertebrate myelinated nerve fibres. Exp. Cell Res. *1:* 309–337.
Findlay, G. P., 1959. Studies of action potentials in the vacuole and cytoplasm of *Nitella*. Aust. J. Biol. Sci. *12:* 412–426.

Findlay, G. P., 1961. Voltage-clamp experiments with *Nitella*. Nature *191:* 812–814.
Finkelstein, A., 1961. Electrical excitability of isolated frog skin. Nature *190:* 1119–1120.
Finkelstein, A., 1964. Electrical excitability of isolated frog skin and toad bladder. J. Gen. Physiol. *47:* 545–565.
FitzHugh, R., 1955. Mathematical models of threshold phenomena in the nerve membrane. Bull. Math. Biophys. *17:* 257–278.
FitzHugh, R., 1960. Thresholds and plateaus in the Hodgkin-Huxley nerve equations. J. Gen. Physiol. *43:* 867–896.
FitzHugh, R., 1961. Impulses and physiological states in theoretical models of nerve membrane. Biophys. J. *1:* 445–466.
FitzHugh, R., 1962. Computation of impulse initiation and saltatory conduction in a myelinated nerve fiber. Biophys. J. *2:* 11–21.
FitzHugh, R., 1965. A kinetic model of the conductance changes in nerve membrane. J. Cell. Comp. Physiol. *66* (Suppl. 2): 111–117.
FitzHugh, R., 1966a. An electronic model of the nerve membrane for demonstration purposes. J. Appl. Physiol. *21:* 305–308.
FitzHugh, R., 1966b. Theoretical effect of temperature on threshold in the Hodgkin-Huxley nerve model. J. Gen. Physiol. *49:* 989–1005.
FitzHugh, R., 1968. Mathematical models of excitation and propagation in nerve. In Schwan, H. P. (ed.) Bioelectronics, New York, McGraw-Hill.
FitzHugh, R., and Cole, K. S., 1964. Theoretical potassium loss from squid axons as a function of temperature. Biophys. J., *4:* 257–265.
Frank, K., and Fuortes, M. G. F., 1956. Stimulation of spinal motoneurones with intracellular electrodes. J. Physiol. *134:* 451–470.
Frank, K., and Tauc, L., 1964. Voltage-clamp studies of molluscan neuron membrane properties. In Hoffman, J. F., (ed.) Society of General Physiologists, The cellular functions of membrane transport, Englewood Cliffs, N. J., Prentice-Hall, p. 113–135.
Frank, K., Fuortes, M. G. F., and Nelson, P. G., 1959. Voltage clamp of motoneuron soma. Science *130:* 38–39.
Frankenhaeuser, B., 1957a. A method for recording resting and action potentials in the isolated myelinated nerve fibre of the frog. J. Physiol. *135:* 550–559.
Frankenhaeuser, B., 1957b. The effect of calcium on the myelinated nerve fibre. J. Physiol. *137:* 245–260.
Frankenhaeuser, B., 1959. Steady state inactivation of sodium permeability in myelinated nerve fibres of *Xenopus laevis*. J. Physiol. *148:* 671–676.
Frankenhaeuser, B., 1960. Quantitative description of sodium currents in myelinated nerve fibres of *Xenopus laevis*. J. Physiol. *151:* 491–501.
Frankenhaeuser, B., 1962a. Delayed currents in myelinated nerve fibres of *Xenopus laevis* investigated with voltage clamp technique. J. Physiol. *160:* 40–45.
Frankenhaeuser, B., 1962b. Instantaneous potassium currents in myelinated nerve fibres of *Xenopus laevis*. J. Physiol. *160:* 46–53.
Frankenhaeuser, B., 1962c. Potassium permeability in myelinated nerve fibres of *Xenopus laevis*. J. Physiol. *160:* 54–61.
Frankenhaeuser, B., 1963a. A quantitative description of potassium currents in myelinated nerve fibres of *Xenopus laevis*. J. Physiol. *169:* 424–430.
Frankenhaeuser, B., 1963b. Inactivation of the sodium-carrying mechanism in myelinated nerve fibers of *Xenopus laevis*. J. Physiol. *169:* 445–451.
Frankenhaeuser, B., 1965. Computed action potentials in nerve from *Xenopus laevis*. J. Physiol. *180:* 780–787.

Frankenhaeuser, B., and Hodgkin, A. L., 1956. The after-effects of impulses in the giant nerve fibres of *Loligo*. J. Physiol. *131:* 341–376.
Frankenhaeuser, B., and Hodgkin, A. L., 1957. The action of calcium on the electrical properties of squid axons. J. Physiol. *137:* 218–244.
Frankenhaeuser, B., and Huxley, A. F., 1964. The action potential in the myelinated nerve fibre of *Xenopus laevis* as computed on the basis of voltage clamp data. J. Physiol. *171:* 302–315.
Frankenhaeuser, B., and Moore, L. E., 1963. The effect of temperature on the sodium and potassium permeability changes in myelinated nerve fibres of *Xenopus laevis*. J. Physiol. *169:* 431–437.
Frankenhaeuser, B., and Persson, A., 1957. Voltage clamp experiments on the myelinated nerve fibre. Acta physiol. Scand. *42* (Suppl. 145): 45–46.
Frankenhaeuser, B., and Vallbo, Å. B., 1965. Accommodation in myelinated nerve fibres of *Xenopus laevis* as computed on the basis of voltage clamp data. Acta physiol. Scand. *63:* 1–20.
Frankenhaeuser, B., Lindley, B. D., and Smith, R. S., 1966. Potentiometric measurement of membrane action potentials in frog muscle fibres. J. Physiol. *183:* 152–166.
Frazier, H. S., 1962. The electrical potential profile of the isolated toad bladder. J. Gen. Physiol. *45:* 515–528.
Gaffey, C. T., and Mullins, L. J., 1958. Ion fluxes during the action potential in *Chara*. J. Physiol. *144:* 505–524.
Gardner, M. F., and Barnes, J. L., 1942. Transients in linear systems studied by the Laplace transformation. Vol. 1, Lumped-constant systems. New York, Wiley.
George, E. P., and Johnson, E. A., 1961. Solutions of the Hodgkin-Huxley equations for squid axon treated with tetraethylammonium and in potassium-rich media. Aust. J. Exp. Biol. Med. Sci. *39:* 275–294.
Gilbert, D. L., and Ehrenstein, G., 1965. Effect of calcium and magnesium on voltage clamped squid axons immersed in isosmotic potassium chloride. In Int. Cong. Physiol. Sci., 23rd., Tokyo, 1965. (Abstracts of papers. #148)
Goldman, D. E., and Blaustein, M. P., 1966. Ions, drugs, and the axon membrane. Ann. N. Y. Acad. Sci., *137:* 967–981.
Goodall, M. C., and Martin, T. E., 1958. Analogue computer solutions of Hodgkin-Huxley type equations. Biophys. Soc. Abstr., I7, p. 23.
Grundfest, H., 1960. A four-factor ionic hypothesis of spike electrogenesis. Biol. Bull. *119:* 284.
Grundfest, H., Kao, C. Y., and Altamirana, M., 1954. Bioelectric effects of ions microinjected into the giant axon of *Loligo*. J. Gen. Physiol. *38:* 245–282.
Guttman, R., 1941. Electrical impedance of muscle at cut and uncut surfaces. J. Cell. Comp. Physiol. *18:* 403–405.
Guttman, R., 1962. Effect of temperature on the potential and current thresholds of axon membrane. J. Gen. Physiol, *46:* 257–266.
Guttman, R., 1966. Temperature characteristics of excitation in space-clamped squid axons. J. Gen. Physiol. *49:* 1007–1018.
Hagiwara, S., 1965. Relation of membrane properties of the giant muscle fiber of a barnacle to internal ionic composition. J. Gen. Physiol. *48:* (5, 2), 55–57.
Hagiwara, S., and Oomura, Y., 1958. The critical depolarization for the spike in the squid. Japan. J. Physiol. *8:* 234–245.
Hagiwara, S., and Saito, N., 1959a. Membrane potential change and membrane current in supramedullary nerve cell of puffer. J. Neurophysiol. *22:* 204–221.

Hagiwara, S., and Saito, N., 1959b. Voltage-current relationships in nerve cell membrane of *Onchidium verruculatum*. J. Physiol. *148:* 161–179.
Hagiwara, S., and Tasaki, I., 1958. A study on the mechanism of impulse transmission across the giant synapse of the squid. J. Physiol. *143:* 114–137.
Hagiwara, S., Naka, K., and Chichibu, S., 1964. Membrane properties of barnacle muscle fiber. Science *143:* 1446–1448.
Hecht, H. H., 1957. Dedication to Frank Norman Wilson. Ann. N. Y. Acad. Sci. *65:* 655.
Hecht, H. H., Hutter, O. F., and Lywood, D. W., 1964. Voltage-current relation of short Purkinje fibres in sodium-deficient solution. J. Physiol. *170:* 5–7P.
Heilbrunn, L. V., 1952. An outline of general physiology. Philadelphia, Saunders.
Hille, B., 1966. Common mode of action of three agents that decrease the transient change in sodium permeability in nerves. Nature, *210:* 1220–1222.
Hille, B., 1967. The selective inhibition of delayed potassium currents in nerve by tetraethylammonium ion. J. Gen. Physiol. *50:* 1287–1302.
Hille, B., Bennett, M. V. L., and Grundfest, H., 1965. Voltage clamp measurements of the Cl-conductance changes in skate electroplaques. Biol. Bull. *129:* 407–408.
Hodgkin, A. L., 1938. The subthreshold potentials in a crustacean nerve fibre. Proc. Roy. Soc. (London) B. *126:* 87–121.
Hodgkin, A. L., 1951. The ionic basis of electrical activity in nerve and muscle. Biol. Rev. *26:* 339–409.
Hodgkin, A. L., 1954. A note on conduction velocity. J. Physiol. *125:* 221–224.
Hodgkin, A. L., 1964. The conduction of the nervous impulse. Springfield, Ill., Thomas.
Hodgkin, A. L., and Huxley, A. F., 1947. Potassium leakage from an active nerve fibre. J. Physiol. *106:* 341–367.
Hodgkin, A. L., and Huxley, A. F., 1952a. Currents carried by sodium and potassium ions through the membrane of the giant axon of *Loligo*. J. Physiol. *116:* 449–472.
Hodgkin, A. L., and Huxley, A. F., 1952b. A quantitative description of membrane current and its application to conduction and excitation in nerve. J. Physiol. *117:* 500-544.
Hodgkin, A. L., and Katz, B., 1949. The effect of sodium ions on the electrical activity of the giant axon of the squid. J. Physiol. *108:* 37–77.
Hodgkin, A. L., and Keynes, R. D., 1956. Experiments on the injection of substances into squid giant axons by means of a microsyringe. J. Physiol. *131:* 592–616.
Hodgkin, A. L., and Keynes, R. D., 1957. Movements of labelled calcium in squid giant axons. J. Physiol. *138:* 253–281.
Hodgkin, A. L., Huxley, A. F., and Katz, B., 1952. Measurement of current-voltage relations in the membrane of the giant axon of *Loligo*. J. Physiol. *116:* 424–448.
Hogg, B. M., Goss, C. M., and Cole, K. S., 1934. Potentials in embryo rat heart muscle cultures. Proc. Soc. Exp. Biol. Med. *32:* 304–307.
Hope, A. B., 1961. The action potential in cells of *Chara*. Nature *191:* 811–812.
Hoyt, R. C., 1963. The squid giant axon. Mathematical models. Biophys. J. *3:* 399–431.
Huneeus-Cox, F., Fernandez, H. L., and Smith, B. H., 1966. Effects of redox and sulfhydryl reagents on the bioelectric properties of the giant axon of the squid. Biophys. J. *6:* 675–689.
Hurlbut, W. P., 1958. Effects of azide and chloretone on the sodium and potassium contents and the respiration of frog sciatic nerves. J. Gen. Physiol. *41:* 959–988.

Hutter, O. F., and Noble, D., 1960. Rectifying properties of heart muscle. Nature *188:* 495.
Huxley, A. F., 1959. Ion movements during nerve activity. Ann. N. Y. Acad. Sci. *81:* 221–246.
Huxley, A. F., and Stämpfli, R., 1949a. Saltatory transmission of the nervous impulse. Arch. Sci. physiol. *3:* 435–448.
Huxley, A. F., and Stämpfli, R., 1949b. Evidence for saltatory conduction in peripheral myelinated nerve fibres. J. Physiol. *108:* 315–339.
Huxley, A. F., and Stämpfli, R., 1951a. Direct determination of membrane resting potential and action potential in single myelinated nerve fibres. J. Physiol. *112:* 476–495.
Huxley, A. F., and Stämpfli, R., 1951b. Effect of potassium and sodium on resting and action potentials of single myelinated nerve fibres. J. Physiol. *112:* 496–508.
Jeans, J. H., 1911. The mathematical theory of electricity and magnetism. 2nd edition, Cambridge, University Press.
Julian, F. J., and Goldman, D. E., 1960. A method for obtaining fullsized resting and action potentials from single axons without internal electrodes. Biophys. Soc. Abst., E3.
Julian, F. J., and Goldman, D. E., 1962. The effects of mechanical stimulation on some electrical properties of axons. J. Gen. Physiol. *46:* 297–313.
Julian, F. J., Moore, J. W., and Goldman, D. E., 1962a. Membrane potentials of the lobster giant axon obtained by use of the sucrose-gap technique. J. Gen. Physiol. *45:* 1195–1216.
Julian, F. J., Moore, J. W., and Goldman, D. E., 1962b. Current-voltage relations in the lobster giant axon membrane under voltage clamp conditions. J. Gen. Physiol. *45:* 1217–1238.
Kato, G., 1934. The microphysiology of nerve. Tokyo, Maruzen.
Katz, B., 1948. The electrical properties of the muscle fibre membrane. Proc. Roy. Soc. (London) B, *135:* 506–534.
Katz, B., 1949. Les constantes électriques de la membrane du muscle. Arch. Sci. physiol. *3:* 285–300.
Kishimoto, U., 1959. Electrical characteristics of *Chara corallina.* Ann. Rep. Scient. Works, Fac. Sci., Osaka Univ. *7:* 115–146.
Kishimoto, U., 1961. Current voltage relations in *Nitella.* Biol. Bull. *121:* 370–371.
Kishimoto, U., 1964. Current voltage relations in *Nitella.* Japanese J. Physiol. *14:* 515–527.
Kishimoto, U., 1965. Voltage clamp and internal perfusion studies on *Nitella* internodes. J. Cell. Comp. Physiol. *66:* (Suppl. 2), 43–53.
Kishimoto, U., and Adelman, W. J., Jr., 1964. Effect of detergent on electrical properties of squid axon membrane. J. Gen. Physiol. *47:* 975–986.
Koefoed-Johnsen, V., and Ussing, H. H., 1958. The nature of the frog skin potential. Acta physiol. Scand. *42:* 298–308.
Lapique, L., 1926. L'Excitabilité en fonction du temp. Paris, Presses Universitaires de France.
Lecar, H., Ehrenstein, G., Binstock, L., and Taylor, R. E., 1967. Removal of potassium negative resistance in perfused squid axons. J. Gen. Physiol. *50:* 1499–1515.
Liebmann, G., 1953. Electrical analogues. Brit. J. Appl. Physics *4:* 193–200.
Lillie, R. S., 1925. Factors affecting transmission and recovery in the passive iron nerve model. J. Gen. Physiol. *7:* 473–507.
Lindemann, B., 1965. Negative slope Na-conductance in the surface structure of frog skin epithelium. Biol. Bull. *129:* 391.

Loewenstein, W. R., Socolar, S. J., Higashino, S., Kanno, Y., and Davidson, N., 1965. Intercellular communication: Renal, urinary bladder, sensory, and salivary gland cells. Science *149:* 295–298.

Lorente de Nó, R., and Condouris, G. A., 1959. Decremental conduction in peripheral nerve. Integration of stimuli. Proc. Natl. Acad. Sci., USA, *45:* 592–617.

Lorente de Nó, R., and Honrubia, V., 1965. Production of action potentials by the internodes of isolated nerve fibers. Proc. Natl. Acad. Sci. USA. *53:* 757–764.

Lorente de Nó, R., and Honrubia, V., 1966. Theory of action currents in isolated myelinated nerve fibers. XII. Proc. Natl. Acad. Sci. USA. *55:* 1118–1125.

Marmont, G., 1949. Studies on the axon membrane; I. A new method. J. Cell. Comp. Physiol. *34:* 351–382.

Moore, J. W., 1958. Temperature and drug effects on squid axon membrane ion conductances. Fed. Proc. *17:* 113.

Moore, J. W., 1959. Excitation of the squid axon membrane in isosmotic potassium chloride. Nature *183:* 265–266.

Moore, J. W., 1965. Voltage clamp studies on internally perfused axons. J. Gen. Physiol. *48:* (5, 2) 11–17.

Moore, J. W., and Cole, K. S., 1963. Voltage clamp techniques. In Nastuk, W. L. (ed.), Physical techniques in biological research Vol. 6, New York, Academic Press, p. 263–321.

Moore, J. W., and del Castillo, J., 1959. An electronic electrode. Inst. Radio Eng. Nat. Conven. Rec., pt. 9, 47–50.

Moore, J. W., and Ulbricht, W., 1963. Effect of ethanol on the squid axon membrane conductance. Biophys. Soc. Abstr. WC 2.

Moore, J. W., Ulbricht, W., and Takata, M., 1964. Effect of ethanol on the sodium and potassium conductances of the squid axon membrane. J. Gen. Physiol., *48:* 279–295.

Mueller, P., 1958. Prolonged action potentials from single nodes of Ranvier. J. Gen. Physiol. *42:* 137–162.

Mullins, L. J., 1962. Efflux of chloride ions during the action potential of *Nitella*. Nature *196:* 986–987.

Mullins, L. J., Adelman, W. J., Jr., and Sjodin, R. A., 1962. Sodium and potassium ion effluxes from squid axons under voltage clamp conditions. Biophys. J. *2:* 257–274.

Nagumo, J., Arimoto, S., and Yoshizawa, S., 1962. An active pulse transmission line simulating nerve axon. Proc. Inst. Radio Eng. *50:* 2061–2070.

Nakamura, Y., Nakajima, S., and Grundfest, H., 1964. The action of tetrodotoxin on electrogenic components of squid giant axons. J. Gen. Physiol. *48:* 985–996.

Nakamura, Y., Nakajima, S., and Grundfest, H., 1965. Analysis of spike electrogenesis and depolarizing K inactivation in electroplaques of *Electrophorus electricus, L.* J. Gen. Physiol. *49:* 321–349.

Narahashi, T., 1963. Dependence of resting and action potentials on internal potassium in perfused squid giant axons. J. Physiol. *169:* 91–115.

Narahashi, T., Anderson, H. C., and Moore, J. W., 1966. Tetrodotoxin does not block excitation from inside the nerve membrane. Science *153:* 765–767.

Narahashi, T., Moore, J. W., and Scott, W. R., 1964. Tetrodotoxin blockage of sodium conductance increase in lobster giant axons. J. Gen. Physiol. *47:* 965–974.

Nastuk, W. L., and Hodgkin, A. L., 1950. The electrical activity of single muscle fibres. J. Cell. Comp. Physiol. *35:* 39–73.

Noble, D., 1962. A modification of the Hodgkin-Huxley equations applicable to Purkinje fibre action and pace-maker potentials. J. Physiol. *160:* 317–352.

Noble, D., 1965. Electrical properties of cardiac muscle attributable to inward going (anomalous) rectification. J. Cell. Comp. Physiol. *66:* (Suppl. 2) 127–135.

Noble, D., 1966. Applications of Hodgkin-Huxley equations to excitable tissues. Physiol. Rev. *46:* 1–50.

Norlund, N. E., 1924. Differenzenrechnung. Berlin, Springer.

Offner, F., Weinberg, A., and Young, G., 1940. Nerve conduction theory: Some mathematical consequences of Bernstein's model. Bull. Math. Biophys. *2:* 89–103.

Oikawa, T., Spyropoulos, C. S., Tasaki, I., and Teorell, T., 1961. Methods for perfusing the giant axon of *Loligo pealii*. Acta physiol. Scand. *52:* 195–196.

Ooyama, H., and Wright, E. B., 1962. Activity of potassium mechanism in single Ranvier node during excitation. J. Neurophysiol. *25:* 67–93.

Osterhout, W. J. V., and Hill, S. E., 1930. Salt bridges and negative variations. J. Gen. Physiol. *13:* 547–552.

Ostwald, W., 1900. Periodische Erscheinungen bei der Auflösung des Chroms in Säuren (Parts I and II). Zeit. phys. Chem. *35:* 33–76 and 204–256.

Parnas, I., and Abbott, B. C., 1964. Deep abdominal muscles of crayfish. Am. Zool. *4:* 284–285.

Ramsey, R. W., and Street, S. F., 1940. The isometric length-tension diagram of isolated skeletal muscle fibers of the frog. J. Cell. Comp. Physiol. *15:* 11–34.

Rashevsky, N., 1933a. Outline of a physico-mathematical theory of excitation and inhibition. Protoplasma *20:* 42–56.

Rashevsky, N., 1933b. Some physico-mathematical aspects of nerve conduction. Physics *4:* 341–349.

Rojas, E., and Ehrenstein, G., 1965. Voltage clamp experiments on axons with potassium as the only internal and external cation. J. Cell. Comp. Physiol. *66:* (Suppl. 2) 71–77.

Rojas, E., and Luxoro, M., 1963. Micro-injection of trypsin into axons of squid Nature *199:* 78–79.

Rosenblueth, A., Wiener, N., Pitts, W., and Garcia Ramos, J., 1948. An account of the spike potential of axons. J. Cell. Comp. Physiol. *32:* 275–318.

Rushton, W. A. H., 1951. A theory of the effects of fibre size in medullated nerve. J. Physiol. *115:* 101–122.

Salmoiraghi, G. C., and Bloom, F. E., 1964. Pharmacology of individual neurons. Science *144:* 493–499.

Salmoiraghi, G. C., and Steiner, F. A., 1963. Acetylcholine sensitivity of cat's medullary neurons. J. Neurophysiol. *26:* 581–597.

Schantz, E. J., 1960. Biochemical studies on paralytic shellfish poisons. Ann. N. Y. Acad. Sci. *90:* 843–855.

Schmitt, F. O., Bear, R. S., and Palmer, K. J., 1941. X-ray diffraction studies on the structure of the nerve myelin sheath. J. Cell. Comp. Physiol. *18:* 31–42.

Segal, J. R., 1958. An anodal threshold phenomenon in the squid giant axon. Nature *182:* 1370.

Shanes, A. M., 1954. Effect of temperature on potassium liberation during nerve activity. Am. J. Physiol. *177:* 377–382.

Shanes, A. M., 1958. Electrochemical aspects of physiological and pharmacological action in excitable cells. Part II. The action potential and excitation. Pharmacol. Rev. *10:* 165–273.

Shanes, A. M., Freygang, W. H., Grundfest, H., and Amatniek, E., 1959. Anesthetic and calcium action in the voltage clamped squid giant axon. J. Gen. Physiol. *42:* 793–802.

Sjodin, R. A., and Mullins, L. J., 1958. Oscillatory behavior of the squid axon membrane potential. J. Gen. Physiol. *42:* 39–47.
Soroka, W. W., 1954. Analog methods in computation and simulation. New York, McGraw-Hill.
Spyropoulos, C. S., 1956. Changes in the duration of the electrical response of single nerve fibers following repetitive stimulation. J. Gen. Physiol. *40:* 19–25.
Spyropoulos, C. S., 1961. Initiation and abolition of electric response of nerve fiber by thermal and chemical means. Am. J. Physiol. *200:* 203–208.
Spyropoulos, C. S., 1965. The role of temperature, potassium, and divalent ions in the current-voltage characteristics of nerve membranes. J. Gen. Physiol. *48:* (5, 2) 49–53.
Spyropoulos, C. S., and Brady, R. O., 1959. Prolongation of response of node of Ranvier by metal ions. Science *129:* 1366–1367.
Stämpfli, R., 1954. A new method for measuring membrane potentials with external electrodes. Experientia *10:* 508–509.
Stämpfli, R., 1958. Die Strom-Spannungs-Charakteristik der erregbaren Membran eines einzelnen Schnürrings und ihre Abhängigkeit von der Ionenkonzentration. Helv. physiol. pharm. Acta. *16:* 127–145.
Strickholm, A., 1961. Impedance of a small electrically isolated area of the muscle cell surface. J. Gen. Physiol. *44:* 1073–1088.
Strickholm, A., 1963. Membrane current of crab muscle. Nature *198:* 393–394.
Tasaki, I., 1939a. The strength-duration relation of the normal, polarized and narcotized nerve fiber. Am. J. Physiol. *125:* 367–379.
Tasaki, I., 1939b. Electric stimulation and the excitatory process in the nerve fiber. Am. J. Physiol. *125:* 380–395.
Tasaki, I., 1939c. The electro-saltatory transmission of the nerve impulse and the effect of narcosis upon the nerve fiber. Am. J. Physiol. *127:* 211–227.
Tasaki, I., 1953. Nervous transmission. Springfield, Ill., Thomas.
Tasaki, I., 1955. New measurements of the capacity and the resistance of the myelin sheath and the nodal membrane of the isolated frog nerve fiber. Am. J. Physiol. *181:* 639–650.
Tasaki, I., 1959. Conduction of the nerve impulse. In Magoun, H. W. (ed.) Handbook of physiology, Section 1: Neurophysiology, Washington, D. C., American Physiological Society, p. 75–121.
Tasaki, I., and Bak, A. F., 1958a. Current-voltage relations in single nodes of Ranvier as examined by voltage-clamp technique. J. Neurophysiol. *21:* 124–137.
Tasaki, I., and Bak, A. F., 1958b. Discrete threshold and repetitive responses in the squid axon under "voltage-clamp." Am. J. Physiol. *193:* 301–308.
Tasaki, I., and Frank, K., 1955. Measurement of the action potential of myelinated nerve fiber. Am. J. Physiol. *182:* 572–578.
Tasaki, I., and Freygang, W. H., Jr., 1955. The parallelism between the action potential, action current, and membrane resistance at a node of Ranvier. J. Gen. Physiol. *39:* 211–223.
Tasaki, I., and Hagiwara, S., 1957. Demonstration of two stable potential states in the squid giant axon under tetraethylammonium chloride. J. Gen. Physiol. *40:* 859–885.
Tasaki, I., and Luxoro, M., 1964. Intracellular perfusion of Chilean giant squid axons. Science *145:* 1313–1315.
Tasaki, I., and Mizuguchi, K., 1948. Response of single Ranvier nodes to electrical stimuli. J. Neurophysiol. *11:* 295–303.

Tasaki, I., and Mizuguchi, K., 1949. The changes in the electric impedance during activity and the effect of alkaloids and polarization upon the bioelectric processes in the myelinated nerve fibre. Biochim. Biophys. Acta *3:* 484–493.

Tasaki, I., and Singer, I., 1965. A macromolecular approach to the excitable membrane. J. Cell. Comp. Physiol. *66:* (Suppl. 2) 137–145.

Tasaki, I., and Spyropoulos, C. S., 1958. Membrane conductance and current-voltage relation in the squid axon under "voltage-clamp." Am. J. Physiol., *193:* 318–327.

Tasaki, I., and Takenaka, T., 1963. Resting and action potential of squid giant axons intracellularly perfused with sodium-rich solutions. Proc. Natl. Acad. Sci., USA, *50:* 619–626.

Tasaki, I., and Takenaka, T., 1964. Effects of various potassium salts and proteases upon excitability of intracellularly perfused squid giant axons. Proc. Natl. Acad. Sci., USA, *52:* 804–810.

Tasaki, I., and Takeuchi, T., 1941. Der am Ranvierschen Knoten entstehende Aktionsstrom und seine Bedeutung für die Erregungsleitung. Arch. ges. Physiol. *244:* 696–711.

Tasaki, I., and Takeuchi, T., 1942. Weitere Studien über den Aktionsstrom der markhaltigen Nervenfasser und über die elektrosaltatorische Übertragung des Nervenimpulses. Arch. ges. Physiol. *245:* 764–782.

Tasaki, I., Singer, I., and Takenaka, T., 1965. Effects of internal and external ionic environment on excitability of squid giant axon. A macromolecular approach. J. Gen. Physiol. *48:* 1095–1123.

Tasaki, I., Watanabe, A., and Singer, I., 1966. Excitability of squid giant axons in the absence of univalent cations in the external medium. Proc. Natl. Acad. Sci., USA, *56:* 1116–1122.

Tauc, L., 1955. Étude de l'activité élémentaire des cellules du ganglion abdominal del'Aplysie. J. Physiol. (Paris) *47:* 769–792.

Taylor, R. E., 1959. Effect of procaine on electrical properties of squid axon membrane. Am. J. Physiol. *196:* 1071–1078.

Taylor, R. E., 1963. Cable Theory. In Nastuck, W. L. (ed.) Physical techniques in biological research Vol. 6, New York, Academic Press, p. 219–262.

Terzuolo, C. A., and Araki, T., 1959. Voltage clamp of the post-synaptic and spike potentials in spinal motoneurons. Fed. Proc. *18:* 158.

Tobias, J. M., 1960. Further studies on the nature of the excitable system in nerve. I. Voltage-induced axoplasm movement in squid axons. II. Penetration of surviving, excitable axons by proteases. III. Effects of proteases and of phospholipases on lobster giant axon resistance and capacity. J. Gen. Physiol. *43:* (5, 2) 57–71.

Tobias, J. M., and Bryant, S. H., 1955. An isolated giant axon preparation from the lobster nerve cord. J. Cell. Comp. Physiol. *46:* 163–182.

Tomita, T., and Wright, E. B., 1965. A study of the crustacean axon repetitive response: I. The effect of membrane potential and resistance. J. Cell. Comp. Physiol. *65:* 195–209.

Trautwein, W., Dudel, J., and Peper, K., 1965. Stationary S-shaped current voltage relation and hysteresis in heart muscle fibers. Excitatory phenomena in Na^+-free bathing solutions. J. Cell. Comp. Physiol. *66:* (Suppl. 2) 79–90.

Ulbricht, W., 1965. Voltage clamp studies of veratrinized frog nodes. J. Cell. Comp. Physiol. *66:* (Suppl. 2) 91–98.

Ussing, H. H., 1952. Some aspects of the application of tracers in permeability studies. Advanc. Enzymol. *13:* 21–65.

Vallbo, Å. B., 1964. Accommodation related to inactivation of the sodium permeability in single myelinated nerve fibres from *Xenopus laevis*. Acta physiol. Scand. *61:* 429–444.

Villegas, R., Caputo, C., and Villegas, L., 1962. Diffusion barriers in the squid nerve fiber. The axolemma and the Schwann layer. J. Gen. Physiol. *46:* 245–255.

von Neumann, J., and Morgenstern, O., 1944. Theory of games and economic behavior. Princeton, Princeton University Press.

Walker, N. A., 1955. Microelectrode experiments on *Nitella*. Aust. J. Biol. Sci. *8:* 476–489.

Weidmann, S., 1952. The electrical constants of Purkinje fibres. J. Physiol. *118:* 348–360.

Weidmann, S., 1955. Effects of calcium ions and local anesthetics on electrical properties of Purkinje fibres. J. Physiol. *129:* 568–582.

Whitcomb, E. R., and Friess, S. L., 1960. Blockade of the action current in single nodes of Ranvier from frog nerve by physostigmine and certain amino derivatives. Arch. Biochem. Biophys. *90:* 260–270.

Wilson, F. N., Macloed, A. G., and Barker, P. S., 1933. The distribution of the currents of action and injury displayed by heart muscle and other excitable tissues. Ann Arbor, University of Michigan Press.

Wright, E. B., and Tomita, T., 1962. Separation of sodium and potassium ion carrier systems in crustacean motor axon. Am. J. Physiol. *202:* 856–864.

Wright, E. B., and Tomita, T., 1965. A study of the crustacean axon repetitive response: II. The effect of cations, sodium, calcium, (magnesium), potassium, and hydrogen (pH) in the external medium. J. Cell. Comp. Physiol. *65:* 211–228.

Wright, E. B., Coleman, P., and Adelman, W. J. Jr., 1955. The effect of potassium chloride on the excitability and conduction of the lobster single nerve fiber. J. Cell. Comp. Physiol. *45:* 273–308.

PART V

Inside the Membrane

Contents to Part V

Introduction

The investigations of living cells based on electrical concepts and using electrical techniques have been amazingly successful. In this way, over a half century, the membranes of cells have been discovered and described and their permeabilities to ions have been found. And for so sophisticated an accomplishment as a nerve impulse these permeabilities have been determined in great detail.

Almost two decades ago an operational description of the black box, the ion-permeable membrane, was found. This was soon nicely partitioned into two amazingly detailed independent channels, two parallel black boxes, one for sodium and one for potassium. No matter what the fallacies or inadequacies of this work, these are a very powerful description of what a normal nerve membrane does—such a powerful description, in fact, that we may not realize for a while that there is, in them, not even a single guess as to *how an axon membrane does what it does*. Much more can be done, and some more has been done along past lines, but the goal has clearly changed.

The new challenge in the new era is to find what happens inside the membrane, the black box of the past. It is certainly possible, if not highly probable, that we may go seriously astray if we fix our attention on only the ion permeability phenomena of any membrane—the more so if it is an excitable membrane. We must look also to the other less complete measurements and the far less concordant interpretations of membrane structure and function. These must all meld together, and in the rather near future. But until they do, there must be many hypotheses as to the structure and function of membranes. Again I can only leave the tracing of the significant developments to the historians of what is to come. I can only sketch the present and I cannot avoid emphasizing what seems important, or perhaps only provoking, to me now.

Membrane Structure

The existence of a cell membrane was long an embattled hypothesis. Before it was in any way conceded to be a reality, Overton (1899) showed that many molecules penetrated into cells with a speed corresponding to their lipid solubility, and he ascribed this to such substances as lecithin and cholesterol near the surface. This result has been vastly extended and refined, as by Jacobs (1935), Collander (1937), and Danielli (Davson and Danielli, 1943), to survive as an important fact of membrane behavior. The effects of solubility and size are seen in figure 5:1 for the single example of *Chara* given by Collander.

The electrical evidence of Höber (1910, 1912) for an electrolytic cytoplasm and the implication of a thin, poorly conducting layer led the way to the membrane as we have come to know it. This membrane was first seen and described for the red blood cell as of molecular dimensions by the theory and experiment of Fricke (1923, 1925). The capacity of 1 $\mu F/cm^2$ would not permit a thickness of less than 10 Å and the assumption of a lipid membrane led

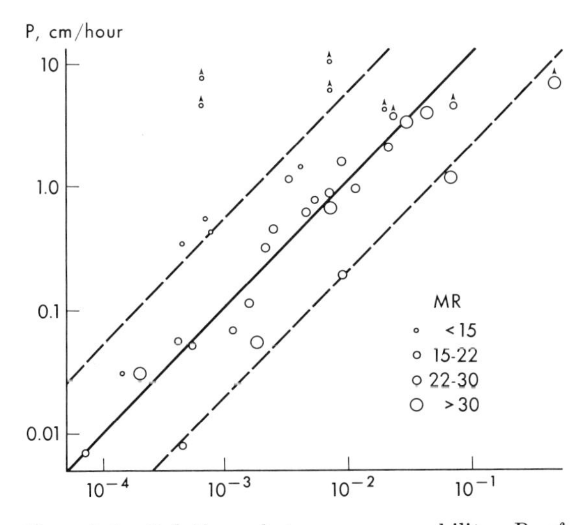

Fig. 5:1. Relations between permeability P of *Chara*, ordinates and olive oil: water partition coefficients, abscissae, for substances classified according to size as given by a molar refraction MR.

to a monomolecular layer of 33 Å. Almost all of the subsequent measurements with many techniques on many living cells, both plant and animal, large and small, show this same membrane capacity within a factor of only a few times (Cole, 1959). We have thus long had basis for the generalization that membranes of all living cells are much the same. The dielectric loss of some of them was characteristic of insulating liquids and solids with strong intermolecular forces. This loss and the capacity were constant, almost independent of the ion permeabilities as they varied a thousand fold and more between rest, activity, injury, and death. So we look upon the membranes as mostly highly organized, inactive structures with seldom more than a few percent of their volume actively involved in the problems of living. Yet there are questions. Why and how do the membrane capacities of some marine eggs increase on fertilization? Why does the capacity of a swollen egg decrease? Is it because the membranes are crinkled—as Rothschild (1957) proposed?

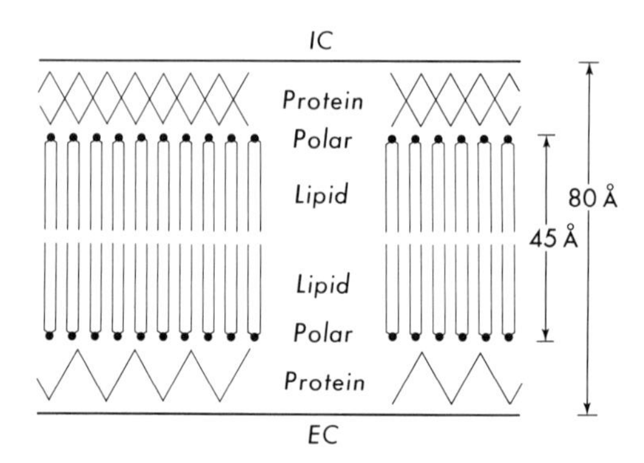

FIG. 5:2. Schematic bimolecular leaflet model for cell membranes. The interior hydrocarbon chains of the lipid may be neither straight nor entirely static, the polar groups may be quite irregular, the proteins at the intracellular *IC* and extracellular *EC* surfaces are probably different in composition, structure, and function.

Soon after Fricke, Gorter and Grendel (1925) spread the lipid extracts from red cell ghosts on water. They found the area of the condensed film was twice that of the original red cells and proposed that the membrane was a bimolecular leaflet as in the core of figure

5:2. Analyses of these lipids and of many since have shown them to be mostly phospholipids and sterols—such as lecithin and cholesterol.

SURFACE FORCE

The next step was made by Harvey (1931a,b) as he showed that the tension at the surface of an egg cell was well below 1 dyne/cm. In my effort to find out why the eggs appeared to be spherical under gravity I confirmed Harvey's result (Cole, 1932) with a value near to 0.1 dyne/cm, which is about the mean of many subsequent measurements with various techniques on numerous cells and protoplasmic surfaces (Harvey and Danielli, 1938; Hiramoto, 1963). The tension to be expected from a bimolecular leaflet or even a single oil-water interface was a hundred times more than this. So Danielli and Davson (1935) postulated and demonstrated that such a high surface energy could be reduced according to Gibbs to about the membrane values by the adsorption of protein on both sides. Of the various possibilities they adopted for their pauci molecular model the phospholipid arrangement with hydrocarbon tails inside and polar groups facing out, and unrolled protein surmounted by globular protein at the water interface. As shown in figure 5:2, this was destined to survive as the bimolecular leaflet, the unit membrane, or simply the bilayer model of living membranes.

I was very surprised to find also that the egg membrane tension was elastic, increasing with increasing surface area rather than of a capillary kind and independent of area. Then the low tension of the undeformed membranes was to be correlated with the slight internal pressure. This, too, became a well-supported generalization; it could be a direct cause and effect relationship or entirely fortuitous, as far as I now know. Several of these measurements, including those of Cole (1932), Mitchison and Schwann (1954), and Hiramoto (1963), have yielded a noncommittal elastic modulus of about 0.4 dyne/cm. This membrane modulus is the surface force required to double the unstretched area of the membrane, and it does not include the estimate of the membrane thickness. To obtain Young's modulus—or the more specific rigidity modulus, if Poisson's ratio is 0.5 as seems uncertain—Mitchison and Schwann and Hiramoto used a microscopic thickness of several microns. Their values were about 10^3

to 10^4 dyne/cm^2 as for a flaccid jelly. I assumed that these and the other similar membrane moduli (unpublished) were to be ascribed almost entirely to a membrane about 100 Å thick, so I arrived at a Young's modulus of the order of 10^6 dyne/cm^2, a tenth of a nominal value for rubber. Interestingly, many of these eggs cytolyze at about twice their normal area to give nearly the same value for an elastic limit. Rather incidentally Danielli and Davson (1935) did show that adsorbed protein changed the tension at an oil-water interface toward that of an elastic film, but I have not obtained a modulus for it.

Once again there are difficulties. Mitchison and Schwann interpret their results as a bending of a thick membrane, while Hiramoto preferred the stretching of a thin membrane. Hiramoto also showed that my original simple analysis was only approximate and that his urchin egg elasticities crept down in five minutes to as little as a third of their initial values. This was confirmed by Yoneda (1964), but he went on to question my measurements of the flattened area of the eggs. His very adroit alternative analysis was that the work done dW in a compression dz against a force F is

$$dW = F\,dz = T\,dS$$

where the surface force is T and the surface area is increased by dS. In this way he arrives at a surface force which is independent of area. Can this be explained by a formation of new membrane? If not, then how?

As usual, things get more complicated when we turn to problems more closely related to mammals and medicine. Rand and Burton (1964) give a "resistance to deformation" for the human red cell of about 0.04 dyne/cm, which might be either bending or stretching. I would like to think of it as mostly bending because this modulus increases many times in the sphered cell to perhaps a stretching tension comparable to the marine eggs. But then the elastic limit must be quite low because the red cell hemolyzes with but little increase of surface area. However, Rand (1964) does measure an elastic relaxation here too, confirming the conclusion of Katchalsky et al. (1960) from hemolysis measurements.

The interpretation and implications of the elastic data are more challenging than informative on the molecular scale. The surface trough measurements on films at an air-water interface seem not yet to have reached a stage comparable to the living membranes.

Phospholipid films usually collapse at barrier pressures much less than that for the surface tension of water, and so they do not correspond to the low surface force for a membrane as indicated in figure 5:3. Cholesterol is described as compacting the phospholipid films, but these also collapse at much too low barrier pressures. Perhaps the addition of protein will produce the desired characteristics. On the other hand, Pethica (personal communication) calculated energy minima for a pattern of dipoles oriented nearly in the surface planes of a bimolecular layer bounded by water. I do not yet know the contribution of the second interface or if such a film has an interesting elasticity. Experimentally, however, Hanai et al. (1965) showed that the surface polar groups of phospholipid bilayers were in the same plane and Thompson (1964) reported a tension of 1.0 dyne/cm.

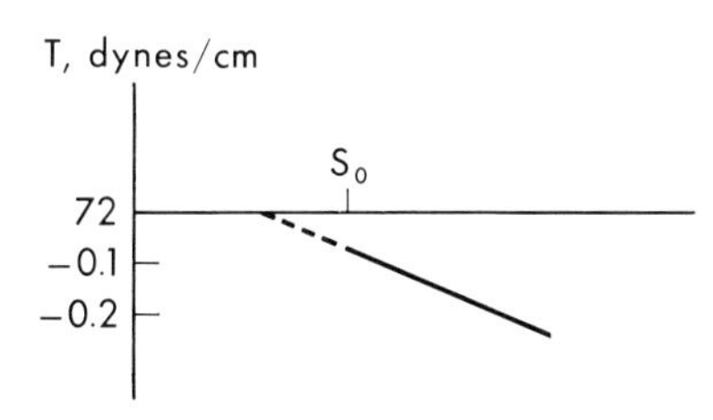

FIG. 5:3. Hypothetical behavior of a monolayer at air water interface analogous to properties of cell membranes. The barrier pressure T should decrease from that of pure water along dashed line to a little less at normal cell membrane area S_0 and continue along solid line up to about 2 S_0.

A most startling behavior of oil drops at the surface of marine eggs was discovered by Chambers and Kopac (1937). A fresh, clean oil drop, 25 μ or more in diameter, could pop inside a naked unfertilized egg of about 100 μ diameter in an unmeasured small fraction of a second. The minimum size of the oil drop that could penetrate corresponded to an energy for the system which was the same for the oil inside as for that outside. Danielli (1954) described a neat kinetics for the process. As the oil touches the egg the protein on the membrane reduces the surfacc tension of the oil. The normal tension over the rest of thc oil gives a high internal pressure which then blows through the weak egg-oil interface. Whatever the explanation, a large hole must be made in the egg membrane, the surface of the membrane increased considerably, and the hole sealed—almost instantaneously. Perhaps a solid membrane can do this, but I expect that it will be found more plausible for a somewhat liquid structure in which the relative positions of the molecules are not fixed. Recent and perhaps similar examples are for the microcapillary electrodes which have been used so successfully on a wide variety of cells. In the range of 0.1–1.0 μ diameter these tips make an enormous hole in the membrane on a molecular scale, yet the membrane very often seems to seal around them quite effectively and to close over the hole when they are withdrawn.

The experiments on the mechanical distortion of numerous cell surfaces are usually interpreted as a stretching of an elastic membrane, with perhaps some contribution by bending. Then these membranes have a Young's modulus only somewhat less than that of a solid such as rubber, and they may show a considerable viscoelasticity. Yet they can be penetrated and can heal with a facility that strongly suggests something approximating a two-dimensional liquid—perhaps a liquid crystal.

OPTICAL, X-RAY, AND ELECTRON INFORMATION

Before and about 1940, F. O. Schmitt and his collaborators gave the basis for much of the biophysics of membrane structure that was to come. Schmitt et al. (1937, 1938) did a polarization analysis of the red cell ghost to show an intrinsic birefringence of lipids caused by their orientation perpendicular to the membrane surface. This interpretation of the observations has been questioned (Mitchison, 1953), but the original conclusion has since become increasingly probable. With Schmitt, Waugh (1940), made a more direct optical attack on the red cell ghost thickness by comparing reflection intensities with a series of standard barium stearate films. From this experiment they proposed that the lipid thickness was about 100 Å, with a layer of protein similar in thickness. Then with the x-ray diffraction approach, Schmitt et al. (1941) showed in myelin the wide angle diffraction at 4.7–4.9 Å, probably representing the spacing of hydrocarbon chains in a plane parallel to the plane of the membrane, and the persistent small angle diffraction of about 75 Å, which was explained as a fundamental bimolecular component. These, somewhat prophetic, descriptions of myelin as many layers of bimolecular cell membranes have since been nicely, and apparently conclusively, confirmed by several other independent lines of investigation.

With lipid extracts, Palmer and Schmitt (1941) showed the decrease of spacing between bilayers to be expected with increasing ionic strength of the medium, except that the interlamellar space was not only reduced with $CaCl_2$ but at no more than 0.1M this space virtually vanished and the entire structure disappeared. This observation is certainly to be remembered and explained.

From the molecular composition of lipids at about phospholipid: cholesterol:cerebroside::2:2:1 and the x-ray diffraction of myelins, Finnean (1962) suggested a tidy candy cane pattern with the phosphate group highly accessible at the boundary and the hydrocarbon turned inside. However, there are species variations of lipid components in both red blood cells and myelin. Much more detailed and complicated organizations have been proposed (Vandenheuvel, 1965); van Deenen (1966) and Chapman (1966), on the basis of infra red and thermal analyses, strongly suggest that the lipids may be usually in the liquid crystal category.

Luzzati and Husson (1962) showed detailed structures of soap and water lamellae with transitions to hexagonal arrays of cylinders with either soap or water for the continuous phase. This is the modern version of the reversible emulsion which Clowes (1918) suggested as a basis for ion permeability changes.

The electron microscope, the improvements in sectioning, following Pease and Baker (1948), and in staining, such as permanganate (Robertson, 1957), have confirmed these conclusions with almost uniquely satisfying visual evidence. On such evidence alone, Robertson (1966) presented the "unit membrane" as a basic, ubiquitous, component of cells, figure 5:4. This, the railroad track

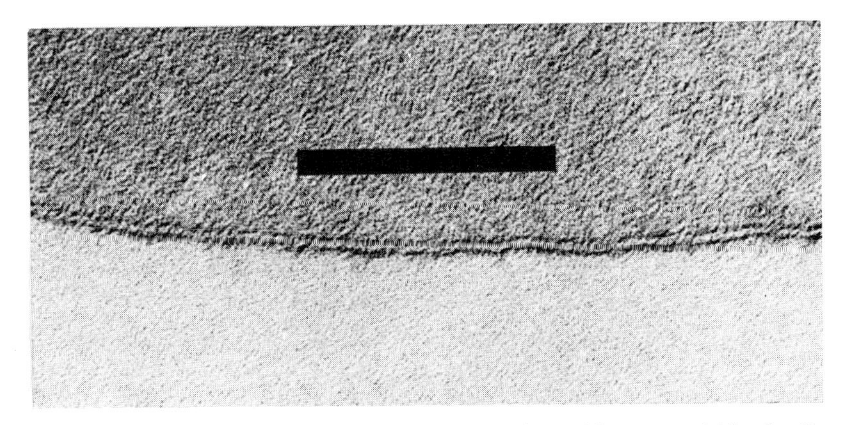

FIG. 5:4. Electron micrograph of bilayer at surface of human red blood cell. The scale marker is 100 Å by 1000 Å.

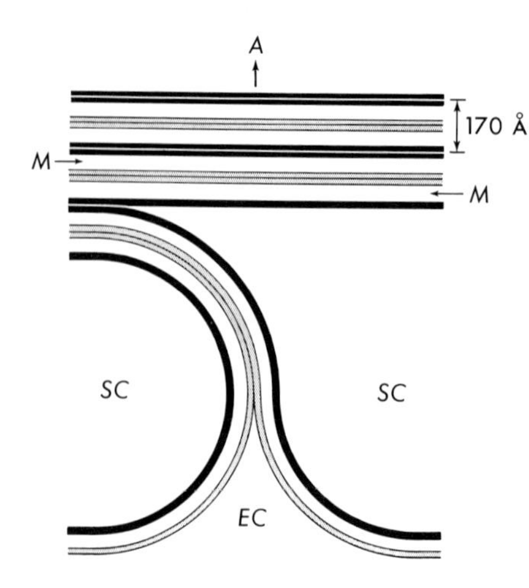

FIG. 5:5. Diagram for formation of myelin sheath. Starting at the axon *A* successive layers *M* have been laid down as a double membrane was taken from the enveloping Schwann cell *SC*. The extracellular *EC* faces of the membranes are shown as stipple and the inner as dark.

of about a 70 Å gauge, looks like a border patrol for each cell. The unit membrane was consistent with the bimolecular membrane theory and with the widespread electrical characteristic of 1 μF/cm^2 for the membranes of living cells. If he had so chosen, Robertson could have strengthened his position by noting the correspondence between the electrical and the electron microscopic observations and using it to support the conclusion that these images were reasonable facsimiles of the living cell membranes in spite of the atrocities of fixation, staining, plastic imbedding, sectioning, and electron bombardment.

The electron microscope also showed the unit membrane in most intracellular organelles, and with a particularly fascinating maze-like plan in the important mitochondria. With it Geren (1954) caught the Schwann cell in the act of laying down its membrane around an axon as a jelly roll to form the myelin on a medullated axon (figure 5:5). But further membrane structure also appeared, with an emphasis on hexagons about 150 Å across in the plane of the membrane as summarized by Kavanau (1963, 1965). The beautiful patterns produced by saponin on red cells were strongly attacked as artifacts characteristic of lecithin and cholesterol. Several membranes with regular arrangements of internal 85 Å spheres were reported; one example was shown most clearly by freeze-fracture (Branton, 1966). I like to think of a "membrane unit" that is like the bathroom tile of figure 5:6, 100–200 Å across and about 100 Å—or a "unit membrane"—in thickness, even though it may not be real.

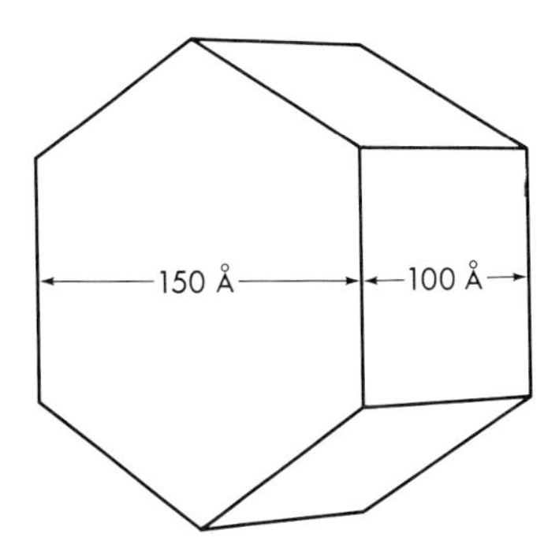

FIG. 5:6. A possible membrane unit.

The late Julian Tobias et al. (1962a) summarized the state of his extensive program which was keyed to optical and mechanical changes accompanying electrical activity, primarily in single axons. These were changes of opacity, diameter, rigidity, axoplasmic flow, surface contour, and length. They were then ascribed to membrane phospholipids because all of the normal electrical membrane characteristics disappeared after treatment with hydrolyzing phospholipases or the destructive digitonin. An essential role was denied to protein because external proteases which break peptide bonds left threshold, conduction velocity, resting and action potentials, and membrane resistance and capacity virtually unchanged. On the basis of experimental evidence that phospholipid was hydrophilic in contact with KCl and hydrophobic in contact with $CaCl_2$, the light-scattering, dimensional, rigidity, and surface contour changes were attributed to hydration changes accompanying changes of potassium vs. calcium concentrations in the membrane surface. These ion concentrations in turn were linked to the ion conductance of the membrane.

The simple bilayer was becoming increasingly unsatisfactory in some respects. Founded as it was in part on the myelin nerve sheath, there was no need, nor space, to accommodate much biochemistry. Green and Perdue (1966) considered the bilayer "a brilliant anticipation . . ." but sought to "make an orderly transition to the biochemically acceptable model." They reviewed the progress of a decade on electron transfer and enzyme systems and on electron microscopic structures, mostly for intracellular organelles. The macromolecular repeating unit shown was like the almost extinct collar button. The bases fit together to form a "membrane continuum" even after removal of the stalk and headpiece which they viewed as the fundamental part in the membrane formation. Thus we may have more and important detail to add to something like the "membrane unit" of figure 5:6. Korn (1966) added criticism of the unit membrane as not proven by the data it was based on and not adequate for recent data or current concepts of genetic control of structure and function.

There seems to be a reasonable certainty of the participation of proteins at the water-lipid interfaces of various membranes. Stoeckenius (1962) made an artificial unit membrane from phospholipid and protein. With the proper protein the appearance in the electron microscope was practically identical with that of the natural membrane. But the structure and function of the protein are even more doubtful than those of the lipid core. As a particular example, Hechter (1965) noted a trend away from the alpha helix and, quite far into the realm of conjecture, he gave a detailed picture of overlapping layers of hexagonal protein units and of ice blockades. Apparently phase modifications might permit impulse transmission along such a membrane without the intermediary of an external current flow, which real excitable membranes seem to require. F. O. Schmitt returned to peripheral nerve, with Davison (1965), to propose a link between the macromolecular protein at the inner surface of the bilayer and the electrical function. These electrogenic protein molecules could be close packed cubes at the inside of a uniform phospholipid bilayer. If a change of electric field distorted perhaps four adjacent cubes into spheres, the phospholipids would be compacted on the reduced available protein surface to leave a pore in the center for ions.

Beginning in 1965 there appeared more objections to the spread out, β configuration, layers of protein such as indicated in figure 5:2. Various spectroscopic measurements by infrared, fluorescence, and rotatory dispersion on isolated membranes of red cells, chloroplasts, and ascites cells were assembled by Wallach and Zahler (1966) in support of a membrane model. They proposed that there were membrane proteins with two widely separated hydrophilic peptide regions. These were located at the hydrated membrane surfaces. Between these regions a hydrophobic rod of helical peptides was to project at right angles through the membrane. Singly or as tubular aggregates, these hydrophobic units were suggested as the paths of transport.

In view of some of the questions being raised probably the somewhat fuzzy electrical generalities should be considered more carefully. Suggesting, as they do, a stable passive structure, such as bilayers might be, it seems to maintain a considerable integrity, in some life processes at least. So perhaps the membrane only supports many and variegated functions in but small parts of its expanse which may be difficult to recognize and explain.

MEMBRANE JUNCTIONS

It has seemed quite fitting and proper that resting membranes should maintain their low conductances even when they were in quite close contact. This has generally been the case in tissues, in densely packed red cells (Robertson, 1960), between two cells in a developing marine egg (Ashman et al., 1964), and apparently across the many layers of membranes wound around an axon to make a myelin sheath (Hodgkin, 1964). In myelin, at least, there was general agreement of the x-ray, visible, and electron microscope evidence for the membranes packed outside to outside and inside to inside as in figure 5:5. There were also more complicated and obviously specialized interfaces in various chemical junctions with internal vesicles and other structures in back of them (figure 5:7a). In some cases, at least, there were channels for the chemical transmission of information—for the "soup" of decades before.

For a long time groups of cells in smooth and cardiac muscle tissues had acted as single units, so protoplasmic bridges, syncytia, and other reasonable connections were postulated, looked for, and even found. But the electron microscope showed each individual cell and organelle to have its own bounding railroad track, with no evidence for sieves, except in some nuclear membranes (Watson, 1959), or even holes connecting adjacent cells. Of course the membrane resistance might be low—of the order of 1–10 Ωcm^2 as found for red cell and active nerve membranes—but such a barrier, corresponding to a millimeter thickness of electrolyte, was considerable.

The idea that a pair of unit membranes might have a negligible resistance—perhaps less than that of a micron thickness of electrolyte was so contrary to past experience as to be quite unbelievable. Yet in a flurry of a few years of intense competition and cooperation, electrophysiology and electron microscopy forced us to believe that two membranes not only can but frequently do join to become essentially perfectly ion-permeable connections between cells.

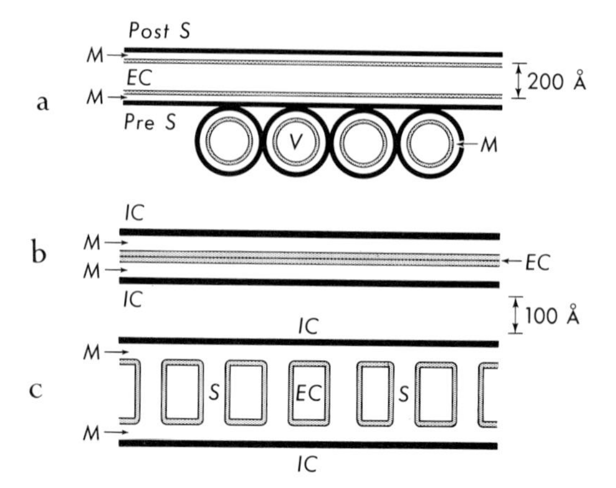

Fig. 5:7. Diagrams for membrane contacts between cells. *EC* and *IC* for extra- and intracellular phases with corresponding light and dark boundaries for the membrane *M*. Membrane bilayers and vesicles *V* associated with chemical synapse transmission *a*. The bilayers correlated with high ion permeability are the tight junctions *b* with little or no *EC* space and the septate junctions *c* with hexagonal structures in the *EC* space.

This was first shown between cells of the same type in several tissues—nervous system, smooth muscle, cardiac muscle, and epithelium. Then similar connections such as an excitatory synapse (Furshpan, 1964; Pappas and Bennett, 1966) appeared between different cell types and more experiments were done. By 1966 there were more than a dozen examples in which electrical coupling between cells was correlated with regions of reduced or negligible extracellular space between the cell membranes. Electrically these were usually known as electrotonic junctions, with resistances from 1 Ωcm^2 on down to practically nothing; morphologically they might be *zonula occludents*, *septate desmosomes*, a nexus, or an intercalated disk, depending on their origin and differences in detail. In general these were the more usual tight junctions of figure 5:7b with no visible extracellular space, and the septate junctions such as that described by Locke (1965) and shown in figure 5:7c.

The dye fluorescein diffused in a salivary gland, crossing the transverse membrane pairs as if they were not there (Loewenstein and Kanno, 1964). Tracer potassium diffused along heart muscle fiber bundles (Weidmann, 1965) as if the intercalated disks had a resistance of no more than 3 Ωcm^2 compared to the outer surface membrane resistance of the order of 1 $k\Omega cm^2$. A nexus was reversibly broken in cardiac tissue with hypertonic shrinking of the cells (Barr et al., 1965). Absence of external calcium irreversibly broke the junction in a nerve tissue (Penn and Loewenstein, 1966); either calcium removal or hypertonic perfusion increased the resistance and separated junctions in an epithelium (Nakas et al., 1966; Palade, 1966); whereas a moderate calcium concentration had broken down *in vitro* bimolecular structures of lipid extracts.

So we had mounting evidence that the external surface was at least considerably involved in the dramatic change from a slightly permeable pair of membranes to the nearly completely permeable combination, and that calcium could be an important factor. Potter et al. (1966) found tight junctions throughout the early squid embryo and these were gradually eliminated during development. So I suspect that our "hot spots" in heart tissue cultures (Hogg et al., 1934) might have been evidence of some such membrane ion path.

In contrast to the primitive use of an axon as a passive cable, these electrotonic junctions are highly developed structural elements which allow much the same electrical performance. As both these and apparent chemical transmitter connections appeared side by side we had an answer to yet another controversy; that of "sparks vs. soup" of three decades before.

From all of the lines of physical and chemical evidence we are led to a bimolecular membrane model with a hydrocarbon central layer about 25 to 50 Å thick and a polar and protein layer of about the same thickness or less on each side, substantially supporting the Fricke, Gorter and Grendel, and Danielli suggestions. There are hopeful indications of a tilelike unit in the plan of the wall that is the membrane. There are also highly permeable contacts between the membranes of some cells. But we seem to be far from an adequate description in molecular terms of the energetics, stability, and distortions, and interactions between the membranes.

Ion Permeabilities

I see no present, definite, and certain indications of the nature of the mechanism by which the ion permeabilities of a membrane are controlled. There is no dearth of proposals. Some of these have been made to describe a single phenomenon and tested only rather casually. For example, a system producing an action potential-like response or a uniform propagation may have no other recognizable similarity to a nerve fiber. The question of criteria to be met is important. There is some minimum, such as electrical instability for excitable membranes; yet to insist that a mechanism is entirely unacceptable because it fails to meet one or more possible criteria may be to exclude a valuable idea in primitive form (Cole, 1966).

At the other extreme are experiments in which both an all-or-none response and its propagation were prevented. It has been quite thoroughly demonstrated that these contain far more information, and more pertinent information, about the normal performance of the squid membrane than does any other single experiment or group of experiments. In these voltage clamp experiments, abrupt changes of the membrane potential produce the membrane ion currents such as figure 3:16, which I first obtained (Cole, 1949) and Hodgkin and Huxley (1952a) analyzed into the sodium and potassium components I_{Na} and I_K, figure 3:29. A model or an analogue which will reproduce this analysis of voltage clamp data will equally well represent most of classical electrophysiology. Similarly, a model or an analogue may be expected to fail to the extent that it does not conform to the voltage clamp data. Few of the models now on hand are developed far enough for such an extensive test, but the clamp results remain a very clear, a quite complete and immediate goal for any mechanism.

We may consider the membrane as an electrolyte with the current density given as the product of a field strength and a conductance. The values of the average and the breakdown field strengths in the membrane are of the order of 10^5 V/cm. This average value for the normal operations of the membrane is certainly striking. It may well be highly significant as more becomes known of the behavior of other systems under such conditions. Membrane conductances range from very small values beyond the negative resting potential to much larger maximum values at positive potentials. The nominal value of 1000 Ωcm^2 at rest corresponds to a column of sea water 50 cm long. A maximum conductance equivalent to 5 Ωcm^2 is approached in a number of membranes under positive voltage clamp, and 0.1 Ωcm^2 has been found for a node. A nominal value of 1 Ωcm^2 is then equivalent to a 0.5 mm of sea water.

In common with squid, other normal excitable membranes have shown a transient ion flow which gives rise to instability; these can be described as quasi steady state negative resistances. It now appears quite certain that there can be no nerve impulse without at least a quasi steady state negative resistance. On the other hand, with only a feeble negative resistance in the membrane, an axon must be excitable and must transmit an impulse. This is then a relatively simple criterion for a model.

But there can also be long lasting or steady state negative resistances, particularly with iso-osmotic potassium at the outside of the membrane. This behavior is contained in the HH axon and was first demonstrated for the squid membrane (Moore, 1959) (figure 4:47), but it is also a characteristic of several other cell membranes, including a lobster axon, the nodes of toad and frog axons, electroplax, and frog skin! Then it was shown that calcium was probably essential for the squid axon (figure 4:51). These performances have given us yet another criterion for models to meet—a particularly important and simple criterion. The model must show a steady state negative resistance between equal concentrations of an ion, such as potassium, on the two sides of the membrane as calcium is added. When there is a good model for such an easy problem as this we should be in a much better position to make a serious attempt at something that may be a bit more difficult—such as the transient sodium ion flow of the normal axon.

Considerable effort was devoted to the expression of membrane parameters in terms of centimeters and seconds but we have reached a stage at which molecular units may be more immediately significant. So it is interesting and perhaps provocative to look at some of the membrane phenomena on an intermediate scale (Cole, 1965a).

First consider the membrane to be 100 Å thick—it is probably more than 50 Å and less than 200 Å—then take the hexagonal faces 150 Å across, which seems a possible unit of structure, as shown in figure 5:6, and use the millisecond as a convenient time interval.

For a resting potential of -60 mV and a capacity of 1 μF/cm^2 there will be an average excess of one negative univalent ion over the inside and one positive ion over the outside of the hexagonal element with face areas of $2 \cdot 10^4$ Å^2 each. When these charges are removed the potential falls to zero and, to the extent that the conventional resting membrane resistance of 1000 Ωcm^2 is linear, this would result in an ion flow through the element of one ion per millisecond. But if an average of only one third of an ion is removed rapidly from each side, to reduce the membrane potential by about 20 mV, the threshold is passed. The resulting impulse goes to a potential where the membrane ion pair is reversed. The ion exchange during the passage of an impulse is interpolated to be 5.6 p mole/cm^2 at 15°C. This becomes the entrance of a mere 7 sodium ions and the later exit of the same number of potassium ions through our membrane unit. Furthermore the loss of sodium from the outside during the impulse corresponds to the average sodium content of a layer of sea water about 1 Å thick, and the potassium of the axoplasm is similarly depleted. Another point of reference is that a single ion in the membrane unit corresponds to a concentration of one millimolar.

However it seems possible, as well as usual and highly desirable in biological systems, that a membrane be able to function normally by using only a small fraction of its ultimate resources. The voltage clamp does produce membrane performance considerably in excess of normal and as such may give a much better measure of the membrane capabilities. For example, if we generously assume a maximum sodium inflow of 10 mA/cm^2 for 1 msec under voltage clamp we would have moved about 100 of these ions through the element. It is a fact that the maximum current densities measured under voltage clamp may be 10 mA/cm^2 for squid and 50 mA/cm^2 for a node. For a figure of 20 mA/cm^2, about 200 monovalent ions will be moving through our hexagonal element in a millisecond. This may seem to be rather dense traffic until we calculate by ordinary kinetic theory that the membrane element is being bombarded inside and out by $6 \cdot 10^9$ ions/msec. Thus only one ion out of $3 \cdot 10^7$ goes through and the membrane still appears as a very formidable barrier, even at such a high current density.

There is so far no direct evidence to suggest the character of a membrane permeability change. Is the membrane as close to being perfectly continuous as is possible with matter and charge in the units of ions? Or are there comparatively large, independent, uniform patches of the membrane area which change either gradually or with an abrupt flip-flop between relatively stable low and high conductance states? Or is the unit of change somewhere in a pathway for a single ion across the membrane? And is the flow through such a channel graded or on and off?

Frankenhaeuser (1960) and with Huxley (1964) mentioned that a principal objective of the voltage clamp work on the nodes was to look for evidence of a unitary response in the smallest membrane area so far available. This did not appear, although an estimated one percent of the maximum inward current could have been detected through the noise on the experimental records. The maximum distance between open or shut sites would then be 1 μ. A positive result was that the permeability to some gauge molecules such as erythritol is increased on activity and less so with lowered external sodium for the Venezuelan squid membrane (Villegas et al., 1966). This suggested that nonelectrolytes and sodium share a common pathway. The data may also be interpreted as an increase in the number of pores in excitation but without change of their diameter.

Until there are more and somewhat consistent indications of the fundamental process of permeability change, the appropriate fields for theoretical investigations are not seriously limited. Personal preference based on experience and hunch are still probably the most important guides.

THEORETICAL MODELS

Continuous parameters

Certainly the oldest and best known cell membrane model is that of Nernst (1888) and Planck (1890) on which Bernstein relied. In this electrodiffusion model the membrane is a thin layer of a uniform nonaqueous electrolyte across which ions move with an electric field and down concentration gradients. A prodigious amount of theoretical work has been done on such a simple system and even more on the problems of added components within the membrane—such as various ion exchangers. It is usual to assume that concentration and electrical variables are constant in planes parallel to the membrane surfaces. Such systems have been classified as in figure 5:8 and their properties reviewed by Eisenman et al. (1966). Beyond this the general problem may still be restricted by considering only systems which are homogeneous across the membrane in the absence of differences of both electrical potential and ionic composition at the two boundaries.

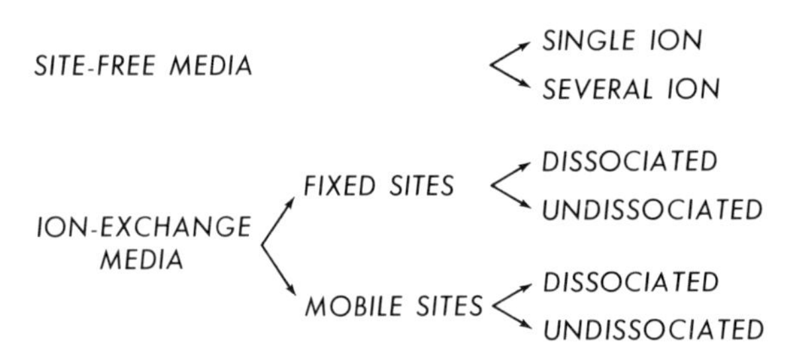

FIG. 5:8. A classification of ion exchange membrane models.

A quarter of a century ago the simple electrodiffusion model was developed to give a very nice potassium explanation of some long-known nerve behavior and for several new and provoking kinetic phenomena that had just appeared in the squid axon. It required the very low mobility and high concentration for this ion that are not yet confirmed or contradicted. However, this model was unable to produce excitation and propagation and has been similarly unable to account for the sodium ion behavior. I reviewed and compared this theory and some squid experiments to conclude that such a simple electrodiffusion mechanism did not provide an adequate explanation of the squid membrane phenomena (Cole, 1965b).

The simple Nernst-Planck electrodiffusion membrane model was approximated by Goldman (1943), it was considerably extended to include fixed charges—such as those of an ion exchanger—by Teorell (1951), it was quite thoroughly generalized by Schlögl (1954), and then arbitrary distributions of fixed charges were considered by Conti and Eisenman (1965). Goldman (1959) gave only a preliminary report of his analogue calculations showing that although the constant field is just an approximate solution, it may sometimes be a rather good one in a steady state. Much more extensive digital computer results have been obtained for a wider variety of steady state problems (Bruner, 1965a,b) and for some transient situations (Cohen and Cooley, 1965).

The fixed charge model with a uniform distribution of negative charges has been a very attractive one, particularly for nerve axon membranes. In this way mobile negative ions could be virtually excluded from the membrane, and positive ions could move relatively freely across it to perform their principal roles in rest and activity. Tasaki et al. (1961) found far larger outfluxes for various tagged positive ions than for any tracer negative ion. They emphasized these results as further support for such a negative ion exchanger membrane model.

Conti and Eisenman (1966) developed an extensive electrodiffusion theory in which a mobile negative component was restricted to the confines of the membrane. Yet on the basis of the first voltage clamp results, Hodgkin and Huxley (1952b) had to reject the earlier hypothesis of lipid-soluble polyvalent carriers for the sodium and potassium ions as they crossed the membrane. Sollner and Shean (1966) consider porous and liquid ion exchange membranes as examples of carrier transport. Meves (1966) reported on a careful search for a sodium carrier. With the relatively impermeable ions choline outside and rubidium inside, no initial carrier movement as large as 10^{-8} coulomb/cm^2 was found in a voltage clamp. This corresponded to a transfer of less than 10^{10} ions/cm^2 for a negative carrier change of six and a spacing of more than 10^3 Å.

The most crushing and probably fatal blow to these nice theories was the direct demonstrations of fewer potassium ions flowing into a membrane as the membrane potential was made more negative, with equal potassium ion concentrations inside and out. This steady state negative resistance region did not appear as a component of any of the simple electrodiffusion models.

In order to produce rhythmic phenomena, which require a negative resistance characteristic, Teorell (1960) added a hydraulic pressure difference across his ion exchanger membrane model. This behavior he explained by the addition of hydraulic and electro-osmotic water flow to the fixed ion extension of the simple Nernst-Planck model. The analysis of this system has been improved by Aranow (1963) and by Kobatake and Fujito (1964a,b). With another ion exchanger membrane Teorell (1962) achieved a negative resistance between identical electrolytes; but this is yet to be explained by theory.

As a possible explanation for their experimental data on crustacean axons, Wright, with Tomita (1965), proposed an additional complication. They suggested an ionic pool between the layers of the bimolecular leaflet. This idea of a transmembrane inhomogeneity had been considered in some detail (Mullins, 1961). Kedem and Katchalsky (1963c) had proposed a model with each side of the bilayer of opposite fixed charge. Later, on the basis of the data of Tasaki et al. (1965a), Katchalsky (1968) elaborated on it. Then as a contrast, the center of the layer was to be a depletion zone to be considered as an analogue of a semiconductor p–n junction.

The second Wien effect of ion pair dissociation in an electric field has often been suggested as a mechanism for the conductance increase of nerve membranes in activity. It seems to have first been presented formally by Monnier et al. (1965), based on the theory of Onsager (1934). The resistance R for a field X is approximately $R = R_0/(1 + qX)$, where R_0 is the resistance for zero field, q is an absolute coefficient and, with a diffusion potential E, $V = E - RI$. As they point out, the field can probably only be high enough over a thin layer of membrane. Bass and Moore (in MS) also invoked the Wien effect.

The problems of membrane permeabilities to ions and molecules were considered in a much more general way by Danielli (Davson and Danielli, 1943) who expressed them in terms of potential barriers. Many small barriers were the modern equivalent of the diffusion process as used by Nernst and Planck. The addition of high barriers, most attractively at the inner and outer boundaries of the membrane, gave interesting models for many cell membrane permeability phenomena, including the characteristic high temperature coefficients.

Some attempt has been made to account for membrane behavior through the formalism of chemistry. This approach was carried through in some detail by Umrath (1927) as an explanation for impulse propagation in *Nitella*. It seems almost as good as—and only slightly more arbitrary than—some recent work which had both the advantage and the disadvantage of forty years of further information.

Polissar (Johnson et al., 1954) proposed a physical chemical model in terms of the membrane potential and a dependent variable, the "potential demand of the membrane." This variable, which seems closely related to the instantaneous Thévenin potential, was to be explained by electrodiffusion and carriers in the membrane. The threshold characteristics were shown on a phase plane by FitzHugh (1955).

An interesting example was one given by Mueller (1958) who calculated the time constant and the displacement of an equilibrium by an electric field in general terms. He then postulated a transfer function by which this displacement modified the membrane potential to provide a feedback circuit. Several suggestions were made as to the possible mechanism of the crucial transfer function, but it remained essentially a process which was undefined except as one might choose to interpret its name—"electromotance." Nonetheless, two analogues based on such a formalism gave good voltage clamp curves and, as a consequence, they incidentally produced quite impressive demonstrations of several other but less demanding performances of living nodes.

On the basis of his experience with the permeabilities of monolayers, Blank (1962, 1965) paid particular attention to the equilibrium between free hydrated and partially hydrated bound positive ions in the regions of fixed negative charges on each side of a bilayer model. This gave high barriers in the aqueous phases rather than in the lipid phase. Depolarization or potassium ions could then release or exchange with bound sodium ions in activity.

The action of anesthetics has long been an attractive problem. There is a voluminous literature on the effects of many compounds and almost as many speculations as to the membrane structure and the location and nature of their actions. The most attractive and useful approaches have come from consideration of the London induced dipole—induced dipole energies. Miller (1961) and Pauling (1961) showed correlations between the expected clathrate formation and anesthetic action. The icelike water structure at a membrane surface was expected to decrease the overall electrical capacity or the intercellular conductance. Each of these changes was then assumed, rather arbitrarily, to block activity. Agin et al. (1965) calculated surface adsorption from London interaction energies which they expressed as the product of molecular polarizability and ionization potential. The minimum external concentration to prevent excitation on isolated muscle or nerve fibers was found to be an exponential function of this product for forty local anesthetics. To the extent that the approximations are valid, the results support the assumption that the minimum effective surface concentration is the same for each of these agents.

In peripheral nerve, and probably other membranes, the nature of the action of local anesthetics and narcotics may be reasonably simple. Normal excitation follows from a prompt flow of inward current, usually sodium, after enough depolarization; the primary action of some blocking agents is to reduce only this current. At least as a first approximation, this action was to reduce the maximum sodium conductance $\bar{g}_{Na}$ in the Hodgkin-Huxley formulation, as was particularly clear for the spectacular tetrodotoxin on the squid membrane. To this powerful toxin Hille (1966) added the local anesthetic, xylocaine, and the tranquillizer, compazine, as examples of substances which act only on $\bar{g}_{Na}$ without changes of any other HH parameters for the frog node.

Tasaki et al. (1965a,b) and with Singer (1965) emphasized the differences between the characteristics of the inner and outer faces of the squid membrane—based on their extensive work on excitability. To the extent that they are confirmed and extended, these results can be expected to help provide a mechanism for the ion conduction process as it has been determined experimentally in the form of electrical conductances of the HH axon.

As an expression of their results, Tasaki and Singer considered the membrane as a macromolecular complex of proteins and lipids held together primarily by saltlike linkages. There was, however, the usual net negative charge throughout, and particularly at the outer surface. This was thought of as a carboxyl colloid while the inner surface was considered to be a phosphate colloid. The order of effectiveness of internal negative ions was explained by ion-fixation processes tending to disrupt the membrane organization. At rest, critical sites were to be occupied primarily by divalent ions from outside and in a stable configuration. In activity, relatively mobile univalent ions filled most of the sites in a second stable state. Such a reversible complex had been proposed as the macromolecular basis for the description of excitability by two-stable states (Tasaki, 1959, 1963). A basis was given for discontinuous changes of the univalent–divalent cation ratio of the outer membrane surface as the external medium or the electric field was changed. There had then to be a population distribution of the exchange parameter to permit the continuous steady state, or quasi steady state, negative conductance characteristic as found in voltage clamps.

As J. G. Kirkwood turned to transport problems he became convinced that the irreversible process theory of Onsager was the most powerful and useful approach to an understanding of the permeabilities of living membranes. He made only an initial step in this direction (Kirkwood, 1954), but it served as a starting point for further developments.

The approach has been limited to conditions of steady states and linear approximations. It has been used generally and with power by Kedem and Katchalsky (1961, 1963a,b,c) who have expressed their results in terms of experimental variables. Extensions and simplifications of the electrodiffusion model considered by Kirkwood were given by Kimizuka and Koketsu (1964), and in this way they arrived at many of the results of other work (Goldman, 1943; Teorell, 1951; Parlin and Eyring, 1954, for example).

An entirely new approach was made by Mauro (1962). He pictured the membrane as a sandwich of negative ion exchanger between two thinner layers of positive exchangers representing the positive ends of the phospholipid polar groups. The equilibrium at the boundaries of a fixed charge membrane had been expressed as a Donnan equilibrium by Teorell (1951) and solved explicitly by Bartlett and Kromhout (1952). However, in order to consider current flow, Mauro resorted to the methods that had been used in semiconductor problems (Shockley, 1949) to obtain approximate solutions to the electrodiffusion equations. He also restricted his treatment to one side of the membrane which was then analogous to a p–n junction. From this he obtained a capacity, largely due to the depletion layer at the junction between the two ion exchangers, of 1 $\mu F/cm^2$. Impedance measurements on such a system (Schwartz and Case, 1964) did indeed give a capacitive reactance. It remains to be seen if this capacity is a function of the potential across the system or if a negative resistance region is to be found. The analysis was extended to high potentials (Coster, 1965) to give a Zener punch-through. This was in reasonable accord with results on *Nitella*. There are probably many examples of "membrane breakdowns" for hyperpolarizations in the range of 200–300 mV (Rudolph and Stämpfli, 1958; Cole and Moore, 1960; Julian et al., 1962;

Taylor and Adelman, unpublished). If indeed these are all punch-through phenomena, the theory will have enough support to make it interesting.

Pores and sites

The second oldest and best known of cell membrane models is that of pores which serve as sieves for molecules and ions. Certainly the long time leader in this field was Traube's membrane of copper ferricyanide. The collodion membrane, investigated extensively by Loeb (1920) and Michaelis (1925), has been widely used as an example of this model; it has been followed by the voluminous work of Sollner (1955) and others over the past twenty years. Here, in simple form, a negative charge on the wall of a pore of radius b would exclude negative ions for $\kappa b < 1$ where $1/\kappa$ is the Debye length. The pore might then sieve positive ions of radius a according as $a >$ or $< 1/\kappa$. All of this field has been interesting and useful physical chemistry although the application to living membranes has not been made clear. Another and quite provocative analysis by Solomon (1959) has been based on the macroscopic analysis for the viscous and diffusion flows of gauge molecules across membranes. In a red blood cell this produces pores about 8 Å in diameter and about 400 Å apart, while Parpart and Ballantine (1952) have estimated pores to be 300 Å apart. Following Solomon, Villegas, and Barnala (1961) arrived at pores for the squid membrane with a similar diameter and spaced 1000 Å apart. Danielli (1954) considered the pore surface of the same structure as, and continuous with, the surfaces of the bimolecular leaflet. C. P. Bean (personal communication) gave reason to believe that such pores are in equilibrium with the rest of a bimolecular membrane—appearing and disappearing from random fluctuations. A more detailed static approach was made by Mullins (1956, 1961) who pointed out that ions with but a single hydration layer might require the highly specific pore diameters of about 8 Å that were interestingly correlated with the biological effectiveness of the more common ions. He also proposed a structure for the membrane pores which was considerably elaborated, from the quite different point of view of membrane stability, by Kavanau (1963, 1965). In a new trend, Onsager (1967) began an analysis of ion passage along a polypeptide chain with proton assistance—as in ice.

As a possible physical model for their kinetic equations, Hodgkin and Huxley (1952b) introduced the idea that three or four activating molecules moving across the membrane to a single site would open a patch to sodium or to potassium ions. However, Hodgkin and Keynes (1955) gave reasons to believe that potassium ions move in single file, possibly along a chain of sites, through the membrane. Agin (1963) called attention to the similarity between the HH functions and mean value functions of statistical mechanics as a possible guide for a physical model. The statistical approaches were further developed by Rosalie Hoyt (1963) who considered particularly the problem of dipole gates in pores, to find that nine such gates best expressed some of the data.

FitzHugh (1965) undertook to set up as simple and general a model as possible, based on chemical kinetics and without relation to a molecular structure. Again, each pore is ion-specific, a single occupied site blocks it, and ions enter and leave sites at random. But in place of 3, 4, 6, 9, or the objectionable 25 sites, FitzHugh considered an infinite number of sites for each pore.

During a transient, after a change from one membrane potential to another, the mean of the Poisson distribution μ varies continuously from μ_0 to μ_∞ to give the fraction of open pores

$$\eta = e^{-\mu}$$

and

$$\mu = \mu_\infty - (\mu_\infty - \mu_0) \exp(-t/\tau).$$

If then, for example, each open pore is ohmic the ionic conductance

$$g = \bar{g}\eta = \bar{g} \exp \{-[\mu_\infty - (\mu_\infty - \mu_0) \exp(-t/\tau)]\}.$$

Some points of one such function are included in the assembly of figure 4:59. The Eyring theory of reaction rates was taken as a simple basis for estimating the steady state mean and the time constant as functions of the membrane potential. It was assumed that the blocking ion crossed a potential barrier to enter or leave a site. The comparisons which FitzHugh made with experimental data showed useful approximations in all cases, with excellent agreement in a few.

At the same time Hoyt (1965) gave reason to believe that coulomb repulsion would permit only a few ions in a pore through the hydrocarbon interior of the membrane. The conductance of each pore would then be ohmic rather than field-dependent, as given by the independence principle. The nonlinear characteristics are produced by the fraction of open channels—which was assumed to depend on time and voltage but not boundary concentrations. She concluded that the Goldman permeabilities were more satisfactory for node data and the ohmic conductances were better for squid data.

Another variant is to assume that there is an interaction between adjacent pores. Tille (1965) considered this preferable to the requirement of a few to many simultaneous but independent events in a restricted region. He found a square array of pores most appropriate, and assumed that a blocking particle could open or close a pore only if it were next to an open pore. These data gave an opening rate

$$dn/dt = P(n)[\alpha(1 - n)n - \beta n^2]$$

where $P(n)$ is a specified cubic and α and β are voltage-dependent rate constants. Then, for a pore density N and

$$g = \bar{g}n/N$$

α and β could be given empirically to represent some voltage clamp data.

Somewhat similar, but far more specific, models are based on the spectacular effect of calcium and some other divalent cations upon ion permeabilities. About the same time as the sodium ion came into the limelight, Karreman (1951), and with Landahl (1953), presented a model in which calcium, reacting with surface sites, made them unavailable as way stations for potassium to move through the membrane. This model was shown to have the characteristics necessary for excitation and repetitive discharge, and FitzHugh (1955) put it on the phase plane. Tobias (1958), and with others (1962a,b), made this specific and gave supporting evidence for calcium association with phospholipid acidic groups, displacement of this calcium by outward moving potassium, and an increase of potassium and/or sodium permeability.

Goldman (1964) outlined a theory based on a detailed molecular organization and on reaction kinetics at the surface polar groups of phospholipids in a bimolecular leaflet (figure 5:9). Sodium and potassium might enter and leave the membrane via the phosphate group, and although the sodium reaction was faster, it was later replaced by one with potassium. However, this route could be blocked by a calcium ion or by a rotation of the positive end group in the electric field. The approximate calculations so far available gave an encouraging agreement with the HH axon. The later evidence for the independence of the sodium and potassium currents—from tetraethylammonium and tetrodotoxin—suggests that the routes for these ions are somewhat distinct and the model can be modified to fit such facts. A preliminary note by Lettvin et al. (1964) contained much the same ingredients for membrane functions, but combined in a propositional form for which symbolic logic may be an appropriate language. This approach may be useful, as well as interesting, when it is stated in a form that can be compared with an extensive body of experimental fact for a living membrane—such as the HH axon.

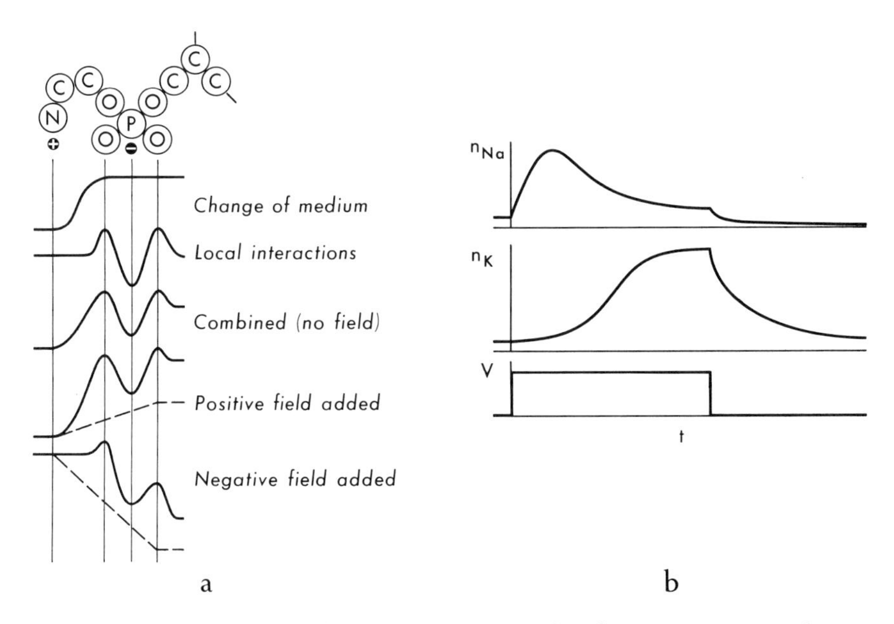

FIG. 5:9. (a.) Energy profile components near phosphate group at membrane surface. Aqueous exterior, left, and hydrocarbon interior, right. (b.) Approximate calculations of sites available for flow of sodium ions, above, and potassium ions, center, during and after a voltage clamp, below.

After his *Loligo* work on TEA injection with Binstock (1965), Armstrong (1966b) continued on *Dosidicus* in the millimolar range, developing the theory for TEA plugging K channels to outward current and presenting evidence of a delay in the subsequent opening to inward current.

Let N_0 inside sites S combine with TEA at concentration T,

$$T + S \underset{\beta}{\overset{\alpha}{\rightleftharpoons}} TS$$

to give open sites N,

$$\frac{dN}{dt} = A(N_\infty - N)$$

and

$$N = N_\infty - (N_0 - N_\infty) \exp(-At)$$

for the equilibrium $N_\infty = \beta N_0/(\alpha T + \beta)$, and the rate constant $A = \alpha T + \beta$.

Assuming in current flow that potassium K, and TEA reach a site in proportion to their concentrations and ignoring dissociation, the probability that the first ion will be TEA and plug the channel is T/K_i. Supposing K to enter at a frequency ν ions/msec, after time t the open channels

$$N = N_0(1 - T/K_i)^{\nu t} = N_0 \exp(-\nu T t/K_i)$$

for T/K_i small. The corresponding mass action expression for β small is $N = N_0 \exp(-\alpha T t)$ so $\nu = \alpha K_i$.

The axons were clamped in $K_e = 440$ mM to +79 and +100 mV above the rest potential and they gave a TS dissociation constant of 0.63 mM, $\beta = 1.1$ msec^{-1} and $\alpha = 2.2$ (msec mM)$^{-1}$. Then at the estimated $K_i = 250$ mM,

$$\nu = 525 \text{ ions/msec}$$

the channel resistance is 10^{12} Ω and, if this parameter is unchanged in sea water, the channels would be about 250 Å apart.

Further, by kinetic theory, the ion bombardment

$$B = nv/4 = 3.8 \cdot 10^{25} \, \gamma/\sqrt{M} \text{ ions/cm}^2 \text{ sec}$$

for n ions/ml or γ moles/l, velocity $v = \sqrt{8kT/\pi m}$ with ion mass m or molecular weight M, and for K = 500 mM, $B = 3 \cdot 10^{24}$ ions/cm² sec. But for a current $I = 10$ mA/cm², the flux is only $F = I/e = 6 \cdot 10^{16}$ and so the transmission F/B is one part in $5 \cdot 10^7$ (Cole, 1965a). Such a transmission and the ion channel frequency then gives site target diameter of 0.05 Å. This seems too small to be a useful or an interesting figure.

It has long been easy and satisfying to assume that ions moved across membranes through channels, pores, or some other paths that could only either be wide open or completely closed. But, in the absence of evidence, this was neither better nor worse than any other of numerous possible guesses. Then, however, Derksen (1965) and with Verveen (1966) measured the spontaneous voltage fluctuations, the "noise," of a single node with highly sophisticated techniques. At rest they found a noise power, $\overline{V^2}$ per cycle of band width, which varied about as 1/f in the range from 1 c to 1 kc. This was swamped at higher frequencies by the usual resistor noise which is independent of frequency. However, after the node was hyperpolarized, by 20 mV to about E_K, the 1/f component disappeared. For this and other reasons they ascribe the "flicker," 1/f, noise to a potassium ion current through the node membrane. This flicker noise is a well-known, but not well-explained, phenomenon in many physical systems when they are carrying current. In a naïve way we may ascribe this node behavior as the Fourier integral (CF No. 415) of random on and off of potassium carrying channels. Then at rest we found (Cole and Lecar, unpublished) such channels to have a resistance of $2 \cdot 10^{12}$ Ω and to be 200 Å apart. In spite of the hazards of interpretation, here at least was an indication of the on-and-off behavior of the elementary membrane conduction process.

Hille (1967) found external TEA to act on only $\bar{g}_K$ for the frog node, in Ringer's solution. He expressed the steady state effect more firmly as a combination of a single ion at a blocking site with a dissociation constant of 0.4 mM. I hope that the agreement between this node and the squid membrane is not fortuitous!

Kao and Nishiyama (1965) pointed out that Lorente de Nó (1949) and Larramendi et al. (1956) found the guanidinium ion to substitute for sodium. So they proposed that tetrodotoxin may block because the guanidinium group is acceptable by the fast mechanism but the whole ion is not. Armstrong (1966a) suggested that unhydrated TEA was similar enough to potassium with a single shell to be accepted at a K site but unsymmetrical enough to block the channel. Also he had found tetra-*n*-propylammonium (Tn–PA) to block the sodium as well as the potassium currents. This was explained on the basis that without hydration the larger Tn–PA could fold enough to compete with potassium and also straighten out to be comparable with double layered sodium.

On a similar basis it may be reasonable to expect that some drugs such as procaine might affect both channels and that some ions such as ammonium could go through both channels.

Although it had not developed formally, a strong trend appeared by 1965 toward the idea that the fast, normally sodium, and slow, normally potassium, channels across the membrane were separate and distinct. Each, however, allowed at least some passage of some other ions and so they were not absolutely specific. Also some blocking agents were not entirely specific. More detailed molecular explanations are a very attractive goal.

Some other views

There are always those who seek the same goals but travel by paths quite different from the currently popular highway. Sometimes the main road is turned to join them promptly—more often belatedly, if at all. I have limited myself almost entirely to what became an outstandingly successful route, largely ignoring the alternatives and the protests. This may be unfair and, since we have no idea where we are going, also unwise. Probably no one of the independents has been more devoted and more insistent than Rafael Lorente de Nó. He proposed his own extensive system of electrophysiology in 1947 and then returned to deny both all-or-none propagation, with Condouris (1959), and the passive role of myelin internodes, with Honrubia (1964–1966). Lund (1928) has long preferred to explain bioelectric potentials, particularly in plants, as oxidation-reduction potentials and T. L. Jahn (1962) extended this to view electron conduction as a source of membrane potentials. There is also David Nachmanson (1959, 1966) who has been determined to have biochemistry, and particularly acetylcholine, play an essential role in impulse conduction—first to bring about the membrane conductance increase and then later to act as the trigger for the sodium permeability increase. The existence of a membrane has been questioned by Gilbert Ling (1962, 1966) who first proposed that cytoplasm be considered as an ion exchanger and has since continually emphasized evidence for transport by diffusion across thick layers. A later and extensive account of a wide variety of membrane phenomena was given by Lee Kavanau (1965). The evidence from many fields of physical chemistry led him to open and closed configurations of lipid protein units. Excitation was then a slight tendency for the closed structure to open up. After his pioneering investigations of the medullated axon, Ichiji Tasaki turned to the squid axons. With the considerable powers at his command, he began by questioning everything about the Hodgkin-Huxley axon—concepts, experiments, and analyses. He continued with imagination and energy, developing his own approaches to the mechanisms of ion permeabilities.

EXPERIMENTAL MODELS

In addition to the various theoretical models there have been numerous experimental analogues and models for membranes of various cells.

There have been many thick analogues, such as oil, glass, and filters as well as the collodion and ion exchangers already mentioned. In this area I have been very much impressed by the work of my erstwhile colleague Tobias and his collaborators (1962a,b). They deposited phospholipid (mostly phosphotidyl serine) and cholesterol on a millipore filter, with KCl on one side and $CaCl_2$ on the other. With an applied field moving potassium into such a thick analogue for the membrane surface, Tobias found a reversible conductance increase in a few seconds. He interpreted this as a ready replacement of blocking calcium on the phospholipid by potassium to give a high conductance. In the reverse direction, as calcium was forced into this structure the conductance is reduced drastically, but over a period of minutes. This is explained as the slow replacement of the lightly bound potassium, which participates in a high conductance, by the relatively tightly bound calcium, which is not easily displaced from the boundary sites. Since these sites are not then accessible to potassium and/or sodium, the conductance decreases as such surface sites, gates, were blocked by the relatively immovable calcium.

Over a decade the Monniers investigated artificial membranes made up of various lipidic derivatives and then turned to oxidized films of unsaturated fats such as linseed oil spread on $KMnO_4$. As summarized by Monnier et al. (1965), these films may be a fixed charge aggregate and represented electrically by a capacity and variable conductance in parallel. During activities analogous to those of living excitable membranes the conductance increased many fold to suggest a negative conductance and its explanation as a Wien effect.

There have also been many artificial models for a bimolecular membrane. Some years ago our attempts to approximate the passive dielectric properties and some of the ion conduction characteristics of cell membranes were partially successful (Dean et al., 1940; Goldman, 1943). These films of protein-stabilized nitrobenzene were, however, of uncertain thickness, structure, and stability. A radically different approach was begun by Langmuir and Waugh (1938) and revived by Mueller and Rudin (1963). Hydrocarbon extracts in a solvent were spread across a small hole in a partition and black films were obtained as the solvent moved into the aqueous phases on each side. Thompson (1964) and with his colleagues Huang et al. (1964), have made films of lecithin that are bimolecular in thickness—as well as having a capacity, conductance, water permeability and tension comparable to living membranes. These films, made of known and well-defined chemical components, are not, however, detectibly nonlinear in their ion current vs. membrane potential relations. Hanai et al. (1964, 1965) have measured a capacity of 0.38 $\mu F/cm^2$, independent of frequency from 0.001 c to 5 Mc, which they ascribe to the hydrocarbon bilayer about 40 Å thick. They further demonstrate by electrophoretic measurements that the polar head groups are parallel to the membrane surface. To account for the capacity and conductance of living membranes they suggest protein lined pores for about 1 percent of the membrane area.

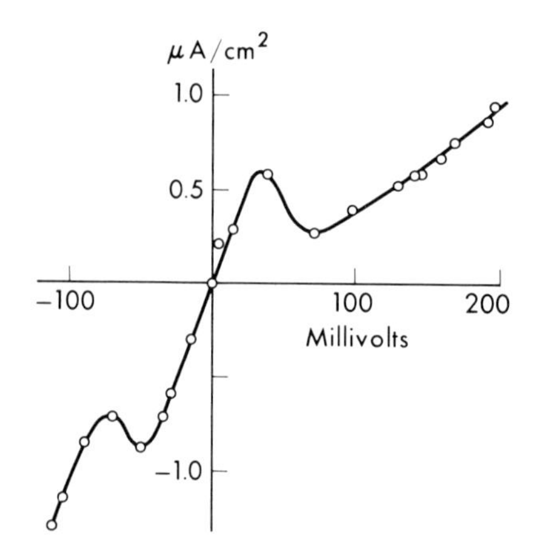

Fig. 5:10. Steady state electrical properties of an artificial bilayer. It was formed from a brain extract and a protein was added.

Yet the films of Mueller and Rudin, made with extracts of brain and somewhat old egg white, were highly nonlinear and even showed the proper *N* shaped negative resistance behavior (figure 5:10) that I think is so important a characteristic of living nerve membranes. Such membranes were found by Bean et al. (1966 and unpublished) to start conducting in steps of $10^9\ \Omega$ and to achieve quite asymmetrical N current-potential characteristics after an activating brain extract was applied to one side. Then a protease which did not penetrate the membrane could reverse the effect if it was added to either side. So Bean proposed that enough of the protein elements reached through the membrane, not only to create the considerable ionic conductance but also to allow enzymatic activity on the far side to remove that conductance.

But, in addition to the mystery proteins, some polypeptides conferred conductance on both red cells and artificial bilayers. In particular, the cyclic valinomycin was highly specific for the potassium ion (Andreoli et al., 1967).

An ingenious new departure was reported by Gregor (1966) as a possible model for a pore. The fracture surfaces of a broken glass slide or cover slip were faced together after the surfaces had been covered with molecular layers by the Langmuir-Blodgett technique. The properties of the opposed molecular layers were measured in the direction of the layer planes. The maximum conductance of a squid axon membrane would be obtained if such structures formed about 1 percent of the membrane area.

Conclusion

As we turn at last to the structure of living membranes and try to come to grips with the passage of ions across it, even a cursory survey is enough to convince us that practically nothing is settled.

The bimolecular leaflet of phospholipids and sterols with a hydrocarbon interior and polar groups at the surfaces is quite well established. Although the structure is not certain, it seems likely to be that of liquid crystals. The original postulate of protein adsorption at the surfaces of the bilayer has several forms of support, yet there is no compelling evidence as to the protein composition, configuration, or function. There is, however, good basis to believe the two surfaces of some membranes are significantly different.

Even if the structure of a single membrane were well defined, from one aqueous solution to the other, there might still not be an adequate basis to decide where and why it is so difficult for any ion to cross in either direction. Thus any theory of ion permeability must assume and support both a structure and a mechanism. It is most unfortunate, yet most challenging, that a possible and reasonable explanation of the most powerful facts of ion permeabilities may be completely wrong. And the concepts gained from artificial models may be entirely misleading.

So it is a struggle in the dark except for a few occasional glimmers of light until the approach of one or more attempts is so close that it cannot be denied.

References

Agin, D., 1963. Some comments on the Hodgkin-Huxley equations. J. Theo. Biol. *5:* 161–170.

Agin, D., Hersh, L., and Holtzman, D., 1965. The action of anesthetics on excitable membranes: Proc. Natl. Acad. Sci. USA, *53:* 952–958.

Andreoli, T. E., Tieffenberg, M., and Tosteson, D. C., 1967. The effect of valinomycin on the ionic permeability of thin lipid membranes. J. Gen. Physiol. *50:* 2527–2545.

Aranow, R. H., 1963. Periodic behavior in charged membranes and its physical and biological implications. Proc. Natl. Acad. Sci. USA, *50:* 1066–1070.

Armstrong, C., 1966a. Interference of injected tetra-n-propylammonium bromide on outward sodium ion current in squid giant axons. Nature *211:* 322–323.

Armstrong, C. M., 1966b. Time course of TEA^+-induced anomalous rectification in squid giant axons. J. Gen. Physiol., *50:* 491–503.

Armstrong, C. M. and Binstock, L., 1965. Anomalous rectification in the squid giant axon injected with tetraethylammonium chloride. J. Gen. Physiol. *48:* 859–872.

Ashman, R. F., Kanno, Y., and Loewenstein, W. R., 1964. Intercellular electrical coupling at a forming membrane junction in a dividing cell. Science *145:* 604–605.

Barr, L., Dewey, M. M. and Berger, W., 1965. Propagation of action potentials and the structure of the nexus in cardiac muscle. J. Gen. Physiol. *48:* 797–823.

Bartlett, J. H., and Kromhout, R. A., 1952. The Donnan equilibrium. Bull. Math. Biophys. *14:* 385–391.

Bean, R. C., 1966. Characterization of a lipid-protein, excitable membrane and its application to problems of molecular biology. Philco Publ., #U3494.

Blank, M., 1962. Monolayer permeability and the properties of natural membranes. J. Phys. Chem. *66:* 1911–1918.

Blank, M., 1965. A physical interpretation of the ionic fluxes in excitable membranes. J. Colloid Sci. *20:* 933–949.

Branton, D., 1966. Fracture faces of frozen membranes. Proc. Natl. Acad. Sci., USA, *55:* 1048–1056.

Bruner, L. J., 1965a. The electrical conductance of semipermeable membranes. I. Biophys. J. *5:* 867–886.

Bruner, L. J., 1965b. The electrical conductance of semipermeable membranes. II. Biophys. J. *5:* 887–908.

Chambers, R., and Kopac, M. J., 1937. The coalescense of living cells with oil drops. I. J. Cell. Comp. Physiol. *9:* 331–341.

Chapman, D., 1966. Liquid crystals and cell membranes. Ann. N. Y. Acad. Sci. *137:* 345–361.

Clowes, G. H. A., 1918. On the action exerted by antagonistic electrolytes on the electrical resistance and permeability on emulsion membranes. Proc. Soc. Exp. Biol. Med. *15:* 108–111.

Cohen, H., and Cooley, J. W., 1965. The numerical solution of the time-dependent Nernst-Planck equations. Biophys. J. *5:* 145–162.

Cole, K. S., 1932. Surface forces of the *Arbacia* egg. J. Cell. Comp. Physiol. *1:* 1–9.

Cole, K. S., 1949. Dynamic electrical characteristics of the squid axon membrane. Arch. Sci. physiol. *3:* 253–258.

Cole, K. S., 1959. The electric structure and function of cells. In Quastler, H. and Morowitz, H. J. (eds.): Proc. First Natl. Biophys. Conf., New Haven, Yale University Press, p. 332–347.

Cole, K. S., 1965a. Theory, experiment and the nerve impulse. In Waterman, T. H. and Morowitz, H. J. (eds.): Theoretical and mathematical biology. New York, Blaisdell, p. 136–171.

Cole, K. S., 1965b. Electrodiffusion models for the membrane of squid giant axon. Physiol. Rev. *45:* 340–379.

Cole, K. S., 1966. The melding of membrane models. Ann. N. Y. Acad. Sci. *137:* 405–408.

Cole, K. S., and Moore, J. W., 1960. Potassium ion current in the squid giant axon: dynamic characteristic. Biophys. J. *1:* 1–14.

Collander, R., 1937. The permeability of plant protoplasts to non-electrolytes. Trans. Faraday Soc. *33:* 958–990.

Conti, F., and Eisenman, G., 1965. The steady state properties of ion exchange membranes with fixed sites. Biophys. J. *5:* 511–530.

Conti, F., and Eisenman, G., 1966. The steady-state properties of an ion exchange membrane with mobile sites. Biophys. J. *6:* 227–246.

Coster, H. G. L., 1965. A quantitative analysis of the voltage-current relationships of fixed charge membranes and the associated property of "punch-through." Biophys. J. *5:* 669–686.

Danielli, J. F., 1954. The present position in the field of facilitated diffusion and selective active transport. In Kitching, J. A. (ed.): Recent developments in cell physiology. New York, Academic Press. *7:* 1–14.

Danielli, J. F., and Davson, H., 1935. A contribution to the theory of permeability of thin films. J. Cell. Comp. Physiol. *5:* 495–508.

Davson, H., and Danielli, J. F., 1943. The permeability of natural membranes. (2nd ed. 1952): Cambridge, University Press.

Dean, R. B., Curtis, H. J., and Cole, K. S., 1940. Impedance of bimolecular films. Science *91:* 50–51.

Derksen, H. E., 1965. Axon membrane voltage fluctuations. Acta Physiol. Pharm. Neerl. *13:* 373–466.

Derksen, H. E., and Verveen, A. A., 1966. Fluctuations of resting neural membrane potential. Science *151:* 1388–1389.

Eisenman, G., Sandblom, J. P., and Walker, J. L., Jr., 1966. Membrane structure and ion permeation. Science, *155:* 965–974.

Finean, J. B., 1962. The nature and stability of the plasma membrane. Circulation *26:* 1151–1162.

FitzHugh, R., 1955. Mathematical models of threshold phenomena in the nerve membrane. Bull. Math. Biophys. *17:* 257–278.

FitzHugh, R., 1965. A kinetic model of the conductance changes in nerve membrane. J. Cell. Comp. Physiol. *66:* (Suppl. 2) 111–117.

Frankenhaeuser, B., 1960. Sodium permeability in toad nerve and squid nerve. J. Physiol. *152:* 159–166.

Frankenhaeuser, B., and Huxley, A. F., 1964. The action potential in the myelinated nerve fibre of *Xenopus laevis* as computed on the basis of voltage clamp data. J. Physiol. *171:* 302–315.

Fricke, H., 1923. The electric capacity of cell suspensions. Physic. Rev. *21:* 708–709.

Fricke, H., 1925. The electric capacity of suspensions with special reference to blood. J. Gen. Physiol. *9:* 137–152.

Furshpan, E. J., 1964. "Electrical transmission" at an excitatory synapse in a vertebrate brain. Science *144:* 878–880.

Geren, B. B., 1954. The formation from the Schwann cell surface of myelin in the peripheral nerves of chick embryos. Exp. Cell Res. *7:* 558–562.

Goldman, D. E., 1943. Potential, impedance, and rectification in membranes. J. Gen. Physiol. *27:* 37–60.

Goldman, D. E., 1959. Analog computer solution of ion flux equations. Biophys. Soc. Abstr., G 3.

Goldman, D. E., 1964. A molecular structural basis for the excitation properties of axons. Biophys. J. *4:* 167–188.

Gorter, E., and Grendel, F., 1925. On bimolecular layers of lipoids on the chromocytes of the blood. J. Exp. Med. *41:* 439–443.

Green, D. E., and Perdue, J. F., 1966. Membranes as expressions of repeating units. Proc. Natl. Acad. Sci. USA *55:* 1295–1302.

Gregor, H. P., 1966. Multilayer electrodes. N. Y. Acad. Sci. Conf.

Hanai, T., Haydon, D. A., and Taylor, J., 1964. An investigation by electrical

methods of lecithin-in-hydrocarbon films in aqueous solutions. Proc. Roy. Soc. (London), A., *281:* 377–391.
Hanai, T., Haydon, D. A., and Taylor, J., 1965. Polar group orientation and the electrical properties of lecithin bimolecular leaflets. J. Theoret. Biol. *9:* 278–296.
Harvey, E. N., 1931a. A determination of the tension at the surface of eggs of the annelid, *Chaetopterus*. Biol. Bull. *60:* 67–71.
Harvey, E. N., 1931b. The tension at the surface of marine eggs, especially those of the sea urchin, *Arbacia*. Biol. Bull. *61:* 273–279.
Harvey, E. N., and Danielli, J. F., 1938. Properties of the cell surface. Biol. Rev. *13:* 319–341.
Hechter, O., 1965. Intracellular water structure and mechanisms of cellular transport. Ann. N. Y. Acad. Sci. *125:* 625–646.
Hille, B., 1966. The common mode of action of three agents that decrease the transient change in sodium permeability in nerves. Nature *210:* 1220–1222.
Hille, B., 1967. The selective inhibition of delayed potassium currents in nerve by tetraethylammonium ion. J. Gen. Physiol. *50:* 1287–1302.
Hiramoto, Y., 1963. Mechanical properties of sea urchin eggs. I. Exp. Cell Res. *32:* 59–75.
Höber, R., 1910. Eine Methode, die elektrische Leitfähigkeit im Innern von Zellen zu messen. Arch. ges. Physiol. *133:* 237–259.
Höber, R., 1912. Ein zweites Verfahren, die Leitfähigkeit im Innern von Zellen zu messen. Arch. ges. Physiol. *148:* 189–221.
Hodgkin, A. L., 1964. The conduction of the nervous impulse. Springfield, Ill., Thomas.
Hodgkin, A. L., and Huxley, A. F., 1952a. Currents carried by sodium and potassium ions through the membrane of the giant axon of *Loligo*. J. Physiol. *116:* 449–472.
Hodgkin, A. L., and Huxley, A. F., 1952b. A quantitative description of membrane current and its application to conduction and excitation in nerve. J. Physiol. *117:* 500–544.
Hodgkin, A. L., and Keynes, R. D., 1955. The potassium permeability of a giant nerve fibre. J. Physiol. *128:* 61–88.
Hogg, B. M., Goss, C. M., and Cole, K. S., 1934. Potentials in embryo rat heart muscle cultures. Proc. Soc. Exp. Biol. Med. *32:* 304–307.
Hoyt, R. C., 1963. The squid giant axon. Mathematical models. Biophys. J. *3:* 399–431.
Hoyt, R. C., 1965. Non-linear membrane currents with ohmic channels. J. Cell. Comp. Physiol. *66:* (Suppl. 2) 119–126.
Huang, C., Wheeldon, L., and Thompson, T. E., 1964. The properties of lipid bilayer membranes separating two aqueous phases. J. Mol. Biol., *8:* 148–160.
Jacobs, M. H., 1935. Diffusion processes. Ergeb. Biol., *12:* 1–160.
Jahn, T. L., 1962. A theory of electronic conduction through membranes, and of active transport of ions, based on redox transmembrane potentials. J. Theo. Biol. *2:* 129–138.
Johnson, F. H., Eyring, H., and Polissar, J. R., 1954. The kinetic basis of molecular biology. New York, Wiley, p. 625–698.
Julian, F. J., Moore, J. W., and Goldman, D. E., 1962. Current-voltage relations of the lobster giant axon membrane under voltage clamp conditions. J. Gen. Physiol. *45:* 1217–1238.
Kao, C. Y., and Nishiyama, A., 1965. Actions of saxitoxin on peripheral neuromuscular systems. J. Physiol. *180:* 50–66.
Karreman, G., 1951. Contributions to the mathematical biology of excitation with

particular emphasis on changes in membrane permeability and on threshold phenomena. Bull. Math. Biophys. *13:* 189–243.

Karreman, G., and Landahl, H. D., 1953. On spontaneous discharges obtained from a physicochemical model of excitation. Bull. Math. Biophys. *15:* 83–91.

Katchalsky, A., 1968. On the nerve membrane. Neurosci. Res. Prog. Bull. (in press).

Katchalsky, A., Kedem, O., Klibansky, C., and de Vries, A., 1960. Rheological considerations of the haemolysing red blood cell. In Copley, A. L. and Stainsby, G. (eds.): Flow properties of blood and other biological systems. New York, Pergamon Press.

Kavanau, J. L., 1963. Structure and functions of biological membranes. Nature *198:* 525–530.

Kavanau, J. L., 1965. Structure and function in biological membranes (2 vols.). San Francisco, Holden-Day.

Kedem, O., and Katchalsky, A., 1961. A physical interpretation of the phenomenological coefficients of membrane permeability. J. Gen. Physiol. *45:* 143–179.

Kedem, O., and Katchalsky, A., 1963a. Permeability of composite membranes. I. Trans. Faraday Soc. *59:* 1918–1930.

Kedem, O., and Katchalsky, A., 1963b. Permeability of composite membranes. II. Trans. Faraday Soc. *59:* 1931–1940.

Kedem, O., and Katchalsky, A., 1963c. Permeability of composite membranes. III. Trans. Faraday Soc., *59:* 1941–1953.

Kimizuka, H., and Koketsu, K., 1964. Ion transport through cell membrane. J. Theo. Biol. *6:* 290–305.

Kirkwood, J. G., 1954. Transport of ions through biological membranes from the standpoint of irreversible thermodynamics. In Clarke, H. T., (ed.): Ion transport across membranes, New York, Academic Press.

Kobatake, Y., and Fujita, H., 1964a. Flows through charged membranes. I. J. Chem. Physics *40:* 2212–2218.

Kobatake, Y., and Fujita, H., 1964b. Flows through charged membranes. II. J. Chem. Physics *40:* 2219–2222.

Korn, E. D., 1966. Structure of biological membranes. Science, *153:* 1491–1498.

Langmuir, I., and Waugh, D. F., 1938. The adsorption of proteins at oil-water interfaces and artificial protein-lipoid membranes. J. Gen. Physiol. *21:* 745–755.

Larramendi, L. M. H., Lorente de Nó, R., and Vidal, F., 1956. Restoration of sodium-deficient frog nerve fibers by an isotonic solution of guanidinium chloride. Nature *178:* 316–317.

Lettvin, J. Y., Pickard, W. F., McCulloch, W. S., and Pitts, W., 1964. A theory of passive ion flux through axon membranes. Nature *202:* 1338–1339.

Ling, G. N., 1962. A physical theory of the living state: the association-induction hypothesis. New York, Blaisdell Publishing Co.

Ling, G. N., 1966. Cell membrane and cell permeability. Ann. N. Y. Acad. Sci. *137:* 837–859.

Locke, M., 1965. The structure of septate desmosomes. J. Cell Biol. *25:* 166–168.

Loeb, J., 1920. Ionic radius and ionic efficiency. J. Gen. Physiol. *2:* 673–687.

Loewenstein, W. R., and Kanno, Y., 1964. Studies on an epithelial (gland) cell junction. I. J. Cell Biol. *22:* 565–586.

Lorente de Nó, R., 1947. A study of nerve physiology. Vols. 131, 132, New York, Rockefeller Institute.

Lorente de Nó, R., 1949. On the effect of certain quaternary ammonium ions upon frog nerve. J. Cell. Comp. Physiol. *33:* (Suppl. 1) 1–231.

Lorente de Nó, R., and Condouris, C. A., 1959. Decremental conduction in peripheral nerve. Integration of stimuli. Proc. Natl. Acad. Sci., USA, *45:* 592–617.

Lorente de Nó, R., and Honrubia, V., 1964. Continuous conduction of action potentials by peripheral myelinated fibers. Proc. Natl. Acad. Sci., USA, *52:* 305–312.
Lorente de Nó, R., and Honrubia, V., 1966. Theory of the flow of action currents in isolated myelinated nerve fibers. XII. Proc. Natl. Acad. Sci., USA, *55:* 1118–1125.
Lund, E. J., 1928. Relation between continuous bioelectric currents and cell respiration. II. J. Exp. Zool., *51:* 265–290.
Luzzati, V., and Husson, F., 1962. The structure of the liquid-crystalline phases of lipid-water systems. J. Cell Biol. *12:* 207–219.
Mauro, A., 1962. Space charge regions in fixed charge membranes and the associated property of capacitance. Biophys. J. *2:* 179–198.
Meves, H., 1966. Experiments on internally perfused squid giant axons. Ann. N. Y. Acad. Sci. *137:* 807–817.
Michaelis, L., 1925. Contribution to the theory of permeability of membranes for electrolytes. J. Gen. Physiol. *8:* 33–59.
Miller, S. L., 1961. A theory of gaseous anesthetics. Proc. Natl. Acad. Sci., USA, *47:* 1515–1524.
Mitchison, J. M., 1953. A polarized light analysis of the human red cell ghost. J. Exp. Biol. *30:* 397–432.
Mitchison, J. M., and Swann, M. M., 1954. The mechanical properties of the cell surface. II. J. Exp. Biol. *31:* 461–472.
Monnier, A. M., Monnier, A., Goudeau, H., and Rebuffel-Reynier, A. M., 1965. Electrically excitable artificial membranes. J. Cell. Comp. Physiol. *66:* (Suppl. 2) 147–154.
Moore, J. W., 1959. Excitation of the squid axon membrane in isosmotic potassium chloride. Nature *183:* 265–266.
Mueller, P., 1958. On the kinetics of potential, electromotance, and chemical change in the excitable system of nerve. J. Gen. Physiol. *42:* 193–229.
Mueller, P., and Rudin, D. O., 1963. Induced excitability in reconstituted cell membrane structure. J. Theo. Biol. *4:* 268–280.
Mullins, L. J., 1956. The structure of nerve cell membranes. In: Molecular structure and functional activity of nerve cell membranes. Am. Inst. Biol. Sci. Publ. No. 1, p. 123–154.
Mullins, L. J., 1961. The macromolecular properties of excitable membranes. Ann. N. Y. Acad. Sci. *94:* 390–404.
Nachmansohn, D., 1959. Chemical and molecular basis of nerve activity. New York, Academic Press.
Nachmansohn, D., 1966. Chemical control of the permeability cycle during the activity of excitable membranes. Ann. N. Y. Acad. Sci. *137:* 877–900.
Nakas, M., Higashino, S., and Loewenstein, W. R., 1966. Uncoupling of an epithelial cell membrane junction by calcium-ion removal. Science *151:* 89–91.
Nernst, W., 1888. Zur Kinctik dcr in Lösung befindlichen Körper; Theorie der Diffusion. Zeit. phys. Chem. *2:* 613–637.
Onsager, L., 1934. Deviations from Ohm's Law in weak electrolytes. J. Chem. Physics *2:* 599–615.
Onsager, L., 1967. Ion passages in lipid bilayers. Science *156:* 541.
Overton, E., 1899. Ueber die allgemeinen osmotischen Eigenschaften der Zelle, ihre vermutlichen Ursachen und ihre Bedeutung für die Physiologie. Vjochr. Naturf. Ges. Zurich. *44:* 88–135.
Palade, G. E., 1965. Cell junctions and functional organization of epithelia. N. Y. Acad. Sci. Conf. on Biol. Memb., Oct. 4.

Palmer, K. J., and Schmitt, F. O., 1941. X-ray diffraction studies of lipide emulsions. J. Cell. Comp. Physiol. *17:* 385–394.

Pappas, G. D., and Bennett, M. L. V., 1966. Specialized junctions involved in electrical transmission between neurons. Ann. N. Y. Acad. Sci. *137:* 495–508.

Parlin, R. B., and Eyring, H., 1954. Membrane permeability and electrical potential. In Clarke, H. T. (ed.): Ion transport across membranes. New York, Academic Press, p. 103–118.

Parpart, A. K., and Ballentine, R., 1952. Molecular anatomy of the red cell plasma membrane. In Barron, E. S. G. (ed.): Trends in physiology and biochemistry. New York, Academic Press, p. 135–148.

Pauling, L., 1961. A molecular theory of general anesthesia. Science *134:* 15–21.

Pease, D. C., and Baker, R. F., 1948. Sectioning techniques for electron microscopy using a conventional microtome. Proc. Soc. Exp. Biol. Med. *67:* 470–474.

Penn, R. D., and Loewenstein, W. R., 1966. Uncoupling of a nerve cell membrane junction by calcium-ion removal. Science *151:* 88–89.

Planck, M., 1890. Ueber die Erregung von Elektricität und Wärme in Elektrolyten. Ann. physik. Chem. *39:* 161–186.

Potter, D. D., Furshpan, E. J., and Lennox, E. S., 1966. Connections between cells of the developing squid as revealed by electrophysiological methods. Proc. Natl. Acad. Sci. USA, *55:* 328–336.

Rand, R. P., 1964. Mechanical properties of the red cell membrane. II. Biophys. J. *4:* 303–316.

Rand, R. P., and Burton, A. C., 1964. Mechanical properties of the red cell membrane. I. Biophys. J. *4:* 115–135.

Robertson, J. D., 1960. The molecular structure and contact relationships of cell membranes. Prog. Biophys. *10:* 343–418.

Robertson, J. D., 1966. Current problems of unit membrane structure and contact relationships. In Rodahl, K., and Issekutz, B., Jr., (eds.) Nerve as a tissue. New York, Harper and Row, p. 11–48.

Rothschild, Lord, 1957. The membrane capacitance of the sea urchin egg. J. Biophys. Biochem. Cytol. *3:* 103–110.

Rudolph, G., and Stämpfli, R., 1958. Anodenöffnungserregungen einzelner Ranvier-Schnürringe. Arch. ges. Physiol. *267:* 524–531.

Schlögl, R., 1954. Elektrodiffusion in freier Lösung und geladenen Membranen. Zeit. physik. Chem., Neue Folge, *1:* 305–339.

Schmitt, F. O., and Davison, P. F., 1965. Role of protein in neural function. Neurosciences Res. Prog. Bull. *3:* 55–76.

Schmitt, F. O., Bear, R. S., and Palmer, K. J., 1941. X-ray diffraction studies on the structure of the nerve myelin sheath. J. Cell. Comp. Physiol. *18:* 31–42.

Schmitt, F. O., Bear, R. S., and Ponder, E., 1937. Optical properties of the red cell membrane. J. Cell. Comp. Physiol. *9:* 89–92.

Schmitt, F. O., Bear, R. S., and Ponder, E., 1938. The red cell envelope considered as a Wiener mixed body. J. Cell. Comp. Physiol. *11:* 309–313.

Schwartz, M., and Case, C. T., 1964. Electric impedance and rectification of fused anion-cation membranes in solution. Biophys. J. *4:* 137–149.

Shockley, W., 1949. The theory of *p-n* junction in semi-conductors and *p-n* junction transistors. Bell Syst. Tech. J. *28:* 435–489.

Sollner, K., 1955. The electrochemistry of porous membranes. In Shedlovsky, T. (ed.): Electrochemistry in biology and medicine. New York, Wiley, p. 33–64.

Sollner, K., and Shean, G., 1966. Carrier mechanisms in the movement of ions across porous and liquid ion exchanger membranes. Ann. N. Y. Acad. Sci. *137:* 759–776.

Solomon, A. K., 1959. Equivalent pore dimensions in cellular membranes. In Quastler, H. and Morowitz, H. J., (eds.): Proc. of the First National Biophysics Conf., New Haven, Yale University Press, p. 314–322.

Stoeckenius, W., 1962. Some electron microscopical observations on liquid-crystalline phase in lipid-water systems. J. Cell Biol. *12:* 221–229.

Tasaki, I., 1959. Demonstration of two stable states of the nerve membrane in potassium-rich media. J. Physiol. *148:* 306–331.

Tasaki, I., 1963. Permeability of squid axon membrane to various ions. J. Gen. Physiol. *46:* 755–772.

Tasaki, I., and Singer, I., 1965. A macromolecular approach to the excitable membrane. J. Cell. Comp. Physiol. *66:* (Suppl. 2) 137–145.

Tasaki, I., Singer, I., and Takenaka, T., 1965a. Effects of internal and external ionic environment on excitability of squid giant axons: A macromolecular approach. J. Gen. Physiol. *48:* 1095–1123.

Tasaki, I., Singer, I., and Watanabe, A., 1965b. Excitation of internally perfused squid giant axons in sodium-free media. Proc. Natl. Acad. Sci., USA, *54:* 763–769.

Tasaki, I., Teorell, T., and Spyropoulos, C. S., 1961. Movement of radioactive tracers across squid axon membrane. Am. J. Physiol. *200:* 11–22.

Teorell, T., 1951. Zur quantitativen Behandlung der Membranpermeabilität. Zeit. Elektrochem. *55:* 460–469.

Teorell, T., 1960. Application of the voltage clamp to the electrohydraulic nerve analog. Acta. Soc. Med. Uppsala *65:* 231–248.

Teorell, T., 1962. Oscillatory electrophoresis in ion exchange membranes. Arkiv Kemi *18:* 401–408.

Thompson, T. E., 1964. The properties of bimolecular phospholipid membranes. In Locke, M. (ed.): Cellular membranes in development. New York, Academic Press, p. 83–96.

Tille, J., 1965. A new interpretation of the dynamic changes of the potassium conductance in the squid giant axon. Biophys. J. *5:* 163–171.

Tobias, J. M., 1958. Experimentally altered structure related to function in the lobster axon with an extrapolation to molecular mechanisms in excitation. J. Cell. Comp. Physiol. *52:* 89–125.

Tobias, J. M., Agin, D. P., and Pawlowski, R., 1962a. The excitable system in the cell surface. Circulation *26:* 1145–1150.

Tobias, J. M., Agin, D. P., and Pawlowski, R., 1962b. Phospholipid-cholesterol membrane model: Control of resistance by ions or current flow. J. Gen. Physiol. *45:* 989–1001.

Umrath, K., 1927. Über die Erregungsleitung bei sensitiven Pflanzen. Planta *5:* 274–324.

van Deenen, L. L. M., 1966. Some structural and dynamic aspects of lipids in biological membranes. Ann. N. Y. Acad. Sci., *137:* 717–730.

Vandenheuvel, F. A., 1965. Structural studies of biological membranes. The structure of myelin. Ann. N. Y. Acad. Sci. *122:* 57–76.

Villegas, R., and Barnola, F. V., 1961. Characterization of the resting axolemma in the giant axon of the squid. J. Gen. Physiol. *44:* 963–977.

Villegas, R., Villegas, G. M., Blei, M., Herrera, F. G., and Villegas, J., 1966. Nonelectrolyte penetration and sodium fluxes through the axolemma of resting and stimulated medium sized axons of the squid *Doryteuthis plei*. J. Gen. Physiol. *50:* 43–59.

Wallach, D. F. H., and Zahler, P. H., 1966. Protein conformations in cellular membranes. Proc. Natl. Acad. Sci. USA, *56:* 1522–1559.

Watson, M. L., 1959. Further observations on the nuclear envelope of the animal cell. J. Biophys. Biochem. Cytol. *6*, 147–156.

Waugh, D. F., and Schmitt, F. O., 1940. Investigations of the thickness and ultrastructure of cellular membranes by the analytical leptoscope. Cold Spring Harbor Symp. Quant. Biol. *8:* 233–241.

Weidmann, S., (with appendix by A. L. Hodgkin) 1966. The diffusion of potassium across intercalated disks of mammalian cardiac muscle. J. Physiol. *187:* 323–342.

Wright, E. B., and Tomita, T., 1965. A study of the crustacean axon repetitive response: II. The effect of cations, sodium, calcium (magnesium), potassium and hydrogen (pH) in the external medium. J. Cell. Comp. Physiol. *65:* 211–228.

Yoneda, M., 1964. Tension at the surface of sea-urchin egg: A critical examination of Cole's experiment. J. Exp. Biol. *41:* 893–906.

Postscript

The development of the concept of the membrane and the description of its ionic behavior, starting early in the century, has been rather simple and straightforward to follow. It came to a dramatic focus for the squid membrane that has been largely supported and clarified in the past decade. Now the principal path of this development is clear and obvious, and little attention has had to be paid to anything else. Hodgkin led us to a powerful statement of what a squid axon membrane normally does, carrying an impulse from start to finish, changing its potential, and letting ions run in and out. This alone was a story worth telling, and it would have been simple and convenient to close it there.

But it was not easy to stop. This was but a description from the outside of the membrane. It was both a challenge and a guide to the next goal—what happens within the membrane. Other membranes have shown many striking similarities and some certainly significant differences. It became increasingly apparent that we had not the solution of any problem but a strong succinct new statement of the old question, "How does an ion get through a membrane?"

There are many new facts and many new hypotheses and these are the necessary introduction to the new era that we have entered. Although I could not avoid it, I have found this new beginning almost impossible to write about because I do not yet know what is to become simple, obvious, and important. There are many models for a membrane, both theoretical and experimental, some apparently naïve and others promising but incomplete. Probably all of these will help toward the goal. Yet I must recall that it was Hodgkin's worry about the single, simple, experimental fact that an action potential was larger than a resting potential—which led us to where we are today.

So it may be that all I can and perhaps should do now is to give the impression and some of the elements of a widespread confusion. And in a decade or so it may be a good beginning for another volume, perhaps under the title of "Membranes, Ions and Impulses: A Chapter of Molecular Biophysics."

Glossary

The following collections are intended as only brief and elementary introductions to some of the terms, concepts, units, abbreviations, and such.

MEASUREMENT AND ANALYSIS

p	pico, (10^{-12}); n nano, (10^{-9}); μ micro, (10^{-6})
m	milli, (10^{-3}); k kilo, (10^{3}); M mega (10^{6})
Å	angstrom (10^{-8}cm); μ micron (μm, 10^{-4}cm)
x, ρ	length and radial variables (cm)
a, b, δ	radii and membrane thickness (cm)
ρ	volume concentration of cells in suspension
λ	characteristic length for an exponential (cm)
t, τ	time and time constant for an exponential (sec)
f	sinusoidal frequency, cycles/sec (c) or cycles/cm
ω	radian frequency $2\pi f$
j	imaginary operation $\sqrt{-1}$ representing $\partial/\partial t$ or a sinusoid phase difference of $\pi/2$
p	general operator $\sigma + j\omega$ which is $j\omega$ for steady state sinusoid
$\dot{F}$	dF/dt
F_x	$\partial F/\partial x$
$\nabla^2 F = 0$	Laplace equation for a potential, F
$\frac{\partial F}{\partial t} = \frac{\partial^2 F}{\partial x^2}$	equation for Kelvin cable and Fick diffusion

ELECTRICAL PARAMETERS AND UNITS

V	potential difference, volts (V), usually referred to ground
V_m	membrane potential (V) usually referred to medium external to a cell or an axon
E	electromotive force (V) measured without current flow
I, i	current, amperes (A), current density (A/cm^2), usually considered positive for current flow from inside to out across a membrane
I_c, I_g	capacitive and ionic conduction membrane current components
Linear, ohmic	passive processes or circuit elements following
Ohm's law	current is proportional to potential difference

R	resistance, ohms (Ω), V/I
r_1, r_2, r_3	specific resistance for external and internal media (Ωcm) and membrane (Ωcm^2)
r_e, r_i, r_m	resistance for unit cable length, external and internal (Ω/cm) and membrane (Ωcm)
$R_0, R_\infty, r_0, r_\infty$	resistances, usually as extrapolated to zero and infinite frequencies
g, etc.	conductance, reciprocal ohms (mho/cm, mho/cm^2)
C, etc.	capacity, farads (F), membrane capacity (F/cm^2, F/cm)
L	inductance, henries (H)
X	reactance (Ω) characteristic of current flow without dissipation as through a capacity or inductance
Z, etc.	complex impedance, (Ω, Ωcm^2, Ωcm) with real, resistive, and imaginary, reactive, components, $R + jX$
g^*	complex admittance (mho) with real and imaginary components $g' + jg''$
ϵ	dielectric constant, may be complex with components $\epsilon' - j\epsilon''$
α, ϕ	frequency and phase parameters of constant phase angle element

Electrodiffusion symbols

k, T, e	Boltzmann constant, absolute temperature and electron charge. $kT/e = 25$ mV at usual biological temperatures.
n, m, c	concentrations for positive, negative and mean number of ions/cm^3
n_e, n_i, etc.	external and internal concentrations.
M	concentration (mole/liter).
D	diffusion coefficient (cm^2/sec) given by ukT for a single particle.
$1/\kappa$	Debye ion atmosphere characteristic thickness.
u, v	mobilities for positive and negative ions (cm/dyne sec)
X	electric field (V/cm) component in x direction.
ue, ve	electrical mobilities (cm^2/V sec)
E_K etc.	concentration equilibrium potentials for K etc. given by $kT/e \ln (K_e/K_i)$
U, V	Planck variables, Σun, Σvm

Physiological terms

Membrane	relatively impermeable structure at surface of living cells, such as red blood cells, marine eggs, and neurons

Cytoplasm, axoplasm	interior of living cell in general, of a neuron axon in particular
Ringer's, Kreb's	artificial, balanced salt solutions approximating normal external media of cells
Osmotic	iso-, hyper-, hypo-, solutions with effective concentrations the same, higher, or less than normal
Q_{10}	variation of a parameter with temperature expressed as a ratio for a change of 10°C
Axon	extension of a neuron, carrying nerve impulses, which may be uniform and smooth in structure or medullated with exposed nodes at regular intervals between heavily myelinated internodes
Absolute Potential	potential at the interior of a membrane relative to the exterior, as measured by reversible electrodes and sometimes corrected for liquid junction potentials—in contrast to changes from a resting potential
Resting Potential	electrical potential of the interior of a cell or axon at rest relative to the exterior—usually -50 to -90 mV
Polarization	de- and hyper-, result of manipulations by which the membrane potential is made less and more negative
Cable theory	as applied to a linear axon expresses
Local circuit	current flows along and into axon in terms of
λ	space constant or characteristic length
Space clamp	experimental arrangements in which potentials and currents are constant over a membrane area, so cable theory does not apply
Excitation	fast process by which a membrane goes from a passive to an active state
Accommodation	slow process tending to prevent excitation from reaching threshold
Two factor theory	an analytical formulation of threshold response to a stimulus in terms of amplitudes and time constants of excitation (V, k) and accommodation (U, λ)
All-or-none law	the usual, impressive observation that there is no intermediate between full blown activity and its absence
Strength-duration curve	relation between a threshold current flow and time of application of stimulus
Action potential	time course of a potential during activity and recovery phases of nerve or other impulse

Spike potential	rapid change of potential from rest to maximum
Overshoot	potential by which a spike exceeds the rest potential
Propagation	conduction process by which activity at one point creates local circuit current to excite adjacent region Saltatory conduction in a medullated axon is from node to node
θ	speed of impulse propagation usually given in m/sec.
Current clamp	control of current, usually electronically and in space clamp, to follow prescribed time course and produce changes of potential. Excitation may occur even if not propagated
Voltage clamp	control of potential, usually electronically and in space clamp, to follow prescribed time course and produce current as the dependent variable. Excitation or other instability and propagation are not normally found
I_p, I_{ss}	membrane currents which appear, usually after a voltage clamp to a depolarizing potential, going first to an early peak value and finally reaching a near to steady state

HODGKIN-HUXLEY DEFINITIONS

I_{K}, I_{Na}, I_L	components of ionic membrane conduction currents as ascribed to K, Na, leakage
V	membrane potential relative to external medium
E_{K}, E_{Na}, E_L	potentials, at which the corresponding current components are zero, and correlated with the Nernst equilibrium emfs for the ions
$g_{\mathrm{K}} = I_{\mathrm{K}}/(V - E_{\mathrm{K}})$, etc.	instantaneous membrane conductance component for K, etc.
$\bar{g}_{\mathrm{K}}$, $\bar{g}_{\mathrm{Na}}$, $\bar{g}_L$	maximum conductance parameters, constants for an axon
n, m, h	dimensionless functions of membrane potential and time to give $g_{\mathrm{K}} = \bar{g}_{\mathrm{K}}n^4$, $g_{\mathrm{Na}} = \bar{g}_{\mathrm{Na}}m^3h$
α_n, β_n, etc.	forward and backward rate constants, functions of potential
$\underline{n}$, τ_n, etc.	steady state and time constant parameters, functions of potential

Acknowledgments

My own participation in the eras of which I have written would have been impossible without the generous help of many organizations and their staffs. First and foremost of these is the Marine Biological Laboratory of Woods Hole, Massachusetts, which provided a continuing education of many kinds over more than forty years. My opportunities at Columbia University made my stay there one of outstanding importance to me. I am further very indebted to Harvard and Princeton Universities, the Universities of Leipzig, Chicago, and California at Berkeley, the Institute for Advanced Study, the Naval Medical Research Institute, the National Institutes of Health, the Rockefeller and Guggenheim Foundations, and the International Education Board.

This book has been produced under the joint auspices of the National Institutes of Health, Bethesda, and the University of California, Berkeley, and the staffs of both have provided many forms of assistance and encouragement from beginning to end. In particular, Grace Ream has contributed far beyond her duties as Secretary of our Laboratory at Bethesda.

More than a dozen of my busy colleagues have very generously taken time to read the manuscript and give me their comments and criticisms; among these Dr. Rita Guttman has been very thorough and also she has prepared the index. All of these suggestions have been most valuable, both in detail and broad outline—as was the opinion of an anonymous reader that the book would be worthless. Many mistakes and oddities of various kinds have been corrected and those that remain, both intentional and irresponsible, are my fault. Most interestingly I have been both criticized and applauded on numerous aspects of the book. Suggestions for more elementary and complete discussions of physics and mathematics have come from those with the most biological background and experience. Pleas for more biological information and discussion along with condensations and consolidation of most of the analysis were made by the more physical scientists.

I have drawn heavily upon unpublished as well as published work for both text and illustrations. Insofar as I have made my own interpretations of the work of others, the responsibility is mine. For reasons of clarity, emphasis, space, and style, most of the figures are new drawings and some reproductions have been modified. The sources of all are identified and acknowledged in the accompanying references and the reproductions are used with the permission of the authors and in the forms required by publishers.

For all of these and the many other forms of help that I have had, I am most grateful and I am glad to put my thanks on record.

Index